Student Solutions Manual for

ELEMENTARY LINEAR ALGEBRA

SIXTH EDITION

Larson

Houghton Mifflin Harcourt Publishing Company

Boston   New York

Publisher: Richard Stratton
Senior Sponsoring Editor: Cathy Cantin
Senior Marketing Manager: Jennifer Jones
Assistant Editor: Janine Tangney
Associate Project Editor: Jill Clark
Senior Manufacturing Buyer: Diane Gibbons
Editorial Assistant: Amy Haines

Printed in Canada.

ISBN 13: 978-0-618-78377-9
ISBN 10: 0-618-78377-6

123456789-TCP-12 11 10 09 08

PREFACE

This *Student Solutions Manual* is designed as a supplement to *Elementary Linear Algebra*, Sixth Edition, by Ron Larson and David C. Falvo. All references to chapters, theorems, and exercises relate to the main text. Solutions to every odd-numbered exercise in the text are given with all essential algebraic steps included. Although this supplement is not a substitute for good study habits, it can be valuable when incorporated into a well-planned course of study.

We have made every effort to see that the solutions are correct. However, we would appreciate hearing about any errors or other suggestions for improvement. Good luck with your study of elementary linear algebra.

Ron Larson

Larson Texts, Inc.

CONTENTS

CHAPTER 1
Systems of Linear Equations

C H A P T E R   1
Systems of Linear Equations

Section 1.1   Introduction to Systems of Linear Equations

1. Because the equation is in the form $a_1x + a_2y = b$, it *is* linear in the variables x and y.

3. Because the equation cannot be written in the form $a_1x + a_2y = b$, it is *not* linear in the variables x and y.

5. Because the equation cannot be written in the form $a_1x + a_2y = b$, it is *not* linear in the variables x and y.

7. Choosing y as the free variable, let $y = t$ and obtain

$$2x - 4t = 0$$
$$2x = 4t$$
$$x = 2t.$$

So, you can describe the solution set as $x = 2t$ and $y = t$, where t is any real number.

9. Choosing y and z as the free variables, let $y = s$ and $z = t$, and obtain $x + s + t = 1$ or $x = 1 - s - t$. So, you can describe the solution set as $x = 1 - s - t$, $y = s$ and $z = t$, where s and t are any real numbers.

11. From Equation 2 you have $x_2 = 3$. Substituting this value into Equation 1 produces $x_1 - 3 = 2$ or $x_1 = 5$. So, the system has exactly one solution: $x_1 = 5$ and $x_2 = 3$.

13. From Equation 3 you can conclude that $z = 0$. Substituting this value into Equation 2 produces

$$2y + 0 = 3$$
$$y = \tfrac{3}{2}.$$

Finally, by substituting $y = \tfrac{3}{2}$ and $z = 0$ into Equation 1, you obtain

$$-x + \tfrac{3}{2} - 0 = 0$$
$$x = \tfrac{3}{2}.$$

So, the system has exactly one solution: $x = \tfrac{3}{2}$, $y = \tfrac{3}{2}$, and $z = 0$.

15. Begin by rewriting the system in row-echelon form. The equations are interchanged.

$$2x_1 + x_2 \qquad = 0$$
$$5x_1 + 2x_2 + x_3 = 0$$

The first equation is multiplied by $\tfrac{1}{2}$.

$$x_1 + \tfrac{1}{2}x_2 \qquad = 0$$
$$5x_1 + 2x_2 + x_3 = 0$$

Adding -5 times the first equation to the second equation produces a new second equation.

$$x_1 + \tfrac{1}{2}x_2 \qquad = 0$$
$$-\tfrac{1}{2}x_2 + x_3 = 0$$

The second equation is multiplied by -2.

$$x_1 + \tfrac{1}{2}x_2 \qquad = 0$$
$$x_2 - 2x_3 = 0$$

To represent the solutions, choose x_3 to be the free variable and represent it by the parameter t. Because $x_2 = 2x_3$ and $x_1 = -\tfrac{1}{2}x_2$, you can describe the solution set as

$$x_1 = -t,\ x_2 = 2t,\ x_3 = t,\text{ where } t \text{ is any real number.}$$

17. $$2x + y = 4$$
$$x - y = 2$$

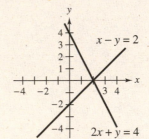

Adding the first equation to the second equation produces a new second equation, $3x = 6$, or $x = 2$. So, $y = 0$, and the solution is $x = 2$, $y = 0$. This is the point where the two lines intersect.

19. $x - y = 1$
$-2x + 2y = 5$

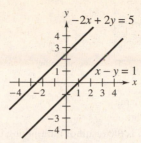

Adding 2 times the first equation to the second equation produces a new second equation.

$x - y = 1$
$\quad 0 = 7$

Because the second equation is a false statement, you can conclude that the original system of equations has no solution. Geometrically, the two lines are parallel.

21. $3x - 5y = 7$
$2x + y = 9$

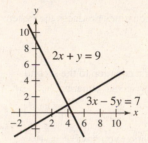

Adding the first equation to 5 times the second equation produces a new second equation, $13x = 52$, or $x = 4$.

So, $2(4) + y = 9$, or $y = 1$, and the solution is: $x = 4$, $y = 1$. This is the point where the two lines intersect.

23. $2x - y = 5$
$5x - y = 11$

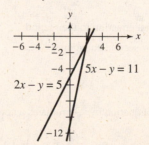

Subtracting the first equation from the second equation produces a new second equation, $3x = 6$, or $x = 2$. So, $2(2) - y = 5$, or $y = -1$, and the solution is: $x = 2$, $y = -1$. This is the point where the two lines intersect.

25. $\dfrac{x + 3}{4} + \dfrac{y - 1}{3} = 1$
$2x - y = 12$

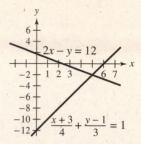

Multiplying the first equation by 12 produces a new first equation.

$3x + 4y = 7$
$2x - y = 12$

Adding the first equation to 4 times the second equation produces a new second equation, $11x = 55$, or $x = 5$. So, $2(5) - y = 12$, or $y = -2$, and the solution is: $x = 5$, $y = -2$. This is the point where the two lines intersect.

27. $0.05x - 0.03y = 0.07$
$0.07x + 0.02y = 0.16$

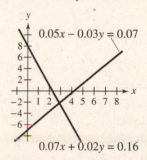

Multiplying the first equation by 200 and the second equation by 300 produces new equations.

$10x - 6y = 14$
$21x + 6y = 48$

Adding the first equation to the second equation produces a new second equation, $31x = 62$, or $x = 2$. So, $10(2) - 6y = 14$, or $y = 1$, and the solution is: $x = 2$, $y = 1$. This is the point where the two lines intersect.

29. $\dfrac{x}{4} + \dfrac{y}{6} = 1$

$x - y = 3$

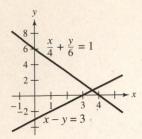

Adding 6 times the first equation to the second equation produces a new second equation, $\frac{5}{2}x = 9$, or $x = \frac{18}{5}$.

So, $\frac{18}{5} - y = 3$, or $y = \frac{3}{5}$, and the solution is: $x = \frac{18}{5}$,

$y = \frac{3}{5}$. This is the point where the two lines intersect.

31. (a)

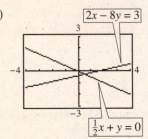

(b) The system is inconsistent.

33. (a)

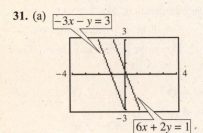

(b) The system is consistent

(c) The solution is approximately $x = \frac{1}{2}$, $y = -\frac{1}{4}$.

(d) Adding $-\frac{1}{4}$ times the first equation to the second equation produces the new second equation
$3y = -\frac{3}{4}$, or $y = -\frac{1}{4}$. So, $x = \frac{1}{2}$, and the solution is $x = \frac{1}{2}$, $y = -\frac{1}{4}$.

(e) The solutions in (c) and (d) are the same.

35. (a)

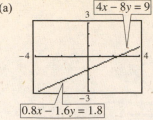

(b) The system is consistent.

(c) There are infinite solutions.

(d) The second equation is the result of multiplying both sides of the first equation by 0.2. A parametric representation of the solution set is given by
$x = \frac{9}{4} + 2t$, $y = t$, where t is any real number.

(e) The solutions in (c) and (d) are consistent.

37. Adding -3 times the first equation to the second equation produces a new second equation.

$$x_1 - x_2 = 0$$
$$x_2 = -1$$

Now, using back-substitution you can conclude that the system has exactly one solution: $x_1 = -1$ and $x_2 = -1$.

39. Interchanging the two equations produces the system

$u + 2v = 120$

$2u + v = 120.$

Adding -2 times the first equation to the second equation produces a new second equation.

$u + 2v = 120$

$-3v = -120$

Solving the second equation you have $v = 40$. Substituting this value into the first equation gives $u + 80 = 120$ or $u = 40$. So, the system has exactly one solution: $u = 40$ and $v = 40$.

41. Dividing the first equation by 9 produces a new first equation.

$x - \frac{1}{3}y = -\frac{1}{9}$

$\frac{1}{5}x + \frac{2}{5}y = -\frac{1}{3}$

Adding $-\frac{1}{5}$ times the first equation to the second equation produces a new second equation.

$x - \frac{1}{3}y = -\frac{1}{9}$

$\frac{7}{15}y = -\frac{14}{45}$

Multiplying the second equation by $\frac{15}{7}$ produces a new second equation.

$x - \frac{1}{3}y = -\frac{1}{9}$

$y = -\frac{2}{3}$

Now, using back-substitution, you can substitute $y = -\frac{2}{3}$ into the first equation to obtain $x + \frac{2}{9} = -\frac{1}{9}$ or $x = -\frac{1}{3}$. So, you can conclude that the system has exactly one solution: $x = -\frac{1}{3}$ and $y = -\frac{2}{3}$.

43. To begin, change the form of the first equation.

$$\tfrac{1}{2}x + \tfrac{1}{3}y = \tfrac{23}{6}$$
$$x - 2y = 5$$

Multiplying the first equation by 2 yields a new first equation.

$$x + \tfrac{2}{3}y = \tfrac{23}{3}$$
$$x - 2y = 5$$

Subtracting the first equation from the second equation yields a new second equation.

$$x + \tfrac{2}{3}y = \tfrac{23}{3}$$
$$-\tfrac{8}{3}y = -\tfrac{8}{3}$$

Dividing the second equation by $-\tfrac{8}{3}$ yields a new second equation.

$$x + \tfrac{2}{3}y = \tfrac{23}{3}$$
$$y = 1$$

Now, using back-substitution you can conclude that the system has exactly one solution:

$$x = 7 \text{ and } y = 1.$$

45. Multiplying the first equation by 50 and the second equation by 100 produces a new system.

$$x_1 - 2.5x_2 = -9.5$$
$$3x_1 + 4x_2 = 52$$

Adding -3 times the first equation to the second equation produces a new second equation.

$$x_1 - 2.5x_2 = -9.5$$
$$11.5x_2 = 80.5$$

Now, using back-substitution, you can conclude that the system has exactly one solution:

$$x_1 = 8 \text{ and } x_2 = 7.$$

47. Adding -2 times the first equation to the second equation yields a new second equation.

$$x + y + z = 6$$
$$-3y - z = -9$$
$$3x - z = 0$$

Adding -3 times the first equation to the third equation yields a new third equation.

$$x + y + z = 6$$
$$-3y - z = -9$$
$$-3y - 4z = -18$$

Dividing the second equation by -3 yields a new second equation.

$$x + y + z = 6$$
$$y + \tfrac{1}{3}z = 3$$
$$-3y - 4z = -18$$

Adding 3 times the second equation to the third equation yields a new third equation.

$$x + y + z = 6$$
$$y + \tfrac{1}{3}z = 3$$
$$-3z = -9$$

Dividing the third equation by -3 yields a new third equation.

$$x + y + z = 6$$
$$y + \tfrac{1}{3}z = 3$$
$$z = 3$$

Now, using back-substitution you can conclude that the system has exactly one solution:

$$x = 1, \ y = 2, \text{ and } z = 3.$$

49. Dividing the first equation by 3 yields a new first equation.

$$x_1 - \tfrac{2}{3}x_2 + \tfrac{4}{3}x_3 = \tfrac{1}{3}$$
$$x_1 + x_2 - 2x_3 = 3$$
$$2x_1 - 3x_2 + 6x_3 = 8$$

Subtracting the first equation from the second equation yields a new second equation.

$$x_1 - \tfrac{2}{3}x_2 + \tfrac{4}{3}x_3 = \tfrac{1}{3}$$
$$\tfrac{5}{3}x_2 - \tfrac{10}{3}x_3 = \tfrac{8}{3}$$
$$2x_1 - 3x_2 + 6x_3 = 8$$

Adding -2 times the first equation to the third equation yields a new third equation.

$$x_1 - \tfrac{2}{3}x_2 + \tfrac{4}{3}x_3 = \tfrac{1}{3}$$
$$\tfrac{5}{3}x_2 - \tfrac{10}{3}x_3 = \tfrac{8}{3}$$
$$-\tfrac{5}{3}x_2 + \tfrac{10}{3}x_3 = \tfrac{22}{3}$$

At this point you should recognize that Equations 2 and 3 cannot both be satisfied. So, the original system of equations has no solution.

51. Dividing the first equation by 2 yields a new first equation.

$$x_1 + \tfrac{1}{2}x_2 - \tfrac{3}{2}x_3 = 2$$
$$4x_1 \qquad + 2x_3 = 10$$
$$-2x_1 + 3x_2 - 13x_3 = -8$$

Adding -4 times the first equation to the second equation produces a new second equation.

$$x_1 + \tfrac{1}{2}x_2 - \tfrac{3}{2}x_3 = 2$$
$$-2x_2 + 8x_3 = 2$$
$$-2x_1 + 3x_2 - 13x_3 = -8$$

Adding 2 times the first equation to the third equation produces a new third equation.

$$x_1 + \tfrac{1}{2}x_2 - \tfrac{3}{2}x_3 = 2$$
$$-2x_2 + 8x_3 = 2$$
$$4x_2 - 16x_3 = -4$$

Dividing the second equation by -2 yields a new second equation.

$$x_1 + \tfrac{1}{2}x_2 - \tfrac{3}{2}x_3 = 2$$
$$x_2 - 4x_3 = -1$$
$$4x_2 - 16x_3 = -4$$

Adding -4 times the second equation to the third equation produces a new third equation.

$$x_1 + \tfrac{1}{2}x_2 - \tfrac{3}{2}x_3 = 2$$
$$x_2 - 4x_3 = -1$$
$$0 = 0$$

Adding $-\tfrac{1}{2}$ times the second equation to the first equation produces a new first equation.

$$x_1 \qquad + \tfrac{1}{2}x_3 = \tfrac{5}{2}$$
$$x_2 - 4x_3 = -1$$

Choosing $x_3 = t$ as the free variable, you can describe the solution as $x_1 = \tfrac{5}{2} - \tfrac{1}{2}t$, $x_2 = 4t - 1$, and $x_3 = t$, where t is any real number.

53. Adding -5 times the first equation to the second equation yields a new second equation.

$$x - 3y + 2z = 18$$
$$0 = -72$$

Because the second equation is a false statement, you can conclude that the original system of equations has no solution.

55. Adding -2 times the first equation to the second, 3 times the first equation to the third, and -1 times the first equation to the fourth, produces

$$x + y + z + w = 6$$
$$y - 2z - 3w = -12$$
$$7y + 4z + 5w = 22$$
$$y - 2z = -6.$$

Adding -7 times the second equation to the third, and -1 times the second equation to the fourth, produces

$$x + y + z + w = 6$$
$$y - 2z - 3w = -12$$
$$18z + 26w = 106$$
$$3w = 6.$$

Using back-substitution, you find the original system has exactly one solution: $x = 1$, $y = 0$, $z = 3$, and $w = 2$.

Answers may vary slightly for Exercises 57–63.

57. Using a computer software program or graphing utility, you obtain

$$x_1 = -15,\; x_2 = 40,\; x_3 = 45,\; x_4 = -75$$

59. Using a computer software program or graphing utility, you obtain

$$x = -1.2,\; y = -0.6,\; z = 2.4.$$

61. Using a computer software program or graphing utility, you obtain

$$x_1 = \tfrac{1}{5},\; x_2 = -\tfrac{4}{5},\; x_3 = \tfrac{1}{2}.$$

63. Using a computer software program or graphing utility, you obtain

$$x = 6.8813,\; y = -163.3111,\; z = -210.2915,$$
$$w = -59.2913.$$

65. $x = y = z = 0$ is clearly a solution.

Dividing the first equation by 4 yields a new first equation.

$x + \frac{3}{4}y + \frac{17}{4}z = 0$

$5x + 4y + 22z = 0$

$4x + 2y + 19z = 0$

Adding -5 times the first equation to the second equation yields a new second equation.

$x + \frac{3}{4}y + \frac{17}{4}z = 0$

$\frac{1}{4}y + \frac{3}{4}z = 0$

$4x + 2y + 19z = 0$

Adding -4 times the first equation to the third equation yields a new third equation.

$x + \frac{3}{4}y + \frac{17}{4}z = 0$

$\frac{1}{4}y + \frac{3}{4}z = 0$

$-y + 2z = 0$

Multiplying the second equation by 4 yields a new second equation.

$x + \frac{3}{4}y + \frac{17}{4}z = 0$

$y + 3z = 0$

$-y + 2z = 0$

Adding the second equation to the third equation yields a new third equation.

$x + \frac{3}{4}y + \frac{17}{4}z = 0$

$y + 3z = 0$

$5z = 0$

Dividing the third equation by 5 yields a new third equation.

$x + \frac{3}{4}y + \frac{17}{4}z = 0$

$y + 3z = 0$

$z = 0$

Now, using back-substitution, you can conclude that the system has exactly one solution: $x = 0$, $y = 0$, and $z = 0$.

67. $x = y = z = 0$ is clearly a solution.

Dividing the first equation by 5 yields a new first equation.

$x + y - \frac{1}{5}z = 0$

$10x + 5y + 2z = 0$

$5x + 15y - 9z = 0$

Adding -10 times the first equation to the second equation yields a new second equation.

$x + y - \frac{1}{5}z = 0$

$-5y + 4z = 0$

$5x + 15y - 9z = 0$

Adding -5 times the first equation to the third equation yields a new third equation.

$x + y - \frac{1}{5}z = 0$

$-5y + 4z = 0$

$10y - 8z = 0$

Dividing the second equation by -5 yields a new second equation.

$x + y - \frac{1}{5}z = 0$

$y - \frac{4}{5}z = 0$

$10y - 8z = 0$

Adding -10 times the second equation to the third equation yields a new third equation.

$x + y - \frac{1}{5}z = 0$

$y - \frac{4}{5}z = 0$

$0 = 0$

Adding -1 times the second equation to the first equation yields a new first equation.

$x + \frac{3}{5}z = 0$

$y - \frac{4}{5}z = 0$

Choosing $z = t$ as the free variable you find the solution to be $x = -\frac{3}{5}t$, $y = \frac{4}{5}t$, and $z = t$, where t is any real number.

69. (a) True. You can describe the entire solution set using parametric representation.

$ax + by = c$

Choosing $y = t$ as the free variable, the solution is

$x = \frac{c}{a} - \frac{b}{a}t$, $y = t$, where t is any real number.

(b) False. For example, consider the system

$x_1 + x_2 + x_3 = 1$

$x_1 + x_2 + x_3 = 2$

which is an inconsistent system.

(c) False. A consistent system may have only one solution.

71. Because $x_1 = t$ and $x_2 = 3t - 4 = 3x_1 - 4$, one answer is the system

$$3x_1 - x_2 = 4$$
$$-3x_1 + x_2 = -4.$$

Letting $x_2 = t$, you get $x_1 = \dfrac{4+t}{3} = \dfrac{4}{3} + \dfrac{t}{3}$.

73. Substituting $A = 1/x$ and $B = 1/y$ into the original system yields

$$12A - 12B = 7$$
$$3A + 4B = 0.$$

Reduce this system to row-echelon form. Dividing the first equation by 12 yields a new first equation.

$$A - B = \tfrac{7}{12}$$
$$3A + 4B = 0$$

Adding -3 times the first equation to the second equation yields a new second equation.

$$A - B = \tfrac{7}{12}$$
$$7B = -\tfrac{7}{4}$$

Dividing the second equation by 7 yields a new second equation.

$$A - B = \tfrac{7}{12}$$
$$B = -\tfrac{1}{4}$$

So, $A = -1/4$ and $B = 1/3$. Because $A = 1/x$ and $B = 1/y$, the solution of the original system of equations is: $x = 3$ and $y = -4$.

75. Substituting $A = 1/x$, $B = 1/y$, and $C = 1/z$, into the original system yields

$$2A + B - 3C = 4$$
$$4A + 2C = 10$$
$$-2A + 3B - 13C = -8.$$

Reduce this system to row-echelon form.

$$A + \tfrac{1}{2}B - \tfrac{3}{2}C = 2$$
$$-2B + 8C = 2$$
$$4B - 16C = -4$$
$$A + \tfrac{1}{2}B - \tfrac{3}{2}C = 2$$
$$B - 4C = -1$$

Letting $t = C$ be the free variable, you have

$C = t$, $B = 4t - 1$ and $A = \dfrac{(-t + 5)}{2}$. So, the solution to the original problem is

$$x = \dfrac{2}{5 - t}, \ y = \dfrac{1}{4t - 1}, \ z = \dfrac{1}{t}, \text{where } t \neq 5, \dfrac{1}{4}, 0.$$

77. Reduce the system to row-echelon form. Dividing the first equation by $\cos \theta$ yields a new second equation.

$$x + \left(\dfrac{\sin \theta}{\cos \theta}\right)y = \dfrac{1}{\cos \theta}$$
$$(-\sin \theta)x + (\cos \theta)y = 0$$

Multiplying the first equation by $\sin \theta$ and adding to the second equation yields a new second equation.

$$x + \left(\dfrac{\sin \theta}{\cos \theta}\right)y = \dfrac{1}{\cos \theta}$$
$$\left(\dfrac{1}{\cos \theta}\right)y = \dfrac{\sin \theta}{\cos \theta}$$

$$\left(\text{Because } \dfrac{\sin^2 \theta}{\cos \theta} + \cos \theta = \dfrac{\sin^2 \theta + \cos^2 \theta}{\cos \theta} = \dfrac{1}{\cos \theta}\right)$$

Multiplying the second equation by $\cos \theta$ yields a new second equation.

$$x + \left(\dfrac{\sin \theta}{\cos \theta}\right)y = \dfrac{1}{\cos \theta}$$
$$y = \sin \theta$$

Substituting $y = \sin \theta$ into the first equation yields

$$x + \left(\dfrac{\sin \theta}{\cos \theta}\right)\sin \theta = \dfrac{1}{\cos \theta}$$
$$x = \dfrac{1 - \sin^2 \theta}{\cos \theta} = \dfrac{\cos^2 \theta}{\cos \theta} = \cos \theta.$$

So, the solution of the original system of equations is: $x = \cos \theta$ and $y = \sin \theta$.

79. For this system to have an infinite number of solutions both equations need to be equivalent.

Multiply the second equation by -2.

$$4x + ky = 6$$
$$-2kx - 2y = 6$$

So, when $k = -2$ the system will have an infinite number of solutions.

81. Reduce the system to row-echelon form.

$$x + ky = 0$$
$$(1 - k^2)y = 0$$

$$x + ky = 0$$
$$y = 0, 1 - k^2 \neq 0$$

$$x = 0$$
$$y = 0, 1 - k^2 \neq 0$$

If $1 - k^2 \neq 0$, that is if $k \neq \pm 1$, the system will have exactly one solution.

83. To begin, reduce the system to row-echelon form.

$$x + 2y + kz = 6$$

$$(8 - 3k)z = -14$$

This system will have no solution if $8 - 3k = 0$, that is, $k = \frac{8}{3}$.

85. Reducing the system to row-echelon form, you have

$$x + \quad y + \quad kz = \quad 3$$

$$(k - 1)y + (1 - k)z = \quad -1$$

$$(1 - k)y + (1 - k^2)z = 1 - 3k$$

$$x + \quad y + \quad kz = \quad 3$$

$$(k - 1)y + \quad (1 - k)z = \quad -1$$

$$(-k^2 - k + 2)z = -3k.$$

If $-k^2 - k + 2 = 0$, then there is no solution. So, if $k = 1$ or $k = -2$, there is not a unique solution.

87. (a) All three of the lines will intersect in exactly one point (corresponding to the solution point).

(b) All three of the lines will coincide (every point on these lines is a solution point).

(c) The three lines have no common point.

89. Answers vary. (*Hint:* Choose three different values for x and solve the resulting system of linear equations in the variables a, b, and c.)

91.

$$x - 4y = -3$$
$$5x - 6y = 13$$

$$x - 4y = -3$$
$$14y = 28$$

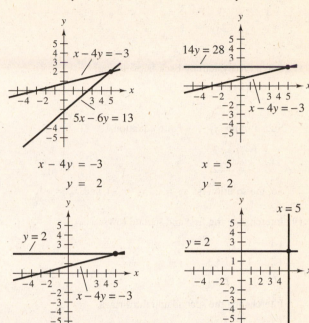

$$x - 4y = -3$$
$$y = 2$$

$$x = 5$$
$$y = 2$$

At each step, the lines always intersect at $(5, 2)$, which is the solution to the system of equations.

93. Solve each equation for y.

$$y = \frac{1}{100}x + 2$$

$$y = \frac{1}{99}x - 2$$

The graphs are misleading because, while they appear parallel, when the equations are solved for y they have slightly different slopes.

Section 1.2   Gaussian Elimination and Gauss-Jordan Elimination

1. Because the matrix has 3 rows and 3 columns, it has size 3×3.

3. Because the matrix has 2 rows and 4 columns, it has size 2×4.

5. Because the matrix has 1 row and 5 columns, it has size 1×5.

7. Because the matrix has 4 rows and 5 columns, it has size 4×5.

9. The matrix satisfies all three conditions in the definition of row-echelon form. Moreover, because each column that has a leading 1 (columns one and two) has zeros elsewhere, the matrix *is* in reduced row-echelon form.

11. Because the matrix has two non-zero rows without leading 1's, it is *not* in row-echelon form.

13. Because the matrix has a non-zero row without a leading 1, it is *not* in row-echelon form.

15. Because the matrix is in reduced row-echelon form, convert back to a system of linear equations

$$x_1 = 0$$

$$x_2 = 2$$

So, the solution is: $x_1 = 0$ and $x_2 = 2$.

17. Because the matrix is in row-echelon form, convert back to a system of linear equations.

$$\begin{aligned} x_1 - x_2 \quad\quad &= 3 \\ x_2 - 2x_3 &= 1 \\ x_3 &= -1 \end{aligned}$$

Solve this system by back-substitution.

$$\begin{aligned} x_2 - 2(-1) &= 1 \\ x_2 &= -1 \end{aligned}$$

Substituting $x_2 = -1$ into equation 1,

$$\begin{aligned} x_1 - (-1) &= 3 \\ x_1 &= 2. \end{aligned}$$

So, the solution is: $x_1 = 2$, $x_2 = -1$, and $x_3 = -1$.

19. Interchange the first and second rows.

$$\begin{bmatrix} 1 & -1 & 1 & 0 \\ 2 & 1 & -1 & 3 \\ 0 & 1 & 2 & 1 \end{bmatrix}$$

Interchange the second and third rows.

$$\begin{bmatrix} 1 & -1 & 1 & 0 \\ 0 & 1 & 2 & 1 \\ 2 & 1 & -1 & 3 \end{bmatrix}$$

Add -2 times the first row to the third row to produce a new third row.

$$\begin{bmatrix} 1 & -1 & 1 & 0 \\ 0 & 1 & 2 & 1 \\ 0 & 3 & -3 & 3 \end{bmatrix}$$

Add -3 times the second row to the third row to produce a new third row.

$$\begin{bmatrix} 1 & -1 & 1 & 0 \\ 0 & 1 & 2 & 1 \\ 0 & 0 & -9 & 0 \end{bmatrix}$$

Divide the third row by -9 to produce a new third row.

$$\begin{bmatrix} 1 & -1 & 1 & 0 \\ 0 & 1 & 2 & 1 \\ 0 & 0 & 1 & 0 \end{bmatrix}$$

Convert back to a system of linear equations.

$$\begin{aligned} x_1 - x_2 + x_3 &= 0 \\ x_2 + 2x_3 &= 1 \\ x_3 &= 0 \end{aligned}$$

Solve this system by back-substitution.

$$x_2 = 1 - 2x_3 = 1 - 2(0) = 1$$

$$x_1 = x_2 - x_3 = 1 - 0 = 1$$

So, the solution is: $x_1 = 1$, $x_2 = 1$, and $x_3 = 0$.

21. Because the matrix is in row-echelon form, convert back to a system of linear equations.

$$\begin{aligned} x_1 + 2x_2 \quad\quad + x_4 &= 4 \\ x_2 + 2x_3 + x_4 &= 3 \\ x_3 + 2x_4 &= 1 \\ x_4 &= 4 \end{aligned}$$

Solve this system by back-substitution.

$$x_3 = 1 - 2x_4 = 1 - 2(4) = -7$$

$$x_2 = 3 - 2x_3 - x_4 = 3 - 2(-7) - 4 = 13$$

$$x_1 = 4 - 2x_2 - x_4 = 4 - 2(13) - 4 = -26$$

So, the solution is: $x_1 = -26$, $x_2 = 13$, $x_3 = -7$, and $x_4 = 4$.

23. The augmented matrix for this system is

$$\begin{bmatrix} 1 & 2 & 7 \\ 2 & 1 & 8 \end{bmatrix}.$$

Adding -2 times the first row to the second row yields a new second row.

$$\begin{bmatrix} 1 & 2 & 7 \\ 0 & -3 & -6 \end{bmatrix}$$

Dividing the second row by -3 yields a new second row.

$$\begin{bmatrix} 1 & 2 & 7 \\ 0 & 1 & 2 \end{bmatrix}$$

Converting back to a system of linear equations produces

$$\begin{aligned} x + 2y &= 7 \\ y &= 2. \end{aligned}$$

Finally, using back-substitution you find that $x = 3$ and $y = 2$.

25. The augmented matrix for this system is

$$\begin{bmatrix} -1 & 2 & 1.5 \\ 2 & -4 & 3 \end{bmatrix}.$$

Gaussian elimination produces the following.

$$\begin{bmatrix} -1 & 2 & 1.5 \\ 2 & -4 & 3 \end{bmatrix} \Rightarrow \begin{bmatrix} 1 & -2 & -\frac{3}{2} \\ 2 & -4 & 3 \end{bmatrix} \Rightarrow \begin{bmatrix} 1 & -2 & -\frac{3}{2} \\ 0 & 0 & 6 \end{bmatrix}$$

Because the second row of this matrix corresponds to the equation $0 = 6$, you can conclude that the original system has no solution.

27. The augmented matrix for this system is

$$\begin{bmatrix} -3 & 5 & -22 \\ 3 & 4 & 4 \\ 4 & -8 & 32 \end{bmatrix}.$$

Dividing the first row by -3 yields a new first row.

$$\begin{bmatrix} 1 & -\frac{5}{3} & \frac{22}{3} \\ 3 & 4 & 4 \\ 4 & -8 & 32 \end{bmatrix}$$

Adding -3 times the first row to the second row yields a new second row.

$$\begin{bmatrix} 1 & -\frac{5}{3} & \frac{22}{3} \\ 0 & 9 & -18 \\ 4 & -8 & 32 \end{bmatrix}$$

Adding -4 times the first row to the third row yields a new third row.

$$\begin{bmatrix} 1 & -\frac{5}{3} & \frac{22}{3} \\ 0 & 9 & -18 \\ 0 & -\frac{4}{3} & \frac{8}{3} \end{bmatrix}$$

Dividing the second row by 9 yields a new second row.

$$\begin{bmatrix} 1 & -\frac{5}{3} & \frac{22}{3} \\ 0 & 1 & -2 \\ 0 & -\frac{4}{3} & \frac{8}{3} \end{bmatrix}$$

Adding $\frac{4}{3}$ times the second row to the third row yields a new third row.

$$\begin{bmatrix} 1 & -\frac{5}{3} & \frac{22}{3} \\ 0 & 1 & -2 \\ 0 & 0 & 0 \end{bmatrix}$$

Converting back to a system of linear equations produces

$$x - \tfrac{5}{3}y = \tfrac{22}{3}$$
$$y = -2.$$

Finally, using back-substitution you find that the solution is: $x = 4$ and $y = -2$.

29. The augmented matrix for this system is

$$\begin{bmatrix} 1 & 0 & -3 & -2 \\ 3 & 1 & -2 & 5 \\ 2 & 2 & 1 & 4 \end{bmatrix}.$$

Gaussian elimination produces the following.

$$\begin{bmatrix} 1 & 0 & -3 & -2 \\ 3 & 1 & -2 & 5 \\ 2 & 2 & 1 & 4 \end{bmatrix} \Rightarrow \begin{bmatrix} 1 & 0 & -3 & -2 \\ 0 & 1 & 7 & 11 \\ 0 & 2 & 7 & 8 \end{bmatrix} \Rightarrow \begin{bmatrix} 1 & 0 & -3 & -2 \\ 0 & 1 & 7 & 11 \\ 0 & 0 & -7 & -14 \end{bmatrix}$$

Back substitution now yields

$$x_3 = 2$$
$$x_2 = 11 - 7x_3 = 11 - (7)2 = -3$$
$$x_1 = -2 + 3x_3 = -2 + 3(2) = 4.$$

So, the solution is: $x_1 = 4$, $x_2 = -3$, and $x_3 = 2$.

31. The augmented matrix for this system is

$$\begin{bmatrix} 1 & 1 & -5 & 3 \\ 1 & 0 & -2 & 1 \\ 2 & -1 & -1 & 0 \end{bmatrix}.$$

Subtracting the first row from the second row yields a new second row.

$$\begin{bmatrix} 1 & 1 & -5 & 3 \\ 0 & -1 & 3 & -2 \\ 2 & -1 & -1 & 0 \end{bmatrix}$$

Adding -2 times the first row to the third row yields a new third row.

$$\begin{bmatrix} 1 & 1 & -5 & 3 \\ 0 & -1 & 3 & -2 \\ 0 & -3 & 9 & -6 \end{bmatrix}$$

Multiplying the second row by -1 yields a new second row.

$$\begin{bmatrix} 1 & 1 & -5 & 3 \\ 0 & 1 & -3 & 2 \\ 0 & -3 & 9 & -6 \end{bmatrix}$$

Adding 3 times the second row to the third row yields a new third row.

$$\begin{bmatrix} 1 & 1 & -5 & 3 \\ 0 & 1 & -3 & 2 \\ 0 & 0 & 0 & 0 \end{bmatrix}$$

Adding -1 times the second row to the first row yields a new first row.

$$\begin{bmatrix} 1 & 0 & -2 & 1 \\ 0 & 1 & -3 & 2 \\ 0 & 0 & 0 & 0 \end{bmatrix}$$

Converting back to a system of linear equations produces

$$x_1 \quad - 2x_3 = 1$$
$$x_2 - 3x_3 = 2.$$

Finally, choosing $x_3 = t$ as the free variable you can describe the solution as $x_1 = 1 + 2t$, $x_2 = 2 + 3t$, and $x_3 = t$, where t is any real number.

33. The augmented matrix for this system is

$$\begin{bmatrix} 4 & 12 & -7 & -20 & 22 \\ 3 & 9 & -5 & -28 & 30 \end{bmatrix}.$$

Dividing the first row by 4 yields a new first row.

$$\begin{bmatrix} 1 & 3 & -\frac{7}{4} & -5 & \frac{11}{2} \\ 3 & 9 & -5 & -28 & 30 \end{bmatrix}$$

Adding -3 times the first row to the second row yields a new second row.

$$\begin{bmatrix} 1 & 3 & -\frac{7}{4} & -5 & \frac{11}{2} \\ 0 & 0 & \frac{1}{4} & -13 & \frac{27}{2} \end{bmatrix}$$

Multiplying the second row by 4 yields a new second row.

$$\begin{bmatrix} 1 & 3 & -\frac{7}{4} & -5 & \frac{11}{2} \\ 0 & 0 & 1 & -52 & 54 \end{bmatrix}$$

Converting back to a system of linear equations produces

$$x + 3y - \tfrac{7}{4}z - 5w = \tfrac{11}{2}$$
$$z - 52w = 54.$$

Choosing $y = s$ and $w = t$ as the free variables you can describe the solution as $x = 100 - 3s + 96t$, $y = s$, $z = 54 + 52t$, and $w = t$, where s and t are any real numbers.

35. The augmented matrix for this system is

$$\begin{bmatrix} 3 & 3 & 12 & 6 \\ 1 & 1 & 4 & 2 \\ 2 & 5 & 20 & 10 \\ -1 & 2 & 8 & 4 \end{bmatrix}.$$

Gaussian elimination produces the following.

$$\begin{bmatrix} 3 & 3 & 12 & 6 \\ 1 & 1 & 4 & 2 \\ 2 & 5 & 20 & 10 \\ -1 & 2 & 8 & 4 \end{bmatrix} \Rightarrow \begin{bmatrix} 1 & 1 & 4 & 2 \\ 1 & 1 & 4 & 2 \\ 2 & 5 & 20 & 10 \\ -1 & 2 & 8 & 4 \end{bmatrix}$$

$$\Rightarrow \begin{bmatrix} 1 & 1 & 4 & 2 \\ 0 & 0 & 0 & 0 \\ 0 & 3 & 12 & 6 \\ 0 & 3 & 12 & 6 \end{bmatrix} \Rightarrow \begin{bmatrix} 1 & 1 & 4 & 2 \\ 0 & 3 & 12 & 6 \\ 0 & 0 & 0 & 0 \\ 0 & 0 & 0 & 0 \end{bmatrix}$$

$$\Rightarrow \begin{bmatrix} 1 & 1 & 4 & 2 \\ 0 & 1 & 4 & 2 \\ 0 & 0 & 0 & 0 \\ 0 & 0 & 0 & 0 \end{bmatrix} \Rightarrow \begin{bmatrix} 1 & 0 & 0 & 0 \\ 0 & 1 & 4 & 2 \\ 0 & 0 & 0 & 0 \\ 0 & 0 & 0 & 0 \end{bmatrix}$$

Letting $z = t$ be the free variable, the solution is: $x = 0$, $y = 2 - 4t$, $z = t$, where t is any real number.

37. Using a computer software program or graphing utility, the augmented matrix reduces to

$$\begin{bmatrix} 1 & 0 & 0 & -0.5278 & 23.5361 \\ 0 & 1 & 0 & -4.1111 & 18.5444 \\ 0 & 0 & 1 & -2.1389 & 7.4306 \end{bmatrix}.$$

Letting $x_4 = t$ be the free variable, you obtain

$$x_1 = 23.5361 + 0.5278t$$
$$x_2 = 18.5444 + 4.1111t$$
$$x_3 = 7.43062 + 2.1389t$$
$$x_4 = t, \text{ where } t \text{ is any real number.}$$

39. Using a computer software program or graphing utility, the augmented matrix reduces to

$$\begin{bmatrix} 1 & 0 & 0 & 0 & 0 & 2 \\ 0 & 1 & 0 & 0 & 0 & -2 \\ 0 & 0 & 1 & 0 & 0 & 3 \\ 0 & 0 & 0 & 1 & 0 & -5 \\ 0 & 0 & 0 & 0 & 1 & 1 \end{bmatrix}.$$

So, the solution is: $x_1 = 2$, $x_2 = -2$, $x_3 = 3$, $x_4 = -5$, and $x_5 = 1$.

41. Using a computer software program or graphing utility, the augmented matrix reduces to

$$\begin{bmatrix} 1 & 0 & 0 & 0 & 0 & 0 & 1 \\ 0 & 1 & 0 & 0 & 0 & 0 & -2 \\ 0 & 0 & 1 & 0 & 0 & 0 & 6 \\ 0 & 0 & 0 & 1 & 0 & 0 & -3 \\ 0 & 0 & 0 & 0 & 1 & 0 & -4 \\ 0 & 0 & 0 & 0 & 0 & 1 & 3 \end{bmatrix}.$$

So, the solution is: $x_1 = 1$, $x_2 = -2$, $x_3 = 6$, $x_4 = -3$, $x_5 = -4$, and $x_6 = 3$.

43. The corresponding system of equations is

$$x_1 = 0$$
$$x_2 + x_3 = 0$$
$$0 = 0$$

Letting $x_3 = t$ be the free variable, the solution is: $x_1 = 0$, $x_2 = -t$, $x_3 = t$, where t is any real number.

45. The corresponding system of equations is

$$x_1 + x_4 = 0$$
$$x_3 = 0$$
$$0 = 0$$

Letting $x_4 = t$ and $x_3 = s$ be the free variables, the solution is: $x_1 = -t$, $x_2 = s$, $x_3 = 0$, $x_4 = t$.

47. (a) Because there are two rows and three columns, there are two equations and two variables.

(b) Gaussian elimination produces the following.

$$\begin{bmatrix} 1 & k & 2 \\ -3 & 4 & 1 \end{bmatrix} \Rightarrow \begin{bmatrix} 1 & k & 2 \\ 0 & 4+3k & 7 \end{bmatrix} \Rightarrow \begin{bmatrix} 1 & k & 2 \\ 0 & 1 & \dfrac{7}{4+3k} \end{bmatrix}$$

The system is consistent if $4 + 3k \neq 0$, or if $k \neq -\frac{4}{3}$.

(c) Because there are two rows and three columns, there are two equations and three variables.

(d) Gaussian elimination produces the following.

$$\begin{bmatrix} 1 & k & 2 & 0 \\ -3 & 4 & 1 & 0 \end{bmatrix} \Rightarrow \begin{bmatrix} 1 & k & 2 & 0 \\ 0 & 4+3k & 7 & 0 \end{bmatrix} \Rightarrow \begin{bmatrix} 1 & k & 2 & 0 \\ 0 & 1 & \dfrac{7}{4+3k} & 0 \end{bmatrix}$$

Notice that $4 + 3k \neq 0$, or $k \neq -\frac{4}{3}$. But if $-\frac{4}{3}$ is substituted for k in the original matrix. Guassian elimination produces the following:

$$\begin{bmatrix} 1 & -\frac{4}{3} & 2 & 0 \\ -3 & 4 & 1 & 0 \end{bmatrix} \Rightarrow \begin{bmatrix} 1 & -\frac{4}{3} & 2 & 0 \\ 0 & 0 & 7 & 0 \end{bmatrix}$$

The system is consistent. So, the original system is consistent for all real k.

49. Begin by forming the augmented matrix for the system

$$\begin{bmatrix} 1 & 1 & 0 & 2 \\ 0 & 1 & 1 & 2 \\ 1 & 0 & 1 & 2 \\ a & b & c & 0 \end{bmatrix}.$$

Then use Gauss-Jordan elimination as follows.

$$\begin{bmatrix} 1 & 1 & 0 & 2 \\ 0 & 1 & 1 & 2 \\ 0 & -1 & 1 & 0 \\ a & b & c & 0 \end{bmatrix} \Rightarrow \begin{bmatrix} 1 & 1 & 0 & 2 \\ 0 & 1 & 1 & 2 \\ 0 & -1 & 1 & 0 \\ 0 & b-a & c & -2a \end{bmatrix}$$

$$\Rightarrow \begin{bmatrix} 1 & 1 & 0 & 2 \\ 0 & 1 & 1 & 2 \\ 0 & 0 & 2 & 2 \\ 0 & b-a & c & -2a \end{bmatrix} \Rightarrow \begin{bmatrix} 1 & 1 & 0 & 2 \\ 0 & 1 & 1 & 2 \\ 0 & 0 & 2 & 2 \\ 0 & 0 & a-b+c & -2b \end{bmatrix}$$

$$\Rightarrow \begin{bmatrix} 1 & 1 & 0 & 2 \\ 0 & 1 & 1 & 2 \\ 0 & 0 & 1 & 1 \\ 0 & 0 & a-b+c & -2b \end{bmatrix} \Rightarrow \begin{bmatrix} 1 & 1 & 0 & 2 \\ 0 & 1 & 1 & 2 \\ 0 & 0 & 1 & 1 \\ 0 & 0 & 0 & a+b+c \end{bmatrix}$$

$$\Rightarrow \begin{bmatrix} 1 & 1 & 0 & 2 \\ 0 & 1 & 0 & 1 \\ 0 & 0 & 1 & 1 \\ 0 & 0 & 0 & a+b+c \end{bmatrix} \Rightarrow \begin{bmatrix} 1 & 0 & 0 & 1 \\ 0 & 1 & 0 & 1 \\ 0 & 0 & 1 & 1 \\ 0 & 0 & 0 & a+b+c \end{bmatrix}$$

Converting back to a system of linear equations

$x = 1$

$y = 1$

$z = 1$

$0 = a + b + c.$

The system

(a) will have a unique solution if $a + b + c = 0$,

(b) will have no solution if $a + b + c \neq 0$, and

(c) cannot have an infinite number of solutions.

51. Solve each pair of equations by Gaussian elimination as follows.

(a) Equations 1 and 2:

$$\begin{bmatrix} 4 & -2 & 5 & 16 \\ 1 & 1 & 0 & 0 \end{bmatrix} \Rightarrow \begin{bmatrix} 1 & 0 & \frac{5}{6} & \frac{8}{3} \\ 0 & 1 & -\frac{5}{6} & -\frac{8}{3} \end{bmatrix} \Rightarrow x = \frac{8}{3} - \frac{5}{6}t,$$
$$y = -\frac{8}{3} + \frac{5}{6}t,$$
$$z = t$$

(b) Equations 1 and 3:

$$\begin{bmatrix} 4 & -2 & 5 & 16 \\ -1 & -3 & 2 & 6 \end{bmatrix} \Rightarrow \begin{bmatrix} 1 & 0 & \frac{11}{14} & \frac{36}{14} \\ 0 & 1 & -\frac{13}{14} & -\frac{40}{14} \end{bmatrix} \Rightarrow x = \frac{18}{7} - \frac{11}{14}t,$$
$$y = -\frac{20}{7} + \frac{13}{14}t,$$
$$z = t$$

(c) Equations 2 and 3:

$$\begin{bmatrix} 1 & 1 & 0 & 0 \\ -1 & -3 & 2 & 6 \end{bmatrix} \Rightarrow \begin{bmatrix} 1 & 0 & 1 & 3 \\ 0 & 1 & -1 & -3 \end{bmatrix} \Rightarrow x = 3 - t,$$
$$y = -3 + t,$$
$$z = t$$

(d) Each of these systems has an infinite number of solutions.

53. Use Gauss-Jordan elimination as follows.

$$\begin{bmatrix} 1 & 2 \\ -1 & 2 \end{bmatrix} \Rightarrow \begin{bmatrix} 1 & 2 \\ 0 & 4 \end{bmatrix} \Rightarrow \begin{bmatrix} 1 & 2 \\ 0 & 1 \end{bmatrix} \Rightarrow \begin{bmatrix} 1 & 0 \\ 0 & 1 \end{bmatrix}$$

55. Begin by finding all possible first rows.

$$\begin{bmatrix} 0 & 0 \end{bmatrix}, \begin{bmatrix} 0 & 1 \end{bmatrix}, \begin{bmatrix} 1 & 0 \end{bmatrix}, \begin{bmatrix} 1 & k \end{bmatrix}.$$

For each of these, examine the possible second rows.

$$\begin{bmatrix} 0 & 0 \\ 0 & 0 \end{bmatrix}, \begin{bmatrix} 0 & 1 \\ 0 & 0 \end{bmatrix}, \begin{bmatrix} 1 & 0 \\ 0 & 1 \end{bmatrix}, \begin{bmatrix} 1 & k \\ 0 & 0 \end{bmatrix}$$

These represent all possible 2×2 reduced row-echelon matrices.

57. (a) True. In the notation $m \times n$, m is the number of rows of the matrix. So, a 6×3 matrix has six rows.

(b) True. On page 19, after Example 4, the sentence reads, "It can be shown that every matrix is row-equivalent to a matrix in row-echelon form."

(c) False. Consider the row-echelon form

$$\begin{bmatrix} 1 & 0 & 0 & 0 & 0 \\ 0 & 1 & 0 & 0 & 1 \\ 0 & 0 & 1 & 0 & 2 \\ 0 & 0 & 0 & 1 & 3 \end{bmatrix}$$

which gives the solution $x_1 = 0$, $x_2 = 1$, $x_3 = 2$, and $x_4 = 3$.

(d) True. Theorem 1.1 states that if a homogeneous system has fewer equations than variables, then it must have an infinite number of solutions.

59. First, a and c cannot both be zero. So, assume $a \neq 0$, and use row reduction as follows.

$$\begin{bmatrix} a & b \\ c & d \end{bmatrix} \Rightarrow \begin{bmatrix} a & b \\ 0 & \frac{-cb}{a} + d \end{bmatrix} \Rightarrow \begin{bmatrix} a & b \\ 0 & ad - bc \end{bmatrix}$$

So, $ad - bc \neq 0$. Similarly, if $c \neq 0$, interchange rows and proceed as above. So the original matrix is row equivalent to the identity if and only if $ad - bc \neq 0$.

61. Form the augmented matrix for this system

$$\begin{bmatrix} \lambda - 2 & 1 & 0 \\ 1 & \lambda - 2 & 0 \end{bmatrix}$$

and reduce the system using elementary row operations.

$$\begin{bmatrix} 1 & \lambda - 2 & 0 \\ \lambda - 2 & 1 & 0 \end{bmatrix} \Rightarrow \begin{bmatrix} 1 & \lambda - 2 & 0 \\ 0 & \lambda^2 - 4\lambda + 3 & 0 \end{bmatrix}$$

To have a nontrivial solution you must have

$$\lambda^2 - 4\lambda + 3 = 0$$
$$(\lambda - 1)(\lambda - 3) = 0.$$

So, if $\lambda = 1$ or $\lambda = 3$, the system will have nontrivial solutions.

63. To show that it is possible you need give only one example, such as

$$x_1 + x_2 + x_3 = 0$$
$$x_1 + x_2 + x_3 = 1$$

which has fewer equations than variables and obviously has no solution.

65. $\begin{bmatrix} a & b \\ c & d \end{bmatrix} \Rightarrow \begin{bmatrix} a-c & b-d \\ c & d \end{bmatrix} \Rightarrow \begin{bmatrix} a-c & b-d \\ a & b \end{bmatrix} \Rightarrow \begin{bmatrix} -c & -d \\ a & b \end{bmatrix} \Rightarrow \begin{bmatrix} c & d \\ a & b \end{bmatrix}$

The rows have been interchanged. In general, the second and third elementary row operations can be used in this manner to interchange two rows of a matrix. So, the first elementary row operation is, in fact, redundant.

67. When a system of linear equations is inconsistent, the row-echelon form of the corresponding augmented matrix will have a row that is all zeros except for the last entry.

69. A matrix in reduced row-echelon form has zeros above each row's leading 1, which may not be true for a matrix in row-echelon form.

Section 1.3   Applications of Systems of Linear Equations

1. (a) Because there are three points, choose a second-degree polynomial, $p(x) = a_0 + a_1 x + a_2 x^2$. Then substitute $x = 2, 3,$ and 4 into $p(x)$ and equate the results to $y = 5, 2,$ and 5, respectively.

$a_0 + a_1(2) + a_2(2)^2 = a_0 + 2a_1 + 4a_2 = 5$
$a_0 + a_1(3) + a_2(3)^2 = a_0 + 3a_1 + 9a_2 = 2$
$a_0 + a_1(4) + a_2(4)^2 = a_0 + 4a_1 + 16a_2 = 5$

Form the augmented matrix

$\begin{bmatrix} 1 & 2 & 4 & 5 \\ 1 & 3 & 9 & 2 \\ 1 & 4 & 16 & 5 \end{bmatrix}$

and use Gauss-Jordan elimination to obtain the equivalent reduced row-echelon matrix

$\begin{bmatrix} 1 & 0 & 0 & 29 \\ 0 & 1 & 0 & -18 \\ 0 & 0 & 1 & 3 \end{bmatrix}$.

So, $p(x) = 29 - 18x + 3x^2$.

(b)

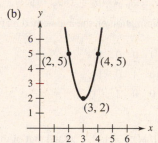

3. (a) Because there are three points, choose a second-degree polynomial, $p(x) = a_0 + a_1 x + a_2 x^2$. Then substitute $x = 2, 3,$ and 5 into $p(x)$ and equate the results to $y = 4, 6,$ and 10, respectively.

$a_0 + a_1(2) + a_2(2)^2 = a_0 + 2a_1 + 4a_2 = 4$
$a_0 + a_1(3) + a_2(3)^2 = a_0 + 3a_1 + 9a_2 = 6$
$a_0 + a_1(5) + a_2(5)^2 = a_0 + 5a_1 + 25a_2 = 10$

Use Gauss-Jordan elimination on the augmented matrix for this system.

$\begin{bmatrix} 1 & 2 & 4 & 4 \\ 1 & 3 & 9 & 6 \\ 1 & 5 & 25 & 10 \end{bmatrix} \Rightarrow \begin{bmatrix} 1 & 0 & 0 & 0 \\ 0 & 1 & 0 & 2 \\ 0 & 0 & 1 & 0 \end{bmatrix}$

So, $p(x) = 2x$.

(b)

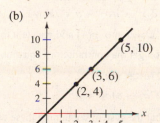

5. (a) Using the translation $z = x - 2007$, the points (z, y) are $(-1, 5)$, $(0, 7)$, and $(1, 12)$. Because there are three points, choose a second-degree polynomial $p(z) = a_0 + a_1 z + a_2 z^2$. Then substitute $z = -1$, 0, and 1 into $p(z)$ and equate the results to $y = 5$, 7, and 12 respectively.

$$a_0 + a_1(-1) + a_2(-1)^2 = a_0 - a_1 + a_2 = 5$$
$$a_0 + a_1(0) + a_2(0)^2 = a_0 \qquad\qquad = 7$$
$$a_0 + a_1(1) + 9_2(1)^2 = a_0 + a_1 + a_2 = 12$$

Use Gauss-Jordan elimination on the augmented matrix for this system.

$$\begin{bmatrix} 1 & -1 & 1 & 5 \\ 1 & 0 & 0 & 7 \\ 1 & 1 & 1 & 12 \end{bmatrix} \Rightarrow \begin{bmatrix} 1 & 0 & 0 & 7 \\ 0 & 1 & 0 & \frac{7}{2} \\ 0 & 0 & 1 & \frac{3}{2} \end{bmatrix}$$

So, $p(z) = 7 + \frac{7}{2}z + \frac{3}{2}z^2$.

Letting $z = x - 2007$, you have $p(x) = 7 + \frac{7}{2}(x - 2007) + \frac{3}{2}(x - 2007)^2$.

(b)

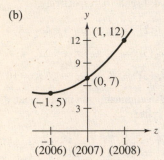

(2006) (2007) (2008)

7. Choose a fourth-degree polynomial and substitute $x = 1, 2, 3$, and 4 into

$p(x) = a_0 + a_1 x + a_2 x^2 + a_3 x^3 + a_4 x^4$. However,

when you substitute $x = 3$ into $p(x)$ and equate it to

$y = 2$ and $y = 3$ you get the contradictory equations

$$a_0 + 3a_1 + 9a_2 + 27a_3 + 81a_4 = 2$$
$$a_0 + 3a_1 + 9a_2 + 27a_3 + 81a_4 = 3$$

and must conclude that the system containing these two equations will have no solution. Also, y is not a function of x because the x-value of 3 is repeated. By similar reasoning, you cannot choose

$p(y) = b_0 + b_1 y + b_2 y^2 + b_3 y^3 + b_4 y^4$ because

$y = 1$ corresponds to both $x = 1$ and $x = 2$.

9. Letting $p(x) = a_0 + a_1 x + a_2 x^2$, substitute $x = 0, 2$,

and 4 into $p(x)$ and equate the results to $y = 1, 3$, and

5, respectively.

$$a_0 + a_1(0) + a_2(0)^2 = a_0 \qquad\qquad = 1$$
$$a_0 + a_1(2) + a_2(2)^2 = a_0 + 2a_1 + 4a_2 = 3$$
$$a_0 + a_1(4) + a_2(4)^2 = a_0 + 4a_1 + 16a_2 = 5$$

Use Gauss-Jordan elimination on the augmented matrix for this system.

$$\begin{bmatrix} 1 & 0 & 0 & 1 \\ 1 & 2 & 4 & 3 \\ 1 & 4 & 16 & 5 \end{bmatrix} \Rightarrow \begin{bmatrix} 1 & 0 & 0 & 1 \\ 0 & 1 & 0 & 1 \\ 0 & 0 & 1 & 0 \end{bmatrix}$$

So, $p(x) = 1 + x$.

The graphs of $y = 1/p(x) = 1/(1 + x)$ and that of the

function $y = 1 - \frac{7}{15}x + \frac{1}{15}x^2$ are shown below.

$$y = \frac{1}{1 + x}$$

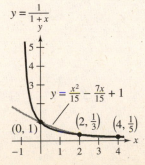

11. To begin, substitute $x = -1$ and $x = 1$ into $p(x) = a_0 + a_1x + a_2x^2 + a_3x^3$ and equate the results

to $y = 2$ and $y = -2$, respectively.

$a_0 - a_1 + a_2 - a_3 = 2$

$a_0 + a_1 + a_2 + a_3 = -2$

Then, differentiate p, yielding $p'(x) = a_1 + 2a_2x + 3a_3x^2$. Substitute $x = -1$ and $x = 1$ into $p'(x)$ and equate the results to 0.

$a_1 - 2a_2 + 3a_3 = 0$

$a_1 + 2a_2 + 3a_3 = 0$

Combining these four equations into one system and forming the augmented matrix, you obtain

$$\begin{bmatrix} 1 & -1 & 1 & -1 & 2 \\ 1 & 1 & 1 & 1 & -2 \\ 0 & 1 & -2 & 3 & 0 \\ 0 & 1 & 2 & 3 & 0 \end{bmatrix}.$$

Use Gauss-Jordan elimination to find the equivalent reduced row-echelon matrix

$$\begin{bmatrix} 1 & 0 & 0 & 0 & 0 \\ 0 & 1 & 0 & 0 & -3 \\ 0 & 0 & 1 & 0 & 0 \\ 0 & 0 & 0 & 1 & 1 \end{bmatrix}.$$

So, $p(x) = -3x + x^3$. The graph of $y = p(x)$ is shown below.

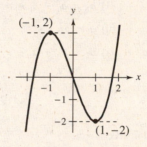

13. (a) Because you are given three points, choose a second-degree polynomial, $p(x) = a_0 + a_1x + a_2x^2$. Because the x-values

are large, use the translation $z = x - 1990$ to obtain $(-10, 227)$, $(0, 249)$, $(10, 281)$. Substituting the given points into

$p(x)$ produces the following system of linear equations.

$a_0 + (-10)a_1 + (-10)^2 a_2 = a_0 - 10a_1 + 100a_2 = 227$

$a_0 + (0)a_1 + (0)^2 a_2 = a_0 \qquad\qquad\quad = 249$

$a_0 + (10)a_1 + (10)^2 a_2 = a_0 + 10a_1 + 100a_2 = 281$

Form the augmented matrix

$$\begin{bmatrix} 1 & -10 & 100 & 227 \\ 1 & 0 & 0 & 249 \\ 1 & 10 & 100 & 281 \end{bmatrix}$$

and use Gauss-Jordan elimination to obtain the equivalent reduced row-echelon matrix

$$\begin{bmatrix} 1 & 0 & 0 & 249 \\ 0 & 1 & 0 & 2.7 \\ 0 & 0 & 1 & 0.05 \end{bmatrix}.$$

So, $p(z) = 249 + 2.7z + 0.05z^2$ and $p(x) = 249 + 2.7(x - 1990) + 0.05(x - 1990)^2$. To predict the population in 2010

and 2020, substitute these values into $p(x)$.

$p(2010) = 249 + 2.7(20) + 0.05(20)^2 = 323$ million

$p(2020) = 249 + 2.7(30) + 0.05(30)^2 = 375$ million

15. (a) Letting $z = x - 2000$ the four points are $(1, 10,003)$, $(3, 10,526)$, $(5, 12,715)$, and $(7, 14,410)$.

Let $p(z) = a_0 + a_1 z + a_2 z^2 + a_3 z^3$.

$$a_0 + a_1(1) + a_2(1)^2 + a_3(1)^3 = a_0 + a_1 + a_2 + a_3 = 10,003$$
$$a_0 + a_1(3) + a_2(3)^2 + a_3(3)^3 = a_0 + 3a_1 + 9a_2 + 27a_3 = 10,526$$
$$a_0 + a_1(5) + a_2(5)^2 + a_3(5)^3 = a_0 + 5a_1 + 25a_2 + 125a_3 = 12,715$$
$$a_0 + a_1(7) + a_2(7)^2 + a_3(7)^3 = a_0 + 7a_1 + 49a_2 + 343a_3 = 14,410$$

(b) Use Gauss-Jordan elimination to solve the system.

$$\begin{bmatrix} 1 & 1 & 1 & 1 & 10,003 \\ 1 & 3 & 9 & 27 & 10,526 \\ 1 & 5 & 25 & 125 & 12,715 \\ 1 & 7 & 49 & 343 & 14,410 \end{bmatrix} \Rightarrow \begin{bmatrix} 1 & 0 & 0 & 0 & 11,041.25 \\ 0 & 1 & 0 & 0 & 1606.5 \\ 0 & 0 & 1 & 0 & 613.25 \\ 0 & 0 & 0 & 1 & -45 \end{bmatrix}$$

So, $p(z) = 11,041.25 + 1606.5z + 613.25z^2 - 45z^3$.

Letting $z = x - 2000$, $p(x) = 11,041.25 + 1606.5(x - 2000) + 613.25(x - 2000)^2 - 45(x - 2000)^3$.

Because the actual net profit increased each year from $2000 - 2007$ except for 2006 and the predicted values decrease each year after 2008, the solution does not produce a reasonable model for predicting future net profits.

17. Choosing a second-degree polynomial approximation,

$p(x) = a_0 + a_1 x + a_2 x^2$, substitute $x = 0, \dfrac{\pi}{2}$, and

π into $p(x)$ and equate the results to $y = 0, 1$, and 0, respectively.

$$a_0 = 0$$
$$a_0 + \frac{\pi}{2}a_1 + \frac{\pi^2}{4}a_2 = 1$$
$$a_0 + \pi a_1 + \pi^2 a_2 = 0$$

Then form the augmented matrix,

$$\begin{bmatrix} 1 & 0 & 0 & 0 \\ 1 & \dfrac{\pi}{2} & \dfrac{\pi^2}{4} & 1 \\ 1 & \pi & \pi^2 & 0 \end{bmatrix}$$

and use Gauss-Jordan elimination to obtain the equivalent reduced row-echelon matrix

$$\begin{bmatrix} 1 & 0 & 0 & 0 \\ 0 & 1 & 0 & \dfrac{4}{\pi} \\ 0 & 0 & 1 & -\dfrac{4}{\pi^2} \end{bmatrix}.$$

So, $p(x) = \dfrac{4}{\pi}x - \dfrac{4}{\pi^2}x^2 = \dfrac{4}{\pi^2}(\pi x - x^2)$.

Furthermore,

$$\sin \frac{\pi}{3} \approx p\left(\frac{\pi}{3}\right) = \frac{4}{\pi^2}\left[\pi\left(\frac{\pi}{3}\right) - \left(\frac{\pi}{3}\right)^2\right]$$

$$= \frac{4}{\pi^2}\left[\frac{2\pi^2}{9}\right] = \frac{8}{9} \approx 0.889.$$

Note that $\sin \pi/3 = 0.866$ to three significant digits.

19. (i) $p(-1) = a_0 + (-1)a_1 + (-1)^2 a_2 = a_0 - a_1 + a_2$

$p(0) = a_0 + (0)a_1 + (0)^2 a_2 = a_0$

$p(1) = a_0 + (1)a_1 + (1)^2 a_2 = a_0 + a_1 + a_2$

(ii) $a_0 - a_1 + a_2 = 0$

$a_0 = 0$

$a_0 + a_1 + a_2 = 0$

(iii) From the augmented matrix

$$\begin{bmatrix} 1 & -1 & 1 & 0 \\ 1 & 0 & 0 & 0 \\ 1 & 1 & 1 & 0 \end{bmatrix}$$

and use Gauss-Jordan elimination to obtain the equivalent reduced row-echelon matrix

$$\begin{bmatrix} 1 & 0 & 0 & 0 \\ 0 & 1 & 0 & 0 \\ 0 & 0 & 1 & 0 \end{bmatrix}.$$

So, $a_0 = 0$, $a_1 = 0$, and $a_2 = 0$.

21. (a) Each of the network's six junctions gives rise to a linear equation as shown below.

input = output

$$600 = x_1 + x_3$$
$$x_1 = x_2 + x_4$$
$$x_2 + x_5 = 500$$
$$x_3 + x_6 = 600$$
$$x_4 + x_7 = x_6$$
$$500 = x_5 + x_7$$

Rearrange these equations, form the augmented matrix, and use Gauss-Jordan elimination.

$$
\begin{bmatrix}
1 & 0 & 1 & 0 & 0 & 0 & 0 & 600 \\
1 & -1 & 0 & -1 & 0 & 0 & 0 & 0 \\
0 & 1 & 0 & 0 & 1 & 0 & 0 & 500 \\
0 & 0 & 1 & 0 & 0 & 1 & 0 & 600 \\
0 & 0 & 0 & 1 & 0 & -1 & 1 & 0 \\
0 & 0 & 0 & 0 & 1 & 0 & 1 & 500
\end{bmatrix}
\Rightarrow
\begin{bmatrix}
1 & 0 & 0 & 0 & 0 & -1 & 0 & 0 \\
0 & 1 & 0 & 0 & 0 & 0 & -1 & 0 \\
0 & 0 & 1 & 0 & 0 & 1 & 0 & 600 \\
0 & 0 & 0 & 1 & 0 & -1 & 1 & 0 \\
0 & 0 & 0 & 0 & 1 & 0 & 1 & 500 \\
0 & 0 & 0 & 0 & 0 & 0 & 0 & 0
\end{bmatrix}
$$

Letting $x_7 = t$ and $x_6 = s$ be the free variables, you have

$$x_1 = s$$
$$x_2 = t$$
$$x_3 = 600 - s$$
$$x_4 = s - t$$
$$x_5 = 500 - t$$
$$x_6 = s$$

$x_7 = t$, where s and t are any real numbers.

(b) If $x_6 = x_7 = 0$, then the solution is $x_1 = 0$, $x_2 = 0$, $x_3 = 600$, $x_4 = 0$, $x_5 = 500$, $x_6 = 0$, $x_7 = 0$.

(c) If $x_5 = 1000$ and $x_6 = 0$, then the solution is $x_1 = 0$, $x_2 = -500$, $x_3 = 600$, $x_4 = 500$, $x_5 = 1000$, $x_6 = 0$, and $x_7 = -500$.

23. (a) Each of the network's four junctions gives rise to a linear equation, as shown below.

input = output

$$200 + x_2 = x_1$$
$$x_4 = x_2 + 100$$
$$x_3 = x_4 + 200$$
$$x_1 + 100 = x_3$$

Rearranging these equations and forming the augmented matrix, you obtain

$$
\begin{bmatrix}
1 & -1 & 0 & 0 & 200 \\
0 & 1 & 0 & -1 & -100 \\
0 & 0 & 1 & -1 & 200 \\
1 & 0 & -1 & 0 & -100
\end{bmatrix}
$$

Gauss-Jordan elimination produces the matrix

$$
\begin{bmatrix}
1 & 0 & 0 & -1 & 100 \\
0 & 1 & 0 & -1 & -100 \\
0 & 0 & 1 & -1 & 200 \\
0 & 0 & 0 & 0 & 0
\end{bmatrix}
$$

Letting $x_4 = t$, you have $x_1 = 100 + t$, $x_2 = -100 + t$, $x_3 = 200 + t$, and $x_4 = t$, where t is a real number.

(b) When $x_4 = t = 0$, then $x_1 = 100$, $x_2 = -100$, $x_3 = 200$.

(c) When $x_4 = t = 100$, then $x_1 = 200$, $x_2 = 0$, $x_3 = 300$.

25. Applying Kirchoff's first law to either junction produces

$$I_1 + I_3 = I_2$$

and applying the second law to the two paths produces

$$R_1 I_1 + R_2 I_2 = 4I_1 + 3I_2 = 3$$
$$R_2 I_2 + R_3 I_3 = 3I_2 + I_3 = 4$$

Rearrange these equations, form the augmented matrix, and use Gauss-Jordan elimination.

$$\begin{bmatrix} 1 & -1 & 1 & 0 \\ 4 & 3 & 0 & 3 \\ 0 & 3 & 1 & 4 \end{bmatrix} \Rightarrow \begin{bmatrix} 1 & 0 & 0 & 0 \\ 0 & 1 & 0 & 1 \\ 0 & 0 & 1 & 1 \end{bmatrix}$$

So, $I_1 = 0$, $I_2 = 1$, and $I_3 = 1$.

27. (a) To find the general solution, let A have a volts and B have b volts. Applying Kirchoff's first law to either junction produces

$$I_1 + I_3 = I_2$$

and applying the second law to the two paths produces

$$R_1 I_1 + R_2 I_2 = I_1 + 2I_2 = a$$
$$R_2 I_2 + R_3 I_3 = 2I_2 + 3I_3 = b$$

Rearrange these three equations and form the augmented matrix.

$$\begin{bmatrix} 1 & -1 & 1 & 0 \\ 1 & 2 & 0 & a \\ 0 & 2 & 4 & b \end{bmatrix}$$

Gauss-Jordan elimination produces the matrix

$$\begin{bmatrix} 1 & 0 & 0 & (3a - b)/7 \\ 0 & 1 & 0 & (4a + b)/14 \\ 0 & 0 & 1 & (3b - 2a)/14 \end{bmatrix}.$$

When $a = 5$ and $b = 8$, then

$$I_1 = 1, I_2 = 2, I_3 = 1.$$

(b) When $a = 2$ and $b = 6$, then

$$I_1 = 0, I_2 = 1, I_3 = 1.$$

29. $$\frac{4x^2}{(x + 1)^2(x - 1)} = \frac{A}{x - 1} + \frac{B}{x + 1} + \frac{C}{(x + 1)^2}$$

$$4x^2 = A(x + 1)^2 + B(x + 1)(x - 1) + C(x - 1)$$

$$4x^2 = Ax^2 + 2Ax + A + Bx^2 - B + Cx - C$$

$$4x^2 = (A + B)x^2 + (2A + C)x + A - B - C$$

So,
$$A + B = 4$$
$$2A \quad + C = 0$$
$$A - B - C = 0.$$

Use Gauss-Jordan elimination to solve the system.

$$\begin{bmatrix} 1 & 1 & 0 & 4 \\ 2 & 0 & 1 & 0 \\ 1 & -1 & -1 & 0 \end{bmatrix} \Rightarrow \begin{bmatrix} 1 & 0 & 0 & 1 \\ 0 & 1 & 0 & 3 \\ 0 & 0 & 1 & -2 \end{bmatrix}$$

The solution is: $A = 1$, $B = 3$, and $C = -2$

So, $$\frac{4x^2}{(x + 1)^2(x - 1)} = \frac{1}{x - 1} + \frac{3}{x + 1} - \frac{2}{(x + 1)^2}.$$

31. $$\frac{20 - x^2}{(x + 2)(x - 2)^2} = \frac{A}{x + 2} + \frac{B}{x - 2} = \frac{C}{(x - 2)^2}$$

$$20 - x^2 = A(x - 2)^2 + B(x + 2)(x - 2) + C(x + 2)$$

$$20 - x^2 = Ax^2 - 4Ax + 4A + Bx^2 - 4B + Cx + 2C$$

$$20 - x^2 = (A + B)x^2 + (-4A + C)x + 4A - 4B + 2C$$

So,
$$A + B = -1$$
$$-4A \quad + C = 0$$
$$4A - 4B + 2C = 20.$$

Use Gauss-Jordan elimination to solve the system.

$$\begin{bmatrix} 1 & 1 & 0 & -1 \\ -4 & 0 & 1 & 0 \\ 4 & -4 & 2 & 20 \end{bmatrix} \Rightarrow \begin{bmatrix} 1 & 0 & 0 & 1 \\ 0 & 1 & 0 & -2 \\ 0 & 0 & 1 & 4 \end{bmatrix}$$

The solution is: $A = 1$, $B = -2$, and $C = 4$.

So, $$\frac{20 - x^2}{(x + 2)(x - 2)^2} = \frac{1}{x + 2} - \frac{2}{x - 2} + \frac{4}{(x - 2)^2}.$$

33. Use Gauss-Jordan elimination to solve the system

$$\begin{bmatrix} 2 & 0 & 1 & 0 \\ 0 & 2 & 1 & 0 \\ 1 & 1 & 0 & 4 \end{bmatrix} \Rightarrow \begin{bmatrix} 1 & 0 & 0 & 2 \\ 0 & 1 & 0 & 2 \\ 0 & 0 & 1 & -4 \end{bmatrix}$$

So, $x = 2$ $y = 2$ and $\lambda = -4$.

35. Let x_1 = number of touchdowns, x_2 = number of extra-point kicks, and x_3 = number of field goals.

$$x_1 + x_2 + x_3 = 13$$
$$6x_1 + x_2 + 3x_3 = 46$$
$$x_2 - x_3 = 0$$

Use Gauss-Jordan elimination to solve the system.

$$\begin{bmatrix} 1 & 1 & 1 & 13 \\ 6 & 1 & 3 & 46 \\ 0 & 1 & -1 & 0 \end{bmatrix} \Rightarrow \begin{bmatrix} 1 & 0 & 0 & 5 \\ 0 & 1 & 0 & 4 \\ 0 & 0 & 1 & 4 \end{bmatrix}$$

Because $x_1 = 5$, $x_2 = 4$, and $x_3 = 4$, there were 5 touchdowns, 4 extra-point kicks, and 4 field goals.

Review Exercises for Chapter 1

1. Because the equation cannot be written in the form $a_1x + a_2y = b$, it is *not* linear in the variables x and y.

3. Because the equation *is* in the form $a_1x + a_2y = b$, it *is* linear in the variables x and y.

5. Because the equation cannot be written in the form $a_1x + a_2y = b$, it is *not* linear in the variables x and y.

7. Because the equation is in the form $a_1x + a_2y = b$, it is linear in the variables x and y.

9. Choosing y and z as the free variables and letting $y = s$ and $z = t$, you have

$$-4x + 2s - 6t = 1$$
$$-4x = 1 - 2s + 6t$$
$$x = -\tfrac{1}{4} + \tfrac{1}{2}s - \tfrac{3}{2}t.$$

So, the solution set can be described as $x = -\tfrac{1}{4} + \tfrac{1}{2}s - \tfrac{3}{2}t$, $y = s$, $z = t$, where s and t are real numbers.

11. Row reduce the augmented matrix for this system.

$$\begin{bmatrix} 1 & 1 & 2 \\ 3 & -1 & 0 \end{bmatrix} \Rightarrow \begin{bmatrix} 1 & 1 & 2 \\ 0 & -4 & -6 \end{bmatrix} \Rightarrow \begin{bmatrix} 1 & 1 & 2 \\ 0 & 1 & \frac{3}{2} \end{bmatrix} \Rightarrow \begin{bmatrix} 1 & 0 & \frac{1}{2} \\ 0 & 1 & \frac{3}{2} \end{bmatrix}$$

Converting back to a linear system, the solution is: $x = \tfrac{1}{2}$ and $y = \tfrac{3}{2}$.

13. Rearrange the equations as shown below.

$$x - y = -4$$
$$2x - 3y = 0$$

Row reduce the augmented matrix for this system.

$$\begin{bmatrix} 1 & -1 & -4 \\ 2 & -3 & 0 \end{bmatrix} \Rightarrow \begin{bmatrix} 1 & -1 & -4 \\ 0 & -1 & 8 \end{bmatrix} \Rightarrow \begin{bmatrix} 1 & -1 & -4 \\ 0 & 1 & -8 \end{bmatrix} \Rightarrow \begin{bmatrix} 1 & 0 & -12 \\ 0 & 1 & -8 \end{bmatrix}$$

Converting back to a linear system, the solution is: $x = -12$ and $y = -8$.

15. Row reduce the augmented matrix for this system.

$$\begin{bmatrix} 1 & 1 & 0 \\ 2 & 1 & 0 \end{bmatrix} \Rightarrow \begin{bmatrix} 1 & 1 & 0 \\ 0 & -1 & 0 \end{bmatrix} \Rightarrow \begin{bmatrix} 1 & 1 & 0 \\ 0 & 1 & 0 \end{bmatrix} \Rightarrow \begin{bmatrix} 1 & 0 & 0 \\ 0 & 1 & 0 \end{bmatrix}$$

Converting back to a linear system, the solution is: $x = 0$ and $y = 0$.

17. The augmented matrix for this system is

$$\begin{bmatrix} 1 & -1 & 9 \\ -1 & 1 & 1 \end{bmatrix}$$

which is equivalent to the reduced row-echelon matrix

$$\begin{bmatrix} 1 & -1 & 0 \\ 0 & 0 & 1 \end{bmatrix}.$$

Because the second row corresponds to $0 = 1$, which is a false statement, you can conclude that the system has no solution.

19. Multiplying both equations by 100 and forming the augmented matrix produces

$$\begin{bmatrix} 20 & 30 & 14 \\ 40 & 50 & 20 \end{bmatrix}.$$

Use Gauss-Jordan elimination as shown below.

$$\begin{bmatrix} 1 & \frac{3}{2} & \frac{7}{10} \\ 40 & 50 & 20 \end{bmatrix} \Rightarrow \begin{bmatrix} 1 & \frac{3}{2} & \frac{7}{10} \\ 0 & -10 & -8 \end{bmatrix} \Rightarrow \begin{bmatrix} 1 & \frac{3}{2} & \frac{7}{10} \\ 0 & 1 & \frac{4}{5} \end{bmatrix} \Rightarrow \begin{bmatrix} 1 & 0 & -\frac{1}{2} \\ 0 & 1 & \frac{4}{5} \end{bmatrix}$$

So, the solution is: $x_1 = -\frac{1}{2}$ and $x_2 = \frac{4}{5}$.

21. Expanding the second equation, $3x + 2y = 0$, the augmented matrix for this system is

$$\begin{bmatrix} \frac{1}{2} & -\frac{1}{3} & 0 \\ 3 & 2 & 0 \end{bmatrix}$$

which is equivalent to the reduced row-echelon matrix

$$\begin{bmatrix} 1 & 0 & 0 \\ 0 & 1 & 0 \end{bmatrix}.$$

So, the solution is: $x = 0$ and $y = 0$.

23. Because the matrix has 2 rows and 3 columns, it has size 2×3.

25. This matrix has the characteristic stair step pattern of leading 1's so it is in row-echelon form. However, the leading 1 in row three of column four has 1's above it, so the matrix is *not* in reduced row-echelon form.

27. Because the first row begins with -1, this matrix is not in row-echelon form.

29. This matrix corresponds to the system

$$x_1 + 2x_2 = 0$$
$$x_3 = 0.$$

Choosing $x_2 = t$ as the free variable you can describe the solution as $x_1 = -2t$, $x_2 = t$, and $x_3 = 0$, where t is a real number.

31. The augmented matrix for this system is

$$\begin{bmatrix} -1 & 1 & 2 & 1 \\ 2 & 3 & 1 & -2 \\ 5 & 4 & 2 & 4 \end{bmatrix}$$

which is equivalent to the reduced row-echelon matrix

$$\begin{bmatrix} 1 & 0 & 0 & 2 \\ 0 & 1 & 0 & -3 \\ 0 & 0 & 1 & 3 \end{bmatrix}.$$

So, the solution is: $x = 2$, $y = -3$, and $z = 3$.

33. Use Gauss-Jordan elimination on the augmented matrix.

$$\begin{bmatrix} 2 & 3 & 3 & 3 \\ 6 & 6 & 12 & 13 \\ 12 & 9 & -1 & 2 \end{bmatrix} \Rightarrow \begin{bmatrix} 1 & 0 & 0 & \frac{1}{2} \\ 0 & 1 & 0 & -\frac{1}{3} \\ 0 & 0 & 1 & 1 \end{bmatrix}$$

So, $x = \frac{1}{2}$, $y = -\frac{1}{3}$, and $z = 1$.

35. The augmented matrix for this system is

$$\begin{bmatrix} 1 & -2 & 1 & -6 \\ 2 & -3 & 0 & -7 \\ -1 & 3 & -3 & 11 \end{bmatrix}$$

which is equivalent to the reduced row-echelon matrix

$$\begin{bmatrix} 1 & 0 & -3 & 4 \\ 0 & 1 & -2 & 5 \\ 0 & 0 & 0 & 0 \end{bmatrix}.$$

Choosing $z = t$ as the free variable you find that the solution set can be described by $x = 4 + 3t$,

$y = 5 + 2t$, and $z = t$, where t is a real number.

37. Use the Gauss-Jordan elimination on the augmented matrix for this system.

$$\begin{bmatrix} 2 & 1 & 2 & 4 \\ 2 & 2 & 0 & 5 \\ 2 & -1 & 6 & 2 \end{bmatrix} \Rightarrow \begin{bmatrix} 1 & 0 & 2 & \frac{3}{2} \\ 0 & 1 & -2 & 1 \\ 0 & 0 & 0 & 0 \end{bmatrix}$$

So, the solution is: $x = \frac{3}{2} - 2t$, $y = 1 + 2t$, $z = t$, where t is any real number.

39. The augmented matrix for this system is

$$\begin{bmatrix} 2 & 1 & 1 & 2 & -1 \\ 5 & -2 & 1 & -3 & 0 \\ -1 & 3 & 2 & 2 & 1 \\ 3 & 2 & 3 & -5 & 12 \end{bmatrix}$$

which is equivalent to the reduced row-echelon matrix

$$\begin{bmatrix} 1 & 0 & 0 & 0 & 1 \\ 0 & 1 & 0 & 0 & 4 \\ 0 & 0 & 1 & 0 & -3 \\ 0 & 0 & 0 & 1 & -2 \end{bmatrix}$$

So, the solution is: $x_1 = 1$, $x_2 = 4$, $x_3 = -3$, and $x_4 = -2$.

41. Using a graphing utility, the augmented matrix reduces to

$$\begin{bmatrix} 1 & 0 & 0 & 0 \\ 0 & 1 & 4 & 2 \\ 0 & 0 & 0 & 0 \\ 0 & 0 & 0 & 0 \end{bmatrix}$$

Choosing $z = t$ as the free variable, you find that the solution set can be described by $x = 0$, $y = z - 4t$, and $z = t$, where t is a real number.

43. Using a graphing utility, the augmented matrix reduces to

$$\begin{bmatrix} 1 & 0 & 0 & 0 & 1 \\ 0 & 1 & 0 & 0 & 0 \\ 0 & 0 & 1 & 0 & 4 \\ 0 & 0 & 0 & 1 & -2 \end{bmatrix}$$

So, the solution is: $x = 1$, $y = 0$, $z = 4$, and $w = -2$.

45. Using a graphing utility, the augmented matrix reduces to

$$\begin{bmatrix} 1 & 0 & 2 & 0 & 0 \\ 0 & 1 & -1 & 0 & 0 \\ 0 & 0 & 0 & 1 & 0 \end{bmatrix}$$

Choosing $z = t$ as the free variable you find that the solution set can be described by $x = -2t$, $y = t$, $z = t$, and $w = 0$, where t is a real number.

47. Use Guass-Jordan elimination on the augmented matrix.

$$\begin{bmatrix} 1 & -2 & -8 & 0 \\ 3 & 2 & 0 & 0 \\ -1 & 1 & 7 & 0 \end{bmatrix} \Rightarrow \begin{bmatrix} 1 & 0 & 0 & 0 \\ 0 & 1 & 0 & 0 \\ 0 & 0 & 1 & 0 \end{bmatrix}$$

So, the solution is $x_1 = x_2 = x_3 = 0$.

49. The augmented matrix for this system is

$$\begin{bmatrix} 2 & -8 & 4 & 0 \\ 3 & -10 & 7 & 0 \\ 0 & 10 & 5 & 0 \end{bmatrix}$$

which is equivalent to the reduced row-echelon matrix

$$\begin{bmatrix} 1 & 0 & 4 & 0 \\ 0 & 1 & \frac{1}{2} & 0 \\ 0 & 0 & 0 & 0 \end{bmatrix}$$

Choosing $x_3 = t$ as the free variable, you find that the solution set can be described by $x_1 = -4t$, $x_2 = -\frac{1}{2}t$, and $x_3 = t$, where t is a real number.

51. Forming the augmented matrix

$$\begin{bmatrix} k & 1 & 0 \\ 1 & k & 1 \end{bmatrix}$$

and using Gauss-Jordan elimination, you obtain

$$\begin{bmatrix} 1 & k & 1 \\ k & 1 & 0 \end{bmatrix} \Rightarrow \begin{bmatrix} 1 & k & 1 \\ 0 & 1-k^2 & -k \end{bmatrix} \Rightarrow \begin{bmatrix} 1 & k & 1 \\ 0 & 1 & \frac{k}{k^2-1} \end{bmatrix} \Rightarrow \begin{bmatrix} 1 & 0 & \frac{-1}{k^2-1} \\ 0 & 1 & \frac{k}{k^2-1} \end{bmatrix}, \ k^2 - 1 \neq 0.$$

So, the system is inconsistent if $k = \pm 1$.

53. Row reduce the augmented matrix.

$$\begin{bmatrix} 1 & 2 & 3 \\ a & b & -9 \end{bmatrix} \Rightarrow \begin{bmatrix} 1 & 2 & 3 \\ 0 & (b-2a) & (-9-3a) \end{bmatrix}$$

(a) There will be no solution if $b - 2a = 0$ *and* $-9 - 3a \neq 0$. That is if $b = 2a$ *and* $a = -3$.

(b) There will be exactly one solution if $b \neq 2a$.

(c) There will be an infinite number of solutions if $b = 2a$ and $a = -3$. That is, if $a = -3$ and $b = -6$.

55. You can show that two matrices of the same size are row equivalent if they both row reduce to the same matrix. The two given matrices are row equivalent because each is row equivalent to the identity matrix.

57. Adding a multiple of row one to each row yields the following matrix.

$$\begin{bmatrix} 1 & 2 & 3 & \cdots & n \\ 0 & -n & -2n & \cdots & -(n-1)n \\ 0 & -2n & -4n & \cdots & -2(n-1)n \\ \vdots & \vdots & \vdots & & \vdots \\ 0 & -(n-1)n & -2(n-1)n & \cdots & -(n-1)(n-1)n \end{bmatrix}$$

Every row below row two is a multiple of row two. Therefore, reduce these rows to zeros.

$$\begin{bmatrix} 1 & 2 & 3 & \cdots & n \\ 0 & -n & -2n & \cdots & -(n-1)n \\ 0 & 0 & 0 & \cdots & 0 \\ \vdots & \vdots & \vdots & \cdots & \vdots \\ 0 & 0 & 0 & & 0 \end{bmatrix}$$

Dividing row two by $-n$ yields a new second row.

$$\begin{bmatrix} 1 & 2 & 3 & \cdots & n \\ 0 & 1 & 2 & \cdots & n-1 \\ 0 & 0 & 0 & \cdots & 0 \\ \vdots & \vdots & \vdots & & \vdots \\ 0 & 0 & 0 & \cdots & 0 \end{bmatrix}$$

Adding -2 times row two to row one yields a new first row.

$$\begin{bmatrix} 1 & 0 & -1 & \cdots & 2-n \\ 0 & 1 & 2 & \cdots & n-1 \\ 0 & 0 & 0 & \cdots & 0 \\ \vdots & \vdots & \vdots & & \vdots \\ 0 & 0 & 0 & \cdots & 0 \end{bmatrix}$$

This matrix is in reduced row-echelon form.

63. $\dfrac{3x^2 - 3x - 2}{(x+2)(x-2)^2} = \dfrac{A}{x+2} + \dfrac{B}{x-2} + \dfrac{C}{(x-2)^2}$

$3x^2 - 3x - 2 = A(x-2)^2 + B(x+2)(x-2) + C(x+2)$

$3x^2 - 3x - 2 = Ax^2 - 4Ax + 4A + Bx^2 - 4B + Cx + 2C$

$3x^2 - 3x - 2 = (A+B)x^2 + (-4A+C)x + 4A - 4B + 2C$

So, $\quad A + B \qquad\qquad = 3$

$\quad -4A + \qquad C = -3$

$\quad 4A - 4B + 2C = -2.$

Use Gauss-Jordan elimination to solve the system.

$$\begin{bmatrix} 1 & 1 & 0 & 3 \\ -4 & 0 & 1 & -3 \\ 4 & -4 & 2 & -2 \end{bmatrix} \Rightarrow \begin{bmatrix} 1 & 0 & 0 & 1 \\ 0 & 1 & 0 & 2 \\ 0 & 0 & 1 & 1 \end{bmatrix}$$

The solution is: $A = 1$, $B = 2$, and $C = 1$. So, $\dfrac{3x^2 - 3x - 2}{(x+2)(x-2)^2} = \dfrac{1}{x+2} + \dfrac{2}{x-2} + \dfrac{1}{(x-2)^2}$.

59. (a) False. See page 3, following Example 2.

(b) True. See page 5, Example 4(b).

61. (a) Let x_1 = number of three-point baskets,
x_2 = number of two-point baskets, and
x_3 = number of one-point free throws.

$3x_1 + 2x_2 + x_3 = 59$

$3x_1 - x_2 \qquad = 0$

$\qquad x_2 - x_3 = 1$

(b) Use Gauss-Jordan elimination to solve the system.

$$\begin{bmatrix} 3 & 2 & 1 & 59 \\ 3 & -1 & 0 & 0 \\ 0 & 1 & -1 & 1 \end{bmatrix} \Rightarrow \begin{bmatrix} 1 & 0 & 0 & 5 \\ 0 & 1 & 0 & 15 \\ 0 & 0 & 1 & 14 \end{bmatrix}$$

Because $x = 5$, $y = 15$, and $z = 14$, there were 5 three-point baskets, 15 two-point baskets, and 14 one-point free throws.

65. (a) Because there are three points, choose a second degree polynomial, $p(x) = a_0 + a_1x + a_2x^2$. By substituting the values at each point into this equation you obtain the system

$$a_0 + 2a_1 + 4a_2 = 5$$
$$a_0 + 3a_1 + 9a_2 = 0$$
$$a_0 + 4a_1 + 16a_2 = 20.$$

Forming the augmented matrix

$$\begin{bmatrix} 1 & 2 & 4 & 5 \\ 1 & 3 & 9 & 0 \\ 1 & 4 & 16 & 20 \end{bmatrix}$$

and using Gauss-Jordan elimination you obtain

$$\begin{bmatrix} 1 & 0 & 0 & 90 \\ 0 & 1 & 0 & -\frac{135}{2} \\ 0 & 0 & 1 & \frac{25}{2} \end{bmatrix}.$$

So, $p(x) = 90 - \frac{135}{2}x + \frac{25}{2}x^2$.

(b)

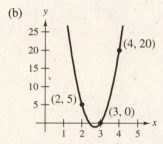

67. Establish the first year as $x = 0$ and substitute the values at each point into $p(x) = a_0 + a_1x + a_2x^2$ to obtain the system

$$a_0 \qquad\qquad = 50$$
$$a_0 + a_1 + a_2 = 60$$
$$a_0 + 2a_1 + 4a_2 = 75.$$

Forming the augmented matrix

$$\begin{bmatrix} 1 & 0 & 0 & 50 \\ 1 & 1 & 1 & 60 \\ 1 & 2 & 4 & 75 \end{bmatrix}$$

and using Gauss-Jordan elimination, you obtain

$$\begin{bmatrix} 1 & 0 & 0 & 50 \\ 0 & 1 & 0 & \frac{15}{2} \\ 0 & 0 & 1 & \frac{5}{2} \end{bmatrix}.$$

So, $p(x) = 50 + \frac{15}{2}x + \frac{5}{2}x^2$. To predict the sales in the fourth year, evaluate $p(x)$ when $x = 3$.

$$p(3) = 50 + \frac{15}{2}(3) + \frac{5}{2}(3)^2 = \$95.$$

69. (a) There are three points: $(0, 80)$, $(4, 68)$, and $(80, 30)$. Because you are given three points, choose a second-degree polynomial, $p(x) = a_0 + a_1x + a_2x^2$. Substituting the given points into $p(x)$ produces the following system of linear equations.

$$a_0 + (0)a_1 + (0)^2 a_2 = a_0 \qquad\qquad = 80$$
$$a_0 + (4)a_1 + (4)^2 a_2 = a_0 + 4a_1 + 16a_2 = 68$$
$$a_0 + (80)a_1 + (80)^2 a_2 = a_0 + 80a_1 + 6400a_2 = 30$$

(b) Form the augmented matrix

$$\begin{bmatrix} 1 & 0 & 0 & 80 \\ 1 & 4 & 16 & 68 \\ 1 & 80 & 6400 & 30 \end{bmatrix}$$

and use Gauss-Jordan elimination to obtain the equivalent reduced row-echelon matrix

$$\begin{bmatrix} 1 & 0 & 0 & 80 \\ 0 & 1 & 0 & -\frac{25}{8} \\ 0 & 0 & 1 & \frac{1}{32} \end{bmatrix}.$$

So, $p(x) = 80 - \frac{25}{8}x + \frac{1}{32}x^2$.

(c) The graphing utility gives $a_0 = 80$, $a_1 = -\frac{25}{8}$, and $a_2 = \frac{1}{32}$. In other words $p(x) = 80 - \frac{25}{8}x + \frac{1}{32}x^2$.

(d) The results to (b) and (c) are the same.

(e) There is precisely one polynomial function of degree $n - 1$ (or less) that fits n distinct points.

71. Applying Kirchoff's first law to either junction produces $I_1 + I_3 = I_2$ and applying the second law to the two paths produces

$$R_1I_1 + R_2I_2 = 3I_1 + 4I_2 = 3$$
$$R_2I_2 + R_3I_3 = 4I_2 + 2I_3 = 2.$$

Rearrange these equations, form the augmented matrix, and use Gauss-Jordan elimination.

$$\begin{bmatrix} 1 & -1 & 1 & 0 \\ 3 & 4 & 0 & 3 \\ 0 & 4 & 2 & 2 \end{bmatrix} \Rightarrow \begin{bmatrix} 1 & 0 & 0 & \frac{5}{13} \\ 0 & 1 & 0 & \frac{6}{13} \\ 0 & 0 & 1 & \frac{1}{13} \end{bmatrix}$$

So, the solution is $I_1 = \frac{5}{13}$, $I_2 = \frac{6}{13}$, and $I_3 = \frac{1}{13}$.

C H A P T E R 2
Matrices

CHAPTER 2
Matrices

Section 2.1 Operations with Matrices

1. (a) $A + B = \begin{bmatrix} 1 & -1 \\ 2 & -1 \end{bmatrix} + \begin{bmatrix} 2 & -1 \\ -1 & 8 \end{bmatrix} = \begin{bmatrix} 1+2 & -1-1 \\ 2-1 & -1+8 \end{bmatrix} = \begin{bmatrix} 3 & -2 \\ 1 & 7 \end{bmatrix}$

(b) $A - B = \begin{bmatrix} 1 & -1 \\ 2 & -1 \end{bmatrix} - \begin{bmatrix} 2 & -1 \\ -1 & 8 \end{bmatrix} = \begin{bmatrix} 1-2 & -1+1 \\ 2+1 & -1-8 \end{bmatrix} = \begin{bmatrix} -1 & 0 \\ 3 & -9 \end{bmatrix}$

(c) $2A = 2\begin{bmatrix} 1 & -1 \\ 2 & -1 \end{bmatrix} = \begin{bmatrix} 2(1) & 2(-1) \\ 2(2) & 2(-1) \end{bmatrix} = \begin{bmatrix} 2 & -2 \\ 4 & -2 \end{bmatrix}$

(d) $2A - B = \begin{bmatrix} 2 & -2 \\ 4 & -2 \end{bmatrix} - \begin{bmatrix} 2 & -1 \\ -1 & 8 \end{bmatrix} = \begin{bmatrix} 0 & -1 \\ 5 & -10 \end{bmatrix}$

(e) $B + \frac{1}{2}A = \begin{bmatrix} 2 & -1 \\ -1 & 8 \end{bmatrix} + \frac{1}{2}\begin{bmatrix} 1 & -1 \\ 2 & -1 \end{bmatrix} = \begin{bmatrix} 2 & -1 \\ -1 & 8 \end{bmatrix} + \begin{bmatrix} \frac{1}{2} & -\frac{1}{2} \\ 1 & -\frac{1}{2} \end{bmatrix} = \begin{bmatrix} \frac{5}{2} & -\frac{3}{2} \\ 0 & \frac{15}{2} \end{bmatrix}$

3. (a) $A + B = \begin{bmatrix} 6 & -1 \\ 2 & 4 \\ -3 & 5 \end{bmatrix} + \begin{bmatrix} 1 & 4 \\ -1 & 5 \\ 1 & 10 \end{bmatrix} = \begin{bmatrix} 6+1 & -1+4 \\ 2+(-1) & 4+5 \\ -3+1 & 5+10 \end{bmatrix} = \begin{bmatrix} 7 & 3 \\ 1 & 9 \\ -2 & 15 \end{bmatrix}$

(b) $A - B = \begin{bmatrix} 6 & -1 \\ 2 & 4 \\ -3 & 5 \end{bmatrix} - \begin{bmatrix} 1 & 4 \\ -1 & 5 \\ 1 & 10 \end{bmatrix} = \begin{bmatrix} 6-1 & -1-4 \\ 2-(-1) & 4-5 \\ -3-1 & 5-10 \end{bmatrix} = \begin{bmatrix} 5 & -5 \\ 3 & -1 \\ -4 & -5 \end{bmatrix}$

(c) $2A = 2\begin{bmatrix} 6 & -1 \\ 2 & 4 \\ -3 & 5 \end{bmatrix} = \begin{bmatrix} 2(6) & 2(-1) \\ 2(2) & 2(4) \\ 2(-3) & 2(5) \end{bmatrix} = \begin{bmatrix} 12 & -2 \\ 4 & 8 \\ -6 & 10 \end{bmatrix}$

(d) $2A - B = \begin{bmatrix} 12 & -2 \\ 4 & 8 \\ -6 & 10 \end{bmatrix} - \begin{bmatrix} 1 & 4 \\ -1 & 5 \\ 1 & 10 \end{bmatrix} = \begin{bmatrix} 12-1 & -2-4 \\ 4-(-1) & 8-5 \\ -6-1 & 10-10 \end{bmatrix} = \begin{bmatrix} 11 & -6 \\ 5 & 3 \\ -7 & 0 \end{bmatrix}$

(e) $B + \frac{1}{2}A = \begin{bmatrix} 1 & 4 \\ -1 & 5 \\ 1 & 10 \end{bmatrix} + \frac{1}{2}\begin{bmatrix} 6 & -1 \\ 2 & 4 \\ -3 & 5 \end{bmatrix} = \begin{bmatrix} 1 & 4 \\ -1 & 5 \\ 1 & 10 \end{bmatrix} + \begin{bmatrix} 3 & -\frac{1}{2} \\ 1 & 2 \\ -\frac{3}{2} & \frac{5}{2} \end{bmatrix} = \begin{bmatrix} 4 & \frac{7}{2} \\ 0 & 7 \\ -\frac{1}{2} & \frac{25}{2} \end{bmatrix}$

5. (a) $A + B = \begin{bmatrix} 3 & 2 & -1 \\ 2 & 4 & 5 \\ 0 & 1 & 2 \end{bmatrix} + \begin{bmatrix} 0 & 2 & 1 \\ 5 & 4 & 2 \\ 2 & 1 & 0 \end{bmatrix} = \begin{bmatrix} 3+0 & 2+2 & -1+1 \\ 2+5 & 4+4 & 5+2 \\ 0+2 & 1+1 & 2+0 \end{bmatrix} = \begin{bmatrix} 3 & 4 & 0 \\ 7 & 8 & 7 \\ 2 & 2 & 2 \end{bmatrix}$

(b) $A - B = \begin{bmatrix} 3 & 2 & -1 \\ 2 & 4 & 5 \\ 0 & 1 & 2 \end{bmatrix} - \begin{bmatrix} 0 & 2 & 1 \\ 5 & 4 & 2 \\ 2 & 1 & 0 \end{bmatrix} = \begin{bmatrix} 3-0 & 2-2 & -1-1 \\ 2-5 & 4-4 & 5-2 \\ 0-2 & 1-1 & 2-0 \end{bmatrix} = \begin{bmatrix} 3 & 0 & -2 \\ -3 & 0 & 3 \\ -2 & 0 & 2 \end{bmatrix}$

(c) $2A = 2\begin{bmatrix} 3 & 2 & -1 \\ 2 & 4 & 5 \\ 0 & 1 & 2 \end{bmatrix} = \begin{bmatrix} 2(3) & 2(2) & 2(-1) \\ 2(2) & 2(4) & 2(5) \\ 2(0) & 2(1) & 2(2) \end{bmatrix} = \begin{bmatrix} 6 & 4 & -2 \\ 4 & 8 & 10 \\ 0 & 2 & 4 \end{bmatrix}$

(d) $2A - B = 2\begin{bmatrix} 3 & 2 & -1 \\ 2 & 4 & 5 \\ 0 & 1 & 2 \end{bmatrix} - \begin{bmatrix} 0 & 2 & 1 \\ 5 & 4 & 2 \\ 2 & 1 & 0 \end{bmatrix} = \begin{bmatrix} 6 & 4 & -2 \\ 4 & 8 & 10 \\ 0 & 2 & 4 \end{bmatrix} - \begin{bmatrix} 0 & 2 & 1 \\ 5 & 4 & 2 \\ 2 & 1 & 0 \end{bmatrix} = \begin{bmatrix} 6 & 2 & -3 \\ -1 & 4 & 8 \\ -2 & 1 & 4 \end{bmatrix}$

(e) $B - \frac{1}{2}A = \begin{bmatrix} 0 & 2 & 1 \\ 5 & 4 & 2 \\ 2 & 1 & 0 \end{bmatrix} + \frac{1}{2}\begin{bmatrix} 3 & 2 & -1 \\ 2 & 4 & 5 \\ 0 & 1 & 2 \end{bmatrix} = \begin{bmatrix} 0 & 2 & 1 \\ 5 & 4 & 2 \\ 2 & 1 & 0 \end{bmatrix} + \begin{bmatrix} \frac{3}{2} & 1 & -\frac{1}{2} \\ 1 & 2 & \frac{5}{2} \\ 0 & \frac{1}{2} & 1 \end{bmatrix} = \begin{bmatrix} \frac{3}{2} & 3 & \frac{1}{2} \\ 6 & 6 & \frac{9}{2} \\ 2 & \frac{3}{2} & 1 \end{bmatrix}$

7. (a) $c_{21} = 2a_{21} - 3b_{21} = 2(-3) - 3(0) = -6$

(b) $c_{13} = 2a_{13} - 3b_{13} = 2(4) - 3(-7) = 29$

9. Expanding both sides of the equation produces

$$\begin{bmatrix} 4x & 4y \\ 4z & -4 \end{bmatrix} = \begin{bmatrix} 2y + 8 & 2z + 2x \\ -2x + 10 & 2 - 2x \end{bmatrix}.$$

By setting corresponding entries equal to each other, you obtain four equations.

$4x = 2y + 8 \implies 4x - 2y = 8$

$4y = 2z + 2x \implies 2x - 4y + 2z = 0$

$4z = -2x + 10 \implies 2x + 4z = 10$

$-4 = 2 - 2x \implies 2x = 6$

Gauss-Jordan elimination produces $x = 3$, $y = 2$, and $z = 1$.

11. (a) $AB = \begin{bmatrix} 1 & 2 \\ 4 & 2 \end{bmatrix}\begin{bmatrix} 2 & -1 \\ -1 & 8 \end{bmatrix} = \begin{bmatrix} 1(2) + 2(-1) & 1(-1) + 2(8) \\ 4(2) + 2(-1) & 4(-1) + 2(8) \end{bmatrix} = \begin{bmatrix} 0 & 15 \\ 6 & 12 \end{bmatrix}$

(b) $BA = \begin{bmatrix} 2 & -1 \\ -1 & 8 \end{bmatrix}\begin{bmatrix} 1 & 2 \\ 4 & 2 \end{bmatrix} = \begin{bmatrix} 2(1) + (-1)(4) & 2(2) + (-1)(2) \\ -1(1) + 8(4) & -1(2) + 8(2) \end{bmatrix} = \begin{bmatrix} -2 & 2 \\ 31 & 14 \end{bmatrix}$

13. (a) AB is not defined because A is 3×2 and B is 3×3.

(b) $BA = \begin{bmatrix} 0 & -1 & 0 \\ 4 & 0 & 2 \\ 8 & -1 & 7 \end{bmatrix}\begin{bmatrix} 2 & 1 \\ -3 & 4 \\ 1 & 6 \end{bmatrix}$

$= \begin{bmatrix} 0(2) + (-1)(-3) + 0(1) & 0(1) + (-1)(4) + 0(6) \\ 4(2) + 0(-3) + 2(1) & 4(1) + 0(4) + 2(6) \\ 8(2) + (-1)(-3) + 7(1) & 8(1) + (-1)(4) + 7(6) \end{bmatrix} = \begin{bmatrix} 3 & -4 \\ 10 & 16 \\ 26 & 46 \end{bmatrix}$

15. (a) $AB = \begin{bmatrix} -1 & 3 \\ 4 & -5 \\ 0 & 2 \end{bmatrix}\begin{bmatrix} 1 & 2 \\ 0 & 7 \end{bmatrix} = \begin{bmatrix} -1(1) + 3(0) & -1(2) + 3(7) \\ 4(1) + (-5)(0) & 4(2) + (-5)(7) \\ 0(1) + 2(0) & 0(2) + 2(7) \end{bmatrix} = \begin{bmatrix} -1 & 19 \\ 4 & -27 \\ 0 & 14 \end{bmatrix}$

(b) BA is not defined because B is a 2×2 matrix and A is a 3×2 matrix.

17. (a) $AB = \begin{bmatrix} 6 \\ -2 \\ 1 \\ 6 \end{bmatrix}\begin{bmatrix} 10 & 12 \end{bmatrix} = \begin{bmatrix} 6(10) & 6(12) \\ -2(10) & -2(12) \\ 1(10) & 1(12) \\ 6(10) & 6(12) \end{bmatrix} = \begin{bmatrix} 60 & 72 \\ -20 & -24 \\ 10 & 12 \\ 60 & 72 \end{bmatrix}$

(b) BA is not defined because B is a 1×2 matrix and A is a 4×1 matrix.

19. Use a graphing utility or computer software program.

(a) $2A + B = \begin{bmatrix} 5 & -2 & 5 & 6 & 1 & 8 \\ 0 & 5 & 5 & -1 & -3 & 8 \\ 6 & -8 & -1 & 4 & 7 & -9 \\ 5 & 0 & 9 & 2 & 3 & 3 \\ 12 & 1 & -7 & -8 & 6 & 8 \\ 5 & 2 & 10 & -10 & -6 & -5 \end{bmatrix}$

(b) $3B - A = \begin{bmatrix} 1 & 8 & -13 & 11 & 3 & 3 \\ 7 & -13 & 1 & 11 & -2 & 3 \\ -3 & -3 & -10 & -2 & 0 & 1 \\ 1 & 7 & 6 & 6 & 2 & -5 \\ 1 & -4 & -7 & 4 & 11 & 3 \\ 1 & -8 & 9 & -2 & -11 & -1 \end{bmatrix}$

(c) $AB = \begin{bmatrix} 2 & -2 & -5 & -2 & -3 & -8 \\ 6 & -27 & 10 & -2 & -21 & 1 \\ 1 & 22 & -34 & 15 & 37 & 7 \\ 4 & -4 & -11 & 1 & 2 & -1 \\ 8 & -4 & -11 & 9 & -10 & 17 \\ -2 & -11 & -30 & 10 & 3 & 9 \end{bmatrix}$

(d) $BA = \begin{bmatrix} 8 & 16 & 21 & -21 & -8 & 28 \\ 15 & -19 & 20 & 6 & 6 & -8 \\ -4 & 0 & -12 & -2 & -5 & 11 \\ 16 & -6 & 12 & 3 & 11 & 6 \\ 20 & 9 & 1 & -26 & -6 & 23 \\ -10 & -26 & 3 & 33 & 9 & -33 \end{bmatrix}$

21. $A + B$ is defined and has size 3×4 because A and B have size 3×4.

23. $\frac{1}{2}D$ is defined and has size 4×2 because D has size 4×2.

25. AC is defined. Because A has size 3×4 and C has size 4×2, the size of AC is 3×2.

27. $E - 2A$ is not defined because E and $2A$ have different sizes.

29. In matrix form $A\mathbf{x} = \mathbf{b}$, the system is

$$\begin{bmatrix} -1 & 1 \\ -2 & 1 \end{bmatrix} \begin{bmatrix} x_1 \\ x_2 \end{bmatrix} = \begin{bmatrix} 4 \\ 0 \end{bmatrix}.$$

Use Gauss-Jordan elimination on the augmented matrix.

$$\begin{bmatrix} -1 & 1 & 4 \\ -2 & 1 & 0 \end{bmatrix} \Rightarrow \begin{bmatrix} 1 & 0 & 4 \\ 0 & 1 & 8 \end{bmatrix}$$

So, the solution is $\begin{bmatrix} x_1 \\ x_2 \end{bmatrix} = \begin{bmatrix} 4 \\ 8 \end{bmatrix}$.

31. In matrix form $A\mathbf{x} = \mathbf{b}$, the system is

$$\begin{bmatrix} -2 & -3 \\ 6 & 1 \end{bmatrix} \begin{bmatrix} x_1 \\ x_2 \end{bmatrix} = \begin{bmatrix} -4 \\ -36 \end{bmatrix}.$$

Use Gauss-Jordan elimination on the augmented matrix.

$$\begin{bmatrix} -2 & -3 & -4 \\ 6 & 1 & -36 \end{bmatrix} \Rightarrow \begin{bmatrix} 1 & 0 & -7 \\ 0 & 1 & 6 \end{bmatrix}$$

So, the solution is $\begin{bmatrix} x_1 \\ x_2 \end{bmatrix} = \begin{bmatrix} -7 \\ 6 \end{bmatrix}$.

33. In matrix form $A\mathbf{x} = \mathbf{b}$, the system is

$$\begin{bmatrix} 1 & -2 & 3 \\ -1 & 3 & -1 \\ 2 & -5 & 5 \end{bmatrix} \begin{bmatrix} x_1 \\ x_2 \\ x_3 \end{bmatrix} = \begin{bmatrix} 9 \\ -6 \\ 17 \end{bmatrix}.$$

Use Gauss-Jordan elimination on the augmented matrix.

$$\begin{bmatrix} 1 & -2 & 3 & 9 \\ -1 & 3 & -1 & -6 \\ 2 & -5 & 5 & 17 \end{bmatrix} \Rightarrow \begin{bmatrix} 1 & 0 & 0 & 1 \\ 0 & 1 & 0 & -1 \\ 0 & 0 & 1 & 2 \end{bmatrix}$$

So, the solution is $\begin{bmatrix} x_1 \\ x_2 \\ x_3 \end{bmatrix} = \begin{bmatrix} 1 \\ -1 \\ 2 \end{bmatrix}$.

35. In matrix form $A\mathbf{x} = \mathbf{b}$, the system is

$$\begin{bmatrix} 1 & -5 & 2 \\ -3 & 1 & -1 \\ 0 & -2 & 5 \end{bmatrix} \begin{bmatrix} x_1 \\ x_2 \\ x_3 \end{bmatrix} = \begin{bmatrix} -20 \\ 8 \\ -16 \end{bmatrix}$$

Use Gauss-Jordan elimination on the augmented matrix.

$$\begin{bmatrix} 1 & -5 & 2 & -20 \\ -3 & 1 & -1 & 8 \\ 0 & -2 & 5 & -16 \end{bmatrix} \Rightarrow \begin{bmatrix} 1 & 0 & 0 & -1 \\ 0 & 1 & 0 & 3 \\ 0 & 0 & 1 & -2 \end{bmatrix}$$

So, the solution is $\begin{bmatrix} x_1 \\ x_2 \\ x_3 \end{bmatrix} = \begin{bmatrix} -1 \\ 3 \\ -2 \end{bmatrix}$.

37. Expanding the left side of the equation produces

$$\begin{bmatrix} 1 & 2 \\ 3 & 5 \end{bmatrix} A = \begin{bmatrix} 1 & 2 \\ 3 & 5 \end{bmatrix} \begin{bmatrix} a_{11} & a_{12} \\ a_{21} & a_{22} \end{bmatrix} = \begin{bmatrix} a_{11} + 2a_{21} & a_{12} + 2a_{22} \\ 3a_{11} + 5a_{21} & 3a_{12} + 5a_{22} \end{bmatrix} = \begin{bmatrix} 1 & 0 \\ 0 & 1 \end{bmatrix}$$

from which you obtain the system

$$\begin{array}{rcl}
a_{11} \quad + 2a_{21} \quad\quad\quad\quad &=& 1 \\
a_{12} \quad\quad + 2a_{22} &=& 0 \\
3a_{11} \quad + 5a_{21} \quad\quad\quad\quad &=& 0 \\
3a_{12} \quad\quad + 5a_{22} &=& 1.
\end{array}$$

Solving by Gauss-Jordan elimination yields $a_{11} = -5$, $a_{12} = 2$, $a_{21} = 3$, and $a_{22} = -1$. So, you have

$$A = \begin{bmatrix} -5 & 2 \\ 3 & -1 \end{bmatrix}.$$

39. Expand the left side of the matrix equation.

$$\begin{bmatrix} 1 & 2 \\ 3 & 4 \end{bmatrix} \begin{bmatrix} a & b \\ c & d \end{bmatrix} = \begin{bmatrix} 6 & 3 \\ 19 & 2 \end{bmatrix}$$

$$\begin{bmatrix} a + 2c & b + 2d \\ 3a + 4c & 3b + 4d \end{bmatrix} = \begin{bmatrix} 6 & 3 \\ 19 & 2 \end{bmatrix}$$

By setting corresponding entries equal to each other, obtain four equations.

$$\begin{array}{rcl}
a \quad + 2c \quad\quad\quad\quad &=& 6 \\
b \quad\quad + 2d &=& 3 \\
3a \quad + 4c \quad\quad\quad\quad &=& 19 \\
3b \quad\quad + 4d &=& 2
\end{array}$$

Gauss-Jordan elimination produces $a = 7$, $b = -4$, $c = -\frac{1}{2}$, and $d = \frac{7}{2}$.

41. Expand $AB = BA$ as follows.

$$\begin{bmatrix} w & x \\ y & z \end{bmatrix}\begin{bmatrix} 1 & 1 \\ -1 & 1 \end{bmatrix} = \begin{bmatrix} 1 & 1 \\ -1 & 1 \end{bmatrix}\begin{bmatrix} w & x \\ y & z \end{bmatrix}$$

$$\begin{bmatrix} w - x & w + x \\ y - z & y + z \end{bmatrix} = \begin{bmatrix} w + y & x + z \\ -w + y & -x + z \end{bmatrix}$$

This yields the system of equations

$$\begin{aligned} -x - y &= 0 \\ w \qquad\quad -z &= 0 \\ w \qquad\quad -z &= 0 \\ x + y &= 0. \end{aligned}$$

Using Gauss-Jordan elimination you can solve this system to obtain $w = t$, $x = -s$, $y = s$, and $z = t$, where s and t are any real numbers. So, $w = z$ and $x = -y$.

43. $AA = \begin{bmatrix} -1 & 0 & 0 \\ 0 & 2 & 0 \\ 0 & 0 & 3 \end{bmatrix}\begin{bmatrix} -1 & 0 & 0 \\ 0 & 2 & 0 \\ 0 & 0 & 3 \end{bmatrix}$

$$= \begin{bmatrix} -1(-1) + 0(0) + 0(0) & -1(0) + 0(2) + 0(0) & -1(0) + 0(0) + 0(3) \\ 0(-1) + 2(0) + 0(0) & 0(0) + 2(2) + 0(0) & 0(0) + 2(0) + 0(3) \\ 0(-1) + 0(0) + 3(0) & 0(0) + 0(2) + 3(0) & 0(0) + 0(0) + 3(3) \end{bmatrix}$$

$$= \begin{bmatrix} 1 & 0 & 0 \\ 0 & 4 & 0 \\ 0 & 0 & 9 \end{bmatrix}$$

45. $AB = \begin{bmatrix} 2 & 0 \\ 0 & -3 \end{bmatrix}\begin{bmatrix} -5 & 0 \\ 0 & 4 \end{bmatrix} = \begin{bmatrix} -10 & 0 \\ 0 & -12 \end{bmatrix}$

$BA = \begin{bmatrix} -5 & 0 \\ 0 & 4 \end{bmatrix}\begin{bmatrix} 2 & 0 \\ 0 & -3 \end{bmatrix} = \begin{bmatrix} -10 & 0 \\ 0 & -12 \end{bmatrix}$

47. Let A and B be diagonal matrices of sizes $n \times n$.

Then, $AB = \begin{bmatrix} c_{ij} \end{bmatrix} = \begin{bmatrix} \sum_{k=1}^{n} a_{ik}b_{kj} \end{bmatrix}$

where $c_{ij} = 0$ if $i \neq j$, and $c_{ii} = a_{ii}b_{ii}$ otherwise. The entries of BA are exactly the same.

49. The trace is the sum of the elements on the main diagonal.

$1 + (-2) + 3 = 2$.

51. The trace is the sum of the elements on the main diagonal.

$1 + 1 + 1 + 1 = 4$

53. (a) $\text{Tr}(A + B) = \text{Tr}(\begin{bmatrix} a_{ij} + b_{ij} \end{bmatrix}) = \sum_{i=1}^{n}(a_{ii} + b_{ii}) = \sum_{i=1}^{n}a_{ii} + \sum_{i=1}^{n}b_{ii}$

$$= \text{Tr}(A) + \text{Tr}(B)$$

(b) $\text{TR}(cA) = \text{Tr}(\begin{bmatrix} ca_{ij} \end{bmatrix}) = \sum_{i=1}^{n}ca_{ii} = c\sum_{i=1}^{n}a_{ii}$

$$= c\text{Tr}(A)$$

55. Let $A = \begin{bmatrix} a_{11} & a_{12} \\ a_{21} & a_{22} \end{bmatrix}$,

then the given matrix equation expands to

$$\begin{bmatrix} a_{11} + a_{21} & a_{12} + a_{22} \\ a_{11} + a_{21} & a_{12} + a_{22} \end{bmatrix} = \begin{bmatrix} 1 & 0 \\ 0 & 1 \end{bmatrix}.$$

Because $a_{11} + a_{21} = 1$ and $a_{11} + a_{21} = 0$ cannot both be true, conclude that there is no solution.

57. (a) $A^2 = \begin{bmatrix} i & 0 \\ 0 & i \end{bmatrix}\begin{bmatrix} i & 0 \\ 0 & i \end{bmatrix} = \begin{bmatrix} i^2 & 0 \\ 0 & i^2 \end{bmatrix} = \begin{bmatrix} -1 & 0 \\ 0 & -1 \end{bmatrix}$

$A^3 = A^2 A = \begin{bmatrix} -1 & 0 \\ 0 & -1 \end{bmatrix}\begin{bmatrix} i & 0 \\ 0 & i \end{bmatrix} = \begin{bmatrix} -i & 0 \\ 0 & -i \end{bmatrix}$

$A^4 = A^3 A = \begin{bmatrix} -i & 0 \\ 0 & -i \end{bmatrix}\begin{bmatrix} i & 0 \\ 0 & i \end{bmatrix} = \begin{bmatrix} -i^2 & 0 \\ 0 & -i^2 \end{bmatrix} = \begin{bmatrix} 1 & 0 \\ 0 & 1 \end{bmatrix}$

A^2, A^3, and A^4 are diagonal matrices with diagonal entries i^2, i^3, and i^4 respectively.

(b) $B^2 = \begin{bmatrix} 0 & -i \\ i & 0 \end{bmatrix}\begin{bmatrix} 0 & -i \\ i & 0 \end{bmatrix} = \begin{bmatrix} 1 & 0 \\ 0 & 1 \end{bmatrix} = A^4 = I$

59. Assume that A is an $m \times n$ matrix and B is a $p \times q$ matrix.

Because AB is defined, you have that $n = p$ and AB is $m \times q$.

Because BA is defined, you have that $m = q$ and so AB is an $m \times m$ square matrix.

Likewise, because BA is defined, $m = q$ and so BA is $p \times n$.

Because AB is defined, you have $n = p$. Therefore BA is an $n \times n$ square matrix.

61. Let $AB = \begin{bmatrix} c_{ij} \end{bmatrix}$, where $c_{ij} = \sum_{k=1}^{n} a_{ik} b_{kj}$. If the ith row of A has all zero entries, then $a_{ik} = 0$ for $k = 1, \ldots, n$. So, $c_{ij} = 0$ for

all $j = 1, \ldots, n$, and the ith row of AB has all zero entries. To show the converse is not true consider

$$AB = \begin{bmatrix} 2 & 1 \\ -2 & -1 \end{bmatrix}\begin{bmatrix} 1 & 1 \\ -2 & -2 \end{bmatrix} = \begin{bmatrix} 0 & 0 \\ 0 & 0 \end{bmatrix}.$$

63. $1.2\begin{bmatrix} 70 & 50 & 25 \\ 35 & 100 & 70 \end{bmatrix} = \begin{bmatrix} 84 & 60 & 30 \\ 42 & 120 & 84 \end{bmatrix}$

65. $BA = \begin{bmatrix} 3.50 & 6.00 \end{bmatrix}\begin{bmatrix} 125 & 100 & 75 \\ 100 & 175 & 125 \end{bmatrix} = \begin{bmatrix} 1037.5 & 1400 & 1012.5 \end{bmatrix}$

The entries of BA represent the profit for both crops at each of the three outlets.

67. (a) True. On page 51, "… for the product of two matrices to be defined, the number of columns of the first matrix must equal the number of rows of the second matrix."

(b) True. On page 55, "… the system $A\mathbf{x} = \mathbf{b}$ is consistent if and only if $\mathbf{b}$ can be expressed as … a linear combination, where the coefficients of the linear combination are a solution of the system."

69. $PP = \begin{bmatrix} 0.75 & 0.15 & 0.10 \\ 0.20 & 0.60 & 0.20 \\ 0.30 & 0.40 & 0.30 \end{bmatrix}\begin{bmatrix} 0.75 & 0.15 & 0.10 \\ 0.20 & 0.60 & 0.20 \\ 0.30 & 0.40 & 0.30 \end{bmatrix} = \begin{bmatrix} 0.6225 & 0.2425 & 0.135 \\ 0.33 & 0.47 & 0.20 \\ 0.395 & 0.405 & 0.20 \end{bmatrix}$

This product represents the changes in party affiliation after *two* elections.

71. $AB = \begin{bmatrix} 1 & 2 & 0 & 0 \\ 0 & 1 & 0 & 0 \\ \hline 0 & 0 & 2 & 1 \end{bmatrix} \begin{bmatrix} 1 & 2 & 0 \\ -1 & 1 & 0 \\ \hline 0 & 0 & 1 \\ 0 & 0 & 3 \end{bmatrix} = \begin{bmatrix} -1 & 4 & 0 \\ -1 & 1 & 0 \\ \hline 0 & 0 & 5 \end{bmatrix}$

73. The augmented matrix row reduces as follows.

$\begin{bmatrix} 1 & -1 & 2 & -1 \\ 3 & -3 & 1 & 7 \end{bmatrix} \Rightarrow \begin{bmatrix} 1 & -1 & 0 & 3 \\ 0 & 0 & 1 & -2 \end{bmatrix}$

There are an infinite number of solutions.

Answers will vary. *Sample answer*: $x_3 = -2$, $x_2 = 0$, $x_1 = 3$.

So,

$\mathbf{b} = \begin{bmatrix} -1 \\ 7 \end{bmatrix} = 3\begin{bmatrix} 1 \\ 3 \end{bmatrix} + 0\begin{bmatrix} -1 \\ -3 \end{bmatrix} - 2\begin{bmatrix} 2 \\ 1 \end{bmatrix}$.

75. The augmented matrix row reduces as follows.

$\begin{bmatrix} 1 & 1 & -5 & 3 \\ 1 & 0 & -1 & 1 \\ 2 & -1 & -1 & 0 \end{bmatrix} \Rightarrow \begin{bmatrix} 1 & 0 & 0 & 1 \\ 0 & 1 & 0 & 2 \\ 0 & 0 & 1 & 0 \end{bmatrix}$

So, $\mathbf{b} = \begin{bmatrix} 3 \\ 1 \\ 0 \end{bmatrix} = 1\begin{bmatrix} 1 \\ 1 \\ 2 \end{bmatrix} + 2\begin{bmatrix} 1 \\ 0 \\ -1 \end{bmatrix} + 0\begin{bmatrix} -5 \\ -1 \\ -1 \end{bmatrix}$.

Section 2.2 Properties of Matrix Operations

1. $3\begin{bmatrix} 1 & 2 \\ 3 & 4 \end{bmatrix} + (-4)\begin{bmatrix} 0 & 1 \\ -1 & 2 \end{bmatrix} = \begin{bmatrix} 3 & 6 \\ 9 & 12 \end{bmatrix} + \begin{bmatrix} 0 & -4 \\ 4 & -8 \end{bmatrix} = \begin{bmatrix} 3 & 2 \\ 13 & 4 \end{bmatrix}$

3. $ab(B) = (3)(-4)\begin{bmatrix} 0 & 1 \\ -1 & 2 \end{bmatrix} = (-12)\begin{bmatrix} 0 & 1 \\ -1 & 2 \end{bmatrix} = \begin{bmatrix} 0 & -12 \\ 12 & -24 \end{bmatrix}$

5. $[3-(-4)]\left(\begin{bmatrix} 1 & 2 \\ 3 & 4 \end{bmatrix} - \begin{bmatrix} 0 & 1 \\ -1 & 2 \end{bmatrix}\right) = 7\begin{bmatrix} 1 & 1 \\ 4 & 2 \end{bmatrix} = \begin{bmatrix} 7 & 7 \\ 28 & 14 \end{bmatrix}$

7. (a)

$3X + 2A = B$

$3X + \begin{bmatrix} -8 & 0 \\ 2 & -10 \\ -6 & 4 \end{bmatrix} = \begin{bmatrix} 1 & 2 \\ -2 & 1 \\ 4 & 4 \end{bmatrix}$

$3X = \begin{bmatrix} 9 & 2 \\ -4 & 11 \\ 10 & 0 \end{bmatrix}$

$X = \begin{bmatrix} 3 & \frac{2}{3} \\ -\frac{4}{3} & \frac{11}{3} \\ \frac{10}{3} & 0 \end{bmatrix}$

(b)
$$2A - 5B = 3X$$

$$\begin{bmatrix} -8 & 0 \\ 2 & -10 \\ -6 & 4 \end{bmatrix} - \begin{bmatrix} 5 & 10 \\ -10 & 5 \\ 20 & 20 \end{bmatrix} = 3X$$

$$\begin{bmatrix} -13 & -10 \\ 12 & -15 \\ -26 & -16 \end{bmatrix} = 3X$$

$$\begin{bmatrix} -\frac{13}{3} & -\frac{10}{3} \\ 4 & -5 \\ -\frac{26}{3} & -\frac{16}{3} \end{bmatrix} = X$$

(c) $X - 3A + 2B = O$

$$X = 3A - 2B$$

$$= \begin{bmatrix} -12 & 0 \\ 3 & -15 \\ -9 & 6 \end{bmatrix} - \begin{bmatrix} 2 & 4 \\ -4 & 2 \\ 8 & 8 \end{bmatrix}$$

$$= \begin{bmatrix} -14 & -4 \\ 7 & -17 \\ -17 & -2 \end{bmatrix}$$

(d) $6X - 4A - 3B = 0$

$$6X = 4A + 3B$$

$$6X = \begin{bmatrix} -16 & 0 \\ 4 & -20 \\ -12 & 8 \end{bmatrix} + \begin{bmatrix} 3 & 6 \\ -6 & 3 \\ 12 & 12 \end{bmatrix}$$

$$6X = \begin{bmatrix} -13 & 6 \\ -2 & -17 \\ 0 & 20 \end{bmatrix}$$

$$X = \begin{bmatrix} -\frac{13}{6} & 1 \\ -\frac{1}{3} & -\frac{17}{6} \\ 0 & \frac{10}{3} \end{bmatrix}$$

9. $B(CA) = \begin{bmatrix} 1 & 3 \\ -1 & 2 \end{bmatrix} \left(\begin{bmatrix} 0 & 1 \\ -1 & 0 \end{bmatrix} \begin{bmatrix} 1 & 2 & 3 \\ 0 & 1 & -1 \end{bmatrix} \right) = \begin{bmatrix} 1 & 3 \\ -1 & 2 \end{bmatrix} \begin{bmatrix} 0 & 1 & -1 \\ -1 & -2 & -3 \end{bmatrix} = \begin{bmatrix} -3 & -5 & -10 \\ -2 & -5 & -5 \end{bmatrix}$

11. $\left(\begin{bmatrix} 1 & 3 \\ -1 & 2 \end{bmatrix} + \begin{bmatrix} 0 & 1 \\ -1 & 0 \end{bmatrix} \right) \begin{bmatrix} 1 & 2 & 3 \\ 0 & 1 & -1 \end{bmatrix} = \begin{bmatrix} 1 & 4 \\ -2 & 2 \end{bmatrix} \begin{bmatrix} 1 & 2 & 3 \\ 0 & 1 & -1 \end{bmatrix} = \begin{bmatrix} 1 & 6 & -1 \\ -2 & -2 & -8 \end{bmatrix}$

13. $\left(-2 \begin{bmatrix} 1 & 3 \\ -1 & 2 \end{bmatrix} \right) \left(\begin{bmatrix} 0 & 1 \\ -1 & 0 \end{bmatrix} + \begin{bmatrix} 0 & 1 \\ -1 & 0 \end{bmatrix} \right) = \begin{bmatrix} -2 & -6 \\ 2 & -4 \end{bmatrix} \begin{bmatrix} 0 & 2 \\ -2 & 0 \end{bmatrix} = \begin{bmatrix} 12 & -4 \\ 8 & 4 \end{bmatrix}$

15. $AC = \begin{bmatrix} 0 & 1 \\ 0 & 1 \end{bmatrix} \begin{bmatrix} 2 & 3 \\ 2 & 3 \end{bmatrix} = \begin{bmatrix} 2 & 3 \\ 2 & 3 \end{bmatrix} = \begin{bmatrix} 1 & 0 \\ 1 & 0 \end{bmatrix} \begin{bmatrix} 2 & 3 \\ 2 & 3 \end{bmatrix} = BC,$ but $A \neq B.$

17. $AB = \begin{bmatrix} 3 & 3 \\ 4 & 4 \end{bmatrix} \begin{bmatrix} 1 & -1 \\ -1 & 1 \end{bmatrix} = \begin{bmatrix} 0 & 0 \\ 0 & 0 \end{bmatrix},$ but $A \neq 0$ and $B \neq 0.$

19. $A^2 = \begin{bmatrix} 1 & 2 \\ 0 & -1 \end{bmatrix} \begin{bmatrix} 1 & 2 \\ 0 & -1 \end{bmatrix} = \begin{bmatrix} 1 & 0 \\ 0 & 1 \end{bmatrix} = I$

21. $\begin{bmatrix} 1 & 2 \\ 0 & -1 \end{bmatrix} \left(\begin{bmatrix} 1 & 0 \\ 0 & 1 \end{bmatrix} + \begin{bmatrix} 1 & 2 \\ 0 & -1 \end{bmatrix} \right) = \begin{bmatrix} 1 & 2 \\ 0 & -1 \end{bmatrix} \begin{bmatrix} 2 & 2 \\ 0 & 0 \end{bmatrix} = \begin{bmatrix} 2 & 2 \\ 0 & 0 \end{bmatrix}$

23. (a) $A^T = \begin{bmatrix} 4 & 2 & 1 \\ 0 & 2 & -1 \end{bmatrix}^T = \begin{bmatrix} 4 & 0 \\ 2 & 2 \\ 1 & -1 \end{bmatrix}$

(b) $A^T A = \begin{bmatrix} 4 & 0 \\ 2 & 2 \\ 1 & -1 \end{bmatrix} \begin{bmatrix} 4 & 2 & 1 \\ 0 & 2 & -1 \end{bmatrix} = \begin{bmatrix} 16 & 8 & 4 \\ 8 & 8 & 0 \\ 4 & 0 & 2 \end{bmatrix}$

(c) $AA^T = \begin{bmatrix} 4 & 2 & 1 \\ 0 & 2 & -1 \end{bmatrix} \begin{bmatrix} 4 & 0 \\ 2 & 2 \\ 1 & -1 \end{bmatrix} = \begin{bmatrix} 21 & 3 \\ 3 & 5 \end{bmatrix}$

25. (a) $A^T = \begin{bmatrix} 2 & 1 & -3 \\ 1 & 4 & 1 \\ 0 & 2 & 1 \end{bmatrix}^T = \begin{bmatrix} 2 & 1 & 0 \\ 1 & 4 & 2 \\ -3 & 1 & 1 \end{bmatrix}$

(b) $A^T A = \begin{bmatrix} 2 & 1 & 0 \\ 1 & 4 & 2 \\ -3 & 1 & 1 \end{bmatrix} \begin{bmatrix} 2 & 1 & -3 \\ 1 & 4 & 1 \\ 0 & 2 & 1 \end{bmatrix} = \begin{bmatrix} 5 & 6 & -5 \\ 6 & 21 & 3 \\ -5 & 3 & 11 \end{bmatrix}$

(c) $AA^T = \begin{bmatrix} 2 & 1 & -3 \\ 1 & 4 & 1 \\ 0 & 2 & 1 \end{bmatrix} \begin{bmatrix} 2 & 1 & 0 \\ 1 & 4 & 2 \\ -3 & 1 & 1 \end{bmatrix} = \begin{bmatrix} 14 & 3 & -1 \\ 3 & 18 & 9 \\ -1 & 9 & 5 \end{bmatrix}$

27. (a) $A^T = \begin{bmatrix} 0 & -4 & 3 & 2 \\ 8 & 4 & 0 & 1 \\ -2 & 3 & 5 & 1 \\ 0 & 0 & -3 & 2 \end{bmatrix}^T = \begin{bmatrix} 0 & 8 & -2 & 0 \\ -4 & 4 & 3 & 0 \\ 3 & 0 & 5 & -3 \\ 2 & 1 & 1 & 2 \end{bmatrix}$

(b) $A^T A = \begin{bmatrix} 0 & 8 & -2 & 0 \\ -4 & 4 & 3 & 0 \\ 3 & 0 & 5 & -3 \\ 2 & 1 & 1 & 2 \end{bmatrix} \begin{bmatrix} 0 & -4 & 3 & 2 \\ 8 & 4 & 0 & 1 \\ -2 & 3 & 5 & 1 \\ 0 & 0 & -3 & 2 \end{bmatrix} = \begin{bmatrix} 68 & 26 & -10 & 6 \\ 26 & 41 & 3 & -1 \\ -10 & 3 & 43 & 5 \\ 6 & -1 & 5 & 10 \end{bmatrix}$

(c) $AA^T = \begin{bmatrix} 0 & -4 & 3 & 2 \\ 8 & 4 & 0 & 1 \\ -2 & 3 & 5 & 1 \\ 0 & 0 & -3 & 2 \end{bmatrix} \begin{bmatrix} 0 & 8 & -2 & 0 \\ -4 & 4 & 3 & 0 \\ 3 & 0 & 5 & -3 \\ 2 & 1 & 1 & 2 \end{bmatrix} = \begin{bmatrix} 29 & 14 & 5 & -5 \\ -14 & 81 & -3 & 2 \\ 5 & -3 & 39 & -13 \\ -5 & 2 & -13 & 13 \end{bmatrix}$

29. In general, $AB \neq BA$ for matrices. So, $(A + B)(A - B) = A^2 + BA - AB - B^2 \neq A^2 - B^2$.

31. $(AB)^T = \left(\begin{bmatrix} -1 & 1 & -2 \\ 2 & 0 & 1 \end{bmatrix} \begin{bmatrix} -3 & 0 \\ 1 & 2 \\ 1 & -1 \end{bmatrix} \right)^T = \begin{bmatrix} 2 & 4 \\ -5 & -1 \end{bmatrix}^T = \begin{bmatrix} 2 & -5 \\ 4 & -1 \end{bmatrix}$

$B^T A^T = \begin{bmatrix} -3 & 0 \\ 1 & 2 \\ 1 & -1 \end{bmatrix}^T \begin{bmatrix} -1 & 1 & -2 \\ 2 & 0 & 1 \end{bmatrix}^T = \begin{bmatrix} -3 & 1 & 1 \\ 0 & 2 & -1 \end{bmatrix} \begin{bmatrix} -1 & 2 \\ 1 & 0 \\ -2 & 1 \end{bmatrix} = \begin{bmatrix} 2 & -5 \\ 4 & -1 \end{bmatrix}$

33. $(AB)^T = \left(\begin{bmatrix} 2 & 1 \\ 0 & 1 \\ -2 & 1 \end{bmatrix} \begin{bmatrix} 2 & 3 & 1 \\ 0 & 4 & -1 \end{bmatrix} \right)^T = \begin{bmatrix} 4 & 10 & 1 \\ 0 & 4 & -1 \\ -4 & -2 & -3 \end{bmatrix}^T = \begin{bmatrix} 4 & 0 & -4 \\ 10 & 4 & -2 \\ 1 & -1 & -3 \end{bmatrix}$

$B^T A^T = \begin{bmatrix} 2 & 3 & 1 \\ 0 & 4 & -1 \end{bmatrix}^T \begin{bmatrix} 2 & 1 \\ 0 & 1 \\ -2 & 1 \end{bmatrix}^T = \begin{bmatrix} 2 & 0 \\ 3 & 4 \\ 1 & -1 \end{bmatrix} \begin{bmatrix} 2 & 0 & -2 \\ 1 & 1 & 1 \end{bmatrix} = \begin{bmatrix} 4 & 0 & -4 \\ 10 & 4 & -2 \\ 1 & -1 & -3 \end{bmatrix}$

35. (a) True. See Theorem 2.1, part 1.

(b) True. See Theorem 2.3, part 1.

(c) False. See Theorem 2.6, part 4, or Example 9.

(d) True. See Example 10.

37. (a) $aX + bY = a\begin{bmatrix} 1 \\ 0 \\ 1 \end{bmatrix} + b\begin{bmatrix} 1 \\ 1 \\ 0 \end{bmatrix} = \begin{bmatrix} 2 \\ -1 \\ 3 \end{bmatrix}$

This matrix equation yields the linear system

$$a + b = 2$$
$$b = -1$$
$$a \quad = 3.$$

The only solution to this system is: $a = 3$ and $b = -1$.

(b) $aX + bY = a\begin{bmatrix} 1 \\ 0 \\ 1 \end{bmatrix} + b\begin{bmatrix} 1 \\ 1 \\ 0 \end{bmatrix} = \begin{bmatrix} 1 \\ 1 \\ 1 \end{bmatrix}$

This matrix equation yields the linear system

$$a + b = 1$$
$$b = 1$$
$$a \quad = 1.$$

The system is inconsistent. So, no values of a and b will satisfy the equation.

(c) $aX + bY + cW = a\begin{bmatrix} 1 \\ 0 \\ 1 \end{bmatrix} + b\begin{bmatrix} 1 \\ 1 \\ 0 \end{bmatrix} + c\begin{bmatrix} 1 \\ 1 \\ 1 \end{bmatrix} = \begin{bmatrix} 0 \\ 0 \\ 0 \end{bmatrix}$

This matrix equation yields the linear system

$$a + b + c = 0$$
$$b + c = 0$$
$$a \quad + c = 0.$$

Then $a = -c$, so $b = 0$. Then $c = 0$, so $a = b = c = 0$.

(d) $aX + bY + cZ = a\begin{bmatrix} 1 \\ 0 \\ 1 \end{bmatrix} + b\begin{bmatrix} 1 \\ 1 \\ 0 \end{bmatrix} + c\begin{bmatrix} 2 \\ -1 \\ 3 \end{bmatrix} = \begin{bmatrix} 0 \\ 0 \\ 0 \end{bmatrix}$

This matrix equation yields the linear system

$$a + b + 2c = 0$$
$$b - c = 0$$
$$a \quad + 3c = 0.$$

Using Gauss-Jordan elimination the solution is $a = -3t$, $b = t$ and $c = t$, where t is any real number.

If $t = 1$, then $a = -3$, $b = 1$, and $c = 1$.

39. $A^{19} = \begin{bmatrix} 1^{19} & 0 & 0 \\ 0 & (-1)^{19} & 0 \\ 0 & 0 & 1^{19} \end{bmatrix} = \begin{bmatrix} 1 & 0 & 0 \\ 0 & -1 & 0 \\ 0 & 0 & 1 \end{bmatrix}$

41. There are four possibilities, such that $A^2 = B$, namely

$$A = \begin{bmatrix} 3 & 0 \\ 0 & 2 \end{bmatrix}, \; A = \begin{bmatrix} -3 & 0 \\ 0 & 2 \end{bmatrix}, \; A = \begin{bmatrix} 3 & 0 \\ 0 & -2 \end{bmatrix}, \; A = \begin{bmatrix} -3 & 0 \\ 0 & -2 \end{bmatrix}.$$

43. $f(A) = \begin{bmatrix} 2 & 0 \\ 4 & 5 \end{bmatrix}^2 - 5\begin{bmatrix} 2 & 0 \\ 4 & 5 \end{bmatrix} + 2\begin{bmatrix} 1 & 0 \\ 0 & 1 \end{bmatrix} = \begin{bmatrix} 2 & 0 \\ 4 & 5 \end{bmatrix}\begin{bmatrix} 2 & 0 \\ 4 & 5 \end{bmatrix} - \begin{bmatrix} 10 & 0 \\ 20 & 25 \end{bmatrix} + \begin{bmatrix} 2 & 0 \\ 0 & 2 \end{bmatrix} = \begin{bmatrix} 4 & 0 \\ 28 & 25 \end{bmatrix} + \begin{bmatrix} -8 & 0 \\ -20 & -23 \end{bmatrix} = \begin{bmatrix} -4 & 0 \\ 8 & 2 \end{bmatrix}$

45. $f(A) = \begin{bmatrix} 2 & 1 \\ -1 & 0 \end{bmatrix}^2 - 3\begin{bmatrix} 2 & 1 \\ -1 & 0 \end{bmatrix} + 2\begin{bmatrix} 1 & 0 \\ 0 & 1 \end{bmatrix} = \begin{bmatrix} 2 & 1 \\ -1 & 0 \end{bmatrix}\begin{bmatrix} 2 & 1 \\ -1 & 0 \end{bmatrix} - \begin{bmatrix} 6 & 3 \\ -3 & 0 \end{bmatrix} + \begin{bmatrix} 2 & 0 \\ 0 & 2 \end{bmatrix} = \begin{bmatrix} 3 & 2 \\ -2 & -1 \end{bmatrix} - \begin{bmatrix} 6 & 3 \\ -3 & 0 \end{bmatrix} + \begin{bmatrix} 2 & 0 \\ 0 & 2 \end{bmatrix} = \begin{bmatrix} -1 & -1 \\ 1 & 1 \end{bmatrix}$

47. $A + (B + C) = [a_{ij}] + ([b_{ij}] + [c_{ij}]) = ([a_{ij}] + [b_{ij}]) + [c_{ij}] = (A + B) + C$

49. $1A = 1[a_{ij}] = [1a_{ij}] = [a_{ij}] = A$

51. (1) $A + O_{mn} = [a_{ij}] + [0] = [a_{ij} + 0] = [a_{ij}] = A$

   (2) $A + (-A) = [a_{ij}] + [-a_{ij}] = [a_{ij} + (-a_{ij})] = [0] = O_{mn}$

   (3) Let $cA = O_{mn}$ and suppose $c \neq 0$. Then, $O_{mn} = cA \Rightarrow c^{-1}O_{mn} = c^{-1}(cA) = (c^{-1}c)A = A$

   and so, $A = O_{mn}$. If $c = 0$, then $OA = [0 \cdot a_{cj}] = [0] = O_{mn}$.

53. (1) The entry in the ith row and jth column of AI_n is
   $a_{i1}0 + \cdots + a_{ij}1 + \cdots + a_{in}0 = a_{ij}$.

   (2) The entry in the ith row and jth column of I_mA is
   $0a_{1j} + \cdots + 1a_{ij} + \cdots + 0a_{nj} = a_{ij}$.

55. $\left(AA^T\right)^T = \left(A^T\right)^T A^T = AA^T$ which implies that AA^T is symmetric.

Similarly, $\left(A^T A\right)^T = A^T\left(A^T\right)^T = A^T A$, which implies that $A^T A$ is symmetric.

57. Because $A^T = \begin{bmatrix} 0 & -2 \\ 2 & 0 \end{bmatrix} = -\begin{bmatrix} 0 & 2 \\ -2 & 0 \end{bmatrix} = -A$ the matrix is skew-symmetric.

59. Because $A^T = \begin{bmatrix} 0 & 2 & 1 \\ 2 & 0 & 3 \\ 1 & 3 & 0 \end{bmatrix} = A$ the matrix is symmetric.

61. Because $A^T = -A$, the diagonal element a_{ii} satisfies $a_{ii} = -a_{ii}$, or $a_{ii} = 0$.

63. (a) If A is a square matrix of order n, then A^T is a square matrix of order n. Let

$$A = \begin{bmatrix} a_{11} & a_{12} & a_{13} & \cdots & a_{1n} \\ a_{21} & a_{22} & a_{23} & \cdots & a_{2n} \\ a_{31} & a_{32} & a_{33} & \cdots & a_{3n} \\ \vdots & \vdots & \vdots & \vdots & \vdots \\ a_{n1} & a_{n2} & a_{n3} & \cdots & a_{nn} \end{bmatrix}. \text{ Then } A^T = \begin{bmatrix} a_{11} & a_{21} & a_{31} & \cdots & a_{n1} \\ a_{12} & a_{22} & a_{32} & \cdots & a_{n2} \\ a_{13} & a_{23} & a_{33} & \cdots & a_{n3} \\ \vdots & \vdots & \vdots & \vdots \\ a_{1n} & a_{2n} & a_{3n} & \cdots & a_{nn} \end{bmatrix}.$$

Now form the sum and scalar multiple, $\frac{1}{2}(A + A^T)$.

$$\frac{1}{2}(A + A^T) = \frac{1}{2}\begin{bmatrix} a_{11} + a_{11} & a_{12} + a_{21} & a_{13} + a_{31} & \cdots & a_{1n} + a_{n1} \\ a_{21} + a_{12} & a_{22} + a_{22} & a_{23} + a_{32} & \cdots & a_{2n} + a_{n2} \\ a_{31} + a_{13} & a_{32} + a_{23} & a_{33} + a_{33} & \cdots & a_{3n} + a_{n3} \\ \vdots & \vdots & \vdots & \vdots \\ a_{n1} + a_{1n} & a_{n2} + a_{2n} & a_{n3} + a_{3n} & \cdots & a_{nn} + a_{nn} \end{bmatrix}.$$

and note that, for the matrix $A + A^T$, the ij^{th} entry is equal to the ji^{th} entry for all $i \neq j$. So, $\frac{1}{2}(A + A^T)$ is symmetric.

(b) Use the matrices in part (a).

$$\frac{1}{2}(A - A^T) = \frac{1}{2}\begin{bmatrix} a_{11} - a_{11} & a_{12} - a_{21} & a_{13} - a_{31} & \cdots & a_{1n} - a_{n1} \\ a_{21} - a_{12} & a_{22} - a_{22} & a_{23} - a_{32} & \cdots & a_{2n} - a_{n2} \\ a_{31} - a_{13} & a_{32} - a_{23} & a_{33} - a_{33} & \cdots & a_{3n} - a_{n3} \\ \vdots & \vdots & \vdots & \vdots \\ a_{n1} - a_{1n} & a_{n2} - a_{2n} & a_{n3} - a_{3n} & \cdots & a_{nn} - a_{nn} \end{bmatrix}.$$

and note that, for the matrix $A - A^T$, the ij^{th} entry is the negative of the ji^{th} entry for all $i \neq j$. So, $\frac{1}{2}(A - A^T)$ is skew symmetric.

(c) For any square matrix A of order n, let $B = \frac{1}{2}\left(A + A^T\right)$ and $C = \frac{1}{2}\left(A - A^T\right)$. By part (a), B is symmetric.

By part (b), C is skew symmetric. And $A = \frac{1}{2}\left(A + A^T\right) + \frac{1}{2}\left(A - A^T\right) = B + C$ as desired.

(d) For the given A, $A^T = \begin{bmatrix} 2 & -3 & 4 \\ 5 & 6 & 1 \\ 3 & 0 & 1 \end{bmatrix}$. Using the notation of part (c),

$$B + C = \frac{1}{2}\left(A + A^T\right) + \frac{1}{2}\left(A - A^T\right) = \begin{bmatrix} 2 & 1 & \frac{7}{2} \\ 1 & 6 & \frac{1}{2} \\ \frac{7}{2} & \frac{1}{2} & 1 \end{bmatrix} + \begin{bmatrix} 0 & 4 & -\frac{1}{2} \\ -4 & 0 & -\frac{1}{2} \\ \frac{1}{2} & \frac{1}{2} & 0 \end{bmatrix} = \begin{bmatrix} 2 & 5 & 3 \\ -3 & 6 & 0 \\ 4 & 1 & 1 \end{bmatrix} = A$$

65. (a) Answers will vary. *Sample answer*: $\begin{bmatrix} 0 & 1 \\ 1 & 0 \end{bmatrix}\begin{bmatrix} -1 & 1 \\ 1 & 0 \end{bmatrix} = \begin{bmatrix} 1 & 0 \\ -1 & 1 \end{bmatrix}$

(b) Let A and B be symmetric.

If $AB = BA$, then $\left(AB\right)^T = B^T A^T = BA = AB$ and AB is symmetric.

If $\left(AB\right)^T = AB$, then $AB = \left(AB\right)^T = B^T A^T = BA$ and $AB = BA$.

Section 2.3 The Inverse of a Matrix

1. $AB = \begin{bmatrix} 1 & 2 \\ 3 & 4 \end{bmatrix}\begin{bmatrix} -2 & 1 \\ \frac{3}{2} & -\frac{1}{2} \end{bmatrix} = \begin{bmatrix} 1 & 0 \\ 0 & 1 \end{bmatrix}$

$BA = \begin{bmatrix} -2 & 1 \\ \frac{3}{2} & -\frac{1}{2} \end{bmatrix}\begin{bmatrix} 1 & 2 \\ 3 & 4 \end{bmatrix} = \begin{bmatrix} 1 & 0 \\ 0 & 1 \end{bmatrix}$

3. $AB = \begin{bmatrix} -2 & 2 & 3 \\ 1 & -1 & 0 \\ 0 & 1 & 4 \end{bmatrix}\left(\frac{1}{3}\right)\begin{bmatrix} -4 & -5 & 3 \\ -4 & -8 & 3 \\ 1 & 2 & 0 \end{bmatrix} = \left(\frac{1}{3}\right)\begin{bmatrix} 3 & 0 & 0 \\ 0 & 3 & 0 \\ 0 & 0 & 3 \end{bmatrix} = \begin{bmatrix} 1 & 0 & 0 \\ 0 & 1 & 0 \\ 0 & 0 & 1 \end{bmatrix}$

$BA = \left(\frac{1}{3}\right)\begin{bmatrix} -4 & -5 & 3 \\ -4 & -8 & 3 \\ 1 & 2 & 0 \end{bmatrix}\begin{bmatrix} -2 & 2 & 3 \\ 1 & -1 & 0 \\ 0 & 1 & 4 \end{bmatrix} = \left(\frac{1}{3}\right)\begin{bmatrix} 3 & 0 & 0 \\ 0 & 3 & 0 \\ 0 & 0 & 3 \end{bmatrix} = \begin{bmatrix} 1 & 0 & 0 \\ 0 & 1 & 0 \\ 0 & 0 & 1 \end{bmatrix}$

5. Use the formula $A^{-1} = \dfrac{1}{ad - bc}\begin{bmatrix} d & -b \\ -c & a \end{bmatrix}$, where

$A = \begin{bmatrix} a & b \\ c & d \end{bmatrix} = \begin{bmatrix} 1 & 2 \\ 3 & 7 \end{bmatrix}$. So, the inverse is

$A^{-1} = \dfrac{1}{(1)(7) - (2)(3)}\begin{bmatrix} 7 & -2 \\ -3 & 1 \end{bmatrix} = \begin{bmatrix} 7 & -2 \\ -3 & 1 \end{bmatrix}$.

7. Use the formula $A^{-1} = \dfrac{1}{ad - bc}\begin{bmatrix} d & -b \\ -c & a \end{bmatrix}$, where

$A = \begin{bmatrix} a & b \\ c & d \end{bmatrix} = \begin{bmatrix} -7 & 33 \\ 4 & -19 \end{bmatrix}$. So, the inverse is

$A^{-1} = \dfrac{1}{(-7)(-19) - 33(4)}\begin{bmatrix} -19 & -33 \\ -4 & -7 \end{bmatrix} = \begin{bmatrix} -19 & -33 \\ -4 & -7 \end{bmatrix}$.

9. Adjoin the identity matrix to form

$$[A \vdots I] = \begin{bmatrix} 1 & 1 & 1 & \vdots & 1 & 0 & 0 \\ 3 & 5 & 4 & \vdots & 0 & 1 & 0 \\ 3 & 6 & 5 & \vdots & 0 & 0 & 1 \end{bmatrix}.$$

Using elementary row operations, rewrite this matrix in reduced row-echelon form.

$$[I \vdots A^{-1}] = \begin{bmatrix} 1 & 0 & 0 & \vdots & 1 & 1 & -1 \\ 0 & 1 & 0 & \vdots & -3 & 2 & -1 \\ 0 & 0 & 1 & \vdots & 3 & -3 & 2 \end{bmatrix}$$

Therefore, the inverse is $A^{-1} = \begin{bmatrix} 1 & 1 & -1 \\ -3 & 2 & -1 \\ 3 & -3 & 2 \end{bmatrix}$.

11. Adjoin the identity matrix to form

$$[A \vdots I] = \begin{bmatrix} 1 & 2 & -1 & \vdots & 1 & 0 & 0 \\ 3 & 7 & -10 & \vdots & 0 & 1 & 0 \\ 7 & 16 & -21 & \vdots & 0 & 0 & 1 \end{bmatrix}.$$

Using elementary row operations, you cannot form the identity matrix on the left side.

$$\begin{bmatrix} 1 & 0 & 13 & \vdots & 0 & -16 & 7 \\ 0 & 1 & -7 & \vdots & 0 & 7 & -3 \\ 0 & 0 & 0 & \vdots & 1 & 2 & -1 \end{bmatrix}.$$

Therefore, the matrix is singular and has no inverse.

13. Adjoin the identity matrix to form

$$[A \vdots I] = \begin{bmatrix} 1 & 1 & 2 & \vdots & 1 & 0 & 0 \\ 3 & 1 & 0 & \vdots & 0 & 1 & 0 \\ -2 & 0 & 3 & \vdots & 0 & 0 & 1 \end{bmatrix}.$$

Using elementary row operations, rewrite this matrix in reduced row-echelon form.

$$[I \vdots A^{-1}] = \begin{bmatrix} 1 & 0 & 0 & \vdots & -\frac{3}{2} & \frac{3}{2} & 1 \\ 0 & 1 & 0 & \vdots & \frac{9}{2} & -\frac{7}{2} & -3 \\ 0 & 0 & 1 & \vdots & -1 & 1 & 1 \end{bmatrix}$$

Therefore, the inverse is

$$A^{-1} = \begin{bmatrix} -\frac{3}{2} & \frac{3}{2} & 1 \\ \frac{9}{2} & -\frac{7}{2} & -3 \\ -1 & 1 & 1 \end{bmatrix}.$$

15. Adjoin the identity matrix to form

$$[A \vdots I] = \begin{bmatrix} 0.1 & 0.2 & 0.3 & \vdots & 1 & 0 & 0 \\ -0.3 & 0.2 & 0.2 & \vdots & 0 & 1 & 0 \\ 0.5 & 0.5 & 0.5 & \vdots & 0 & 0 & 1 \end{bmatrix}.$$

Using elementary row operations, reduce the matrix as follows.

$$[I \vdots A^{-1}] = \begin{bmatrix} 1 & 0 & 0 & \vdots & 0 & -2 & 0.8 \\ 0 & 1 & 0 & \vdots & -10 & 4 & 4.4 \\ 0 & 0 & 1 & \vdots & 10 & -2 & -3.2 \end{bmatrix}$$

Therefore, the inverse is

$$A^{-1} = \begin{bmatrix} 0 & -2 & 0.8 \\ -10 & 4 & 4.4 \\ 10 & -2 & -3.2 \end{bmatrix}.$$

17. Adjoin the identity matrix to form

$$[A \vdots I] = \begin{bmatrix} 1 & 0 & 0 & \vdots & 1 & 0 & 0 \\ 3 & 4 & 0 & \vdots & 0 & 1 & 0 \\ 2 & 5 & 5 & \vdots & 0 & 0 & 1 \end{bmatrix}.$$

Using elementary row operations, rewrite this matrix in reduced row-echelon form.

$$[I \vdots A^{-1}] = \begin{bmatrix} 1 & 0 & 0 & \vdots & 1 & 0 & 0 \\ 0 & 1 & 0 & \vdots & -\frac{3}{4} & \frac{1}{4} & 0 \\ 0 & 0 & 1 & \vdots & \frac{7}{20} & -\frac{1}{4} & \frac{1}{5} \end{bmatrix}$$

Therefore, the inverse is $A^{-1} = \begin{bmatrix} 1 & 0 & 0 \\ -\frac{3}{4} & \frac{1}{4} & 0 \\ \frac{7}{20} & -\frac{1}{4} & \frac{1}{5} \end{bmatrix}.$

19. Adjoin the identity matrix to form

$$[A \vdots I] = \begin{bmatrix} -8 & 0 & 0 & 0 & \vdots & 1 & 0 & 0 & 0 \\ 0 & 1 & 0 & 0 & \vdots & 0 & 1 & 0 & 0 \\ 0 & 0 & 0 & 0 & \vdots & 0 & 0 & 1 & 0 \\ 0 & 0 & 0 & -5 & \vdots & 0 & 0 & 0 & 1 \end{bmatrix}.$$

Using elementary row operations, you cannot form the identity matrix on the left side. Therefore, the matrix A is singular and has no inverse.

21. Use a graphing utility or a computer software program. The inverse is

$$A^{-1} = \begin{bmatrix} -24 & 7 & 1 & -2 \\ -10 & 3 & 0 & -1 \\ -29 & 7 & 3 & -2 \\ 12 & -3 & -1 & 1 \end{bmatrix}.$$

23. Using a graphing utility or a computer software program, you find that the matrix is singular and therefore has no inverse.

25. The coefficient matrix for each system is $A = \begin{bmatrix} 1 & 2 \\ 1 & -2 \end{bmatrix}$

and the formula for the inverse of a 2×2 matrix produces

$$A^{-1} = \frac{1}{-2-2}\begin{bmatrix} -2 & -2 \\ -1 & 1 \end{bmatrix} = \begin{bmatrix} \frac{1}{2} & \frac{1}{2} \\ \frac{1}{4} & -\frac{1}{4} \end{bmatrix}.$$

(a) $\mathbf{x} = A^{-1}\mathbf{b} = \begin{bmatrix} \frac{1}{2} & \frac{1}{2} \\ \frac{1}{4} & -\frac{1}{4} \end{bmatrix}\begin{bmatrix} -1 \\ 3 \end{bmatrix} = \begin{bmatrix} 1 \\ -1 \end{bmatrix}$

The solution is: $x = 1$ and $y = -1$.

(b) $\mathbf{x} = A^{-1}\mathbf{b} = \begin{bmatrix} \frac{1}{2} & \frac{1}{2} \\ \frac{1}{4} & -\frac{1}{4} \end{bmatrix}\begin{bmatrix} 10 \\ -6 \end{bmatrix} = \begin{bmatrix} 2 \\ 4 \end{bmatrix}$

The solution is: $x = 2$ and $y = 4$.

(c) $\mathbf{x} = A^{-1}\mathbf{b} = \begin{bmatrix} \frac{1}{2} & \frac{1}{2} \\ \frac{1}{4} & -\frac{1}{4} \end{bmatrix}\begin{bmatrix} -3 \\ 0 \end{bmatrix} = \begin{bmatrix} -\frac{3}{2} \\ -\frac{3}{4} \end{bmatrix}$

The solution is: $x = -\frac{3}{2}$ and $y = -\frac{3}{4}$.

27. The coefficient matrix for each system is
$$A = \begin{bmatrix} 1 & 2 & 1 \\ 1 & 2 & -1 \\ 1 & -2 & 1 \end{bmatrix}.$$

Using the algorithm to invert a matrix, you find that the inverse is
$$A^{-1} = \begin{bmatrix} 0 & \frac{1}{2} & \frac{1}{2} \\ \frac{1}{4} & 0 & -\frac{1}{4} \\ \frac{1}{2} & -\frac{1}{2} & 0 \end{bmatrix}$$

(a) $\mathbf{x} = A^{-1}\mathbf{b} = \begin{bmatrix} 0 & \frac{1}{2} & \frac{1}{2} \\ \frac{1}{4} & 0 & -\frac{1}{4} \\ \frac{1}{2} & -\frac{1}{2} & 0 \end{bmatrix}\begin{bmatrix} 2 \\ 4 \\ -2 \end{bmatrix} = \begin{bmatrix} 1 \\ 1 \\ -1 \end{bmatrix}$

The solution is: $x_1 = 1$, $x_2 = 1$, and $x_3 = -1$.

(b) $\mathbf{x} = A^{-1}\mathbf{b} = \begin{bmatrix} 0 & \frac{1}{2} & \frac{1}{2} \\ \frac{1}{4} & 0 & -\frac{1}{4} \\ \frac{1}{2} & -\frac{1}{2} & 0 \end{bmatrix}\begin{bmatrix} 1 \\ 3 \\ -3 \end{bmatrix} = \begin{bmatrix} 0 \\ 1 \\ -1 \end{bmatrix}$

The solution is: $x_1 = 0$, $x_2 = 1$, and $x_3 = -1$.

29. Using a graphing utility or a computer software program, you have
$$A\mathbf{x} = \mathbf{b}$$
$$\mathbf{x} = A^{-1}\mathbf{b} = \begin{bmatrix} 0 \\ 1 \\ 2 \\ -1 \\ 0 \end{bmatrix} \text{ where }$$

$$A = \begin{bmatrix} 1 & 2 & -1 & 3 & -1 \\ 1 & -3 & 1 & 2 & -1 \\ 2 & 1 & 1 & -3 & 1 \\ 1 & -1 & 2 & 1 & -1 \\ 1 & 1 & -1 & 2 & 1 \end{bmatrix}, \mathbf{x} = \begin{bmatrix} x_1 \\ x_2 \\ x_3 \\ x_4 \\ x_5 \end{bmatrix}, \text{ and } \mathbf{b} = \begin{bmatrix} -3 \\ -3 \\ 6 \\ 2 \\ -3 \end{bmatrix}.$$

The solution is: $x_1 = 0$, $x_2 = 1$, $x_3 = 2$, $x_4 = -1$, and $x_5 = 0$.

31. Using a graphing utility or a computer software program you have
$$A\mathbf{x} = \mathbf{b}$$
$$\mathbf{x} = A^{-1}\mathbf{b} = \begin{bmatrix} 1 \\ -2 \\ 3 \\ 0 \\ 1 \\ -2 \end{bmatrix} \text{ where }$$

$$A = \begin{bmatrix} 2 & -3 & 1 & -2 & 1 & -4 \\ 3 & 1 & -4 & 1 & -1 & 2 \\ 4 & 1 & -3 & 4 & -1 & 2 \\ -5 & -1 & 4 & 2 & -5 & 3 \\ 1 & 1 & -3 & 4 & -3 & 1 \\ 3 & -1 & 2 & -3 & 2 & -6 \end{bmatrix},$$

$$\mathbf{x} = \begin{bmatrix} x_1 \\ x_2 \\ x_3 \\ x_4 \\ x_5 \\ x_6 \end{bmatrix}, \text{ and } \mathbf{b} = \begin{bmatrix} 20 \\ -16 \\ -12 \\ -2 \\ -15 \\ 25 \end{bmatrix}.$$

The solution is: $x_1 = 1$, $x_2 = -2$, $x_3 = 3$, $x_4 = 0$, $x_5 = 1$, $x_6 = -2$.

33. (a) $(AB)^{-1} = B^{-1}A^{-1} = \begin{bmatrix} 7 & -3 \\ 2 & 0 \end{bmatrix}\begin{bmatrix} 2 & 5 \\ -7 & 6 \end{bmatrix} = \begin{bmatrix} 35 & 17 \\ 4 & 10 \end{bmatrix}$

(b) $(A^T)^{-1} = (A^{-1})^T = \begin{bmatrix} 2 & 5 \\ -7 & 6 \end{bmatrix}^T = \begin{bmatrix} 2 & -7 \\ 5 & 6 \end{bmatrix}$

(c) $A^{-2} = (A^{-1})^2 = \begin{bmatrix} 2 & 5 \\ -7 & 6 \end{bmatrix}\begin{bmatrix} 2 & 5 \\ -7 & 6 \end{bmatrix} = \begin{bmatrix} -31 & 40 \\ -56 & 1 \end{bmatrix}$

(d) $(2A)^{-1} = \frac{1}{2}A^{-1} = \frac{1}{2}\begin{bmatrix} 2 & 5 \\ -7 & 6 \end{bmatrix} = \begin{bmatrix} 1 & \frac{5}{2} \\ -\frac{7}{2} & 3 \end{bmatrix}$

35. (a) $(AB)^{-1} = B^{-1}A^{-1} = \begin{bmatrix} 2 & 4 & \frac{5}{2} \\ -\frac{3}{4} & 2 & \frac{1}{4} \\ \frac{1}{4} & \frac{1}{2} & 2 \end{bmatrix}\begin{bmatrix} 1 & -\frac{1}{2} & \frac{3}{4} \\ \frac{3}{2} & \frac{1}{2} & -2 \\ \frac{1}{4} & 1 & \frac{1}{2} \end{bmatrix} = \begin{bmatrix} \frac{69}{8} & \frac{7}{2} & -\frac{21}{4} \\ \frac{37}{16} & \frac{13}{8} & -\frac{71}{16} \\ \frac{3}{2} & \frac{17}{8} & \frac{3}{16} \end{bmatrix} = \frac{1}{16}\begin{bmatrix} 138 & 56 & -84 \\ 37 & 26 & -71 \\ 24 & 34 & 3 \end{bmatrix}$

(b) $(A^T)^{-1} = (A^{-1})^T = \begin{bmatrix} 1 & \frac{3}{2} & \frac{1}{4} \\ -\frac{1}{2} & \frac{1}{2} & 1 \\ \frac{3}{4} & -2 & \frac{1}{2} \end{bmatrix} = \frac{1}{4}\begin{bmatrix} 4 & 6 & 1 \\ -2 & 2 & 4 \\ 3 & -8 & 2 \end{bmatrix}$

(c) $A^{-2} = (A^{-1})^2 = \begin{bmatrix} 1 & -\frac{1}{2} & \frac{3}{4} \\ \frac{3}{2} & \frac{1}{2} & -2 \\ \frac{1}{4} & 1 & \frac{1}{2} \end{bmatrix}^2 = \begin{bmatrix} \frac{7}{16} & 0 & \frac{17}{8} \\ \frac{7}{4} & -\frac{5}{2} & -\frac{7}{8} \\ \frac{15}{8} & \frac{7}{8} & -\frac{25}{16} \end{bmatrix} = \frac{1}{16}\begin{bmatrix} 7 & 0 & 34 \\ 28 & -40 & -14 \\ 30 & 14 & -25 \end{bmatrix}$

(d) $(2A)^{-1} = \frac{1}{2}A^{-1} = \begin{bmatrix} \frac{1}{2} & -\frac{1}{4} & \frac{3}{8} \\ \frac{3}{4} & \frac{1}{4} & -1 \\ \frac{1}{8} & \frac{1}{2} & \frac{1}{4} \end{bmatrix} = \frac{1}{8}\begin{bmatrix} 4 & -2 & 3 \\ 6 & 2 & -8 \\ 1 & 4 & 2 \end{bmatrix}$

37. Using the formula for the inverse of a 2×2 matrix, you have

$$A^{-1} = \frac{1}{2x - 9}\begin{bmatrix} -3 & -x \\ 2 & 3 \end{bmatrix}.$$

Letting $A^{-1} = A$, you find that $1/(2x - 9) = -1$. So, $x = 4$.

39. The matrix $\begin{bmatrix} 4 & x \\ -2 & -3 \end{bmatrix}$ will be singular if

$ad - bc = (4)(-3) - (x)(-2) = 0$, which implies that $2x = 12$ or $x = 6$.

41. First find $2A$.

$$2A = \left[(2A)^{-1}\right]^{-1} = \frac{1}{4 - 6}\begin{bmatrix} 4 & -2 \\ -3 & 1 \end{bmatrix} = \begin{bmatrix} -2 & 1 \\ \frac{3}{2} & -\frac{1}{2} \end{bmatrix}$$

Then divide by 2 to obtain

$$A = \tfrac{1}{2}(2A) = \tfrac{1}{2}\begin{bmatrix} -2 & 1 \\ \frac{3}{2} & -\frac{1}{2} \end{bmatrix} = \begin{bmatrix} -1 & \frac{1}{2} \\ \frac{3}{4} & -\frac{1}{4} \end{bmatrix}.$$

43. Using the formula for the inverse of a 2×2 matrix, you have

$$A^{-1} = \frac{1}{ad - bc}\begin{bmatrix} d & -b \\ -c & a \end{bmatrix}$$

$$= \frac{1}{\sin^2 \theta + \cos^2 \theta}\begin{bmatrix} \sin \theta & -\cos \theta \\ \cos \theta & \sin \theta \end{bmatrix}$$

$$= \begin{bmatrix} \sin \theta & -\cos \theta \\ \cos \theta & \sin \theta \end{bmatrix}.$$

45. (a) True. See Theorem 2.7.

(b) True. See Theorem 2.10, part 1.

(c) False. See Theorem 2.9

(d) True. See "Finding the Inverse of a Matrix by Gauss-Jordan Elimination," part 2, on page 76.

47. Use mathematical induction. The property is clearly true if $k = 1$. Suppose the property is true for $k = n$, and consider the case for $k = n + 1$.

$$\left(A^{n+1}\right)^{-1} = \left(AA^n\right)^{-1}$$

$$= \left(A^n\right)^{-1}A^{-1} = \left(\underbrace{A^{-1} \cdots A^{-1}}_{n \text{ times}}\right)A^{-1} = \underbrace{A^{-1} \cdots A^{-1}}_{n+1 \text{ times}}$$

49. Let A be symmetric and nonsingular. Then $A^T = A$ and $\left(A^{-1}\right)^T = \left(A^T\right)^{-1} = A^{-1}$. Therefore A^{-1} is symmetric.

51. $(I - 2A)(I - 2A) = I^2 - 2IA - 2AI + 4A^2$

$$= I - 4A + 4A^2$$

$$= I - 4A + 4A \quad \left(\text{because } A = A^2\right)$$

$$= I$$

So, $(I - 2A)^{-1} = I - 2A$.

53. Because A is invertible, you can multiply both sides of the equation $AB = O$ by A^{-1} to obtain the following.

$$A^{-1}(AB) = A^{-1}O$$

$$\left(A^{-1}A\right)B = O$$

$$B = O$$

55. No. For instance, $\begin{bmatrix} 1 & 0 \\ 0 & 1 \end{bmatrix} + \begin{bmatrix} -1 & 0 \\ 0 & -1 \end{bmatrix} = \begin{bmatrix} 0 & 0 \\ 0 & 0 \end{bmatrix}$.

57. To find the inverses, take the reciprocals of the diagonal entries.

(a) $A^{-1} = \begin{bmatrix} -1 & 0 & 0 \\ 0 & \frac{1}{3} & 0 \\ 0 & 0 & \frac{1}{2} \end{bmatrix}$

(b) $A^{-1} = \begin{bmatrix} 2 & 0 & 0 \\ 0 & 3 & 0 \\ 0 & 0 & 4 \end{bmatrix}$

59. (a) Let $H = I - 2\mathbf{u}\mathbf{u}^T$, where $\mathbf{u}^T\mathbf{u} = 1$. Then

$$H^T = \left(I - 2\mathbf{u}\mathbf{u}^T\right)^T$$

$$= I^T - 2\left(\mathbf{u}\mathbf{u}^T\right)^T$$

$$= I - 2\left[\left(\mathbf{u}^T\right)^T\mathbf{u}^T\right] = I - 2\mathbf{u}\mathbf{u}^T = H.$$

So, H is symmetric. Furthermore,

$$HH = \left(I - 2\mathbf{u}\mathbf{u}^T\right)\left(I - 2\mathbf{u}\mathbf{u}^T\right)$$

$$= I^2 - 4\mathbf{u}\mathbf{u}^T + 4\left(\mathbf{u}\mathbf{u}^T\right)^2$$

$$= I - 4\mathbf{u}\mathbf{u}^T + 4\mathbf{u}\mathbf{u}^T$$

$$= I.$$

So, H is nonsingular.

(b) $\mathbf{u}^T\mathbf{u} = \begin{bmatrix} \dfrac{\sqrt{2}}{2} & \dfrac{\sqrt{2}}{2} & 0 \end{bmatrix} \begin{bmatrix} \dfrac{\sqrt{2}}{2} \\ \dfrac{\sqrt{2}}{2} \\ 0 \end{bmatrix} = \begin{bmatrix} \dfrac{2}{4} + \dfrac{2}{4} + 0 \end{bmatrix} = 1.$

$$H = I_n - 2\mathbf{u}\mathbf{u}^T$$

$$= \begin{bmatrix} 1 & 0 & 0 \\ 0 & 1 & 0 \\ 0 & 0 & 1 \end{bmatrix} - 2\begin{bmatrix} \dfrac{\sqrt{2}}{2} \\ \dfrac{\sqrt{2}}{2} \\ 0 \end{bmatrix} \begin{bmatrix} \dfrac{\sqrt{2}}{2} & \dfrac{\sqrt{2}}{2} & 0 \end{bmatrix}$$

$$= \begin{bmatrix} 1 & 0 & 0 \\ 0 & 1 & 0 \\ 0 & 0 & 1 \end{bmatrix} - 2\begin{bmatrix} \dfrac{1}{2} & \dfrac{1}{2} & 0 \\ \dfrac{1}{2} & \dfrac{1}{2} & 0 \\ 0 & 0 & 0 \end{bmatrix} = \begin{bmatrix} 0 & -1 & 0 \\ -1 & 0 & 0 \\ 0 & 0 & 1 \end{bmatrix}$$

61. If P is nonsingular then P^{-1} exists. Use matrix multiplication to solve for A. Because all of the matrices are $n \times n$ matrices

$$AP = PD$$

$$APP^{-1} = PDP^{-1}$$

$$A = PDP^{-1}.$$

No, it is not necessarily true that $A = D$.

63. Answers will vary. *Sample answer*:

$$A = \begin{bmatrix} 1 & 0 \\ -1 & 0 \end{bmatrix} \text{ or } A = \begin{bmatrix} 1 & 0 \\ 1 & 0 \end{bmatrix}.$$

Section 2.4   Elementary Matrices

1. The matrix *is* elementary. It can be obtained by multiplying the second row of I_2 by 2.

3. The matrix *is* elementary. Two times the first row was added to the second row.

5. The matrix is *not* elementary. The first row was multiplied by 2 and the second and third rows were interchanged.

7. This matrix *is* elementary. It can be obtained by multiplying the second row of I_4 by -5, and adding the result to the third row.

9. B is obtained by interchanging the first and third rows of A. So,

$$E = \begin{bmatrix} 0 & 0 & 1 \\ 0 & 1 & 0 \\ 1 & 0 & 0 \end{bmatrix}.$$

11. A is obtained by interchanging the first and third rows of B. So,

$$E = \begin{bmatrix} 0 & 0 & 1 \\ 0 & 1 & 0 \\ 1 & 0 & 0 \end{bmatrix}.$$

13. To obtain the inverse matrix, reverse the elementary row operation that produced it. So, interchange the first and second rows of I_2 to obtain

$$E^{-1} = \begin{bmatrix} 0 & 1 \\ 1 & 0 \end{bmatrix}.$$

15. To obtain the inverse matrix, reverse the elementary row operation that produced it. So, interchange the first and third rows to obtain

$$E^{-1} = \begin{bmatrix} 0 & 0 & 1 \\ 0 & 1 & 0 \\ 1 & 0 & 0 \end{bmatrix}.$$

17. To obtain the inverse matrix, reverse the elementary row operation that produced it. So, divide the first row by k to obtain

$$E^{-1} = \begin{bmatrix} \frac{1}{k} & 0 \\ 0 & 1 \end{bmatrix}, \, k \neq 0.$$

19. To obtain the inverse matrix, reverse the elementary row operation that produced it. So, interchange the second and third row to obtain

$$E^{-1} = \begin{bmatrix} 1 & 0 & 0 \\ 0 & 0 & 1 \\ 0 & 1 & 0 \end{bmatrix}.$$

21. Find a sequence of elementary row operations that can be used to rewrite A in reduced row-echelon form.

$$\begin{bmatrix} 1 & 0 \\ 3 & -2 \end{bmatrix} R_1 \leftrightarrow R_2 \qquad E_1 = \begin{bmatrix} 0 & 1 \\ 1 & 0 \end{bmatrix}$$

$$\begin{bmatrix} 1 & 0 \\ 0 & -2 \end{bmatrix} R_2 - 3R_1 \rightarrow R_2 \qquad E_2 = \begin{bmatrix} 1 & 0 \\ -3 & 1 \end{bmatrix}$$

$$\begin{bmatrix} 1 & 0 \\ 0 & 1 \end{bmatrix} \left(-\tfrac{1}{2}\right) R_2 \rightarrow R_2 \qquad E_3 = \begin{bmatrix} 1 & 0 \\ 0 & -\frac{1}{2} \end{bmatrix}$$

Use the elementary matrices to find the inverse.

$$A^{-1} = E_3 E_2 E_1$$

$$= \begin{bmatrix} 1 & 0 \\ 0 & -\frac{1}{2} \end{bmatrix} \begin{bmatrix} 1 & 0 \\ -3 & 1 \end{bmatrix} \begin{bmatrix} 0 & 1 \\ 1 & 0 \end{bmatrix}$$

$$= \begin{bmatrix} 1 & 0 \\ \frac{3}{2} & -\frac{1}{2} \end{bmatrix} \begin{bmatrix} 0 & 1 \\ 1 & 0 \end{bmatrix} = \begin{bmatrix} 0 & 1 \\ -\frac{1}{2} & \frac{3}{2} \end{bmatrix}$$

23. Find a sequence of elementary row operations that can be used to rewrite A in reduced row-echelon form.

$$\begin{bmatrix} 1 & 0 & -1 \\ 0 & 1 & -\frac{1}{6} \\ 0 & 0 & 4 \end{bmatrix} \left(\tfrac{1}{6}\right) R_2 \rightarrow R_2 \qquad E_1 = \begin{bmatrix} 1 & 0 & 0 \\ 0 & \frac{1}{6} & 0 \\ 0 & 0 & 1 \end{bmatrix}$$

$$\begin{bmatrix} 1 & 0 & -1 \\ 0 & 1 & -\frac{1}{6} \\ 0 & 0 & 1 \end{bmatrix} \left(\tfrac{1}{4}\right) R_3 \rightarrow R_3 \qquad E_2 = \begin{bmatrix} 1 & 0 & 0 \\ 0 & 1 & 0 \\ 0 & 0 & \frac{1}{4} \end{bmatrix}$$

$$\begin{bmatrix} 1 & 0 & 0 \\ 0 & 1 & -\frac{1}{6} \\ 0 & 0 & 1 \end{bmatrix} R_1 + R_3 \rightarrow R_1 \qquad E_3 = \begin{bmatrix} 1 & 0 & 1 \\ 0 & 1 & 0 \\ 0 & 0 & 1 \end{bmatrix}$$

$$\begin{bmatrix} 1 & 0 & 0 \\ 0 & 1 & 0 \\ 0 & 0 & 1 \end{bmatrix} R_2 + \left(\tfrac{1}{6}\right) R_3 \rightarrow R_2 \quad E_4 = \begin{bmatrix} 1 & 0 & 0 \\ 0 & 1 & \frac{1}{6} \\ 0 & 0 & 1 \end{bmatrix}$$

Use the elementary matrices to find the inverse.

$$A^{-1} = E_4 E_3 E_2 E_1$$

$$= \begin{bmatrix} 1 & 0 & 0 \\ 0 & 1 & \frac{1}{6} \\ 0 & 0 & 1 \end{bmatrix} \begin{bmatrix} 1 & 0 & 1 \\ 0 & 1 & 0 \\ 0 & 0 & 1 \end{bmatrix} \begin{bmatrix} 1 & 0 & 0 \\ 0 & 1 & 0 \\ 0 & 0 & \frac{1}{4} \end{bmatrix} \begin{bmatrix} 1 & 0 & 0 \\ 0 & \frac{1}{6} & 0 \\ 0 & 0 & 1 \end{bmatrix}$$

$$= \begin{bmatrix} 1 & 0 & 1 \\ 0 & 1 & \frac{1}{6} \\ 0 & 0 & 1 \end{bmatrix} \begin{bmatrix} 1 & 0 & 0 \\ 0 & 1 & 0 \\ 0 & 0 & \frac{1}{4} \end{bmatrix} \begin{bmatrix} 1 & 0 & 0 \\ 0 & \frac{1}{6} & 0 \\ 0 & 0 & 1 \end{bmatrix}$$

$$= \begin{bmatrix} 1 & 0 & \frac{1}{4} \\ 0 & 1 & \frac{1}{24} \\ 0 & 0 & \frac{1}{4} \end{bmatrix} \begin{bmatrix} 1 & 0 & 0 \\ 0 & \frac{1}{6} & 0 \\ 0 & 0 & 1 \end{bmatrix} = \begin{bmatrix} 1 & 0 & \frac{1}{4} \\ 0 & \frac{1}{6} & \frac{1}{24} \\ 0 & 0 & \frac{1}{4} \end{bmatrix}$$

For Exercises 25–31, answers will vary. Sample answers are shown below.

25. <u>Matrix</u>   <u>Elementary Row Operation</u>   <u>Elementary Matrix</u>

$$\begin{bmatrix} 1 & 2 \\ 0 & -2 \end{bmatrix} \qquad \text{Add } -1 \text{ times row one to row two.} \qquad E_1 = \begin{bmatrix} 1 & 0 \\ -1 & 1 \end{bmatrix}$$

$$\begin{bmatrix} 1 & 0 \\ 0 & -2 \end{bmatrix} \qquad \text{Add row two to row one.} \qquad E_2 = \begin{bmatrix} 1 & 1 \\ 0 & 1 \end{bmatrix}$$

$$\begin{bmatrix} 1 & 0 \\ 0 & 1 \end{bmatrix} \qquad \text{Divide row two by } -2. \qquad E_3 = \begin{bmatrix} 1 & 0 \\ 0 & -\frac{1}{2} \end{bmatrix}$$

Because $E_3 E_2 E_1 A = I_2$, factor A as follows.

$$A = E_1^{-1} E_2^{-1} E_3^{-1} = \begin{bmatrix} 1 & 0 \\ 1 & 1 \end{bmatrix}\begin{bmatrix} 1 & -1 \\ 0 & 1 \end{bmatrix}\begin{bmatrix} 1 & 0 \\ 0 & -2 \end{bmatrix}$$

Note that this factorization is not unique. For example, another factorization is

$$A = \begin{bmatrix} 0 & 1 \\ 1 & 0 \end{bmatrix}\begin{bmatrix} 1 & 0 \\ 1 & 1 \end{bmatrix}\begin{bmatrix} 1 & 0 \\ 0 & 2 \end{bmatrix}.$$

27. <u>Matrix</u>   <u>Elementary Row Operation</u>   <u>Elementary Matrix</u>

$$\begin{bmatrix} 1 & 0 \\ 3 & -1 \end{bmatrix} \qquad \text{Add } (-1) \text{ times row two to row one.} \qquad E_1 = \begin{bmatrix} 1 & -1 \\ 0 & 1 \end{bmatrix}$$

$$\begin{bmatrix} 1 & 0 \\ 0 & -1 \end{bmatrix} \qquad \text{Add } -3 \text{ times row one to row two.} \qquad E_2 = \begin{bmatrix} 1 & 0 \\ -3 & 1 \end{bmatrix}$$

$$\begin{bmatrix} 1 & 0 \\ 0 & 1 \end{bmatrix} \qquad \text{Multiply row two by } -1. \qquad E_3 = \begin{bmatrix} 1 & 0 \\ 0 & -1 \end{bmatrix}$$

Because $E_3 E_2 E_1 A = I_2$, one way to factor A is as follows.

$$A = E_1^{-1} E_2^{-1} E_3^{-1} = \begin{bmatrix} 1 & 1 \\ 0 & 1 \end{bmatrix}\begin{bmatrix} 1 & 0 \\ 3 & 1 \end{bmatrix}\begin{bmatrix} 1 & 0 \\ 0 & -1 \end{bmatrix}$$

29. <u>Matrix</u>   <u>Elementary Row Operation</u>   <u>Elementary Matrix</u>

$$\begin{bmatrix} 1 & -2 & 0 \\ 0 & 1 & 0 \\ 0 & 0 & 1 \end{bmatrix} \qquad \text{Add row one to row two.} \qquad E_1 = \begin{bmatrix} 1 & 0 & 0 \\ 1 & 1 & 0 \\ 0 & 0 & 1 \end{bmatrix}$$

$$\begin{bmatrix} 1 & 0 & 0 \\ 0 & 1 & 0 \\ 0 & 0 & 1 \end{bmatrix} \qquad \text{Add 2 times row two to row one.} \qquad E_2 = \begin{bmatrix} 1 & 2 & 0 \\ 0 & 1 & 0 \\ 0 & 0 & 1 \end{bmatrix}$$

Because $E_2 E_1 A = I_3$, one way to factor A is

$$A = E_1^{-1} E_2^{-1} = \begin{bmatrix} 1 & 0 & 0 \\ -1 & 1 & 0 \\ 0 & 0 & 1 \end{bmatrix}\begin{bmatrix} 1 & -2 & 0 \\ 0 & 1 & 0 \\ 0 & 0 & 1 \end{bmatrix}.$$

31. Find a sequence of elementary row operations that can be used to rewrite A in reduced row-echelon form.

$$\begin{bmatrix} 1 & 0 & 0 & 1 \\ 0 & 1 & -3 & 0 \\ 0 & 0 & 2 & 0 \\ 0 & 0 & 1 & -1 \end{bmatrix} \, -R_2 \to R_2 \qquad\qquad E_1 = \begin{bmatrix} 1 & 0 & 0 & 0 \\ 0 & -1 & 0 & 0 \\ 0 & 0 & 1 & 0 \\ 0 & 0 & 0 & 1 \end{bmatrix}$$

$$\begin{bmatrix} 1 & 0 & 0 & 1 \\ 0 & 1 & -3 & 0 \\ 0 & 0 & 1 & 0 \\ 0 & 0 & 1 & -1 \end{bmatrix} \, \left(\tfrac{1}{2}\right)R_3 \to R_3 \qquad\qquad E_2 = \begin{bmatrix} 1 & 0 & 0 & 0 \\ 0 & 1 & 0 & 0 \\ 0 & 0 & \tfrac{1}{2} & 0 \\ 0 & 0 & 0 & 1 \end{bmatrix}$$

$$\begin{bmatrix} 1 & 0 & 0 & 1 \\ 0 & 1 & -3 & 0 \\ 0 & 0 & 1 & 0 \\ 0 & 0 & -1 & 1 \end{bmatrix} \, -R_4 \to R_4 \qquad\qquad E_3 = \begin{bmatrix} 1 & 0 & 0 & 0 \\ 0 & 1 & 0 & 0 \\ 0 & 0 & 1 & 0 \\ 0 & 0 & 0 & -1 \end{bmatrix}$$

$$\begin{bmatrix} 1 & 0 & 0 & 1 \\ 0 & 1 & -3 & 0 \\ 0 & 0 & 1 & 0 \\ 0 & 0 & 0 & 1 \end{bmatrix} \, R_4 + R_3 \to R_4 \qquad\qquad E_4 = \begin{bmatrix} 1 & 0 & 0 & 0 \\ 0 & 1 & 0 & 0 \\ 0 & 0 & 1 & 0 \\ 0 & 0 & 1 & 1 \end{bmatrix}$$

$$\begin{bmatrix} 1 & 0 & 0 & 1 \\ 0 & 1 & 0 & 0 \\ 0 & 0 & 1 & 0 \\ 0 & 0 & 0 & 1 \end{bmatrix} \, R_2 + 3R_3 \to R_2 \qquad\qquad E_5 = \begin{bmatrix} 1 & 0 & 0 & 0 \\ 0 & 1 & 3 & 0 \\ 0 & 0 & 1 & 0 \\ 0 & 0 & 0 & 1 \end{bmatrix}$$

$$\begin{bmatrix} 1 & 0 & 0 & 0 \\ 0 & 1 & 0 & 0 \\ 0 & 0 & 1 & 0 \\ 0 & 0 & 0 & 1 \end{bmatrix} \, R_1 - R_4 \to R_1 \qquad\qquad E_6 = \begin{bmatrix} 1 & 0 & 0 & -1 \\ 0 & 1 & 0 & 0 \\ 0 & 0 & 1 & 0 \\ 0 & 0 & 0 & 1 \end{bmatrix}$$

So, one way to factor A is

$$A = E_1^{-1} E_2^{-1} E_3^{-1} E_4^{-1} E_5^{-1} E_6^{-1}$$

$$= \begin{bmatrix} 1 & 0 & 0 & 0 \\ 0 & -1 & 0 & 0 \\ 0 & 0 & 1 & 0 \\ 0 & 0 & 0 & 1 \end{bmatrix} \begin{bmatrix} 1 & 0 & 0 & 0 \\ 0 & 1 & 0 & 0 \\ 0 & 0 & 2 & 0 \\ 0 & 0 & 0 & 1 \end{bmatrix} \begin{bmatrix} 1 & 0 & 0 & 0 \\ 0 & 1 & 0 & 0 \\ 0 & 0 & 1 & 0 \\ 0 & 0 & 0 & -1 \end{bmatrix} \begin{bmatrix} 1 & 0 & 0 & 0 \\ 0 & 1 & 0 & 0 \\ 0 & 0 & 1 & 0 \\ 0 & 0 & -1 & 1 \end{bmatrix} \begin{bmatrix} 1 & 0 & 0 & 0 \\ 0 & 1 & -3 & 0 \\ 0 & 0 & 1 & 0 \\ 0 & 0 & 0 & 1 \end{bmatrix} \begin{bmatrix} 1 & 0 & 0 & 1 \\ 0 & 1 & 0 & 0 \\ 0 & 0 & 1 & 0 \\ 0 & 0 & 0 & 1 \end{bmatrix}.$$

33. (a) True. See "Remark" following the "Definition of an Elementary Matrix" on page 87.

    (b) False. Multiplication of a matrix by a scalar is not a single elementary row operation so it cannot be represented by a corresponding elementary matrix.

    (c) True. See "Definition of Row Equivalence," on page 90.

    (d) True. See Theorem 2.13.

35. (a) EA has the same rows as A except the two rows that are interchanged in E will be interchanged in EA.

    (b) Multiplying a matrix on the left by E interchanges the same two rows that are interchanged from I_n in E. So, multiplying E by itself interchanges the rows twice and so $E^2 = I_n$.

37. $A^{-1} = \begin{bmatrix} 1 & 0 & 0 \\ 0 & 1 & 0 \\ 0 & 0 & c \end{bmatrix}^{-1} \begin{bmatrix} 1 & 0 & 0 \\ b & 1 & 0 \\ 0 & 0 & 1 \end{bmatrix}^{-1} \begin{bmatrix} 1 & a & 0 \\ 0 & 1 & 0 \\ 0 & 0 & 1 \end{bmatrix}^{-1} = \begin{bmatrix} 1 & 0 & 0 \\ 0 & 1 & 0 \\ 0 & 0 & \frac{1}{c} \end{bmatrix} \begin{bmatrix} 1 & 0 & 0 \\ -b & 1 & 0 \\ 0 & 0 & 1 \end{bmatrix} \begin{bmatrix} 1 & -a & 0 \\ 0 & 1 & 0 \\ 0 & 0 & 1 \end{bmatrix} = \begin{bmatrix} 1 & -a & 0 \\ -b & 1+ab & 0 \\ 0 & 0 & \frac{1}{c} \end{bmatrix}.$

39. No. For example, $\begin{bmatrix} 1 & 0 \\ 2 & 1 \end{bmatrix} \begin{bmatrix} 1 & 1 \\ 0 & 1 \end{bmatrix} = \begin{bmatrix} 1 & 1 \\ 2 & 3 \end{bmatrix}$, which is not elementary.

For Exercises 41–45, answers will vary. Sample answers are shown below.

41. Because the matrix is lower triangular, an *LU*-factorization is

$$\begin{bmatrix} 1 & 0 \\ -2 & 1 \end{bmatrix} \begin{bmatrix} 1 & 0 \\ 0 & 1 \end{bmatrix}$$

43.   Matrix   Elementary Matrix

$\begin{bmatrix} 3 & 0 & 1 \\ 6 & 1 & 1 \\ -3 & 1 & 0 \end{bmatrix} = A$

$\begin{bmatrix} 3 & 0 & 1 \\ 0 & 1 & -1 \\ -3 & 1 & 0 \end{bmatrix}$ $E_1 = \begin{bmatrix} 1 & 0 & 0 \\ -2 & 1 & 0 \\ 0 & 0 & 1 \end{bmatrix}$

$\begin{bmatrix} 3 & 0 & 1 \\ 0 & 1 & -1 \\ 0 & 1 & 1 \end{bmatrix}$ $E_2 = \begin{bmatrix} 1 & 0 & 0 \\ 0 & 1 & 0 \\ 1 & 0 & 1 \end{bmatrix}$

$\begin{bmatrix} 3 & 0 & 1 \\ 0 & 1 & -1 \\ 0 & 0 & 2 \end{bmatrix} = U$ $E_3 = \begin{bmatrix} 1 & 0 & 0 \\ 0 & 1 & 0 \\ 0 & -1 & 1 \end{bmatrix}$

$E_3 E_2 E_1 A = U$

$A = E_1^{-1} E_2^{-1} E_3^{-1} U = \begin{bmatrix} 1 & 0 & 0 \\ 2 & 1 & 0 \\ -1 & 1 & 1 \end{bmatrix} \begin{bmatrix} 3 & 0 & 1 \\ 0 & 1 & -1 \\ 0 & 0 & 2 \end{bmatrix} = LU$

45. (a)   Matrix   Elementary Matrix

$\begin{bmatrix} 2 & 1 & 0 \\ 0 & 1 & -1 \\ -2 & 1 & 1 \end{bmatrix} = A$

$\begin{bmatrix} 2 & 1 & 0 \\ 0 & 1 & -1 \\ 0 & 2 & 1 \end{bmatrix}$ $E_1 = \begin{bmatrix} 1 & 0 & 0 \\ 0 & 1 & 0 \\ 1 & 0 & 1 \end{bmatrix}$

$\begin{bmatrix} 2 & 1 & 0 \\ 0 & 1 & -1 \\ 0 & 0 & 3 \end{bmatrix} = U$ $E_2 = \begin{bmatrix} 1 & 0 & 0 \\ 0 & 1 & 0 \\ 0 & -2 & 1 \end{bmatrix}$

$E_2 E_1 A = U$

$A = E_1^{-1} E_2^{-1} U = \begin{bmatrix} 1 & 0 & 0 \\ 0 & 1 & 0 \\ -1 & 2 & 1 \end{bmatrix} \begin{bmatrix} 2 & 1 & 0 \\ 0 & 1 & -1 \\ 0 & 0 & 3 \end{bmatrix} = LU$

(b) $L\mathbf{y} = \mathbf{b}$: $\begin{bmatrix} 1 & 0 & 0 \\ 0 & 1 & 0 \\ -1 & 2 & 1 \end{bmatrix} \begin{bmatrix} y_1 \\ y_2 \\ y_3 \end{bmatrix} = \begin{bmatrix} 1 \\ 2 \\ -2 \end{bmatrix}$

$y_1 = 1, \; y_2 = 2,$

and

$-y_1 + 2y_2 + y_3 = -2 \Rightarrow y_3 = -5$

(c) $U\mathbf{x} = \mathbf{y}$: $\begin{bmatrix} 2 & 1 & 0 \\ 0 & 1 & -1 \\ 0 & 0 & 3 \end{bmatrix} \begin{bmatrix} x_1 \\ x_2 \\ x_3 \end{bmatrix} = \begin{bmatrix} 1 \\ 2 \\ -5 \end{bmatrix}$

$x_3 = -\frac{5}{3}, \; x_2 - x_3 = 2 \Rightarrow x_2 = \frac{1}{3}$

and $2x_1 + x_2 = 1 \Rightarrow x_1 = \frac{1}{3}$.

So, the solution to the system $A\mathbf{x} = \mathbf{b}$ is

$x_1 = \frac{1}{3}, \; x_2 = \frac{1}{3}, \; x_3 = -\frac{5}{3}.$

47. You could first factor the matrix $A = LU$. Then for each right hand side $\mathbf{b}_i$, solve $L\mathbf{y} = \mathbf{b}_i$ and $U\mathbf{x} = \mathbf{y}$.

49. $A^2 = \begin{bmatrix} 1 & 0 \\ 0 & 0 \end{bmatrix}^2 = \begin{bmatrix} 1 & 0 \\ 0 & 0 \end{bmatrix} = A.$

Because $A^2 = A$, A is idempotent.

51. $A^2 = \begin{bmatrix} 2 & 3 \\ -1 & -2 \end{bmatrix}\begin{bmatrix} 2 & 3 \\ -1 & -2 \end{bmatrix} = \begin{bmatrix} 1 & 0 \\ 0 & 1 \end{bmatrix} \neq A.$

Because $A^2 \neq A$, A is *not* idempotent.

53. $A^2 = \begin{bmatrix} 0 & 0 & 1 \\ 0 & 1 & 0 \\ 1 & 0 & 0 \end{bmatrix}\begin{bmatrix} 0 & 0 & 1 \\ 0 & 1 & 0 \\ 1 & 0 & 0 \end{bmatrix} = \begin{bmatrix} 1 & 0 & 0 \\ 0 & 1 & 0 \\ 0 & 0 & 1 \end{bmatrix}.$

Because $A^2 \neq A$, A is *not* idempotent.

55. Begin by finding A^2.

$A^2 = \begin{bmatrix} 1 & 0 \\ a & b \end{bmatrix}\begin{bmatrix} 1 & 0 \\ a & b \end{bmatrix} = \begin{bmatrix} 1 & 0 \\ a(1+b) & b^2 \end{bmatrix}$

Setting $A^2 = A$ yields the equations $a = a(1+b)$ and $b = b^2$. The second equation is satisfied when $b = 1$ or $b = 0$. If $b = 1$, then $a = 0$, and if $b = 0$, then a can be any real number.

57. Because A is idempotent and invertible, you have

$A^2 = A$

$A^{-1}A^2 = A^{-1}A$

$(A^{-1}A)A = I$

$A = I.$

59. $(AB)^2 = (AB)(AB)$

$= A(BA)B = A(AB)B = (AA)(BB) = AB.$

So, $(AB)^2 = AB$, and AB is idempotent.

61. $A = E_k \cdots E_2 E_1 B$, for $E_1, \ldots, E_k$ elementary.

$B = F_t \cdots F_2 F_1 C$, for $F_1, \ldots, F_t$ elementary.

Then

$A = E_k \cdots E_2 E_1 B = (E_k \cdots E_2 E_1)(F_t \cdots F_2 F_1 C)$

which shows that A is row equivalent to C.

Section 2.5   Applications of Matrix Operations

1. The matrix is *not* stochastic because every entry of a stochastic matrix must satisfy the inequality
$0 \leq a_{ij} \leq 1.$

3. This matrix *is* stochastic because each entry is between 0 and 1, and each column adds up to 1.

5. The matrix *is* stochastic because $0 \leq a_{ij} \leq 1$ and each column adds up to 1.

7. Form the matrix representing the given transition probabilities. Let A represent people who purchased the product and B represent people who did not.

$$P = \begin{array}{c} \\ \end{array}\overset{\text{From}}{\underset{}{\begin{array}{cc} A & B \end{array}}}$$

$P = \begin{bmatrix} 0.80 & 0.30 \\ 0.20 & 0.70 \end{bmatrix}\begin{matrix} A \\ B \end{matrix}\Big\}$ To

The state matrix representing the current population is

$X = \begin{bmatrix} 100 \\ 900 \end{bmatrix}\begin{matrix} A \\ B \end{matrix}$

The state matrix for next month is

$PX = \begin{bmatrix} 0.80 & 0.30 \\ 0.20 & 0.70 \end{bmatrix}\begin{bmatrix} 100 \\ 900 \end{bmatrix} = \begin{bmatrix} 350 \\ 650 \end{bmatrix}.$

The state matrix for the month after next is

$P(PX) = \begin{bmatrix} 0.80 & 0.30 \\ 0.20 & 0.70 \end{bmatrix}\begin{bmatrix} 350 \\ 650 \end{bmatrix} = \begin{bmatrix} 475 \\ 525 \end{bmatrix}.$

So, next month 350 people will purchase the product. In two months 475 people will purchase the product.

9. Form the matrix representing the given transition probabilities. Let N represent nonsmokers, S_0 represent those who smoke one pack or less, and S_1 represent those who smoke more than one pack.

$$\overset{\text{From}}{\begin{array}{ccc} N & S_0 & S_1 \end{array}}$$

$P = \begin{bmatrix} 0.93 & 0.10 & 0.05 \\ 0.05 & 0.80 & 0.10 \\ 0.02 & 0.10 & 0.85 \end{bmatrix}\begin{matrix} N \\ S_0 \\ S_1 \end{matrix}\Big\}$ To

The state matrix representing the current population is

$X = \begin{bmatrix} 5000 \\ 2500 \\ 2500 \end{bmatrix}\begin{matrix} N \\ S_0 \\ S_1 \end{matrix}$

The state matrix for the next month is

$PX = \begin{bmatrix} 0.93 & 0.10 & 0.05 \\ 0.05 & 0.80 & 0.10 \\ 0.02 & 0.10 & 0.85 \end{bmatrix}\begin{bmatrix} 5000 \\ 2500 \\ 2500 \end{bmatrix} = \begin{bmatrix} 5025 \\ 2500 \\ 2475 \end{bmatrix}.$

The state matrix for the month after next is

$P(PX) = \begin{bmatrix} 0.93 & 0.10 & 0.05 \\ 0.05 & 0.80 & 0.10 \\ 0.02 & 0.10 & 0.85 \end{bmatrix}\begin{bmatrix} 5025 \\ 2500 \\ 2475 \end{bmatrix} = \begin{bmatrix} 5047 \\ 2498.75 \\ 2454.25 \end{bmatrix}.$

So, next month the population will be grouped as follows: 5025 nonsmokers, 2500 smokers of one pack or less per day, and 2475 smokers of more than one pack per day. In two months the population will be grouped as follows: 5047 nonsmokers, 2499 smokers of one pack or less per day, and 2454 smokers of more than one pack per day.

11. Form the matrix representing the given transition probabilities. Let A represent an hour or more of TV and B less than one hour.

$$\overset{\text{From}}{\underset{A \quad B}{}}$$

$$P = \begin{bmatrix} 0 & 0.25 \\ 1 & 0.75 \end{bmatrix} \!\! \begin{matrix} A \\ B \end{matrix} \Bigg\} \text{To}$$

The state matrix representing the current distribution is $X = \begin{bmatrix} 100 \\ 100 \end{bmatrix}$.

The state matrix for one day later is $PX = \begin{bmatrix} 0 & 0.25 \\ 1 & 0.75 \end{bmatrix}\begin{bmatrix} 100 \\ 100 \end{bmatrix} = \begin{bmatrix} 25 \\ 175 \end{bmatrix}$.

In two days the state matrix is $P(PX) = \begin{bmatrix} 0 & 0.25 \\ 1 & 0.75 \end{bmatrix}\begin{bmatrix} 25 \\ 175 \end{bmatrix} \approx \begin{bmatrix} 44 \\ 156 \end{bmatrix}$.

In thirty days, the state matrix will be $P^{30}X = \begin{bmatrix} 40 \\ 160 \end{bmatrix}$.

So, 25 will watch TV for an hour or more tomorrow, 44 the day after tomorrow, and 40 in thirty days.

13. Because the columns in a stochastic matrix add up to 1, you can represent two stochastic matrices as

$$P\begin{bmatrix} a & b \\ 1-a & 1-b \end{bmatrix} \text{and } Q = \begin{bmatrix} c & d \\ 1-c & 1-d \end{bmatrix}$$

Then,

$$PQ = \begin{bmatrix} a & b \\ 1-a & 1-b \end{bmatrix}\begin{bmatrix} c & d \\ 1-c & 1-d \end{bmatrix}$$

$$= \begin{bmatrix} ac + b(1-c) & ad + b(1-d) \\ c(1-a) + (1-b)(1-c) & d(1-a) + (1-b)(1-d) \end{bmatrix} = \begin{bmatrix} ac + b - bc & ad + b - bd \\ 1 - (ac + b - bc) & 1 - (ad + b - bd) \end{bmatrix}$$

The columns of PQ add up to 1, and the entries are non-negative, because those of P and Q are non-negative. So, PQ is stochastic.

15. Divide the message into groups of three and form the uncoded matrices.

$$\begin{matrix} \text{S} & \text{E} & \text{L} \\ [19 & 5 & 12] \end{matrix} \quad \begin{matrix} \text{L} & \_ & \text{C} \\ [12 & 0 & 3] \end{matrix} \quad \begin{matrix} \text{O} & \text{N} & \text{S} \\ [15 & 14 & 19] \end{matrix} \quad \begin{matrix} \text{O} & \text{L} & \text{I} \\ [15 & 12 & 9] \end{matrix} \quad \begin{matrix} \text{D} & \text{A} & \text{T} \\ [4 & 1 & 20] \end{matrix} \quad \begin{matrix} \text{E} & \text{D} & \_ \\ [5 & 4 & 0] \end{matrix}$$

Multiplying each uncoded row matrix on the right by A yields the coded row matrices

$$[19 \;\; 5 \;\; 12]A = [19 \;\; 5 \;\; 12]\begin{bmatrix} 1 & -1 & 0 \\ 1 & 0 & -1 \\ -6 & 2 & 3 \end{bmatrix} = [-48 \;\; 5 \;\; 31]$$

$$[12 \;\; 0 \;\; 3]A = [-6 \;\; -6 \;\; 9]$$

$$[15 \;\; 14 \;\; 19]A = [-85 \;\; 23 \;\; 43]$$

$$[15 \;\; 12 \;\; 9]A = [-27 \;\; 3 \;\; 15]$$

$$[4 \;\; 1 \;\; 20]A = [-115 \;\; 36 \;\; 59]$$

$$[5 \;\; 4 \;\; 0]A = [9 \;\; -5 \;\; -4]$$

So, the coded message is $-48, 5, 31, -6, -6, 9, -85, 23, 43, -27, 3, 15, -115, 36, 59, 9, -5, -4$.

17. Divide the message into pairs of letters and form the coded matrices.

C   O   M   E   _   H   O   M   E   _   S   O   O   N

$[3 \ 15] \quad [13 \ 5] \quad [0 \ 8] \quad [15 \ 13] \quad [5 \ 0] \quad [19 \ 15] \quad [15 \ 14]$

Multiplying each uncoded row matrix on the right by A yields the coded row matrices.

$$[3 \ 15]\begin{bmatrix} 1 & 2 \\ 3 & 5 \end{bmatrix} = [48 \ 81]$$

$$[13 \ 5]\begin{bmatrix} 1 & 2 \\ 3 & 5 \end{bmatrix} = [28 \ 51]$$

$$[0 \ 8]\begin{bmatrix} 1 & 2 \\ 3 & 5 \end{bmatrix} = [24 \ 40]$$

$$[15 \ 13]\begin{bmatrix} 1 & 2 \\ 3 & 5 \end{bmatrix} = [54 \ 95]$$

$$[5 \ 0]\begin{bmatrix} 1 & 2 \\ 3 & 5 \end{bmatrix} = [5 \ 10]$$

$$[19 \ 15]\begin{bmatrix} 1 & 2 \\ 3 & 5 \end{bmatrix} = [64 \ 113]$$

$$[15 \ 14]\begin{bmatrix} 1 & 2 \\ 3 & 5 \end{bmatrix} = [57 \ 100]$$

So, the coded message is 48, 81, 28, 51, 24, 40, 54, 95, 5, 10, 64, 113, 57, 100.

19. Find $A^{-1} = \begin{bmatrix} -5 & 2 \\ 3 & -1 \end{bmatrix}$

and multiply each coded row matrix on the right by A^{-1} to find the associated uncoded row matrix.

$$[11 \ 21]\begin{bmatrix} -5 & 2 \\ 3 & -1 \end{bmatrix} = [8 \ 1] \Rightarrow \text{H, A}$$

$$[64 \ 112]\begin{bmatrix} -5 & 2 \\ 3 & -1 \end{bmatrix} = [16 \ 16] \Rightarrow \text{P, P}$$

$$[25 \ 50]\begin{bmatrix} -5 & 2 \\ 3 & -1 \end{bmatrix} = [25 \ 0] \Rightarrow \text{Y, \_}$$

$$[29 \ 53]\begin{bmatrix} -5 & 2 \\ 3 & -1 \end{bmatrix} = [14 \ 5] \Rightarrow \text{N, E}$$

$$[23 \ 46]\begin{bmatrix} -5 & 2 \\ 3 & -1 \end{bmatrix} = [23 \ 0] \Rightarrow \text{W, \_}$$

$$[40 \ 75]\begin{bmatrix} -5 & 2 \\ 3 & -1 \end{bmatrix} = [25 \ 5] \Rightarrow \text{Y, E}$$

$$[55 \ 92]\begin{bmatrix} -5 & 2 \\ 3 & -1 \end{bmatrix} = [1 \ 18] \Rightarrow \text{A, R}$$

So, the message is HAPPY_NEW_YEAR.

21. Find $A^{-1} = \begin{bmatrix} -13 & 6 & 4 \\ 12 & -5 & -3 \\ -5 & 2 & 1 \end{bmatrix}$

and multiply each coded row matrix on the right by A^{-1} to find the associated uncoded row matrix.

$$[13 \ \ 19 \ \ 10]A^{-1} = [13 \ \ 19 \ \ 10]\begin{bmatrix} -13 & 6 & 4 \\ 12 & -5 & -3 \\ -5 & 2 & 1 \end{bmatrix} = [9 \ \ 3 \ \ 5] \Rightarrow \text{I, C, E}$$

$$[-1 \ \ -33 \ \ -77]A^{-1} = [2 \ \ 5 \ \ 18] \Rightarrow \text{B, E, R}$$
$$[3 \ \ -2 \ \ -14]A^{-1} = [7 \ \ 0 \ \ 4] \Rightarrow \text{G, \_, D}$$
$$[4 \ \ 1 \ \ -9]A^{-1} = [5 \ \ 1 \ \ 4] \Rightarrow \text{E, A, D}$$
$$[-5 \ \ -25 \ \ -47]A^{-1} = [0 \ \ 1 \ \ 8] \Rightarrow \text{\_, A, H}$$
$$[4 \ \ 1 \ \ -9]A^{-1} = [5 \ \ 1 \ \ 4] \Rightarrow \text{E, A, D}$$

The message is ICEBERG_DEAD_AHEAD.

23. Let a $A^{-1} = \begin{bmatrix} a & b \\ c & d \end{bmatrix}$ and find that

$$[-18 \ \ -18]\begin{bmatrix} a & b \\ c & d \end{bmatrix} = [0 \ \ 18] \quad \overset{\_ \ R}{}$$

$$[-18(a + c) - 18(b + d)] = [0 \ \ 18].$$

So, $c = -a$ and $d = -1 - b$. Using these values you find that

$$[1 \ \ 16]\begin{bmatrix} a & b \\ -a & -(1 + b) \end{bmatrix} = [15 \ \ 14] \quad \overset{\text{O} \quad \text{N}}{}$$

$$[-15a \ \ -15b - 16] = [15 \ \ 14].$$

So, $a = -1$, $b = -2$, $c = 1$, and $d = 1$. Using the matrix

$$A^{-1} = \begin{bmatrix} -1 & -2 \\ 1 & 1 \end{bmatrix}$$

multiply each coded row matrix to yield the uncoded row matrices

$[13 \ \ 5]$, $[5 \ \ 20]$, $[0 \ \ 13]$, $[5 \ \ 0]$, $[20 \ \ 15]$, $[14 \ \ 9]$, $[7 \ \ 8]$, $[20 \ \ 0]$, $[0 \ \ 18]$, $[15 \ \ 14]$.

This corresponds to the message MEET_ME_TONIGHT_RON.

25. For $A = \begin{bmatrix} 1 & 0 & 2 \\ 2 & -1 & 1 \\ 0 & 1 & 2 \end{bmatrix}$ you have $A^{-1} = \begin{bmatrix} -3 & 2 & 2 \\ -4 & 2 & 3 \\ 2 & -1 & -1 \end{bmatrix}$.

Multiply each coded row matrix on the right by A^{-1} to find the associated uncoded row matrix.

$\begin{bmatrix} 38 & -14 & 29 \end{bmatrix} A^{-1} = \begin{bmatrix} 0 & 19 & 5 \end{bmatrix} \Rightarrow \_,\ \text{S},\ \text{E}$

$\begin{bmatrix} 56 & -15 & 62 \end{bmatrix} A^{-1} = \begin{bmatrix} 16 & 20 & 5 \end{bmatrix} \Rightarrow \text{P},\ \text{T},\ \text{E}$

$\begin{bmatrix} 17 & 3 & 38 \end{bmatrix} A^{-1} = \begin{bmatrix} 13 & 2 & 5 \end{bmatrix} \Rightarrow \text{M},\ \text{B},\ \text{E}$

$\begin{bmatrix} 18 & 20 & 76 \end{bmatrix} A^{-1} = \begin{bmatrix} 18 & 0 & 20 \end{bmatrix} \Rightarrow \text{R},\ \_,\ \text{T}$

$\begin{bmatrix} 18 & -5 & 21 \end{bmatrix} A^{-1} = \begin{bmatrix} 8 & 5 & 0 \end{bmatrix} \Rightarrow \text{H},\ \text{E},\ \_$

$\begin{bmatrix} 29 & -7 & 32 \end{bmatrix} A^{-1} = \begin{bmatrix} 5 & 12 & 5 \end{bmatrix} \Rightarrow \text{E},\ \text{L},\ \text{E}$

$\begin{bmatrix} 32 & 9 & 77 \end{bmatrix} A^{-1} = \begin{bmatrix} 22 & 5 & 14 \end{bmatrix} \Rightarrow \text{V},\ \text{E},\ \text{N}$

$\begin{bmatrix} 36 & -8 & 48 \end{bmatrix} A^{-1} = \begin{bmatrix} 20 & 8 & 0 \end{bmatrix} \Rightarrow \text{T},\ \text{H},\ \_$

$\begin{bmatrix} 33 & -5 & 51 \end{bmatrix} A^{-1} = \begin{bmatrix} 23 & 5 & 0 \end{bmatrix} \Rightarrow \text{W},\ \text{E},\ \_$

$\begin{bmatrix} 41 & 3 & 79 \end{bmatrix} A^{-1} = \begin{bmatrix} 23 & 9 & 12 \end{bmatrix} \Rightarrow \text{W},\ \text{I},\ \text{L}$

$\begin{bmatrix} 12 & 1 & 26 \end{bmatrix} A^{-1} = \begin{bmatrix} 12 & 0 & 1 \end{bmatrix} \Rightarrow \text{L},\ \_,\ \text{A}$

$\begin{bmatrix} 58 & -22 & 49 \end{bmatrix} A^{-1} = \begin{bmatrix} 12 & 23 & 1 \end{bmatrix} \Rightarrow \text{L},\ \text{W},\ \text{A}$

$\begin{bmatrix} 63 & -19 & 69 \end{bmatrix} A^{-1} = \begin{bmatrix} 25 & 19 & 0 \end{bmatrix} \Rightarrow \text{Y},\ \text{S},\ \_$

$\begin{bmatrix} 28 & 8 & 67 \end{bmatrix} A^{-1} = \begin{bmatrix} 18 & 5 & 13 \end{bmatrix} \Rightarrow \text{R},\ \text{E},\ \text{M}$

$\begin{bmatrix} 31 & -11 & 27 \end{bmatrix} A^{-1} = \begin{bmatrix} 5 & 13 & 2 \end{bmatrix} \Rightarrow \text{E},\ \text{M},\ \text{B}$

$\begin{bmatrix} 41 & -18 & 28 \end{bmatrix} A^{-1} = \begin{bmatrix} 5 & 18 & 0 \end{bmatrix} \Rightarrow \text{E},\ \text{R},\ \_$

The message is _SEPTEMBER_THE_ELEVENTH_ WE_WILL_ALWAYS_REMEMBER_

27. Use the given information to find D.

$$D = \begin{bmatrix} 0.10 & 0.20 \\ 0.80 & 0.10 \end{bmatrix} \begin{matrix} \text{Coal} \\ \text{Steel} \end{matrix} \Big\} \text{Supplier}$$

with column headers: User — Coal, Steel.

The equation $X = DX + E$ may be rewritten in the form $(I - D)X = E$; that is,

$$\begin{bmatrix} 0.90 & -0.20 \\ -0.80 & 0.90 \end{bmatrix} X = \begin{bmatrix} 10,000 \\ 20,000 \end{bmatrix}.$$

Solve this system by using Gauss-Jordan elimination to obtain $X = \begin{bmatrix} 20,000 \\ 40,000 \end{bmatrix}$.

29. From the given matrix D, form the linear system $X = DX + E$, which can be written as

$(I - D)X = E$

$$\begin{bmatrix} 0.6 & -0.5 & -0.5 \\ -0.3 & 1 & -0.3 \\ -0.2 & -0.2 & 1.0 \end{bmatrix} X = \begin{bmatrix} 1000 \\ 1000 \\ 1000 \end{bmatrix}.$$

Solving this system, $X = \begin{bmatrix} 8622.0 \\ 4685.0 \\ 3661.4 \end{bmatrix}$.

31. (a) The line that best fits the given points is shown on the graph.

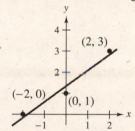

(b) Using the matrices $X = \begin{bmatrix} 1 & -2 \\ 1 & 0 \\ 1 & 2 \end{bmatrix}$ and $Y = \begin{bmatrix} 0 \\ 1 \\ 3 \end{bmatrix}$, you have

$$X^T X = \begin{bmatrix} 1 & 1 & 1 \\ -2 & 0 & 2 \end{bmatrix} \begin{bmatrix} 1 & -2 \\ 1 & 0 \\ 1 & 2 \end{bmatrix} = \begin{bmatrix} 3 & 0 \\ 0 & 8 \end{bmatrix}.$$

$$X^T Y = \begin{bmatrix} 1 & 1 & 1 \\ -2 & 0 & 2 \end{bmatrix} \begin{bmatrix} 0 \\ 1 \\ 3 \end{bmatrix} = \begin{bmatrix} 4 \\ 6 \end{bmatrix}.$$

$$A = (X^T X)^{-1} X^T Y = \begin{bmatrix} \frac{1}{3} & 0 \\ 0 & \frac{1}{8} \end{bmatrix} \begin{bmatrix} 4 \\ 6 \end{bmatrix} = \begin{bmatrix} \frac{4}{3} \\ \frac{3}{4} \end{bmatrix}.$$

So, the least square regression line is $y = \frac{3}{4}x + \frac{4}{3}$.

(c) Solving $Y = XA + E$ for E,

$$E = Y - XA = \begin{bmatrix} 0 \\ 1 \\ 3 \end{bmatrix} - \begin{bmatrix} 1 & -2 \\ 1 & 0 \\ 1 & 2 \end{bmatrix}\begin{bmatrix} \frac{4}{3} \\ \frac{3}{4} \end{bmatrix} = \begin{bmatrix} \frac{1}{6} \\ -\frac{1}{3} \\ \frac{1}{6} \end{bmatrix}.$$

So, the sum of the squared error is

$$E^T E = \begin{bmatrix} \frac{1}{6} & -\frac{1}{3} & \frac{1}{6} \end{bmatrix}\begin{bmatrix} \frac{1}{6} \\ -\frac{1}{3} \\ \frac{1}{6} \end{bmatrix} = \frac{1}{6}.$$

33. (a) The line that best fits the given points is shown on the graph.

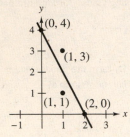

(b) Using the matrices $X = \begin{bmatrix} 1 & 0 \\ 1 & 1 \\ 1 & 1 \\ 1 & 2 \end{bmatrix}$ and $Y = \begin{bmatrix} 4 \\ 3 \\ 1 \\ 0 \end{bmatrix}$, you have

$$X^T X = \begin{bmatrix} 1 & 1 & 1 & 1 \\ 0 & 1 & 1 & 2 \end{bmatrix}\begin{bmatrix} 1 & 0 \\ 1 & 1 \\ 1 & 1 \\ 1 & 2 \end{bmatrix} = \begin{bmatrix} 4 & 4 \\ 4 & 6 \end{bmatrix}.$$

$$X^T Y = \begin{bmatrix} 1 & 1 & 1 & 1 \\ 0 & 1 & 1 & 2 \end{bmatrix}\begin{bmatrix} 4 \\ 3 \\ 1 \\ 0 \end{bmatrix} = \begin{bmatrix} 8 \\ 4 \end{bmatrix}.$$

$$A = \left(X^T X\right)^{-1} X^T Y = \frac{1}{8}\begin{bmatrix} 6 & -4 \\ -4 & 4 \end{bmatrix}\begin{bmatrix} 8 \\ 4 \end{bmatrix} = \begin{bmatrix} 4 \\ -2 \end{bmatrix}.$$

So, the least squares regression line is $y = 4 - 2x$.

(c) Solving $Y = XA + E$ for E,

$$E = Y - XA = \begin{bmatrix} 4 \\ 3 \\ 1 \\ 0 \end{bmatrix} - \begin{bmatrix} 1 & 0 \\ 1 & 1 \\ 1 & 1 \\ 1 & 2 \end{bmatrix}\begin{bmatrix} 4 \\ -2 \end{bmatrix} = \begin{bmatrix} 4 \\ 3 \\ 1 \\ 0 \end{bmatrix} - \begin{bmatrix} 4 \\ 2 \\ 2 \\ 0 \end{bmatrix} = \begin{bmatrix} 0 \\ 1 \\ -1 \\ 0 \end{bmatrix}.$$

So, the sum of the squared error is

$$E^T E = \begin{bmatrix} 0 & 1 & -1 & 0 \end{bmatrix}\begin{bmatrix} 0 \\ 1 \\ -1 \\ 0 \end{bmatrix} = 2.$$

35. Using the matrices

$$X = \begin{bmatrix} 1 & 0 \\ 1 & 1 \\ 1 & 2 \end{bmatrix} \text{ and } Y = \begin{bmatrix} 0 \\ 1 \\ 4 \end{bmatrix} \text{ you have}$$

$$X^T X = \begin{bmatrix} 3 & 3 \\ 3 & 5 \end{bmatrix} \text{ and } X^T Y = \begin{bmatrix} 5 \\ 9 \end{bmatrix}$$

$$A = \left(X^T X\right)^{-1}\left(X^T Y\right) = \begin{bmatrix} -\frac{1}{3} \\ 2 \end{bmatrix}.$$

So, the least squares regression line is $y = 2x - \frac{1}{3}$.

37. Using the matrices

$$X = \begin{bmatrix} 1 & -2 \\ 1 & -1 \\ 1 & 0 \\ 1 & 1 \end{bmatrix} \text{ and } Y = \begin{bmatrix} 0 \\ 1 \\ 1 \\ 2 \end{bmatrix} \text{ you have}$$

$$X^T X = \begin{bmatrix} 4 & -2 \\ -2 & 6 \end{bmatrix} \text{ and } X^T Y = \begin{bmatrix} 4 \\ 1 \end{bmatrix}.$$

$$A = \left(X^T X\right)^{-1} X^T Y = \begin{bmatrix} 0.3 & 0.1 \\ 0.1 & 0.2 \end{bmatrix}\begin{bmatrix} 4 \\ 1 \end{bmatrix} = \begin{bmatrix} 1.3 \\ 0.6 \end{bmatrix}.$$

So, the least squares regression line is $y = 0.6x + 1.3$.

39. Using the four given points, the matrices X and Y are

$$X = \begin{bmatrix} 1 & -5 \\ 1 & 1 \\ 1 & 2 \\ 1 & 2 \end{bmatrix} \text{ and } Y = \begin{bmatrix} 1 \\ 3 \\ 3 \\ 5 \end{bmatrix}. \text{ This means that}$$

$$X^T X = \begin{bmatrix} 1 & 1 & 1 & 1 \\ -5 & 1 & 2 & 2 \end{bmatrix}\begin{bmatrix} 1 & -5 \\ 1 & 1 \\ 1 & 2 \\ 1 & 2 \end{bmatrix} = \begin{bmatrix} 4 & 0 \\ 0 & 34 \end{bmatrix} \text{ and } X^T Y = \begin{bmatrix} 1 & 1 & 1 & 1 \\ -5 & 1 & 2 & 2 \end{bmatrix}\begin{bmatrix} 1 \\ 3 \\ 3 \\ 5 \end{bmatrix} = \begin{bmatrix} 12 \\ 14 \end{bmatrix}.$$

Now, using $\left(X^T X\right)^{-1}$ to find the coefficient matrix A, you have

$$A = \left(X^T X\right)^{-1} X^T Y = \begin{bmatrix} 4 & 0 \\ 0 & 34 \end{bmatrix}^{-1}\begin{bmatrix} 12 \\ 14 \end{bmatrix} = \begin{bmatrix} \frac{1}{4} & 0 \\ 0 & \frac{1}{34} \end{bmatrix}\begin{bmatrix} 12 \\ 14 \end{bmatrix} = \begin{bmatrix} 3 \\ \frac{7}{17} \end{bmatrix}.$$

So, the least squares regression line is $y = \frac{7}{17}x + 3 = 0.412x + 3$.

41. Using the five given points, the matrices X and Y are

$$X = \begin{bmatrix} 1 & -5 \\ 1 & -1 \\ 1 & 3 \\ 1 & 7 \\ 1 & 5 \end{bmatrix} \text{ and } Y = \begin{bmatrix} 10 \\ 8 \\ 6 \\ 4 \\ 5 \end{bmatrix}. \text{ This means that}$$

$$X^T X = \begin{bmatrix} 1 & 1 & 1 & 1 & 1 \\ -5 & -1 & 3 & 7 & 5 \end{bmatrix}\begin{bmatrix} 1 & -5 \\ 1 & -1 \\ 1 & 3 \\ 1 & 7 \\ 1 & 5 \end{bmatrix} = \begin{bmatrix} 5 & 9 \\ 9 & 109 \end{bmatrix} \text{ and } X^T Y = \begin{bmatrix} 1 & 1 & 1 & 1 & 1 \\ -5 & -1 & 3 & 7 & 5 \end{bmatrix}\begin{bmatrix} 10 \\ 8 \\ 6 \\ 4 \\ 5 \end{bmatrix} = \begin{bmatrix} 33 \\ 13 \end{bmatrix}.$$

Now, using $\left(X^T X\right)^{-1}$ to find the coefficient matrix A, you have

$$A = \left(X^T X\right)^{-1} X^T Y = \begin{bmatrix} 5 & 9 \\ 9 & 109 \end{bmatrix}^{-1}\begin{bmatrix} 33 \\ 13 \end{bmatrix} = \begin{bmatrix} \frac{109}{464} & -\frac{9}{464} \\ -\frac{9}{464} & \frac{5}{464} \end{bmatrix}\begin{bmatrix} 33 \\ 13 \end{bmatrix} = \begin{bmatrix} \frac{15}{2} \\ -\frac{1}{2} \end{bmatrix}.$$

So, the least squares regression line is $y = -\frac{1}{2}x + \frac{15}{2} = -0.5x + 7.5$.

43. (a) Using the matrices $X = \begin{bmatrix} 1 & 3.00 \\ 1 & 3.25 \\ 1 & 3.50 \end{bmatrix}$ and $Y = \begin{bmatrix} 4500 \\ 3750 \\ 3300 \end{bmatrix}$, you have

$$X^T X = \begin{bmatrix} 1 & 1 & 1 \\ 3.00 & 3.25 & 3.50 \end{bmatrix}\begin{bmatrix} 1 & 3.00 \\ 1 & 3.25 \\ 1 & 3.50 \end{bmatrix} = \begin{bmatrix} 3 & 9.75 \\ 9.75 & 31.8125 \end{bmatrix} \text{ and } X^T Y = \begin{bmatrix} 1 & 1 & 1 \\ 3.00 & 3.25 & 3.50 \end{bmatrix}\begin{bmatrix} 4500 \\ 3750 \\ 3300 \end{bmatrix} = \begin{bmatrix} 11{,}550 \\ 37{,}237.5 \end{bmatrix}.$$

Now, using $\left(X^T X\right)^{-1}$ to find the coefficient matrix A, you have

$$A = \left(X^T X\right)^{-1} X^T Y = \frac{1}{0.375}\begin{bmatrix} 31.8125 & -9.75 \\ -9.75 & 3 \end{bmatrix}\begin{bmatrix} 11{,}550 \\ 37{,}237.5 \end{bmatrix} = \frac{1}{0.375}\begin{bmatrix} 4368.75 \\ -900 \end{bmatrix} = \begin{bmatrix} 11{,}650 \\ -2400 \end{bmatrix}.$$

So, the least squares regression line is $y = 11{,}650 - 2400x$.

(b) When $x = 3.40$, $y = 11{,}650 - 2400(3.40) = 3490$. So, the demand is 3490 gallons.

45. (a) Using the matrices $X = \begin{bmatrix} 1 & 0 \\ 1 & 1 \\ 1 & 2 \\ 1 & 3 \\ 1 & 4 \end{bmatrix}$ and $Y = \begin{bmatrix} 221.5 \\ 230.4 \\ 229.6 \\ 231.4 \\ 237.2 \end{bmatrix}$ you have

$$X^T X = \begin{bmatrix} 1 & 1 & 1 & 1 & 1 \\ 0 & 1 & 2 & 3 & 4 \end{bmatrix}\begin{bmatrix} 1 & 0 \\ 1 & 1 \\ 1 & 2 \\ 1 & 3 \\ 1 & 4 \end{bmatrix} = \begin{bmatrix} 5 & 10 \\ 10 & 30 \end{bmatrix} \text{ and } X^T Y = \begin{bmatrix} 1 & 1 & 1 & 1 & 1 \\ 0 & 1 & 2 & 3 & 4 \end{bmatrix}\begin{bmatrix} 221.5 \\ 230.4 \\ 229.6 \\ 231.4 \\ 237.2 \end{bmatrix} = \begin{bmatrix} 1150.1 \\ 2332.6 \end{bmatrix}.$$

Now, using $\left(X^T X\right)^{-1}$ to find the coefficient matrix A, you have

$$A = \left(X^T X\right)^{-1} X^T Y = \tfrac{1}{50}\begin{bmatrix} 30 & -10 \\ -10 & 5 \end{bmatrix}\begin{bmatrix} 1150.1 \\ 2332.6 \end{bmatrix} = \tfrac{1}{50}\begin{bmatrix} 11{,}177 \\ 162 \end{bmatrix} = \begin{bmatrix} 223.54 \\ 3.24 \end{bmatrix}.$$

So, the least squares regression line is $y = 3.24t + 223.54$.

(b) Using a graphing utility with $L_1 = \{0, 1, 2, 3, 4\}$ and $L_2 = \{221.5, 230.4, 229.6, 231.4, 237.2\}$ gives the same least squares regression line: $y = 3.24x + 223.54$.

Review Exercises for Chapter 2

1. $\begin{bmatrix} 2 & 1 & 0 \\ 0 & 5 & -4 \end{bmatrix} - 3\begin{bmatrix} 5 & 3 & -6 \\ 0 & -2 & 5 \end{bmatrix} = \begin{bmatrix} 2 & 1 & 0 \\ 0 & 5 & -4 \end{bmatrix} - \begin{bmatrix} 15 & 9 & -18 \\ 0 & -6 & 15 \end{bmatrix} = \begin{bmatrix} -13 & -8 & 18 \\ 0 & 11 & -19 \end{bmatrix}$

3. $\begin{bmatrix} 1 & 2 \\ 5 & -4 \\ 6 & 0 \end{bmatrix}\begin{bmatrix} 6 & -2 & 8 \\ 4 & 0 & 0 \end{bmatrix} = \begin{bmatrix} 1(6) + 2(4) & 1(-2) + 2(0) & 1(8) + 2(0) \\ 5(6) - 4(4) & 5(-2) - 4(0) & 5(8) - 4(0) \\ 6(6) + 0(4) & 6(-2) + 0(0) & 6(8) + 0(0) \end{bmatrix} = \begin{bmatrix} 14 & -2 & 8 \\ 14 & -10 & 40 \\ 36 & -12 & 48 \end{bmatrix}$

5. $\begin{bmatrix} 1 & 3 & 2 \\ 0 & 2 & -4 \\ 0 & 0 & 3 \end{bmatrix}\begin{bmatrix} 4 & -3 & 2 \\ 0 & 3 & -1 \\ 0 & 0 & 2 \end{bmatrix} = \begin{bmatrix} 1(4) & 1(-3) + 3(3) & 1(2) + 3(-1) + 2(2) \\ 0 & 2(3) & 2(-1) + (-4)(2) \\ 0 & 0 & 3(2) \end{bmatrix} = \begin{bmatrix} 4 & 6 & 3 \\ 0 & 6 & -10 \\ 0 & 0 & 6 \end{bmatrix}$

7. Multiplying the left side of the equation yields

$$\begin{bmatrix} 5x + 4y \\ -x + y \end{bmatrix} = \begin{bmatrix} 2 \\ -22 \end{bmatrix}.$$

So, the corresponding system of linear equations is

$$5x + 4y = 2$$
$$-x + y = -22.$$

9. Multiplying the left side of the equation yields

$$\begin{bmatrix} x_2 - 2x_3 \\ -x_1 + 3x_2 + x_3 \\ 2x_1 - 2x_2 + 4x_3 \end{bmatrix} = \begin{bmatrix} -1 \\ 0 \\ 2 \end{bmatrix}.$$

So, the corresponding system of linear equations is

$$x_2 - 2x_3 = -1$$
$$-x_1 + 3x_2 + x_3 = 0$$
$$2x_1 - 2x_2 + 4x_3 = 2.$$

11. Letting

$$A = \begin{bmatrix} 2 & -1 \\ 3 & 2 \end{bmatrix}, \quad \mathbf{x} = \begin{bmatrix} x \\ y \end{bmatrix}, \text{ and } \mathbf{b} = \begin{bmatrix} 5 \\ -4 \end{bmatrix}, \text{ the given}$$

system can be written in matrix form

$$A\mathbf{x} = \mathbf{b}$$

$$\begin{bmatrix} 2 & -1 \\ 3 & 2 \end{bmatrix} \begin{bmatrix} x \\ y \end{bmatrix} = \begin{bmatrix} 5 \\ -4 \end{bmatrix}.$$

13. Letting

$$A = \begin{bmatrix} 2 & 3 & 1 \\ 2 & -3 & -3 \\ 4 & -2 & 3 \end{bmatrix}, \quad \mathbf{x} = \begin{bmatrix} x_1 \\ x_2 \\ x_3 \end{bmatrix} \text{ and } \mathbf{b} = \begin{bmatrix} 10 \\ 22 \\ -2 \end{bmatrix}$$

the given system can be written in matrix form

$$A\mathbf{x} = \mathbf{b}$$

$$\begin{bmatrix} 2 & 3 & 1 \\ 2 & -3 & -3 \\ 4 & -2 & 3 \end{bmatrix} \begin{bmatrix} x_1 \\ x_2 \\ x_3 \end{bmatrix} = \begin{bmatrix} 10 \\ 22 \\ -2 \end{bmatrix}.$$

15. $A^T = \begin{bmatrix} 1 & 0 \\ 2 & 1 \\ -3 & 2 \end{bmatrix},$

$$A^T A = \begin{bmatrix} 1 & 0 \\ 2 & 1 \\ -3 & 2 \end{bmatrix} \begin{bmatrix} 1 & 2 & -3 \\ 0 & 1 & 2 \end{bmatrix} = \begin{bmatrix} 1 & 2 & -3 \\ 2 & 5 & -4 \\ -3 & -4 & 13 \end{bmatrix}$$

$$AA^T = \begin{bmatrix} 1 & 2 & -3 \\ 0 & 1 & 2 \end{bmatrix} \begin{bmatrix} 1 & 0 \\ 2 & 1 \\ -3 & 2 \end{bmatrix} = \begin{bmatrix} 14 & -4 \\ -4 & 5 \end{bmatrix}$$

17. $A^T = \begin{bmatrix} 1 & 3 & -1 \end{bmatrix}$

$$A^T A = \begin{bmatrix} 1 & 3 & -1 \end{bmatrix} \begin{bmatrix} 1 \\ 3 \\ -1 \end{bmatrix} = \begin{bmatrix} 11 \end{bmatrix}$$

$$AA^T = \begin{bmatrix} 1 \\ 3 \\ -1 \end{bmatrix} \begin{bmatrix} 1 & 3 & -1 \end{bmatrix} = \begin{bmatrix} 1 & 3 & -1 \\ 3 & 9 & -3 \\ -1 & -3 & 1 \end{bmatrix}$$

19. Use the formula for the inverse of a 2×2 matrix.

$$A^{-1} = \frac{1}{ad - bc} \begin{bmatrix} d & -b \\ -c & a \end{bmatrix} = \frac{1}{3(-1) - (-1)(2)} \begin{bmatrix} -1 & 1 \\ -2 & 3 \end{bmatrix}$$

$$= \begin{bmatrix} 1 & -1 \\ 2 & -3 \end{bmatrix}$$

21. Begin by adjoining the identity matrix to the given matrix.

$$[A \vdots I] = \begin{bmatrix} 2 & 3 & 1 & \vdots & 1 & 0 & 0 \\ 2 & -3 & -3 & \vdots & 0 & 1 & 0 \\ 4 & 0 & 3 & \vdots & 0 & 0 & 1 \end{bmatrix}$$

This matrix reduces to

$$[I \vdots A^{-1}] = \begin{bmatrix} 1 & 0 & 0 & \vdots & \frac{3}{20} & \frac{3}{20} & \frac{1}{10} \\ 0 & 1 & 0 & \vdots & \frac{3}{10} & -\frac{1}{30} & -\frac{2}{15} \\ 0 & 0 & 1 & \vdots & -\frac{1}{5} & -\frac{1}{5} & \frac{1}{5} \end{bmatrix}.$$

So, the inverse matrix is

$$A^{-1} = \begin{bmatrix} \frac{3}{20} & \frac{3}{20} & \frac{1}{10} \\ \frac{3}{20} & -\frac{1}{30} & -\frac{2}{15} \\ -\frac{1}{5} & -\frac{1}{5} & \frac{1}{5} \end{bmatrix}.$$

23. $A\mathbf{x} = \mathbf{b}$

$$\begin{bmatrix} 5 & 4 \\ -1 & 1 \end{bmatrix} \begin{bmatrix} x_1 \\ x_2 \end{bmatrix} = \begin{bmatrix} 2 \\ -22 \end{bmatrix}$$

Because

$$A^{-1} = \frac{1}{5(1) - 4(-1)} \begin{bmatrix} 1 & -4 \\ 1 & 5 \end{bmatrix} = \begin{bmatrix} \frac{1}{9} & -\frac{4}{9} \\ \frac{1}{9} & \frac{5}{9} \end{bmatrix}$$

solve the equation $A\mathbf{x} = \mathbf{b}$ as follows.

$$\mathbf{x} = A^{-1}\mathbf{b} = \begin{bmatrix} \frac{1}{9} & -\frac{4}{9} \\ \frac{1}{9} & \frac{5}{9} \end{bmatrix} \begin{bmatrix} 2 \\ -22 \end{bmatrix} = \begin{bmatrix} 10 \\ -12 \end{bmatrix}$$

25.
$$A\mathbf{x} = \mathbf{b}$$

$$\begin{bmatrix} -1 & 1 & 2 \\ 2 & 3 & 1 \\ 5 & 4 & 2 \end{bmatrix} \begin{bmatrix} x_1 \\ x_2 \\ x_3 \end{bmatrix} = \begin{bmatrix} 1 \\ -2 \\ 4 \end{bmatrix}$$

Using Gauss-Jordan elimination, you find that

$$A^{-1} = \begin{bmatrix} -\frac{2}{15} & -\frac{2}{5} & \frac{1}{3} \\ -\frac{1}{15} & \frac{4}{5} & -\frac{1}{3} \\ \frac{7}{15} & -\frac{3}{5} & \frac{1}{3} \end{bmatrix}.$$

So, solve the equation $A\mathbf{x} = \mathbf{b}$ as follows.

$$\mathbf{x} = A^{-1}\mathbf{b} = \begin{bmatrix} -\frac{2}{15} & -\frac{2}{5} & \frac{1}{3} \\ -\frac{1}{15} & \frac{4}{5} & -\frac{1}{3} \\ \frac{7}{15} & -\frac{3}{5} & \frac{1}{3} \end{bmatrix} \begin{bmatrix} 1 \\ -2 \\ 4 \end{bmatrix} = \begin{bmatrix} 2 \\ -3 \\ 3 \end{bmatrix}$$

27. Because

$$(3A)^{-1} = \begin{bmatrix} 4 & -1 \\ 2 & 3 \end{bmatrix},$$

you can use the formula for the inverse of a 2×2 matrix to obtain

$$3A = \begin{bmatrix} 4 & -1 \\ 2 & 3 \end{bmatrix}^{-1} = \frac{1}{4(3) - (-1)(2)} \begin{bmatrix} 3 & 1 \\ -2 & 4 \end{bmatrix}$$

$$= \frac{1}{14} \begin{bmatrix} 3 & 1 \\ -2 & 4 \end{bmatrix}.$$

So, $A = \dfrac{1}{42} \begin{bmatrix} 3 & 1 \\ -2 & 4 \end{bmatrix} = \begin{bmatrix} \frac{1}{14} & \frac{1}{42} \\ -\frac{1}{21} & \frac{2}{21} \end{bmatrix}.$

29. A is nonsingular if and only if the second row is not a multiple of the first. That is, A is nonsingular if and only if $x \neq -3$.

Alternatively, you could use the formula for the inverse of a 2×2 matrix to show that A is nonsingular if $ad - bc = 3(-1) - 1(x) \neq 0$. That is, $x \neq -3$.

31. Because the given matrix represents the addition of 4 times the third row to the first row of I_3, reverse the operation and subtract 4 times the third row from the first row.

$$E^{-1} = \begin{bmatrix} 1 & 0 & -4 \\ 0 & 1 & 0 \\ 0 & 0 & 1 \end{bmatrix}$$

For Exercises 33–39, answers will vary. Sample answers are shown below.

33. Begin by finding a sequence of elementary row operations that can be used to write A in reduced row-echelon form.

Matrix	Elementary Row Operation	Elementary Matrix
$\begin{bmatrix} 1 & \frac{3}{2} \\ 0 & 1 \end{bmatrix}$	Divide row one by 2.	$E_1 = \begin{bmatrix} \frac{1}{2} & 0 \\ 0 & 1 \end{bmatrix}$
$\begin{bmatrix} 1 & 0 \\ 0 & 1 \end{bmatrix}$	Subtract $\frac{3}{2}$ times row two from row one.	$E_2 = \begin{bmatrix} 1 & -\frac{3}{2} \\ 0 & 1 \end{bmatrix}$

So, factor A as follows.

$$A = E_1^{-1}E_2^{-1} = \begin{bmatrix} 2 & 0 \\ 0 & 1 \end{bmatrix} \begin{bmatrix} 1 & \frac{3}{2} \\ 0 & 1 \end{bmatrix}$$

35. Begin by finding a sequence of elementary row operations to write A in reduced row-echelon form.

Matrix	Elementary Row Operation	Elementary Matrix
$\begin{bmatrix} 1 & 0 & 1 \\ 0 & 1 & -2 \\ 0 & 0 & 1 \end{bmatrix}$	$\frac{1}{4}$ times row 3.	$E_1 = \begin{bmatrix} 1 & 0 & 0 \\ 0 & 1 & 0 \\ 0 & 0 & \frac{1}{4} \end{bmatrix}$
$\begin{bmatrix} 1 & 0 & 1 \\ 0 & 1 & 0 \\ 0 & 0 & 1 \end{bmatrix}$	Add two times row three to row two.	$E_2 = \begin{bmatrix} 1 & 0 & 0 \\ 0 & 1 & 2 \\ 0 & 0 & 1 \end{bmatrix}$
$\begin{bmatrix} 1 & 0 & 0 \\ 0 & 1 & 0 \\ 0 & 0 & 1 \end{bmatrix}$	Add -1 times row three to row one.	$E_3 = \begin{bmatrix} 1 & 0 & -1 \\ 0 & 1 & 0 \\ 0 & 0 & 1 \end{bmatrix}$

So, factor A as follows.

$$A = E_1^{-1}E_2^{-1}E_3^{-1} = \begin{bmatrix} 1 & 0 & 0 \\ 0 & 1 & 0 \\ 0 & 0 & 4 \end{bmatrix}\begin{bmatrix} 1 & 0 & 0 \\ 0 & 1 & -2 \\ 0 & 0 & 1 \end{bmatrix}\begin{bmatrix} 1 & 0 & 1 \\ 0 & 1 & 0 \\ 0 & 0 & 1 \end{bmatrix}.$$

37. Let $A = \begin{bmatrix} a & b \\ c & d \end{bmatrix}$ then $A^2 = \begin{bmatrix} a & b \\ c & d \end{bmatrix}\begin{bmatrix} a & b \\ c & d \end{bmatrix} = \begin{bmatrix} a^2 + bc & ad + bd \\ ca + dc & cb + d^2 \end{bmatrix} = \begin{bmatrix} 1 & 0 \\ 0 & 1 \end{bmatrix}.$

So, many answers are possible. $\begin{bmatrix} 1 & 0 \\ 0 & 1 \end{bmatrix}, \begin{bmatrix} -1 & 0 \\ 0 & -1 \end{bmatrix}, \begin{bmatrix} -1 & 0 \\ 0 & 1 \end{bmatrix}$, etc.

39. Let $A = \begin{bmatrix} a & b \\ c & d \end{bmatrix}$ then $A^2 = \begin{bmatrix} a & b \\ c & d \end{bmatrix}\begin{bmatrix} a & b \\ c & d \end{bmatrix} = \begin{bmatrix} a^2 + bc & b(a + d) \\ c(a + d) & bc + d^2 \end{bmatrix}.$

Solving $A^2 = A$ gives the system of nonlinear equations

$a^2 + bc = a$

$d^2 + bc = d$

$b(a + d) = b$

$c(a + d) = c.$

From this system, conclude that any of the following matrices are solutions to the equation $A^2 = A$.

$\begin{bmatrix} 0 & 0 \\ 0 & 0 \end{bmatrix}, \begin{bmatrix} 0 & 0 \\ t & 1 \end{bmatrix}, \begin{bmatrix} 0 & t \\ 0 & 1 \end{bmatrix}, \begin{bmatrix} 1 & 0 \\ t & 0 \end{bmatrix}, \begin{bmatrix} 1 & t \\ 0 & 0 \end{bmatrix}, \begin{bmatrix} 1 & 0 \\ 0 & 1 \end{bmatrix}$

41. (a) Letting $W = aX + bY + cZ$ yields the system of linear equations

$\begin{aligned} a - b + 3c &= 3 \\ 2a \quad\quad + 4c &= 2 \\ 3b - c &= -4 \\ a + 2b + 2c &= -1 \end{aligned}$

which has the solution: $a = -1, b = -1, c = 1.$

(b) Letting $Z = aX + bY$ yields the system of linear equations

$\begin{aligned} a - b &= 3 \\ 2a \quad\quad &= 4 \\ 3b &= -1 \\ a + 2b &= 2 \end{aligned}$

which has no solution.

43. Because $\left(A^{-1} + B^{-1}\right)\left(A^{-1} + B^{-1}\right)^{-1} = I$, if $\left(A^{-1} + B^{-1}\right)^{-1}$ exists, it is sufficient to show that $\left(A^{-1} + B^{-1}\right)\left(A(A + B)^{-1}B\right) = I$ for equality of the second factors in each equation.

$$\left(A^{-1} + B^{-1}\right)\left(A(A + B)^{-1}B\right) = A^{-1}\left(A(A + B)^{-1}B\right) + B^{-1}\left(A(A + B)^{-1}B\right)$$

$$= A^{-1}A(A + B)^{-1}B + B^{-1}A(A + B)^{-1}B$$

$$= I(A + B)^{-1}B + B^{-1}A(A + B)^{-1}B$$

$$= \left(I + B^{-1}A\right)\left((A + B)^{-1}B\right)$$

$$= \left(B^{-1}B + B^{-1}A\right)\left((A + B)^{-1}B\right)$$

$$= B^{-1}(B + A)(A + B)^{-1}B$$

$$= B^{-1}(A + B)(A + B)^{-1}B$$

$$= B^{-1}IB$$

$$= B^{-1}B$$

$$= I.$$

Therefore, $\left(A^{-1} + B^{-1}\right)^{-1} = A(A + B)^{-1}B$.

45. Answers will vary. *Sample answer*:

Matrix	Elementary Matrix

$$\begin{bmatrix} 2 & 5 \\ 6 & 14 \end{bmatrix} = A$$

$$\begin{bmatrix} 2 & 5 \\ 0 & -1 \end{bmatrix} = U \qquad E = \begin{bmatrix} 1 & 0 \\ -3 & 1 \end{bmatrix}$$

$$EA = U$$

$$A = E^{-1}U = \begin{bmatrix} 1 & 0 \\ 3 & 1 \end{bmatrix}\begin{bmatrix} 2 & 5 \\ 0 & -1 \end{bmatrix} = LU$$

47.

Matrix	Elementary Matrix

$$\begin{bmatrix} 1 & 0 & 1 \\ 2 & 1 & 2 \\ 3 & 2 & 6 \end{bmatrix} = A$$

$$\begin{bmatrix} 1 & 0 & 1 \\ 0 & 1 & 0 \\ 3 & 2 & 6 \end{bmatrix} \qquad E_1 = \begin{bmatrix} 1 & 0 & 0 \\ -2 & 1 & 0 \\ 0 & 0 & 1 \end{bmatrix}$$

$$\begin{bmatrix} 1 & 0 & 1 \\ 0 & 1 & 0 \\ 0 & 2 & 3 \end{bmatrix} \qquad E_2 = \begin{bmatrix} 1 & 0 & 0 \\ 0 & 1 & 0 \\ -3 & 0 & 1 \end{bmatrix}$$

$$\begin{bmatrix} 1 & 0 & 1 \\ 0 & 1 & 0 \\ 0 & 0 & 3 \end{bmatrix} = U \qquad E_3 = \begin{bmatrix} 1 & 0 & 0 \\ 0 & 1 & 0 \\ 0 & -2 & 1 \end{bmatrix}$$

$$E_3E_2E_1A = U$$

$$A = E_1^{-1}E_2^{-1}E_3^{-1}U = LU = \begin{bmatrix} 1 & 0 & 0 \\ 2 & 1 & 0 \\ 3 & 2 & 1 \end{bmatrix}\begin{bmatrix} 1 & 0 & 1 \\ 0 & 1 & 0 \\ 0 & 0 & 3 \end{bmatrix}.$$

$$L\mathbf{y} = \mathbf{b}: \begin{bmatrix} 1 & 0 & 0 \\ 2 & 1 & 0 \\ 3 & 2 & 1 \end{bmatrix}\begin{bmatrix} y_1 \\ y_2 \\ y_3 \end{bmatrix} = \begin{bmatrix} 3 \\ 7 \\ 8 \end{bmatrix} \Rightarrow \mathbf{y} = \begin{bmatrix} 3 \\ 1 \\ -3 \end{bmatrix}$$

$$U\mathbf{x} = \mathbf{y}: \begin{bmatrix} 1 & 0 & 1 \\ 0 & 1 & 0 \\ 0 & 0 & 3 \end{bmatrix}\begin{bmatrix} x_1 \\ x_2 \\ x_3 \end{bmatrix} = \begin{bmatrix} 3 \\ 1 \\ -3 \end{bmatrix} \Rightarrow \mathbf{x} = \begin{bmatrix} 3 \\ 1 \\ -1 \end{bmatrix}$$

So, $x = 4$, $y = 1$, and $z = -1$.

49. (a) False. See Theorem 2.1, part 1, page 61

    (b) True. See Theorem 2.6, part 2, page 68.

51. (a) False. The matrix $\begin{bmatrix} 1 & 0 \\ 0 & 0 \end{bmatrix}$ is not invertible.

(b) False. See Exercise 55, page 72.

53. (a) $AB = \begin{bmatrix} 580 & 840 & 320 \\ 560 & 420 & 160 \\ 860 & 1020 & 540 \end{bmatrix} \begin{bmatrix} 3.05 & 0.05 \\ 3.15 & 0.08 \\ 3.25 & 0.10 \end{bmatrix} = \begin{bmatrix} 5455 & 128.2 \\ 3551 & 77.6 \\ 7591 & 178.6 \end{bmatrix}$

This matrix shows the total sales of gas each day in the first column and the total profit each day in the second column.

(b) The gasoline sales profit for Friday through Sunday is the sum of the elements in the second column of AB,
$128.2 + 77.6 + 178.6 = \$384.40$.

55. (a)
$\begin{array}{ccc} \text{Bicycling} & \text{Jogging} & \text{Walking} \end{array}$
$\begin{array}{l} \text{20-min} \\ \text{periods} \end{array} \begin{bmatrix} 2 & \frac{1}{2} & 3 \end{bmatrix} = B$

(b) $BA = \begin{bmatrix} 2 & \frac{1}{2} & 3 \end{bmatrix} \begin{bmatrix} 109 & 136 \\ 127 & 159 \\ 64 & 79 \end{bmatrix} = \begin{bmatrix} 473.5 & 588.5 \end{bmatrix}$

(c) The matrix BA represents the number of calories each person burned during the exercises.

57. The given matrix is not stochastic because the entries in columns two and three do not add up to one.

59. $PX = \begin{bmatrix} \frac{1}{2} & \frac{1}{4} \\ \frac{1}{2} & \frac{3}{4} \end{bmatrix} \begin{bmatrix} 128 \\ 64 \end{bmatrix} = \begin{bmatrix} 80 \\ 112 \end{bmatrix}$

$P^2 X = P \begin{bmatrix} 80 \\ 112 \end{bmatrix} = \begin{bmatrix} 68 \\ 124 \end{bmatrix}$

$P^3 X = P \begin{bmatrix} 68 \\ 124 \end{bmatrix} = \begin{bmatrix} 65 \\ 127 \end{bmatrix}$

61. Begin by forming the matrix of transition probabilities.

$$\begin{array}{c} \text{From Region} \\ \begin{array}{ccc} 1 & 2 & 3 \end{array} \end{array}$$

$P = \begin{bmatrix} 0.85 & 0.15 & 0.10 \\ 0.10 & 0.80 & 0.10 \\ 0.05 & 0.05 & 0.80 \end{bmatrix} \begin{array}{l} 1 \\ 2 \\ 3 \end{array} \Biggr\} \text{ To Region}$

(a) The population in each region after one year is given by

$PX = \begin{bmatrix} 0.85 & 0.15 & 0.10 \\ 0.10 & 0.80 & 0.10 \\ 0.05 & 0.05 & 0.80 \end{bmatrix} \begin{bmatrix} 100{,}000 \\ 100{,}000 \\ 100{,}000 \end{bmatrix}$

$= \begin{bmatrix} 110{,}000 \\ 100{,}000 \\ 90{,}000 \end{bmatrix} \begin{array}{l} \text{Region 1} \\ \text{Region 2} \\ \text{Region 3} \end{array}$

(b) The population in each region after three years is given by

$P^3 X = \begin{bmatrix} 0.665375 & 0.322375 & 0.2435 \\ 0.219 & 0.562 & 0.219 \\ 0.115625 & 0.115625 & 0.5375 \end{bmatrix} \begin{bmatrix} 100{,}000 \\ 100{,}000 \\ 100{,}000 \end{bmatrix} = \begin{bmatrix} 123{,}125 \\ 100{,}000 \\ 76{,}875 \end{bmatrix} \begin{array}{l} \text{Region 1} \\ \text{Region 2} \\ \text{Region 3} \end{array}$

63. The uncoded row matrices are

$\begin{array}{ccccccc} \text{O} & \text{N} & \text{E} \_ & \text{I} & \text{F} \_ & \text{B} & \text{Y} \_ & \text{L} & \text{A} & \text{N} & \text{D} \end{array}$
$\begin{bmatrix} 15 & 14 \end{bmatrix} \quad \begin{bmatrix} 5 & 0 \end{bmatrix} \quad \begin{bmatrix} 9 & 6 \end{bmatrix} \quad \begin{bmatrix} 0 & 2 \end{bmatrix} \quad \begin{bmatrix} 25 & 0 \end{bmatrix} \quad \begin{bmatrix} 12 & 1 \end{bmatrix} \quad \begin{bmatrix} 14 & 4 \end{bmatrix}$.

Multiplying each 1×2 matrix on the right by A yields the coded row matrices

$\begin{bmatrix} 103 & 44 \end{bmatrix}, \begin{bmatrix} 25 & 10 \end{bmatrix}, \begin{bmatrix} 57 & 24 \end{bmatrix}, \begin{bmatrix} 4 & 2 \end{bmatrix}, \begin{bmatrix} 125 & 50 \end{bmatrix}, \begin{bmatrix} 62 & 25 \end{bmatrix}, \begin{bmatrix} 78 & 32 \end{bmatrix}$.

So, the coded message is 103, 44, 25, 10, 57, 24, 4, 2, 125, 50, 62, 25, 78, 32.

65. You can find A^{-1} to be $\begin{bmatrix} 3 & 2 \\ 4 & 3 \end{bmatrix}$ and the coded row matrices are

$\begin{bmatrix} -45 & 34 \end{bmatrix}$, $\begin{bmatrix} 36 & -24 \end{bmatrix}$, $\begin{bmatrix} -43 & 37 \end{bmatrix}$, $\begin{bmatrix} -23 & 22 \end{bmatrix}$, $\begin{bmatrix} -37 & 29 \end{bmatrix}$, $\begin{bmatrix} 57 & -38 \end{bmatrix}$, $\begin{bmatrix} -39 & 31 \end{bmatrix}$.

Multiplying each coded row matrix on the right by A^{-1} yields the uncoded row matrices

$\begin{array}{ccccccc} \text{A} \quad \text{L} & \text{L} \quad \_ & \text{S} \quad \text{Y} & \text{S} \quad \text{T} & \text{E} \quad \text{M} & \text{S} \quad \_ & \text{G} \quad \text{O} \end{array}$

$\begin{bmatrix} 1 & 12 \end{bmatrix}$ $\begin{bmatrix} 12 & 0 \end{bmatrix}$ $\begin{bmatrix} 19 & 25 \end{bmatrix}$ $\begin{bmatrix} 19 & 20 \end{bmatrix}$ $\begin{bmatrix} 5 & 13 \end{bmatrix}$ $\begin{bmatrix} 19 & 0 \end{bmatrix}$ $\begin{bmatrix} 7 & 15 \end{bmatrix}$.

The decoded message is ALL_SYSTEMS_GO.

67. You can find A^{-1} to be $\begin{bmatrix} -2 & -1 & 0 \\ 0 & 1 & 1 \\ -5 & -3 & -3 \end{bmatrix}$ and the coded row matrices are

$\begin{bmatrix} 58 & -3 & -25 \end{bmatrix}$, $\begin{bmatrix} -48 & 28 & 19 \end{bmatrix}$, $\begin{bmatrix} -40 & 13 & 13 \end{bmatrix}$, $\begin{bmatrix} -98 & 39 & 39 \end{bmatrix}$, $\begin{bmatrix} 118 & -25 & -48 \end{bmatrix}$, $\begin{bmatrix} 28 & -14 & -14 \end{bmatrix}$.

Multiplying each coded row matrix on the right by A^{-1} yields the uncoded row matrices

$\begin{array}{cccccc} \text{I} \quad \text{N} \quad \text{V} & \text{A} \quad \text{S} \quad \text{I} & \text{O} \quad \text{N} \quad \_ & \text{A} \quad \text{T} \quad \_ & \text{D} \quad \text{A} \quad \text{W} & \text{N} \quad \_ \quad \_ \end{array}$

$\begin{bmatrix} 9 & 14 & 22 \end{bmatrix}$ $\begin{bmatrix} 1 & 19 & 9 \end{bmatrix}$ $\begin{bmatrix} 15 & 14 & 0 \end{bmatrix}$ $\begin{bmatrix} 1 & 20 & 0 \end{bmatrix}$ $\begin{bmatrix} 4 & 1 & 23 \end{bmatrix}$ $\begin{bmatrix} 14 & 0 & 0 \end{bmatrix}$.

So, the message is INVASION_AT_DAWN.

69. Find $A^{-1} = \begin{bmatrix} -1 & -10 & -8 \\ -1 & -6 & -5 \\ 0 & -1 & -1 \end{bmatrix}$,

and multiply each coded row matrix on the right by A^{-1} to find the associated uncoded row matrix.

$\begin{bmatrix} -2 & 2 & 5 \end{bmatrix} A^{-1} = \begin{bmatrix} -2 & 2 & 5 \end{bmatrix} \begin{bmatrix} -1 & -10 & -8 \\ -1 & -6 & -5 \\ 0 & -1 & -1 \end{bmatrix} = \begin{bmatrix} 0 & 3 & 1 \end{bmatrix} \Rightarrow \_, \; \text{C}, \; \text{A}$

$\begin{bmatrix} 39 & -53 & -72 \end{bmatrix} A^{-1} = \begin{bmatrix} 14 & 0 & 25 \end{bmatrix} \Rightarrow \text{N}, \; \_, \; \text{Y}$

$\begin{bmatrix} -6 & -9 & 93 \end{bmatrix} A^{-1} = \begin{bmatrix} 15 & 21 & 0 \end{bmatrix} \Rightarrow \text{O}, \; \text{U}, \; \_$

$\begin{bmatrix} 4 & -12 & 27 \end{bmatrix} A^{-1} = \begin{bmatrix} 8 & 5 & 1 \end{bmatrix} \Rightarrow \text{H}, \; \text{E}, \; \text{A}$

$\begin{bmatrix} 31 & -49 & -16 \end{bmatrix} A^{-1} = \begin{bmatrix} 18 & 0 & 13 \end{bmatrix} \Rightarrow \text{R}, \; \_, \; \text{M}$

$\begin{bmatrix} 19 & -24 & -46 \end{bmatrix} A^{-1} = \begin{bmatrix} 5 & 0 & 14 \end{bmatrix} \Rightarrow \text{E}, \; \_, \; \text{N}$

$\begin{bmatrix} -8 & -7 & 99 \end{bmatrix} A^{-1} = \begin{bmatrix} 15 & 23 & 0 \end{bmatrix} \Rightarrow \text{O}, \; \text{W}, \; \_$

The message is _CAN_YOU_HEAR_ME_NOW_.

71. First find the input-output matrix D.

$$\begin{array}{c} \overbrace{\text{User Industry}}^{} \\ \overbrace{A \quad B}^{} \end{array}$$

$D = \begin{bmatrix} 0.20 & 0.50 \\ 0.30 & 0.10 \end{bmatrix} \begin{matrix} A \\ B \end{matrix} \Big\} \text{Supplier Industry}$

Then solve the equation $X = DX + E$ for X to obtain $(I - D)X = E$, which corresponds to solving the augmented matrix

$\begin{bmatrix} 0.80 & -0.50 & : & 40{,}000 \\ -0.30 & 0.90 & : & 80{,}000 \end{bmatrix}$. The solution to this system gives you $X \approx \begin{bmatrix} 133{,}333 \\ 133{,}333 \end{bmatrix}$.

73. Using the matrices

$$X = \begin{bmatrix} 1 & 1 \\ 1 & 2 \\ 1 & 3 \end{bmatrix} \text{ and } Y = \begin{bmatrix} 5 \\ 4 \\ 2 \end{bmatrix} \text{ you have}$$

$$X^T X = \begin{bmatrix} 3 & 6 \\ 6 & 14 \end{bmatrix} \text{ and } X^T Y = \begin{bmatrix} 11 \\ 19 \end{bmatrix}$$

$$A = \left(X^T X \right)^{-1} X^T Y = \begin{bmatrix} \frac{7}{3} & -1 \\ -1 & \frac{1}{2} \end{bmatrix} \begin{bmatrix} 11 \\ 19 \end{bmatrix} = \begin{bmatrix} \frac{20}{3} \\ -\frac{3}{2} \end{bmatrix}.$$

So, the least squares regression line is $y = -\frac{3}{2}x + \frac{20}{3}$.

75. Using the matrices

$$X = \begin{bmatrix} 1 & -2 \\ 1 & -1 \\ 1 & 0 \\ 1 & 1 \\ 1 & 2 \end{bmatrix} \text{ and } Y = \begin{bmatrix} 4 \\ 2 \\ 1 \\ -2 \\ -3 \end{bmatrix} \text{ you have}$$

$$X^T X = \begin{bmatrix} 5 & 0 \\ 0 & 10 \end{bmatrix} \text{ and } X^T Y = \begin{bmatrix} 2 \\ -18 \end{bmatrix}$$

$$A = \left(X^T X \right)^{-1} X^T Y = \begin{bmatrix} \frac{1}{5} & 0 \\ 0 & \frac{1}{10} \end{bmatrix} \begin{bmatrix} 2 \\ -18 \end{bmatrix} \begin{bmatrix} 0.4 \\ -1.8 \end{bmatrix}.$$

So, the least squares regression line is

$y = -1.8x + 0.4$, or $y = -\frac{9}{5}x + \frac{2}{5}$.

77. (a) Begin by finding the matrices X and Y.

$$X = \begin{bmatrix} 1 & 1.0 \\ 1 & 1.5 \\ 1 & 2.0 \\ 1 & 2.5 \end{bmatrix} \text{ and } Y = \begin{bmatrix} 32 \\ 41 \\ 48 \\ 53 \end{bmatrix} \text{Then}$$

$$X^T X = \begin{bmatrix} 1 & 1 & 1 & 1 \\ 1.0 & 1.5 & 2.0 & 2.5 \end{bmatrix} \begin{bmatrix} 1 & 1.0 \\ 1 & 1.5 \\ 1 & 2.0 \\ 1 & 2.5 \end{bmatrix} = \begin{bmatrix} 4 & 7 \\ 7 & 13.5 \end{bmatrix}$$

and

$$X^T Y = \begin{bmatrix} 1 & 1 & 1 & 1 \\ 1.0 & 1.5 & 2.0 & 2.5 \end{bmatrix} \begin{bmatrix} 32 \\ 41 \\ 48 \\ 53 \end{bmatrix} = \begin{bmatrix} 174 \\ 322 \end{bmatrix}.$$

The matrix of coefficients is

$$A = \left(X^T X \right)^{-1} X^T Y = \frac{1}{5} \begin{bmatrix} 13.5 & -7 \\ -7 & 4 \end{bmatrix} \begin{bmatrix} 174 \\ 322 \end{bmatrix} = \begin{bmatrix} 19 \\ 14 \end{bmatrix}.$$

So, the least squares regression line is $y = 19 + 14x$.

(b) When

$x = 1.6(160$ kilograms per square kilometer$)$,

$y = 41.4$ (kilograms per square kilometer).

79. (a) Using the matrices $X = \begin{bmatrix} 1 & 0 \\ 1 & 1 \\ 1 & 2 \\ 1 & 3 \\ 1 & 4 \\ 1 & 5 \end{bmatrix}$ and $Y = \begin{bmatrix} 30.37 \\ 32.87 \\ 34.71 \\ 36.59 \\ 38.14 \\ 39.63 \end{bmatrix}$ you have

$$X^T X = \begin{bmatrix} 1 & 1 & 1 & 1 & 1 & 1 \\ 0 & 1 & 2 & 3 & 4 & 5 \end{bmatrix} \begin{bmatrix} 1 & 0 \\ 1 & 1 \\ 1 & 2 \\ 1 & 3 \\ 1 & 4 \\ 1 & 5 \end{bmatrix} = \begin{bmatrix} 6 & 15 \\ 15 & 55 \end{bmatrix} \text{ and } X^T Y = \begin{bmatrix} 1 & 1 & 1 & 1 & 1 & 1 \\ 0 & 1 & 2 & 3 & 4 & 5 \end{bmatrix} \begin{bmatrix} 30.37 \\ 32.87 \\ 34.71 \\ 36.59 \\ 38.14 \\ 39.63 \end{bmatrix} = \begin{bmatrix} 212.31 \\ 562.77 \end{bmatrix}.$$

Now, using $\left(X^T X \right)^{-1}$ to find the coefficient matrix A, you have

$$A = \left(X^T X \right)^{-1} X^T Y = \frac{1}{105} \begin{bmatrix} 55 & -15 \\ -15 & 6 \end{bmatrix} \begin{bmatrix} 212.31 \\ 562.77 \end{bmatrix} = \frac{1}{105} \begin{bmatrix} 3235.5 \\ 191.97 \end{bmatrix} \approx \begin{bmatrix} 30.81 \\ 1.828 \end{bmatrix}$$

So, the least squares regression line is $y = 1.828x + 30.81$.

(b) Using a graphing utility with $L_1 = \{0, 1, 2, 3, 4, 5\}$ and $L_2 = \{30.37, 32.87, 34.71, 36.59, 38.14, 39.63\}$ gives the same least squares regression line: $y = 1.828x + 30.81$.

(c)

Year	2000	2001	2002	2003	2004	2005
Actual	30.37	32.87	34.71	36.59	38.14	39.63
Estimated	30.81	32.64	34.47	36.29	38.12	39.95

The estimated values are close to the actual values.

(d) The average monthly rate in 2010 is $y = 1.828(10) + 30.81 = \$49.09$.

(e) $51 = 1.828x + 30.81$

$20.19 = 1.828x$

$11 \approx x$

The average monthly rate will be \$50.00 in 2011.

81. (a) Using the matrices $X = \begin{bmatrix} 1 & 0 \\ 1 & 1 \\ 1 & 2 \\ 1 & 3 \\ 1 & 4 \\ 1 & 5 \end{bmatrix}$ and $Y = \begin{bmatrix} 1.8 \\ 2.1 \\ 2.3 \\ 2.4 \\ 2.3 \\ 2.5 \end{bmatrix}$ you have

$$X^T X = \begin{bmatrix} 1 & 1 & 1 & 1 & 1 & 1 \\ 0 & 1 & 2 & 3 & 4 & 5 \end{bmatrix} \begin{bmatrix} 1 & 0 \\ 1 & 1 \\ 1 & 2 \\ 1 & 3 \\ 1 & 4 \\ 1 & 5 \end{bmatrix} = \begin{bmatrix} 6 & 15 \\ 15 & 55 \end{bmatrix} \text{ and } X^T Y = \begin{bmatrix} 1 & 1 & 1 & 1 & 1 & 1 \\ 0 & 1 & 2 & 3 & 4 & 5 \end{bmatrix} \begin{bmatrix} 1.8 \\ 2.1 \\ 2.3 \\ 2.4 \\ 2.3 \\ 2.5 \end{bmatrix} = \begin{bmatrix} 13.4 \\ 35.6 \end{bmatrix}.$$

Now, using $\left(X^T X\right)^{-1}$ to find the coefficient matrix A, you have

$$A = \left(X^T X\right)^{-1} X^T Y = \tfrac{1}{105}\begin{bmatrix} 55 & -15 \\ -15 & 6 \end{bmatrix}\begin{bmatrix} 13.4 \\ 35.6 \end{bmatrix} = \tfrac{1}{105}\begin{bmatrix} 203 \\ 12.6 \end{bmatrix} = \begin{bmatrix} 1.93 \\ 0.12 \end{bmatrix}$$

So, the least squares regression line is $y = 0.12x + 1.9$.

(b) Using a graphing utility with $L_1 = \{0, 1, 2, 3, 4, 5\}$ and $L_2 = \{1.8, 2.1, 2.3, 2.4, 2.3, 2.5\}$

gives the same least squares regression line: $y = 0.12x + 1.9$.

(c)

Year	2000	2001	2002	2003	2004	2005
Actual	1.8	2.1	2.3	2.4	2.3	2.5
Estimated	1.9	2.0	2.1	2.3	2.4	2.5

The estimated values are close to the actual values.

(d) The average salary in 2010 is $y = 0.12(10) + 1.9 = \$3.1$ million.

(e) $3.7 = 0.12x + 1.9$

$1.8 = 0.12x$

$15 = x$

The average salary will be \$3.7 million in 2015.

CHAPTER 3
Determinants

CHAPTER 3
Determinants

Section 3.1   The Determinant of a Matrix

1. The determinant of a matrix of order 1 is the entry in the matrix. So, $\det[1] = 1$.

3. $\begin{vmatrix} 2 & 1 \\ 3 & 4 \end{vmatrix} = 2(4) - 3(1) = 5$

5. $\begin{vmatrix} 5 & 2 \\ -6 & 3 \end{vmatrix} = 5(3) - (-6)(2) = 27$

7. $\begin{vmatrix} -7 & 6 \\ \frac{1}{2} & 3 \end{vmatrix} = -7(3) - \left(\frac{1}{2}\right)(6) = -24$

9. $\begin{vmatrix} 2 & 6 \\ 0 & 3 \end{vmatrix} = 2(3) - 0(6) = 6$

11. $\begin{vmatrix} \lambda - 3 & 2 \\ 4 & \lambda - 1 \end{vmatrix} = (\lambda - 3)(\lambda - 1) - 4(2)$

$= \lambda^2 - 4\lambda - 5$

13. (a) The minors of the matrix are shown below.

$M_{11} = |4| = 4 \qquad M_{12} = |3| = 3$

$M_{21} = |2| = 2 \qquad M_{22} = |1| = 1$

(b) The cofactors of the matrix are shown below.

$C_{11} = (-1)^2 M_{11} = 4 \qquad C_{12} = (-1)^3 M_{12} = -3$

$C_{21} = (-1)^3 M_{21} = -2 \qquad C_{22} = (-1)^4 M_{22} = 1$

15. (a) The minors of the matrix are shown below.

$M_{11} = \begin{vmatrix} 5 & 6 \\ -3 & 1 \end{vmatrix} = 23 \quad M_{12} = \begin{vmatrix} 4 & 6 \\ 2 & 1 \end{vmatrix} = -8 \quad M_{13} = \begin{vmatrix} 4 & 5 \\ 2 & -3 \end{vmatrix} = -22$

$M_{21} = \begin{vmatrix} 2 & 1 \\ -3 & 1 \end{vmatrix} = 5 \quad M_{22} = \begin{vmatrix} -3 & 1 \\ 2 & 1 \end{vmatrix} = -5 \quad M_{23} = \begin{vmatrix} -3 & 2 \\ 2 & -3 \end{vmatrix} = 5$

$M_{31} = \begin{vmatrix} 2 & 1 \\ 5 & 6 \end{vmatrix} = 7 \quad M_{32} = \begin{vmatrix} -3 & 1 \\ 4 & 6 \end{vmatrix} = -22 \quad M_{33} = \begin{vmatrix} -3 & 2 \\ 4 & 5 \end{vmatrix} = -23$

(b) The cofactors of the matrix are shown below.

$C_{11} = (-1)^2 M_{11} = 23 \qquad C_{12} = (-1)^3 M_{12} = 8 \qquad C_{13} = (-1)^4 M_{13} = -22$

$C_{21} = (-1)^3 M_{21} = -5 \qquad C_{22} = (-1)^4 M_{22} = -5 \qquad C_{23} = (-1)^5 M_{23} = -5$

$C_{31} = (-1)^4 M_{31} = 7 \qquad C_{32} = (-1)^5 M_{32} = 22 \qquad C_{33} = (-1)^6 M_{33} = -23$

17. (a) You found the cofactors of the matrix in Exercise 15. Now find the determinant by expanding along the second row.

$\begin{vmatrix} -3 & 2 & 1 \\ 4 & 5 & 6 \\ 2 & -3 & 1 \end{vmatrix} = 4C_{21} + 5C_{22} + 6C_{23} = 4(-5) + 5(-5) + 6(-5) = -75$

(b) Expanding along the second column,

$\begin{vmatrix} -3 & 2 & 1 \\ 4 & 5 & 6 \\ 2 & -3 & 1 \end{vmatrix} = 2C_{12} + 5C_{22} - 3C_{32} = 2(8) + 5(-5) - 3(22) = -75.$

19. Expand along the second row because it has a zero.

$$\begin{vmatrix} 1 & 4 & -2 \\ 3 & 2 & 0 \\ -1 & 4 & 3 \end{vmatrix} = -3\begin{vmatrix} 4 & -2 \\ 4 & 3 \end{vmatrix} + 2\begin{vmatrix} 1 & -2 \\ -1 & 3 \end{vmatrix} - 0\begin{vmatrix} 1 & 4 \\ -1 & 4 \end{vmatrix} = -3(20) + 2(1) = -58$$

21. Expand along the first column because it has two zeros.

$$\begin{vmatrix} 2 & 4 & 6 \\ 0 & 3 & 1 \\ 0 & 0 & -5 \end{vmatrix} = 2\begin{vmatrix} 3 & 1 \\ 0 & -5 \end{vmatrix} - 0\begin{vmatrix} 4 & 6 \\ 0 & -5 \end{vmatrix} + 0\begin{vmatrix} 4 & 6 \\ 3 & 1 \end{vmatrix} = 2(-15) = -30$$

23. Expand along the first row.

$$\begin{vmatrix} 0.1 & 0.2 & 0.3 \\ -0.3 & 0.2 & 0.2 \\ 0.5 & 0.4 & 0.4 \end{vmatrix} = 0.1\begin{vmatrix} 0.2 & 0.2 \\ 0.4 & 0.4 \end{vmatrix} - 0.2\begin{vmatrix} -0.3 & 0.2 \\ 0.5 & 0.4 \end{vmatrix} + 0.3\begin{vmatrix} -0.3 & 0.2 \\ 0.5 & 0.4 \end{vmatrix}$$

$$= 0.1(0) - 0.2(-0.22) + 0.3(-0.22)$$

$$= -0.022$$

25. Use the third row because it has a zero.

$$\begin{vmatrix} x & y & 1 \\ 2 & 3 & 1 \\ 0 & -1 & 1 \end{vmatrix} = -(-1)\begin{vmatrix} x & 1 \\ 2 & 1 \end{vmatrix} + 1\begin{vmatrix} x & y \\ 2 & 3 \end{vmatrix}$$

$$= x - 2 + 3x - 2y$$

$$= 4x - 2y - 2$$

27. Expand along the third column because it has two zeros.

$$\begin{vmatrix} 2 & 6 & 6 & 2 \\ 2 & 7 & 3 & 6 \\ 1 & 5 & 0 & 1 \\ 3 & 7 & 0 & 7 \end{vmatrix} = 6\begin{vmatrix} 2 & 7 & 6 \\ 1 & 5 & 1 \\ 3 & 7 & 7 \end{vmatrix} - 3\begin{vmatrix} 2 & 6 & 2 \\ 1 & 5 & 1 \\ 3 & 7 & 7 \end{vmatrix}$$

The 3×3 determinants are:

$$\begin{vmatrix} 2 & 7 & 6 \\ 1 & 5 & 1 \\ 3 & 7 & 7 \end{vmatrix} = 2\begin{vmatrix} 5 & 1 \\ 7 & 7 \end{vmatrix} - 7\begin{vmatrix} 1 & 1 \\ 3 & 7 \end{vmatrix} + 6\begin{vmatrix} 1 & 5 \\ 3 & 7 \end{vmatrix}$$

$$= 2(28) - 7(4) + 6(-8)$$

$$= -20$$

$$\begin{vmatrix} 2 & 6 & 2 \\ 1 & 5 & 1 \\ 3 & 7 & 7 \end{vmatrix} = 2\begin{vmatrix} 5 & 1 \\ 7 & 7 \end{vmatrix} - 6\begin{vmatrix} 1 & 1 \\ 3 & 7 \end{vmatrix} + 2\begin{vmatrix} 1 & 5 \\ 3 & 7 \end{vmatrix}$$

$$= 2(28) - 6(4) + 2(-8)$$

$$= 16$$

So, the determinant of the original matrix is

$$6(-20) - 3(16) = -168.$$

29. Use the first column because it has two zeros.

$$\begin{vmatrix} 5 & 3 & 0 & 6 \\ 4 & 6 & 4 & 12 \\ 0 & 2 & -3 & 4 \\ 0 & 1 & -2 & 2 \end{vmatrix} = 5\begin{vmatrix} 6 & 4 & 12 \\ 2 & -3 & 4 \\ 1 & -2 & 2 \end{vmatrix} - 4\begin{vmatrix} 3 & 0 & 6 \\ 2 & -3 & 4 \\ 1 & -2 & 2 \end{vmatrix}$$

The determinants of the two 3×3 matrices are:

$$\begin{vmatrix} 6 & 4 & 12 \\ 2 & -3 & 4 \\ 1 & -2 & 2 \end{vmatrix} = 6\begin{vmatrix} -3 & 4 \\ -2 & 2 \end{vmatrix} - 2\begin{vmatrix} 4 & 12 \\ -2 & 2 \end{vmatrix} + 1\begin{vmatrix} 4 & 12 \\ -3 & 4 \end{vmatrix}$$

$$= 6(2) - 2(32) + 52$$

$$= 0$$

$$\begin{vmatrix} 3 & 0 & 6 \\ 2 & -3 & 4 \\ 1 & -2 & 2 \end{vmatrix} = 3\begin{vmatrix} -3 & 4 \\ -2 & 2 \end{vmatrix} - 0\begin{vmatrix} 2 & 4 \\ 1 & 2 \end{vmatrix} + 6\begin{vmatrix} 2 & -3 \\ 1 & -2 \end{vmatrix}$$

$$= 3(2) + 6(-1)$$

$$= 0$$

So,

$$\begin{vmatrix} 5 & 3 & 0 & 6 \\ 4 & 6 & 4 & 12 \\ 0 & 2 & -3 & 4 \\ 0 & 1 & -2 & 2 \end{vmatrix} = 5(0) - 4(0) = 0.$$

31. Using the first row you have

$$\begin{vmatrix} w & x & y & z \\ 21 & -15 & 24 & 30 \\ -10 & 24 & -32 & 18 \\ -40 & 22 & 32 & -35 \end{vmatrix} = w\begin{vmatrix} -15 & 24 & 30 \\ 24 & -32 & 18 \\ 22 & 32 & -35 \end{vmatrix} - x\begin{vmatrix} 21 & 24 & 30 \\ -10 & -32 & 18 \\ -40 & 32 & -35 \end{vmatrix} + y\begin{vmatrix} 21 & -15 & 30 \\ -10 & 24 & 18 \\ -40 & 22 & -35 \end{vmatrix} - z\begin{vmatrix} 21 & -15 & 24 \\ -10 & 24 & -32 \\ -40 & 22 & 32 \end{vmatrix}.$$

The determinants of the 3×3 matrices are:

$$\begin{vmatrix} -15 & 24 & 30 \\ 24 & -32 & 18 \\ 22 & 32 & -35 \end{vmatrix} = -15\begin{vmatrix} -32 & 18 \\ 32 & -35 \end{vmatrix} - 24\begin{vmatrix} 24 & 18 \\ 22 & -35 \end{vmatrix} + 30\begin{vmatrix} 24 & -32 \\ 22 & 32 \end{vmatrix}$$

$$= -15(544) - 24(-1236) + 30(1472)$$

$$= 65,664$$

$$\begin{vmatrix} 21 & 24 & 30 \\ -10 & -32 & 18 \\ -40 & 32 & -35 \end{vmatrix} = 21\begin{vmatrix} -32 & 18 \\ 32 & -35 \end{vmatrix} - 24\begin{vmatrix} -10 & 18 \\ -40 & -35 \end{vmatrix} + 30\begin{vmatrix} -10 & -32 \\ -40 & 32 \end{vmatrix}$$

$$= 21(544) - 24(1070) + 30(-1600)$$

$$= -62,256$$

$$\begin{vmatrix} 21 & -15 & 30 \\ -10 & 24 & 18 \\ -40 & 22 & -35 \end{vmatrix} = 21\begin{vmatrix} 24 & 18 \\ 22 & -35 \end{vmatrix} + 15\begin{vmatrix} -10 & 18 \\ -40 & -35 \end{vmatrix} + 30\begin{vmatrix} -10 & 24 \\ -40 & 22 \end{vmatrix}$$

$$= 21(-1236) + 15(1070) + 30(740)$$

$$= 12,294$$

$$\begin{vmatrix} 21 & -15 & 24 \\ -10 & 24 & -32 \\ -40 & 22 & 32 \end{vmatrix} = 21\begin{vmatrix} 24 & -32 \\ 22 & 35 \end{vmatrix} + 15\begin{vmatrix} -10 & -32 \\ -40 & 32 \end{vmatrix} + 24\begin{vmatrix} -10 & 24 \\ -40 & 22 \end{vmatrix}$$

$$= 21(1472) + 15(-1600) + 24(740)$$

$$= 24,672$$

So, $\begin{vmatrix} w & x & y & z \\ 21 & -15 & 24 & 30 \\ -10 & 24 & -32 & 18 \\ -40 & 22 & 32 & -35 \end{vmatrix} = 65,664w + 62,256x + 12,294y - 24,672z$

33. Expand along the first column, and then along the first column of the 4×4 matrix.

$$\begin{vmatrix} 5 & 2 & 0 & 0 & -2 \\ 0 & 1 & 4 & 3 & 2 \\ 0 & 0 & 2 & 6 & 3 \\ 0 & 0 & 3 & 4 & 1 \\ 0 & 0 & 0 & 0 & 2 \end{vmatrix} = 5\begin{vmatrix} 1 & 4 & 3 & 2 \\ 0 & 2 & 6 & 3 \\ 0 & 3 & 4 & 1 \\ 0 & 0 & 0 & 2 \end{vmatrix} = 5(1)\begin{vmatrix} 2 & 6 & 3 \\ 3 & 4 & 1 \\ 0 & 0 & 2 \end{vmatrix}$$

Now expand along the third row, and obtain

$$5(1)\begin{vmatrix} 2 & 6 & 3 \\ 3 & 4 & 1 \\ 0 & 0 & 2 \end{vmatrix} = 5(2)\begin{vmatrix} 2 & 6 \\ 3 & 4 \end{vmatrix} = 10(-10) = -100.$$

35. $\begin{vmatrix} \frac{1}{2} & 1 & -5 \\ 4 & -\frac{1}{4} & -4 \\ 3 & 2 & -2 \end{vmatrix} = -43.5$

37. $\begin{vmatrix} 4 & 3 & 2 & 5 \\ 1 & 6 & -1 & 2 \\ -3 & 2 & 4 & 5 \\ 6 & 1 & 3 & -2 \end{vmatrix} = -1098$

39.
$$\begin{vmatrix} 1 & 2 & 1 & 4 & 2 & -1 \\ 0 & 1 & 2 & -2 & -3 & 1 \\ 0 & 3 & 2 & -1 & 3 & -2 \\ 1 & 2 & 0 & -2 & 3 & 2 \\ 1 & -2 & 3 & 1 & 2 & -1 \\ 2 & 0 & 2 & 3 & 1 & 1 \end{vmatrix} = 329$$

41. The determinant of a triangular matrix is the product of the elements on its main diagonal.
$$\begin{vmatrix} -2 & 0 & 0 \\ 4 & 6 & 0 \\ -3 & 7 & 2 \end{vmatrix} = -2(6)(2) = -24$$

43. The determinant of a triangular matrix is the product of the elements on its main diagonal.
$$\begin{vmatrix} 5 & 8 & -4 & 2 \\ 0 & 0 & 6 & 0 \\ 0 & 0 & 2 & 2 \\ 0 & 0 & 0 & -1 \end{vmatrix} = 5(0)(2)(-1) = 0$$

45. The determinant of a triangular matrix is the product of the elements on its main diagonal.
$$\begin{vmatrix} -1 & 4 & 2 & 1 & -3 \\ 0 & 3 & -4 & 5 & 2 \\ 0 & 0 & 2 & 7 & 0 \\ 0 & 0 & 0 & 5 & -1 \\ 0 & 0 & 0 & 0 & 1 \end{vmatrix} = (-1)(3)(2)(5)(1) = -30$$

47. (a) False. See "Definition of the Determinant of a 2×2 Matrix," page 123.

(b) True. See the first line, page 124.

(c) False. See "Definition of Minors and Cofactors of a Matrix," page 124.

49. $(x + 3)(x + 2) - 1(2) = 0$
$$x^2 + 5x + 6 - 2 = 0$$
$$x^2 + 5x + 4 = 0$$
$$(x + 4)(x + 1) = 0$$
$$x = -4, -1$$

51. $(x + 1)(x - 2) - 1(-2) = 0$
$$x^2 - x - 2 + 2 = 0$$
$$x^2 - x = 0$$
$$x(x - 1) = 0$$
$$x = 0, 1$$

53. $(x - 1)(x - 2) - 3(2) = 0$
$$x^2 - 3x + 2 - 6 = 0$$
$$x^2 - 3x - 4 = 0$$
$$(x - 4)(x + 1) = 0$$
$$x = 4, -1$$

55. $\begin{vmatrix} \lambda + 2 & 2 \\ 1 & \lambda \end{vmatrix} = (\lambda + 2)(\lambda) - 1(2) = \lambda^2 + 2\lambda - 2$

The determinant is zero when $\lambda^2 + 2\lambda - 2 = 0$.
Use the Quadratic Formula to find λ.
$$\lambda = \frac{-2 \pm \sqrt{2^2 - 4(1)(-2)}}{2(1)}$$
$$= \frac{-2 \pm \sqrt{12}}{2}$$
$$= \frac{-2 \pm 2\sqrt{3}}{2}$$
$$= -1 \pm \sqrt{3}$$

57. $\begin{vmatrix} \lambda & 2 & 0 \\ 0 & \lambda + 1 & 2 \\ 0 & 1 & \lambda \end{vmatrix} = \lambda \begin{vmatrix} \lambda + 1 & 2 \\ 1 & \lambda \end{vmatrix}$
$$= \lambda(\lambda^2 + \lambda - 2) = \lambda(\lambda + 2)(\lambda - 1)$$

The determinant is zero when $\lambda(\lambda + 2)(\lambda - 1) = 0$.
So, $\lambda = 0, -2, 1$.

59. $\begin{vmatrix} 4u & -1 \\ -1 & 2v \end{vmatrix} = (4u)(2v) - (-1)(-1) = 8uv - 1$

61. $\begin{vmatrix} e^{2x} & e^{3x} \\ 2e^{2x} & 3e^{3x} \end{vmatrix} = e^{2x}(3e^{3x}) - 2e^{2x}(e^{3x})$
$$= 3e^{5x} - 2e^{5x} = e^{5x}$$

63. $\begin{vmatrix} x & \ln x \\ 1 & 1/x \end{vmatrix} = x(1/x) - 1(\ln x) = 1 - \ln x$

65. Expanding along the first row, the determinant of a 4×4 matrix involves four 3×3 determinants. Each of these 3×3 determinants requires 6 triple products. So, there are $4(6) = 24$ quadruple products.

67. Evaluating the left side yields
$$\begin{vmatrix} w & x \\ y & z \end{vmatrix} = wz - xy.$$

Evaluating the right side yields
$$-\begin{vmatrix} y & z \\ w & x \end{vmatrix} = -(xy - wz) = wz - xy.$$

69. Evaluating the left side yields

$$\begin{vmatrix} w & x \\ y & z \end{vmatrix} = wz - xy.$$

Evaluating the right side yields

$$\begin{vmatrix} w & x + cw \\ y & z + cy \end{vmatrix} = w(z + cy) - y(x + cw)$$

$$= wz + cwy - xy - cwy$$

$$= wz - xy.$$

71. Evaluating the left side yields

$$\begin{vmatrix} 1 & x & x^2 \\ 1 & y & y^2 \\ 1 & z & z^2 \end{vmatrix} = \begin{vmatrix} y & y^2 \\ z & z^2 \end{vmatrix} - \begin{vmatrix} x & x^2 \\ z & z^2 \end{vmatrix} + \begin{vmatrix} x & x^2 \\ y & y^2 \end{vmatrix}$$

$$= yz^2 - y^2z - \left(xz^2 - x^2z\right) + xy^2 - x^2y$$

$$= xy^2 - xz^2 + yz^2 - x^2y + x^2z - y^2z.$$

Expanding the right side yields

$$(y - x)(z - x)(z - y) = \left(yz - xy - xz + x^2\right)(z - y)$$

$$= yz^2 - y^2z - xyz + xy^2 - xz^2 + xyz + x^2z - x^2y$$

$$= xy^2 - xz^2 + yz^2 - x^2y + x^2z - y^2z.$$

73. Expanding the determinant along the first row yields

$$\begin{vmatrix} 1 & 1 & 1 \\ a & b & c \\ a^2 & b^2 & c^2 \end{vmatrix} = \left(bc^2 - cb^2\right) - \left(ac^2 - ca^2\right) + \left(ab^2 - a^2b\right) = bc^2 + ca^2 + ab^2 - ba^2 - ac^2 - cb^2.$$

Expanding the right side yields

$$(a - b)(b - c)(c - a) = \left(ab - b^2 - ac + bc\right)(c - a)$$

$$= abc - cb^2 - ac^2 + bc^2 - ba^2 + ab^2 + ca^2 - abc$$

$$= bc^2 + ca^2 + ab^2 - ba^2 - ac^2 - cb^2.$$

75. (a) Expanding along the first row,

$$\begin{vmatrix} x & 0 & c \\ -1 & x & b \\ 0 & -1 & a \end{vmatrix} = x \begin{vmatrix} x & b \\ -1 & a \end{vmatrix} + c \begin{vmatrix} -1 & x \\ 0 & -1 \end{vmatrix} = x(ax + b) + c(1)$$

$$= ax^2 + bx + c.$$

(b) The right column contains the coefficients a, b, c. So,

$$\begin{vmatrix} x & 0 & 0 & d \\ -1 & x & 0 & c \\ 0 & -1 & x & b \\ 0 & 0 & -1 & a \end{vmatrix} = x \begin{vmatrix} x & 0 & c \\ -1 & x & b \\ 0 & -1 & a \end{vmatrix} + 1 \begin{vmatrix} 0 & 0 & d \\ -1 & x & b \\ 0 & -1 & a \end{vmatrix}$$

$$= x\left(ax^2 + bx + c\right) + d$$

$$= ax^3 + bx^2 + cx + d.$$

Section 3.2   Evaluation of a Determinant Using Elementary Operations

1. Because the first row is a multiple of the second row, the determinant is zero.

3. Because the second row is composed of all zeros, the determinant is zero.

5. Because the second and third columns are interchanged, the sign of the determinant is changed.

7. Because 5 has been factored out of the first row, the first determinant is 5 times the second one.

9. Because 4 has been factored out of the second column, and 3 factored out of the third column, the first determinant is 12 times the second one.

11. Because each row in the matrix on the left is divided by 5 to yield the matrix on the right, the determinant of the matrix on the left is 5^3 times the determinant of the matrix on the right.

13. Because a multiple of the first row of the matrix on the left was added to the second row to produce the matrix on the right, the determinants are equal.

15. Because a multiple of the first row of the matrix on the left was added to the second row to produce the matrix on the right, the determinants are equal.

17. Because the second row of the matrix on the left was multiplied by (-1), the sign of the determinant is changed.

19. Because the sixth column is a multiple of the first column, the determinant is zero.

21. Expand by cofactors along the second column.

$$\begin{vmatrix} 1 & 0 & 2 \\ -1 & 1 & 4 \\ 2 & 0 & 3 \end{vmatrix} = 1\begin{vmatrix} 1 & 2 \\ 2 & 3 \end{vmatrix} = 3 - 4 = -1$$

A graphing utility or computer software program produces the same determinant, -1.

23. Rewrite the matrix in triangular form.

$$\begin{vmatrix} 1 & 2 & 1 & -1 \\ 0 & 1 & 0 & 2 \\ 0 & 3 & -1 & 1 \\ 0 & 0 & 4 & 1 \end{vmatrix} = \begin{vmatrix} 1 & 2 & 1 & -1 \\ 0 & 1 & 0 & 2 \\ 0 & 0 & -1 & -5 \\ 0 & 0 & 4 & 1 \end{vmatrix} = \begin{vmatrix} 1 & 2 & 1 & -1 \\ 0 & 1 & 0 & 2 \\ 0 & 0 & -1 & -5 \\ 0 & 0 & 0 & -19 \end{vmatrix}$$

$$= 1(1)(-1)(-19) = 19$$

A graphing utility or computer software program produces the same determinant, 19.

25. $\begin{vmatrix} 1 & 7 & -3 \\ 1 & 3 & 1 \\ 4 & 8 & 1 \end{vmatrix} = \begin{vmatrix} 1 & 7 & -3 \\ 0 & -4 & 4 \\ 4 & 8 & 1 \end{vmatrix}$

$$= \begin{vmatrix} 1 & 7 & -3 \\ 0 & -4 & 4 \\ 0 & -20 & 13 \end{vmatrix}$$

$$= \begin{vmatrix} 1 & 7 & -3 \\ 0 & -4 & 4 \\ 0 & 0 & -7 \end{vmatrix} = 1(-4)(-7) = 28$$

27. $\begin{vmatrix} 2 & -1 & -1 \\ 1 & 3 & 2 \\ 1 & 1 & 3 \end{vmatrix} = -\begin{vmatrix} 1 & 3 & 2 \\ 2 & -1 & -1 \\ 1 & 1 & 3 \end{vmatrix}$

$$= -\begin{vmatrix} 1 & 3 & 2 \\ 0 & -7 & -5 \\ 0 & -2 & 1 \end{vmatrix} = (-1)(-7 - 10) = 17$$

29. $\begin{vmatrix} 4 & 3 & -2 \\ 5 & 4 & 1 \\ -2 & 3 & 4 \end{vmatrix} = 4\begin{vmatrix} 14 & 11 & 0 \\ 5 & 4 & 1 \\ -22 & -13 & 0 \end{vmatrix}$

$$= (-1)\begin{vmatrix} 14 & 11 \\ -22 & -13 \end{vmatrix}$$

$$= (-1)\big[(14(-13) - (-22)(11))\big]$$

$$= (-1)(60) = -60$$

31. $\begin{vmatrix} 5 & -8 & 0 \\ 9 & 7 & 4 \\ -8 & 7 & 1 \end{vmatrix} = \begin{vmatrix} 5 & -8 & 0 \\ 41 & -21 & 0 \\ -8 & 7 & 1 \end{vmatrix} = 1(-105 + 328) = 223$

33. $\begin{vmatrix} 4 & -7 & 9 & 1 \\ 6 & 2 & 7 & 0 \\ 3 & 6 & -3 & 3 \\ 0 & 7 & 4 & -1 \end{vmatrix} = \begin{vmatrix} 4 & -7 & 9 & 1 \\ 6 & 2 & 7 & 0 \\ -9 & 27 & -30 & 0 \\ 4 & 0 & 13 & 0 \end{vmatrix}$

$$= -\begin{vmatrix} 6 & 2 & 7 \\ -9 & 27 & -30 \\ 4 & 0 & 13 \end{vmatrix}$$

$$= 3\begin{vmatrix} 6 & 2 & 7 \\ 3 & -9 & 10 \\ 4 & 0 & 13 \end{vmatrix}$$

$$= 3\begin{vmatrix} 0 & 20 & -13 \\ 3 & -9 & 10 \\ 4 & 0 & 13 \end{vmatrix}$$

$$= 3\big[(-3)260 + 4(200 - 117)\big]$$

$$= -1344$$

35. $\begin{vmatrix} 1 & -2 & 7 & 9 \\ 3 & -4 & 5 & 5 \\ 3 & 6 & 1 & -1 \\ 4 & 5 & 3 & 2 \end{vmatrix} = \begin{vmatrix} 1 & -2 & 7 & 9 \\ 0 & 2 & -16 & -22 \\ 0 & 12 & -20 & -28 \\ 0 & 13 & -25 & -34 \end{vmatrix}$

$\qquad = 2 \begin{vmatrix} 1 & -2 & 7 & 9 \\ 0 & 1 & -8 & -11 \\ 0 & 12 & -20 & -28 \\ 0 & 13 & -25 & -34 \end{vmatrix}$

$\qquad = 2 \begin{vmatrix} 1 & -2 & 7 & 9 \\ 0 & 1 & -8 & -11 \\ 0 & 0 & 76 & 104 \\ 0 & 0 & 79 & 109 \end{vmatrix}$

$\qquad = 2(1)(1) \begin{vmatrix} 76 & 104 \\ 79 & 109 \end{vmatrix}$

$\qquad = 2 \big[76(109) - 79(104) \big] = 136$

37. $\begin{vmatrix} 1 & -1 & 8 & 4 & 2 \\ 2 & 6 & 0 & -4 & 3 \\ 2 & 0 & 2 & 6 & 2 \\ 0 & 2 & 8 & 0 & 0 \\ 0 & 1 & 1 & 2 & 2 \end{vmatrix} = \begin{vmatrix} 1 & -1 & 8 & 4 & 2 \\ 0 & 8 & -16 & -12 & -1 \\ 0 & 2 & -14 & -2 & -2 \\ 0 & 2 & 8 & 0 & 0 \\ 0 & 1 & 1 & 2 & 2 \end{vmatrix}$

$\qquad = \begin{vmatrix} 1 & -1 & 8 & 4 & 2 \\ 0 & 0 & -24 & -28 & -17 \\ 0 & 0 & -16 & -6 & -6 \\ 0 & 0 & 6 & -4 & -4 \\ 0 & 1 & 1 & 2 & 2 \end{vmatrix}$

$\qquad = (-1) \begin{vmatrix} -24 & -28 & -17 \\ -16 & -6 & -6 \\ 6 & -4 & -4 \end{vmatrix}$

$\qquad = 2 \begin{vmatrix} -24 & -28 & -17 \\ -16 & -6 & -6 \\ -3 & 2 & 2 \end{vmatrix}$

$\qquad = 2 \begin{vmatrix} -24 & -28 & -17 \\ -25 & 0 & 0 \\ -3 & 2 & 2 \end{vmatrix} = 50(-56 + 34) = -1100$

39. (a) True. See Theorem 3.3, part 1, page 134.

(b) True. See Theorem 3.3, part 3, page 134.

(c) True. See Theorem 3.4, part 2, page 136.

41. $\begin{vmatrix} 1 & 0 & 0 \\ 0 & k & 0 \\ 0 & 0 & 1 \end{vmatrix} = k \begin{vmatrix} 1 & 0 & 0 \\ 0 & 1 & 0 \\ 0 & 0 & 1 \end{vmatrix} = k$

43. $\begin{vmatrix} 0 & 1 & 0 \\ 1 & 0 & 0 \\ 0 & 0 & 1 \end{vmatrix} = - \begin{vmatrix} 1 & 0 & 0 \\ 0 & 1 & 0 \\ 0 & 0 & 1 \end{vmatrix} = -1$

45. $\begin{vmatrix} 1 & 0 & 0 \\ k & 1 & 0 \\ 0 & 0 & 1 \end{vmatrix} = \begin{vmatrix} 1 & 0 & 0 \\ 0 & 1 & 0 \\ 0 & 0 & 1 \end{vmatrix} = 1$

47. Expand the two determinants on the left.

$$\begin{vmatrix} a_{11} & a_{12} & a_{13} \\ a_{21} & a_{22} & a_{23} \\ a_{31} & a_{32} & a_{33} \end{vmatrix} + \begin{vmatrix} b_{11} & a_{12} & a_{13} \\ b_{21} & a_{22} & a_{23} \\ b_{31} & a_{32} & a_{33} \end{vmatrix}$$

$$= a_{11}\begin{vmatrix} a_{22} & a_{23} \\ a_{32} & a_{33} \end{vmatrix} - a_{21}\begin{vmatrix} a_{12} & a_{13} \\ a_{32} & a_{33} \end{vmatrix} + a_{31}\begin{vmatrix} a_{12} & a_{13} \\ a_{22} & a_{23} \end{vmatrix} + b_{11}\begin{vmatrix} a_{22} & a_{23} \\ a_{32} & a_{33} \end{vmatrix} - b_{21}\begin{vmatrix} a_{12} & a_{13} \\ a_{32} & a_{33} \end{vmatrix} + b_{31}\begin{vmatrix} a_{12} & a_{13} \\ a_{22} & a_{23} \end{vmatrix}$$

$$= (a_{11} + b_{11})\begin{vmatrix} a_{22} & a_{23} \\ a_{32} & a_{33} \end{vmatrix} - (a_{21} + b_{21})\begin{vmatrix} a_{12} & a_{13} \\ a_{32} & a_{33} \end{vmatrix} + (a_{31} + b_{31})\begin{vmatrix} a_{12} & a_{13} \\ a_{22} & a_{23} \end{vmatrix}$$

$$= \begin{vmatrix} (a_{11} + b_{11}) & a_{12} & a_{13} \\ (a_{21} + b_{21}) & a_{22} & a_{23} \\ (a_{31} + b_{31}) & a_{32} & a_{33} \end{vmatrix}$$

49. $\begin{vmatrix} \cos \theta & \sin \theta \\ -\sin \theta & \cos \theta \end{vmatrix} = \cos \theta(\cos \theta) - (-\sin \theta)(\sin \theta) = \cos^2 \theta + \sin^2 \theta = 1$

51. $\begin{vmatrix} \sin \theta & 1 \\ 1 & \sin \theta \end{vmatrix} = (\sin \theta)(\sin \theta) - 1(1) = \sin^2 \theta - 1 = -\cos^2 \theta$

53. $\begin{vmatrix} \cos x & 0 & \sin x \\ \sin x & 0 & -\cos x \\ \sin x - \cos x & 1 & \sin x - \cos x \end{vmatrix} = -1\begin{vmatrix} \cos x & \sin x \\ \sin x & -\cos x \end{vmatrix} = -1(-\cos^2 x - \sin^2 x) = \cos^2 x + \sin^2 x = 1$

The value of the determinant is 1 for all x and therefore there is no value of x such that the determinant has a value of zero.

55. If B is obtained from A by multiplying a row of A by a nonzero constant c, then

$$\det(B) = \det\begin{bmatrix} a_{11} & \cdots & a_{1n} \\ \vdots & & \\ ca_{i1} & \cdots & ca_{in} \\ \vdots & & \\ a_{n1} & \cdots & a_{nn} \end{bmatrix} = ca_{i1}C_{i1} + \ldots + ca_{in}C_{in} = c(a_{i1}C_{i1} + \ldots + ca_{in}C_{in}) = c\det(A)$$

Section 3.3   Properties of Determinants

1. (a) $|A| = \begin{vmatrix} -2 & 1 \\ 4 & -2 \end{vmatrix} = 0$

(b) $|B| = \begin{vmatrix} 1 & 1 \\ 0 & -1 \end{vmatrix} = -1$

(c) $AB = \begin{bmatrix} -2 & 1 \\ 4 & -2 \end{bmatrix}\begin{bmatrix} 1 & 1 \\ 0 & -1 \end{bmatrix} = \begin{bmatrix} -2 & -3 \\ 4 & 6 \end{bmatrix}$

(d) $|AB| = \begin{vmatrix} -2 & -3 \\ 4 & 6 \end{vmatrix} = 0$

Notice that $|A||B| = 0(-1) = 0 = |AB|$.

3. (a) $|A| = \begin{vmatrix} -1 & 2 & 1 \\ 1 & 0 & 1 \\ 0 & 1 & 0 \end{vmatrix} = 2$

(b) $|B| = \begin{vmatrix} -1 & 0 & 0 \\ 0 & 2 & 0 \\ 0 & 0 & 3 \end{vmatrix} = -6$

(c) $AB = \begin{bmatrix} -1 & 2 & 1 \\ 1 & 0 & 1 \\ 0 & 1 & 0 \end{bmatrix}\begin{bmatrix} -1 & 0 & 0 \\ 0 & 2 & 0 \\ 0 & 0 & 3 \end{bmatrix} = \begin{bmatrix} 1 & 4 & 3 \\ -1 & 0 & 3 \\ 0 & 2 & 0 \end{bmatrix}$

(d) $|AB| = \begin{vmatrix} 1 & 4 & 3 \\ -1 & 0 & 3 \\ 0 & 2 & 0 \end{vmatrix} = -12$

Notice that $|A||B| = 2(-6) = -12 = |AB|$.

5. (a) $|A| = \begin{vmatrix} 2 & 0 & 1 & 1 \\ 1 & -1 & 0 & 1 \\ 2 & 3 & 1 & 0 \\ 1 & 2 & 3 & 0 \end{vmatrix} = \begin{vmatrix} 1 & 1 & 1 & 0 \\ 1 & -1 & 0 & 1 \\ 2 & 3 & 1 & 0 \\ 1 & 2 & 3 & 0 \end{vmatrix} = \begin{vmatrix} 1 & 1 & 1 \\ 2 & 3 & 1 \\ 1 & 2 & 3 \end{vmatrix} = \begin{vmatrix} 1 & 1 & 1 \\ 0 & 1 & -1 \\ 1 & 2 & 3 \end{vmatrix} = \begin{vmatrix} 1 & 1 & 1 \\ 0 & 1 & -1 \\ 0 & 1 & 2 \end{vmatrix} = 3$

(b) $|B| = \begin{vmatrix} 1 & 0 & -1 & 1 \\ 2 & 1 & 0 & 2 \\ 1 & 1 & -1 & 0 \\ 3 & 2 & 1 & 0 \end{vmatrix} = \begin{vmatrix} 1 & 0 & -1 & 1 \\ 0 & 1 & 2 & 0 \\ 1 & 1 & -1 & 0 \\ 3 & 2 & 1 & 0 \end{vmatrix} = -\begin{vmatrix} 0 & 1 & 2 \\ 1 & 1 & -1 \\ 3 & 2 & 1 \end{vmatrix} = -\begin{vmatrix} 0 & 1 & 2 \\ 1 & 1 & -1 \\ 0 & -1 & 4 \end{vmatrix} = 6$

(c) $AB = \begin{bmatrix} 2 & 0 & 1 & 1 \\ 1 & -1 & 0 & 1 \\ 2 & 3 & 1 & 0 \\ 1 & 2 & 3 & 0 \end{bmatrix}\begin{bmatrix} 1 & 0 & -1 & 1 \\ 2 & 1 & 0 & 2 \\ 1 & 1 & -1 & 0 \\ 3 & 2 & 1 & 0 \end{bmatrix} = \begin{bmatrix} 6 & 3 & -2 & 2 \\ 2 & 1 & 0 & -1 \\ 9 & 4 & -3 & 8 \\ 8 & 5 & -4 & 5 \end{bmatrix}$

(d) $|AB| = \begin{vmatrix} 6 & 3 & -2 & 2 \\ 2 & 1 & 0 & -1 \\ 9 & 4 & -3 & 8 \\ 8 & 5 & -4 & 5 \end{vmatrix} = \begin{vmatrix} 0 & 0 & -2 & 5 \\ 2 & 1 & 0 & -1 \\ 9 & 4 & -3 & 8 \\ 8 & 5 & -4 & 5 \end{vmatrix} = -2\begin{vmatrix} 2 & 1 & -1 \\ 9 & 4 & 8 \\ 8 & 5 & 5 \end{vmatrix} - 5\begin{vmatrix} 2 & 1 & 0 \\ 9 & 4 & -3 \\ 8 & 5 & -4 \end{vmatrix} = -2\begin{vmatrix} 2 & 1 & -1 \\ 1 & 0 & 12 \\ 8 & 5 & 5 \end{vmatrix} - 5\begin{vmatrix} 2 & 1 & 0 \\ 3 & \frac{1}{4} & 0 \\ 8 & 5 & -4 \end{vmatrix}$

$= -2\begin{vmatrix} 2 & 1 & -1 \\ 1 & 0 & 12 \\ -2 & 0 & 10 \end{vmatrix} - 5(-4)\left(\frac{1}{2} - 3\right)$

$= -2(-34) - 50 = 18$

Notice that $|A||B| = 3 \cdot 6 = 18 = |AB|$.

7. $|A| = \begin{vmatrix} 4 & 2 \\ 6 & -8 \end{vmatrix} = 2^2\begin{vmatrix} 2 & 1 \\ 3 & -4 \end{vmatrix} = 4(-11) = -44$

9. $|A| = \begin{vmatrix} -3 & 6 & 9 \\ 6 & 9 & 12 \\ 9 & 12 & 15 \end{vmatrix} = 3^3\begin{vmatrix} -1 & 2 & 3 \\ 2 & 3 & 4 \\ 3 & 4 & 5 \end{vmatrix} = 3^3\begin{vmatrix} -1 & 2 & 3 \\ 0 & 7 & 10 \\ 0 & 10 & 14 \end{vmatrix}$

$= (-27)(-2) = 54$

11. (a) $|A| = \begin{vmatrix} -1 & 1 \\ 2 & 0 \end{vmatrix} = -2$

(b) $|B| = \begin{vmatrix} 1 & -1 \\ -2 & 0 \end{vmatrix} = -2$

(c) $|A + B| = \left\|\begin{bmatrix} -1 & 1 \\ 2 & 0 \end{bmatrix} + \begin{bmatrix} 1 & -1 \\ -2 & 0 \end{bmatrix}\right\| = \begin{vmatrix} 0 & 0 \\ 0 & 0 \end{vmatrix} = 0$

Notice that
$|A| + |B| = -2 + (-2) = -4 \neq |A + B|$.

13. (a) $|A| = \begin{vmatrix} 1 & 0 & 1 \\ -1 & 2 & 1 \\ 0 & 1 & 1 \end{vmatrix} = \begin{vmatrix} 1 & 0 & 1 \\ 0 & 2 & 2 \\ 0 & 1 & 1 \end{vmatrix} = \begin{vmatrix} 1 & 0 & 1 \\ 0 & 2 & 2 \\ 0 & 0 & 0 \end{vmatrix} = 0$

(b) $|B| = \begin{vmatrix} -1 & 0 & 2 \\ 0 & 1 & 2 \\ 1 & 1 & 1 \end{vmatrix} = \begin{vmatrix} 0 & 1 & 3 \\ 0 & 1 & 2 \\ 1 & 1 & 1 \end{vmatrix} = -1$

(c) $|A + B| = \left\|\begin{bmatrix} 1 & 0 & 1 \\ -1 & 2 & 1 \\ 0 & 1 & 1 \end{bmatrix} + \begin{bmatrix} -1 & 0 & 2 \\ 0 & 1 & 2 \\ 1 & 1 & 1 \end{bmatrix}\right\| = \begin{vmatrix} 0 & 0 & 3 \\ -1 & 3 & 3 \\ 1 & 2 & 2 \end{vmatrix}$

$= -15$

Notice that $|A| + |B| = 0 + (-1) = -1 \neq |A + B|$.

15. First observe that $|A| = \begin{vmatrix} 6 & -11 \\ 4 & -5 \end{vmatrix} = 14$.

(a) $|A^T| = |A| = 14$

(b) $|A^2| = |A||A| = |A|^2 = 196$

(c) $|AA^T| = |A||A^T| = 14(14) = 196$

(d) $|2A| = 4|A| = 56$

(e) $|A^{-1}| = \dfrac{1}{|A|} = \dfrac{1}{14}$

17. First observe that $|A| = \begin{vmatrix} 2 & 0 & 5 \\ 4 & -1 & 6 \\ 3 & 2 & 1 \end{vmatrix} = 29.$

(a) $|A^T| = |A| = 29$

(b) $|A^2| = |A||A| = 29^2 = 841$

(c) $|AA^T| = |A||A^T| = 29(29) = 841$

(d) $|2A| = 2^3|A| = 8(29) = 232$

(e) $|A^{-1}| = \dfrac{1}{|A|} = \dfrac{1}{29}$

19. (a) $|A| = \begin{vmatrix} 4 & 2 \\ -1 & 5 \end{vmatrix} = 22$

(b) $|A^T| = \begin{vmatrix} 4 & -1 \\ 2 & 5 \end{vmatrix} = 22$

(c) $|A^2| = \begin{vmatrix} 14 & 18 \\ -9 & 23 \end{vmatrix} = 484$

(d) $|2A| = \begin{vmatrix} 8 & 4 \\ -2 & 10 \end{vmatrix} = 88$

(e) $|A^{-1}| = \begin{vmatrix} \frac{5}{22} & -\frac{1}{11} \\ \frac{1}{22} & \frac{2}{11} \end{vmatrix} = \dfrac{1}{22}$

21. (a) $|A| = \begin{vmatrix} 4 & -2 & 1 & 5 \\ 3 & 8 & 2 & -1 \\ 6 & 8 & 9 & 2 \\ 2 & 3 & -1 & 0 \end{vmatrix} = -115$

(b) $|A^T| = \begin{vmatrix} 4 & 3 & 6 & 2 \\ -2 & 8 & 8 & 3 \\ 1 & 2 & 9 & -1 \\ 5 & -1 & 2 & 0 \end{vmatrix} = -115$

(c) $|A^2| = \begin{vmatrix} 26 & -1 & 4 & 24 \\ 46 & 71 & 38 & 11 \\ 106 & 130 & 101 & 40 \\ 11 & 12 & -1 & 5 \end{vmatrix} = 13{,}225$

(d) $|2A| = \begin{vmatrix} 8 & -4 & 2 & 10 \\ 6 & 16 & 4 & -2 \\ 12 & 16 & 18 & 4 \\ 4 & 6 & -2 & 0 \end{vmatrix} = -1840$

(e) $|A^{-1}| = \begin{vmatrix} -\frac{63}{115} & -\frac{173}{115} & \frac{71}{115} & 2 \\ \frac{38}{115} & \frac{108}{115} & -\frac{41}{115} & -1 \\ -\frac{12}{115} & -\frac{22}{115} & \frac{19}{115} & 0 \\ \frac{91}{115} & \frac{186}{115} & -\frac{77}{115} & -2 \end{vmatrix} = -\dfrac{1}{115}$

23. (a) $|AB| = |A||B| = -5(3) = -15$

(b) $|A^3| = |A|^3 = (-5)^3 = -125$

(c) $|3B| = 3^4|B| = 81(3) = 243$

(d) $\left|(AB)^T\right| = |AB| = -15$

(e) $|A^{-1}| = \dfrac{1}{|A|} = -\dfrac{1}{5}$

25. (a) $|BA| = |B||A| = 2 \cdot 4 = 8$

(b) $|B^2| = |B|^2 = 2^2 = 4$

(c) $|2A| = 2^4|A| = 16 \cdot 4 = 64$

(d) $\left|(AB)^T\right| = |B^T A^T| = |B^T||A^T| = |B||A| = 2 \cdot 4 = 8$

Equivalently,

$$\left|(AB)^T\right| = |AB| = |A||B| = 4 \cdot 2 = 8$$

(e) $|B^{-1}| = \dfrac{1}{|B|} = \dfrac{1}{2}$

27. Because

$$\begin{vmatrix} 5 & 4 \\ 10 & 8 \end{vmatrix} = 0,$$

the matrix is singular.

29. Because

$$\begin{vmatrix} 14 & 5 & 7 \\ -2 & 0 & 3 \\ 1 & -5 & -10 \end{vmatrix} = 195 \neq 0,$$

the matrix is nonsingular.

31. Because

$$\begin{vmatrix} \frac{1}{2} & \frac{3}{2} & 2 \\ \frac{2}{3} & -\frac{1}{3} & 0 \\ 1 & 1 & 1 \end{vmatrix} = \frac{5}{6} \neq 0,$$

the matrix is nonsingular.

33. Because

$$\begin{vmatrix} 1 & 0 & -8 & 2 \\ 0 & 8 & -1 & 10 \\ 0 & 0 & 0 & 1 \\ 0 & 0 & 0 & 2 \end{vmatrix} = 0,$$

the matrix is singular.

35. $A^{-1} = \dfrac{1}{5}\begin{bmatrix} 4 & -3 \\ -1 & 2 \end{bmatrix} = \begin{bmatrix} \dfrac{4}{5} & -\dfrac{3}{5} \\ -\dfrac{1}{5} & \dfrac{2}{5} \end{bmatrix}$

$\left|A^{-1}\right| = \dfrac{4}{5}\left(\dfrac{2}{5}\right) - \left(-\dfrac{1}{5}\right)\left(-\dfrac{3}{5}\right) = \dfrac{8}{25} - \dfrac{3}{25} = \dfrac{1}{5}$

Notice that $|A| = 5$, so $\left|A^{-1}\right| = \dfrac{1}{|A|} = \dfrac{1}{5}$.

37. $A^{-1} = \begin{bmatrix} -2 & 2 & -1 \\ \dfrac{1}{2} & 0 & -\dfrac{1}{2} \\ \dfrac{3}{2} & -1 & \dfrac{1}{2} \end{bmatrix}$

$\left|A^{-1}\right| = \begin{vmatrix} -2 & 2 & -1 \\ \dfrac{1}{2} & 0 & -\dfrac{1}{2} \\ \dfrac{3}{2} & -1 & \dfrac{1}{2} \end{vmatrix} = \begin{vmatrix} 1 & 0 & 0 \\ \dfrac{1}{2} & 0 & -\dfrac{1}{2} \\ \dfrac{3}{2} & -1 & \dfrac{1}{2} \end{vmatrix} = -\dfrac{1}{2}$

Notice that $|A| = \begin{vmatrix} 1 & 0 & 2 \\ 2 & -1 & 3 \\ 1 & -2 & 2 \end{vmatrix} = \begin{vmatrix} 1 & 0 & 2 \\ 2 & -1 & 3 \\ -3 & 0 & -4 \end{vmatrix} = -2$, so

$\left|A^{-1}\right| = \dfrac{1}{|A|} = -\dfrac{1}{2}$.

39. $A^{-1} = \begin{bmatrix} -\dfrac{1}{8} & -\dfrac{5}{8} & \dfrac{7}{8} & 0 \\ \dfrac{5}{12} & \dfrac{5}{12} & -\dfrac{1}{4} & -\dfrac{1}{3} \\ \dfrac{3}{8} & \dfrac{7}{8} & -\dfrac{5}{8} & 0 \\ \dfrac{1}{2} & \dfrac{1}{2} & -\dfrac{1}{2} & 0 \end{bmatrix}$

$\left|A^{-1}\right| = \begin{vmatrix} -\dfrac{1}{8} & -\dfrac{5}{8} & \dfrac{7}{8} & 0 \\ \dfrac{5}{12} & \dfrac{5}{12} & -\dfrac{1}{4} & -\dfrac{1}{3} \\ \dfrac{3}{8} & \dfrac{7}{8} & -\dfrac{5}{8} & 0 \\ \dfrac{1}{2} & \dfrac{1}{2} & -\dfrac{1}{2} & 0 \end{vmatrix} = -\dfrac{1}{3}\begin{vmatrix} -\dfrac{1}{8} & -\dfrac{5}{8} & \dfrac{7}{8} \\ \dfrac{3}{8} & \dfrac{7}{8} & -\dfrac{5}{8} \\ \dfrac{1}{2} & \dfrac{1}{2} & -\dfrac{1}{2} \end{vmatrix} = -\dfrac{1}{3}\begin{vmatrix} -\dfrac{1}{8} & -\dfrac{5}{8} & \dfrac{7}{8} \\ 0 & -1 & 2 \\ \dfrac{1}{2} & \dfrac{1}{2} & -\dfrac{1}{2} \end{vmatrix} = -\dfrac{1}{3}\begin{vmatrix} -\dfrac{1}{8} & -\dfrac{5}{8} & \dfrac{7}{8} \\ 0 & -1 & 2 \\ 0 & -2 & 3 \end{vmatrix} = \dfrac{1}{24}$

Notice that $|A| = \begin{vmatrix} 1 & 0 & -1 & 3 \\ 1 & 0 & 3 & -2 \\ 2 & 0 & 2 & -1 \\ 1 & -3 & 1 & 2 \end{vmatrix} = -3\begin{vmatrix} 1 & -1 & 3 \\ 1 & 3 & -2 \\ 2 & 2 & -1 \end{vmatrix} = -3\begin{vmatrix} 1 & -1 & 3 \\ 0 & 4 & -5 \\ 2 & 2 & -1 \end{vmatrix} = -3\begin{vmatrix} 1 & -1 & 3 \\ 0 & 4 & -5 \\ 0 & 4 & -7 \end{vmatrix} = 24.$

So, $\left|A^{-1}\right| = \dfrac{1}{|A|} = \dfrac{1}{24}$.

41. The coefficient matrix of the system is

$$\begin{bmatrix} 1 & -1 & 1 \\ 2 & -1 & 1 \\ 3 & -2 & 2 \end{bmatrix}$$

which has a determinant of

$$\begin{vmatrix} 1 & -1 & 1 \\ 2 & -1 & 1 \\ 3 & -2 & 2 \end{vmatrix} = 0.$$

Because the determinant is zero, the system does not have a unique solution.

43. The coefficient matrix of the system is

$$\begin{vmatrix} 2 & 1 & 5 & -1 \\ 1 & 1 & -3 & -4 \\ 2 & 2 & 2 & -3 \\ 1 & 5 & -6 & 0 \end{vmatrix}.$$

Because the determinant of this matrix is 115, and not zero, the system has a unique solution.

45. Find the values of k necessary to make A singular by setting $|A| = 0$.

$$|A| = \begin{vmatrix} k-1 & 3 \\ 2 & k-2 \end{vmatrix}$$

$$= (k-1)(k-2) - 6$$

$$= k^2 - 3k - 4$$

$$= (k-4)(k+1) = 0$$

So, $|A| = 0$ when $k = -1, 4$.

47. Find the value of k necessary to make A singular by setting $|A| = 0$.

$$|A| = \begin{vmatrix} 1 & 0 & 3 \\ 2 & -1 & 0 \\ 4 & 2 & k \end{vmatrix} = 1(-k) + 3(8) = 0$$

So, $k = 24$.

49. $AB = I$, which implies that $|AB| = |A||B| = |I| = 1$.

So, both $|A|$ and $|B|$ must be nonzero, because their product is 1.

51. Let

$$A = \begin{bmatrix} 1 & 0 \\ 0 & 0 \end{bmatrix} \quad \text{and} \quad B = \begin{bmatrix} 0 & 1 \\ 0 & 0 \end{bmatrix}.$$

Then

$$|A| + |B| = 0 + 0 = 0, \text{ and } |A + B| = \begin{vmatrix} 1 & 1 \\ 0 & 0 \end{vmatrix} = 0.$$

(The answer is not unique.)

53. For each i, $i = 1, 2, ..., n$, the ith row of A can be written as

$$a_{i1}, \quad a_{i2}, \quad ..., \quad a_{in-1}, \quad -\sum_{j=1}^{n-1} a_{ij}.$$

Therefore, the last column can be reduced to all zeros by adding the other columns of A to it. Because A can be reduced to a matrix with a column of zeros, $|A| = 0$.

55. Let $\det(A) = x$ and $\det(A^{-1}) = y$. First note that

$$xy = \det(A) \cdot \det(A^{-1})$$

$$= \det(AA^{-1})$$

$$= \det(I)$$

$$= 1.$$

Assume that all of the entries of A and A^{-1} are integers. Because a determinant is a product of the entries of a matrix, $x = \det(A)$ and $y = \det(A^{-1})$ are integers.

Therefore it must be that x and y are each ± 1 because these are the only integer solutions to $xy = 1$.

57. $P^{-1}AP \neq A$ in general. For example,

$$P = \begin{bmatrix} 1 & 2 \\ 3 & 5 \end{bmatrix}, P^{-1} = \begin{bmatrix} -5 & 2 \\ 3 & -1 \end{bmatrix}, A = \begin{bmatrix} 2 & 1 \\ -1 & 0 \end{bmatrix},$$

$$P^{-1}AP = \begin{bmatrix} -27 & -49 \\ 16 & 29 \end{bmatrix} \neq A.$$

However, the determinants $|A|$ and $|P^{-1}AP|$ are equal.

$$|P^{-1}AP| = |P^{-1}||A||P| = |P^{-1}||P||A| = \frac{1}{|P|}|P||A| = |A|$$

59. (a) False. See Theorem 3.6, page 144.

(b) True. See Theorem 3.8, page 146.

(c) True. See "Equivalent Conditions for a Nonsingular Matrix," parts 1 and 2, page 147.

61. Let A be an $n \times n$ matrix satisfying $A^T = -A$. Then,

$$|A| = |A^T| = |-A| = (-1)^n |A|.$$

63. The inverse of this matrix is

$$\begin{bmatrix} 0 & 1 \\ 1 & 0 \end{bmatrix}^{-1} = \begin{bmatrix} 0 & 1 \\ 1 & 0 \end{bmatrix}.$$

Because $A^T = A^{-1}$, $\begin{bmatrix} 0 & 1 \\ 1 & 0 \end{bmatrix}$ is orthogonal.

65. Because the matrix does not have an inverse (its determinant is 0), it is *not* orthogonal.

67. The inverse of this elementary matrix is

$$A^{-1} = \begin{bmatrix} 1 & 0 & 0 \\ 0 & 0 & 1 \\ 0 & 1 & 0 \end{bmatrix}.$$

Because $A^{-1} = A^T$, the matrix *is* orthogonal.

69. If $A^T = A^{-1}$, then $\left| A^T \right| = \left| A^{-1} \right|$ and so

$$\left| I \right| = \left| AA^{-1} \right| = \left| A \right| \left| A^{-1} \right| = \left| A \right| \left| A^T \right| = \left| A \right|^2 = 1 \Rightarrow \left| A \right| = \pm 1.$$

71. $A = \begin{bmatrix} \frac{2}{3} & -\frac{2}{3} & \frac{1}{3} \\ \frac{2}{3} & \frac{1}{3} & -\frac{2}{3} \\ \frac{1}{3} & \frac{2}{3} & \frac{2}{3} \end{bmatrix}$

Using a graphing utility you have

(a), (b) $A^{-1} = \begin{bmatrix} \frac{2}{3} & \frac{2}{3} & \frac{1}{3} \\ -\frac{2}{3} & \frac{1}{3} & \frac{2}{3} \\ \frac{1}{3} & -\frac{2}{3} & \frac{2}{3} \end{bmatrix} = A^T$

(c) As shown in Exercise 69, if A is an orthogonal matrix then $\left| A \right| = \pm 1$. For this given A you have $\left| A \right| = 1$. Because $A^{-1} = A^T$, A is an orthogonal matrix.

73. $\left| SB \right| = \left| S \right| \left| B \right| = 0 \left| B \right| = 0 \Rightarrow SB$ is singular.

Section 3.4   Introduction to Eigenvalues

1. $A\mathbf{x}_1 = \begin{bmatrix} 1 & 2 \\ 0 & -3 \end{bmatrix} \begin{bmatrix} 1 \\ 0 \end{bmatrix} = \begin{bmatrix} 1 \\ 0 \end{bmatrix} = \lambda_1 \mathbf{x}_1$

$A\mathbf{x}_2 = \begin{bmatrix} 1 & 2 \\ 0 & -3 \end{bmatrix} \begin{bmatrix} -1 \\ 2 \end{bmatrix} = \begin{bmatrix} 3 \\ -6 \end{bmatrix} = -3 \begin{bmatrix} -1 \\ 2 \end{bmatrix} = \lambda_2 \mathbf{x}_2$

3. $A\mathbf{x}_1 = \begin{bmatrix} 1 & 1 & 1 \\ 0 & 1 & 0 \\ 1 & 1 & 1 \end{bmatrix} \begin{bmatrix} 1 \\ 0 \\ 1 \end{bmatrix} = \begin{bmatrix} 2 \\ 0 \\ 2 \end{bmatrix} = 2 \begin{bmatrix} 1 \\ 0 \\ 1 \end{bmatrix} = \lambda_1 \mathbf{x}_1$

$A\mathbf{x}_2 = \begin{bmatrix} 1 & 1 & 1 \\ 0 & 1 & 0 \\ 1 & 1 & 1 \end{bmatrix} \begin{bmatrix} -1 \\ 0 \\ 1 \end{bmatrix} = \begin{bmatrix} 0 \\ 0 \\ 0 \end{bmatrix} = 0 \begin{bmatrix} -1 \\ 0 \\ 1 \end{bmatrix} = \lambda_2 \mathbf{x}_2$

$A\mathbf{x}_3 = \begin{bmatrix} 1 & 1 & 1 \\ 0 & 1 & 0 \\ 1 & 1 & 1 \end{bmatrix} \begin{bmatrix} -1 \\ 1 \\ -1 \end{bmatrix} = \begin{bmatrix} -1 \\ 1 \\ -1 \end{bmatrix} = 1 \begin{bmatrix} -1 \\ 1 \\ -1 \end{bmatrix} = \lambda_3 \mathbf{x}_3$

5. (a) $\left| \lambda I - A \right| = \left| \begin{bmatrix} \lambda & 0 \\ 0 & \lambda \end{bmatrix} - \begin{bmatrix} 4 & -5 \\ 2 & -3 \end{bmatrix} \right|$

$= \begin{bmatrix} \lambda - 4 & 5 \\ -2 & \lambda + 3 \end{bmatrix} = \lambda^2 - \lambda - 2$

The characteristic equation is $\lambda^2 - \lambda - 2 = 0$.

(b) Solve the characteristic equation.

$$\lambda^2 - \lambda - 2 = 0$$
$$(\lambda - 2)(\lambda + 1) = 0$$
$$\lambda = 2, -1$$

The eigenvalues are $\lambda_1 = 2$ and $\lambda_2 = -1$.

(c) For $\lambda_1 = 2$: $\begin{bmatrix} -2 & 5 \\ -2 & 5 \end{bmatrix} \Rightarrow \begin{bmatrix} -2 & 5 \\ 0 & 0 \end{bmatrix}$

The corresponding eigenvectors are nonzero scalar multiples of $\begin{bmatrix} 5 \\ 2 \end{bmatrix}$.

For $\lambda_2 = -1$: $\begin{bmatrix} -5 & 5 \\ -2 & 2 \end{bmatrix} \Rightarrow \begin{bmatrix} 1 & -1 \\ 0 & 0 \end{bmatrix}$

The corresponding eigenvectors are nonzero scalar multiples of $\begin{bmatrix} 1 \\ 1 \end{bmatrix}$.

7. (a) $|\lambda I - A| = \left|\begin{bmatrix} \lambda & 0 \\ 0 & \lambda \end{bmatrix} - \begin{bmatrix} 2 & 1 \\ 3 & 0 \end{bmatrix}\right| = \begin{vmatrix} \lambda - 2 & -1 \\ -3 & \lambda \end{vmatrix}$

$$= \lambda^2 - 2\lambda - 3$$

The characteristic equation is $\lambda^2 - 2\lambda - 3 = 0$.

(b) Solve the characteristic equation.

$$\lambda^2 - 2\lambda - 3 = 0$$
$$(\lambda - 3)(\lambda + 1) = 0$$
$$\lambda = 3, -1$$

The eigenvalues are $\lambda_1 = 3$ and $\lambda_2 = -1$.

(c) For $\lambda_1 = 3$: $\begin{bmatrix} 1 & -1 \\ -3 & 3 \end{bmatrix} \Rightarrow \begin{bmatrix} 1 & -1 \\ 0 & 0 \end{bmatrix}$

The corresponding eigenvectors are the nonzero

multiples of $\begin{bmatrix} 1 \\ 1 \end{bmatrix}$.

For $\lambda_2 = -1$: $\begin{bmatrix} -3 & -1 \\ -3 & -1 \end{bmatrix} \Rightarrow \begin{bmatrix} 1 & \frac{1}{3} \\ -3 & -1 \end{bmatrix} \Rightarrow \begin{bmatrix} 1 & \frac{1}{3} \\ 0 & 0 \end{bmatrix}$

The corresponding eigenvectors are the nonzero

multiples of $\begin{bmatrix} -1 \\ 3 \end{bmatrix}$.

9. (a) $|\lambda I - A| = \left|\begin{bmatrix} \lambda & 0 \\ 0 & \lambda \end{bmatrix} - \begin{bmatrix} -2 & 4 \\ 2 & 5 \end{bmatrix}\right| = \begin{vmatrix} \lambda + 2 & -4 \\ -2 & \lambda - 5 \end{vmatrix}$

$$= \lambda^2 - 3\lambda - 18$$

The characteristic equation is $\lambda^2 - 3\lambda - 18 = 0$.

(b) Solve the characteristic equation.

$$\lambda^2 - 3\lambda - 18 = 0$$
$$(\lambda - 6)(\lambda + 3) = 0$$
$$\lambda = 6, -3$$

The eigenvalues are $\lambda_1 = 6$ and $\lambda_2 = -3$.

(c) For $\lambda_1 = 6$: $\begin{bmatrix} 8 & -4 \\ -2 & 1 \end{bmatrix} \Rightarrow \begin{bmatrix} 1 & -\frac{1}{2} \\ 0 & 0 \end{bmatrix}$

The corresponding eigenvectors are the nonzero

multiples of $\begin{bmatrix} 1 \\ 2 \end{bmatrix}$.

For $\lambda_2 = -3$: $\begin{bmatrix} -1 & -4 \\ -2 & -8 \end{bmatrix} \Rightarrow \begin{bmatrix} 1 & 4 \\ 0 & 0 \end{bmatrix}$

The corresponding eigenvectors are the nonzero

multiples of $\begin{bmatrix} -4 \\ 1 \end{bmatrix}$.

11. (a) $|\lambda I - A| = \begin{vmatrix} \lambda - 1 & 1 & 1 \\ -1 & \lambda - 3 & -1 \\ 3 & -1 & \lambda + 1 \end{vmatrix} = (\lambda - 1)(\lambda^2 - 2\lambda - 4) + 1(\lambda + 2) + 3(2 - \lambda) = \lambda^3 - 3\lambda^2 - 4\lambda + 12$

The characteristic equation is $\lambda^3 - 3\lambda^2 - 4\lambda + 12 = 0$.

(b) Solve the characteristic equation.

$$\lambda^3 - 3\lambda^2 - 4\lambda + 12 = 0$$
$$(\lambda - 2)(\lambda + 2)(\lambda - 3) = 0$$

The eigenvalues are $\lambda_1 = 2, \lambda_2 = -2, \lambda_3 = 3$.

(c) For $\lambda_1 = 2$: $\begin{bmatrix} 1 & 1 & 1 \\ -1 & -1 & -1 \\ 3 & -1 & 3 \end{bmatrix} \Rightarrow \begin{bmatrix} 1 & 1 & 1 \\ 0 & -4 & 0 \\ 0 & 0 & 0 \end{bmatrix} \Rightarrow \begin{bmatrix} 1 & 0 & 1 \\ 0 & 1 & 0 \\ 0 & 0 & 0 \end{bmatrix}$

The corresponding eigenvectors are the nonzero multiples of $\begin{bmatrix} -1 \\ 0 \\ 1 \end{bmatrix}$.

For $\lambda_2 = -2$: $\begin{bmatrix} -3 & 1 & 1 \\ -1 & -5 & -1 \\ 3 & -1 & -1 \end{bmatrix} \Rightarrow \begin{bmatrix} 1 & 5 & 1 \\ 0 & 16 & 4 \\ 0 & 0 & 0 \end{bmatrix} \Rightarrow \begin{bmatrix} 1 & 5 & 1 \\ 0 & 4 & 1 \\ 0 & 0 & 0 \end{bmatrix}$

The corresponding eigenvectors are the nonzero multiples of $\begin{bmatrix} 1 \\ -1 \\ 4 \end{bmatrix}$.

For $\lambda_3 = 3$: $\begin{bmatrix} 2 & 1 & 1 \\ -1 & 0 & -1 \\ 3 & -1 & 4 \end{bmatrix} \Rightarrow \begin{bmatrix} 1 & 0 & 1 \\ 0 & 1 & -1 \\ 0 & -1 & 1 \end{bmatrix} \Rightarrow \begin{bmatrix} 1 & 0 & 1 \\ 0 & 1 & -1 \\ 0 & 0 & 0 \end{bmatrix}$

The corresponding eigenvectors are the nonzero multiples of $\begin{bmatrix} -1 \\ 1 \\ 1 \end{bmatrix}$.

13. (a) $|\lambda I - A| = \left| \begin{bmatrix} \lambda & 0 & 0 \\ 0 & \lambda & 0 \\ 0 & 0 & \lambda \end{bmatrix} - \begin{bmatrix} 1 & 2 & 1 \\ 0 & 1 & 0 \\ 4 & 0 & 1 \end{bmatrix} \right| = \begin{vmatrix} \lambda - 1 & -2 & -1 \\ 0 & \lambda - 1 & 0 \\ -4 & 0 & \lambda - 1 \end{vmatrix}$

$$= (\lambda - 1)(\lambda^2 - 2\lambda + 1) - 4(\lambda - 1)$$

$$= \lambda^3 - 3\lambda^2 - \lambda + 3$$

The characteristic equation is $\lambda^3 - 3\lambda^2 - \lambda + 3 = 0$.

(b) Solve the characteristic equation.

$$\lambda^3 - 3\lambda^2 - \lambda + 3 = 0$$

$$\lambda^2(\lambda - 3) - 1(\lambda - 3) = 0$$

$$(\lambda^2 - 1)(\lambda - 3) = 0$$

The eigenvalues are $\lambda_1 = -1, \lambda_2 = 1, \lambda_3 = 3$.

(c) For $\lambda_1 = -1$: $\begin{bmatrix} -2 & -2 & -1 \\ 0 & -2 & 0 \\ -4 & 0 & -2 \end{bmatrix} \Rightarrow \begin{bmatrix} -2 & -2 & -1 \\ 0 & -2 & 0 \\ 0 & 4 & 0 \end{bmatrix} \Rightarrow \begin{bmatrix} 2 & 2 & 1 \\ 0 & 1 & 0 \\ 0 & 0 & 0 \end{bmatrix} \Rightarrow \begin{bmatrix} 1 & 0 & \frac{1}{2} \\ 0 & 1 & 0 \\ 0 & 0 & 0 \end{bmatrix}$

The corresponding eigenvectors are the nonzero multiples of $\begin{bmatrix} -1 \\ 0 \\ 2 \end{bmatrix}$.

For $\lambda_2 = 1$: $\begin{bmatrix} 0 & -2 & -1 \\ 0 & 0 & 0 \\ -4 & 0 & 0 \end{bmatrix} \Rightarrow \begin{bmatrix} 0 & 1 & \frac{1}{2} \\ 0 & 0 & 0 \\ 1 & 0 & 0 \end{bmatrix}$

The corresponding eigenvectors are the nonzero multiples of $\begin{bmatrix} 0 \\ -1 \\ 2 \end{bmatrix}$.

For $\lambda_3 = 3$: $\begin{bmatrix} 2 & -2 & -1 \\ 0 & 2 & 0 \\ -4 & 0 & 2 \end{bmatrix} \Rightarrow \begin{bmatrix} 2 & -2 & -1 \\ 0 & 1 & 0 \\ 0 & -4 & 0 \end{bmatrix} \Rightarrow \begin{bmatrix} 1 & 0 & -\frac{1}{2} \\ 0 & 1 & 0 \\ 0 & 0 & 0 \end{bmatrix}$

The corresponding eigenvectors are the nonzero multiples of $\begin{bmatrix} 1 \\ 0 \\ 2 \end{bmatrix}$.

15. Using a graphing utility or computer software program with $A = \begin{bmatrix} 2 & 5 \\ -1 & -4 \end{bmatrix}$ produces the eigenvalues $\{1 \quad -3\}$.

So, $\lambda_1 = 1$ and $\lambda_2 = -3$.

$\lambda_1 = 1$: $\begin{bmatrix} -1 & -5 \\ 1 & 5 \end{bmatrix} \Rightarrow \begin{bmatrix} 1 & 5 \\ 0 & 0 \end{bmatrix}$ and $\mathbf{x}_1 = \begin{bmatrix} -5 \\ 1 \end{bmatrix}$

$\lambda_2 = -3$: $\begin{bmatrix} -5 & -5 \\ 1 & 1 \end{bmatrix} \Rightarrow \begin{bmatrix} 1 & 1 \\ 0 & 0 \end{bmatrix}$ and $\mathbf{x}_2 = \begin{bmatrix} -1 \\ 1 \end{bmatrix}$

17. Using a graphing utility or computer software program with $A = \begin{bmatrix} 4 & -2 & -2 \\ 0 & 1 & 0 \\ 1 & 0 & 1 \end{bmatrix}$ produces the eigenvalues $\{3 \quad 2 \quad 1\}$.

So, $\lambda_1 = 3, \lambda_2 = 2,$ and $\lambda_3 = 1.$

$\lambda_1 = 3$: $\begin{bmatrix} -1 & 2 & 2 \\ 0 & 2 & 0 \\ -1 & 0 & 2 \end{bmatrix} \Rightarrow \begin{bmatrix} 1 & 0 & -2 \\ 0 & 1 & 0 \\ 0 & 0 & 0 \end{bmatrix}$ and $\mathbf{x}_1 = \begin{bmatrix} 2 \\ 0 \\ 1 \end{bmatrix}$

$\lambda_2 = 2$: $\begin{bmatrix} -2 & 2 & 2 \\ 0 & 1 & 0 \\ -1 & 0 & 1 \end{bmatrix} \Rightarrow \begin{bmatrix} 1 & 0 & -1 \\ 0 & 1 & 0 \\ 0 & 0 & 0 \end{bmatrix}$ and $\mathbf{x}_2 = \begin{bmatrix} 1 \\ 0 \\ 1 \end{bmatrix}$

$\lambda_3 = 1$: $\begin{bmatrix} -3 & 2 & 2 \\ 0 & 0 & 0 \\ -1 & 0 & 0 \end{bmatrix} \Rightarrow \begin{bmatrix} 1 & 0 & 0 \\ 0 & 1 & 1 \\ 0 & 0 & 0 \end{bmatrix}$ and $\mathbf{x}_3 = \begin{bmatrix} 0 \\ -1 \\ 1 \end{bmatrix}$

19. Using a graphing utility or computer software program with $A = \begin{bmatrix} 1 & 0 & -1 \\ 0 & -2 & 0 \\ 0 & -2 & -2 \end{bmatrix}$ produces the eigenvalues $\{1 \ -2\}$.

So, $\lambda_1 = 1$ and $\lambda_2 = -2.$

$\lambda_1 = 1$: $\begin{bmatrix} 0 & 0 & 1 \\ 0 & 1 & 0 \\ 0 & 2 & 1 \end{bmatrix} \Rightarrow \begin{bmatrix} 0 & 1 & 0 \\ 0 & 0 & 1 \\ 0 & 0 & 0 \end{bmatrix}$ and $\mathbf{x}_1 = \begin{bmatrix} 1 \\ 0 \\ 0 \end{bmatrix}$

$\lambda_2 = -2$: $\begin{bmatrix} -3 & 0 & 1 \\ 0 & 0 & 0 \\ 0 & 2 & 0 \end{bmatrix} \Rightarrow \begin{bmatrix} 1 & 0 & -\frac{1}{3} \\ 0 & 1 & 0 \\ 0 & 0 & 0 \end{bmatrix}$ and $\mathbf{x}_2 = \begin{bmatrix} 1 \\ 0 \\ 3 \end{bmatrix}$

21. Using a graphing utility or computer software program with $A = \begin{bmatrix} 3 & 0 & 0 & 0 \\ 0 & -1 & 0 & 0 \\ 0 & 0 & 2 & 5 \\ 0 & 0 & 3 & 0 \end{bmatrix}$ produces the eigenvalues $\{5 \ -3 \ 3 \ -1\}$.

So, $\lambda_1 = 5, \lambda_2 = -3, \lambda_3 = 3,$ and $\lambda_4 = -1.$

$\lambda_1 = 5$: $\begin{bmatrix} 2 & 0 & 0 & 0 \\ 0 & 6 & 0 & 0 \\ 0 & 0 & 3 & -5 \\ 0 & 0 & -3 & 5 \end{bmatrix} \Rightarrow \begin{bmatrix} 1 & 0 & 0 & 0 \\ 0 & 1 & 0 & 0 \\ 0 & 0 & 1 & -\frac{5}{3} \\ 0 & 0 & 0 & 0 \end{bmatrix}$ and $\mathbf{x}_1 = \begin{bmatrix} 0 \\ 0 \\ 5 \\ 3 \end{bmatrix}$

$\lambda_2 = -3$: $\begin{bmatrix} -6 & 0 & 0 & 0 \\ 0 & -2 & 0 & 0 \\ 0 & 0 & -5 & -5 \\ 0 & 0 & -3 & -3 \end{bmatrix} \Rightarrow \begin{bmatrix} 1 & 0 & 0 & 0 \\ 0 & 1 & 0 & 0 \\ 0 & 0 & 1 & 1 \\ 0 & 0 & 0 & 0 \end{bmatrix}$ and $\mathbf{x}_2 = \begin{bmatrix} 0 \\ 0 \\ -1 \\ 1 \end{bmatrix}$

$\lambda_3 = 3$: $\begin{bmatrix} 0 & 0 & 0 & 0 \\ 0 & 4 & 0 & 0 \\ 0 & 0 & 1 & -5 \\ 0 & 0 & -3 & 3 \end{bmatrix} \Rightarrow \begin{bmatrix} 0 & 1 & 0 & 0 \\ 0 & 0 & 1 & 0 \\ 0 & 0 & 0 & 1 \\ 0 & 0 & 0 & 0 \end{bmatrix}$ and $\mathbf{x}_3 = \begin{bmatrix} 1 \\ 0 \\ 0 \\ 0 \end{bmatrix}$

$\lambda_4 = -1$: $\begin{bmatrix} -4 & 0 & 0 & 0 \\ 0 & 0 & 0 & 0 \\ 0 & 0 & -3 & -5 \\ 0 & 0 & -3 & -1 \end{bmatrix} \Rightarrow \begin{bmatrix} 1 & 0 & 0 & 0 \\ 0 & 0 & 1 & 0 \\ 0 & 0 & 0 & 1 \\ 0 & 0 & 0 & 0 \end{bmatrix}$ and $\mathbf{x}_4 = \begin{bmatrix} 0 \\ 1 \\ 0 \\ 0 \end{bmatrix}$

23. Using a graphing utility or computer software program with $A = \begin{bmatrix} 1 & 0 & 1 & 0 \\ 0 & -2 & 0 & 0 \\ 0 & 0 & 2 & 1 \\ 0 & 0 & 3 & 0 \end{bmatrix}$ produces the eigenvalues $\{-2\ -1\ 1\ 3\}$.

So, $\lambda_1 = -2$, $\lambda_2 = -1$, $\lambda_3 = 1$, and $\lambda_4 = 3$.

$\lambda_1 = -2$: $\begin{bmatrix} -3 & 0 & -1 & 0 \\ 0 & 0 & 0 & 0 \\ 0 & 0 & -4 & -1 \\ 0 & 0 & -3 & -2 \end{bmatrix} \Rightarrow \begin{bmatrix} 1 & 0 & 0 & 0 \\ 0 & 0 & 1 & 0 \\ 0 & 0 & 0 & 1 \\ 0 & 0 & 0 & 0 \end{bmatrix}$ and $\mathbf{x}_1 = \begin{bmatrix} 0 \\ 1 \\ 0 \\ 0 \end{bmatrix}$

$\lambda_2 = -1$: $\begin{bmatrix} -2 & 0 & -1 & 0 \\ 0 & 1 & 0 & 0 \\ 0 & 0 & -3 & -1 \\ 0 & 0 & -3 & -1 \end{bmatrix} \Rightarrow \begin{bmatrix} 1 & 0 & 0 & -\frac{1}{6} \\ 0 & 1 & 0 & 0 \\ 0 & 0 & 1 & \frac{1}{3} \\ 0 & 0 & 0 & 0 \end{bmatrix}$ and $\mathbf{x}_2 = \begin{bmatrix} 1 \\ 0 \\ -2 \\ 6 \end{bmatrix}$

$\lambda_3 = 1$: $\begin{bmatrix} 0 & 0 & -1 & 0 \\ 0 & 3 & 0 & 0 \\ 0 & 0 & -1 & -1 \\ 0 & 0 & -3 & 1 \end{bmatrix} \Rightarrow \begin{bmatrix} 0 & 1 & 0 & 0 \\ 0 & 0 & 1 & 0 \\ 0 & 0 & 0 & 1 \\ 0 & 0 & 0 & 0 \end{bmatrix}$ and $\mathbf{x}_3 = \begin{bmatrix} 1 \\ 0 \\ 0 \\ 0 \end{bmatrix}$

$\lambda_4 = 3$: $\begin{bmatrix} 2 & 0 & -1 & 0 \\ 0 & 5 & 0 & 0 \\ 0 & 0 & 1 & -1 \\ 0 & 0 & -3 & 3 \end{bmatrix} \Rightarrow \begin{bmatrix} 1 & 0 & 0 & -\frac{1}{2} \\ 0 & 1 & 0 & 0 \\ 0 & 0 & 1 & -1 \\ 0 & 0 & 0 & 0 \end{bmatrix}$ and $\mathbf{x}_4 = \begin{bmatrix} 1 \\ 0 \\ 2 \\ 2 \end{bmatrix}$

25. (a) False. The statement should read "... any *nonzero* multiple ...". See paragraph following Example 1 on page 153.

(b) False. Eigenvalues are solutions to the characteristic equation $|\lambda I - A| = 0$.

Section 3.5 Applications of Determinants

1. The matrix of cofactors is

$\begin{bmatrix} 4 & -3 \\ -2 & 1 \end{bmatrix} = \begin{bmatrix} 4 & -3 \\ -2 & 1 \end{bmatrix}.$

So, the adjoint of A is

$\text{adj}(A) = \begin{bmatrix} 4 & -3 \\ -2 & 1 \end{bmatrix}^T = \begin{bmatrix} 4 & -2 \\ -3 & 1 \end{bmatrix}.$

Because $|A| = -2$, the inverse of A is

$A^{-1} = \frac{1}{|A|}\text{adj}(A) = -\frac{1}{2}\begin{bmatrix} 4 & -2 \\ -3 & 1 \end{bmatrix} = \begin{bmatrix} -2 & 1 \\ \frac{3}{2} & -\frac{1}{2} \end{bmatrix}.$

3. The matrix of cofactors is

$\begin{bmatrix} \begin{vmatrix} 2 & 6 \\ -4 & -12 \end{vmatrix} & -\begin{vmatrix} 0 & 6 \\ 0 & -12 \end{vmatrix} & \begin{vmatrix} 0 & 2 \\ 0 & -4 \end{vmatrix} \\ -\begin{vmatrix} 0 & 0 \\ -4 & -12 \end{vmatrix} & \begin{vmatrix} 1 & 0 \\ 0 & -12 \end{vmatrix} & -\begin{vmatrix} 1 & 0 \\ 0 & -4 \end{vmatrix} \\ \begin{vmatrix} 0 & 0 \\ 2 & 6 \end{vmatrix} & -\begin{vmatrix} 1 & 0 \\ 0 & 6 \end{vmatrix} & \begin{vmatrix} 1 & 0 \\ 0 & 2 \end{vmatrix} \end{bmatrix}$

$= \begin{bmatrix} 0 & 0 & 0 \\ 0 & -12 & 4 \\ 0 & -6 & 2 \end{bmatrix}.$

So, the adjoint of A is $\text{adj}(A) = \begin{bmatrix} 0 & 0 & 0 \\ 0 & -12 & -6 \\ 0 & 4 & 2 \end{bmatrix}.$

Because row 3 of A is a multiple of row 2, the determinant is zero, and A has no inverse.

5. The matrix of cofactors is

$$
\begin{bmatrix}
\begin{vmatrix} 4 & 3 \\ 1 & -1 \end{vmatrix} & -\begin{vmatrix} 2 & 3 \\ 0 & -1 \end{vmatrix} & \begin{vmatrix} 2 & 4 \\ 0 & 1 \end{vmatrix} \\[6pt]
-\begin{vmatrix} -5 & -7 \\ 1 & -1 \end{vmatrix} & \begin{vmatrix} -3 & -7 \\ 0 & -1 \end{vmatrix} & -\begin{vmatrix} -3 & -5 \\ 0 & 1 \end{vmatrix} \\[6pt]
\begin{vmatrix} -5 & -7 \\ 4 & 3 \end{vmatrix} & -\begin{vmatrix} -3 & -7 \\ 2 & 3 \end{vmatrix} & \begin{vmatrix} -3 & -5 \\ 2 & 4 \end{vmatrix}
\end{bmatrix}
=
\begin{bmatrix}
-7 & 2 & 2 \\
-12 & 3 & 3 \\
13 & -5 & -2
\end{bmatrix}.
$$

So, the adjoint is $\operatorname{adj}(A) = \begin{bmatrix} -7 & -12 & 13 \\ 2 & 3 & -5 \\ 2 & 3 & -2 \end{bmatrix}$. Because $|A| = -3$, the inverse of A is

$$
A^{-1} = \frac{1}{|A|}\operatorname{adj}(A) = -\frac{1}{3}\begin{bmatrix} -7 & -12 & 13 \\ 2 & 3 & -5 \\ 2 & 3 & -2 \end{bmatrix} = \begin{bmatrix} \dfrac{7}{3} & 4 & -\dfrac{13}{3} \\[6pt] -\dfrac{2}{3} & -1 & \dfrac{5}{3} \\[6pt] -\dfrac{2}{3} & -1 & \dfrac{2}{3} \end{bmatrix}.
$$

7. The matrix of cofactors is

$$
\begin{bmatrix}
\begin{vmatrix} -1 & 4 & 1 \\ 0 & 1 & 2 \\ 1 & 1 & 2 \end{vmatrix} & -\begin{vmatrix} 3 & 4 & 1 \\ 0 & 1 & 2 \\ -1 & 1 & 2 \end{vmatrix} & \begin{vmatrix} 3 & -1 & 1 \\ 0 & 0 & 2 \\ -1 & 1 & 2 \end{vmatrix} & -\begin{vmatrix} 3 & -1 & 4 \\ 0 & 0 & 1 \\ -1 & 1 & 1 \end{vmatrix} \\[10pt]
-\begin{vmatrix} 2 & 0 & 1 \\ 0 & 1 & 2 \\ 1 & 1 & 2 \end{vmatrix} & \begin{vmatrix} -1 & 0 & 1 \\ 0 & 1 & 2 \\ -1 & 1 & 2 \end{vmatrix} & -\begin{vmatrix} -1 & 2 & 1 \\ 0 & 0 & 2 \\ -1 & 1 & 2 \end{vmatrix} & \begin{vmatrix} -1 & 2 & 0 \\ 0 & 0 & 1 \\ -1 & 1 & 1 \end{vmatrix} \\[10pt]
\begin{vmatrix} 2 & 0 & 1 \\ -1 & 4 & 1 \\ 1 & 1 & 2 \end{vmatrix} & -\begin{vmatrix} -1 & 0 & 1 \\ 3 & 4 & 1 \\ -1 & 1 & 2 \end{vmatrix} & \begin{vmatrix} -1 & 2 & 1 \\ 3 & -1 & 1 \\ -1 & 1 & 2 \end{vmatrix} & -\begin{vmatrix} -1 & 2 & 0 \\ 3 & -1 & 4 \\ -1 & 1 & 1 \end{vmatrix} \\[10pt]
-\begin{vmatrix} 2 & 0 & 1 \\ -1 & 4 & 1 \\ 0 & 1 & 2 \end{vmatrix} & \begin{vmatrix} -1 & 0 & 1 \\ 3 & 4 & 1 \\ 0 & 1 & 2 \end{vmatrix} & -\begin{vmatrix} -1 & 2 & 1 \\ 3 & -1 & 1 \\ 0 & 0 & 2 \end{vmatrix} & \begin{vmatrix} -1 & 2 & 0 \\ 3 & -1 & 4 \\ 0 & 0 & 1 \end{vmatrix}
\end{bmatrix}
=
\begin{bmatrix}
7 & 7 & -4 & 2 \\
1 & 1 & 2 & -1 \\
9 & 0 & -9 & 9 \\
-13 & -4 & 10 & -5
\end{bmatrix}.
$$

So, the adjoint of A is $\operatorname{adj}(A) = \begin{bmatrix} 7 & 1 & 9 & -13 \\ 7 & 1 & 0 & -4 \\ -4 & 2 & -9 & 10 \\ 2 & -1 & 9 & -5 \end{bmatrix}$. Because $\det(A) = 9$, the inverse of A is

$$
A^{-1} = \frac{1}{|A|}\operatorname{adj}(A) = \begin{bmatrix} \dfrac{7}{9} & \dfrac{1}{9} & 1 & -\dfrac{13}{9} \\[6pt] \dfrac{7}{9} & \dfrac{1}{9} & 0 & -\dfrac{4}{9} \\[6pt] -\dfrac{4}{9} & \dfrac{2}{9} & -1 & \dfrac{10}{9} \\[6pt] \dfrac{2}{9} & -\dfrac{1}{9} & 1 & -\dfrac{5}{9} \end{bmatrix}.
$$

9. If all the entries of A are integers, then so are those of the adjoint of A.

Because $A^{-1} = \dfrac{1}{|A|} \operatorname{adj}(A)$, and $|A| = 1$, the entries of A^{-1} must be integers.

11. Because $\operatorname{adj}(A) = |A| A^{-1}$,

$$\left| \operatorname{adj}(A) \right| = \left| |A| A^{-1} \right| = |A|^n \left| A^{-1} \right| = |A|^n \dfrac{1}{|A|} = |A|^{n-1}.$$

13. $\left| \operatorname{adj}(A) \right| = \begin{vmatrix} -2 & 0 \\ -1 & 1 \end{vmatrix} = -2$

$|A| = \begin{vmatrix} 1 & 0 \\ 1 & -2 \end{vmatrix} = -2$

So, $\left| \operatorname{adj}(A) \right| = |A|$.

15. Because $\operatorname{adj}\left(A^{-1} \right) = \left| A^{-1} \right| A$ and

$\left(\operatorname{adj}(A) \right)^{-1} = \left(|A| A^{-1} \right)^{-1} = \dfrac{1}{|A|} A,$

you have $\operatorname{adj}\left(A^{-1} \right) = \left(\operatorname{adj}(A) \right)^{-1}.$

17. The coefficient matrix is

$A = \begin{bmatrix} 1 & 2 \\ -1 & 1 \end{bmatrix}$, and $|A| = 3.$

Because $|A| \neq 0$, you can use Cramer's Rule. Replace column one with the column of constants to obtain

$A_1 = \begin{bmatrix} 5 & 2 \\ 1 & 1 \end{bmatrix}$, $|A_1| = 3.$

Similarly, replace column two with the column of constants to obtain

$A_2 = \begin{bmatrix} 1 & 5 \\ -1 & 1 \end{bmatrix}$, $|A_2| = 6.$

Then solve for x_1 and x_2.

$x_1 = \dfrac{|A_1|}{|A|} = \dfrac{3}{3} = 1$

$x_2 = \dfrac{|A_2|}{|A|} = \dfrac{6}{3} = 2$

19. The coefficient matrix is

$A = \begin{bmatrix} 3 & 4 \\ 5 & 3 \end{bmatrix}$, and $|A| = -11.$

Because $|A| \neq 0$, you can use Cramer's Rule.

$A_1 = \begin{bmatrix} -2 & 4 \\ 4 & 3 \end{bmatrix}$, $|A_1| = -22$

$A_2 = \begin{bmatrix} 3 & -2 \\ 5 & 4 \end{bmatrix}$, $|A_2| = 22$

The solution is as follows.

$x_1 = \dfrac{|A_1|}{|A|} = \dfrac{-22}{-11} = 2$

$x_2 = \dfrac{|A_2|}{|A|} = \dfrac{22}{-11} = -2$

21. The coefficient matrix is

$A = \begin{bmatrix} 20 & 8 \\ 12 & -24 \end{bmatrix}$, and $|A| = -576.$

Because $|A| \neq 0$, you can use Cramer's Rule.

$A_1 = \begin{bmatrix} 11 & 8 \\ 21 & -24 \end{bmatrix}$, $|A_1| = -432$

$A_2 = \begin{bmatrix} 20 & 11 \\ 12 & 21 \end{bmatrix}$, $|A_2| = 288$

The solution is

$x_1 = \dfrac{|A_1|}{|A|} = \dfrac{-432}{-576} = \dfrac{3}{4}$

$x_2 = \dfrac{|A_2|}{|A|} = \dfrac{288}{-576} = -\dfrac{1}{2}.$

23. The coefficient matrix is

$A = \begin{bmatrix} -0.4 & 0.8 \\ 2 & -4 \end{bmatrix}$, and $|A| = 0.$

Because $|A| = 0$, Cramer's Rule cannot be applied. (The system does not have a solution.)

25. The coefficient matrix is

$A = \begin{bmatrix} 3 & 6 \\ 6 & 12 \end{bmatrix}$, and $|A| = 0.$

Because $|A| = 0$, Cramer's Rule cannot be applied. (The system has an infinite number of solutions.)

27. The coefficient matrix is

$$A = \begin{bmatrix} 4 & -1 & -1 \\ 2 & 2 & 3 \\ 5 & -2 & -2 \end{bmatrix}, \quad \text{and } |A| = 3.$$

Because $|A| \neq 0$, you can use Cramer's Rule.

$$A_1 = \begin{bmatrix} 1 & -1 & -1 \\ 10 & 2 & 3 \\ -1 & -2 & -2 \end{bmatrix}, \quad |A_1| = 3$$

$$A_2 = \begin{bmatrix} 4 & 1 & -1 \\ 2 & 10 & 3 \\ 5 & -1 & -2 \end{bmatrix}, \quad |A_2| = 3$$

$$A_3 = \begin{bmatrix} 4 & -1 & 1 \\ 2 & 2 & 10 \\ 5 & -2 & -1 \end{bmatrix}, \quad |A_3| = 6$$

The solution is

$$x_1 = \frac{|A_1|}{|A|} = \frac{3}{3} = 1$$

$$x_2 = \frac{|A_2|}{|A|} = \frac{3}{3} = 1$$

$$x_3 = \frac{|A_3|}{|A|} = \frac{6}{3} = 2.$$

29. The coefficient matrix is

$$A = \begin{bmatrix} 3 & 4 & 4 \\ 4 & -4 & 6 \\ 6 & -6 & 0 \end{bmatrix}, \quad \text{and } |A| = 252.$$

Because $|A| \neq 0$, you can use Cramer's Rule.

$$A_1 = \begin{bmatrix} 11 & 4 & 4 \\ 11 & -4 & 6 \\ 3 & -6 & 0 \end{bmatrix}, \quad |A_1| = 252$$

$$A_2 = \begin{bmatrix} 3 & 11 & 4 \\ 4 & 11 & 6 \\ 6 & 3 & 0 \end{bmatrix}, \quad |A_2| = 126$$

$$A_3 = \begin{bmatrix} 3 & 4 & 11 \\ 4 & -4 & 11 \\ 6 & -6 & 3 \end{bmatrix}, \quad |A_3| = 378$$

The solution is

$$x_1 = \frac{|A_1|}{|A|} = \frac{252}{252} = 1$$

$$x_2 = \frac{|A_2|}{|A|} = \frac{126}{252} = \frac{1}{2}$$

$$x_3 = \frac{|A_3|}{|A|} = \frac{378}{252} = \frac{3}{2}.$$

31. The coefficient matrix is

$$A = \begin{bmatrix} 3 & 3 & 5 \\ 3 & 5 & 9 \\ 5 & 9 & 17 \end{bmatrix}, \quad \text{and } |A| = 4.$$

Because $|A| \neq 0$, you can use Cramer's Rule.

$$A_1 = \begin{bmatrix} 1 & 3 & 5 \\ 2 & 5 & 9 \\ 4 & 9 & 17 \end{bmatrix}, \quad |A_1| = 0$$

$$A_2 = \begin{bmatrix} 3 & 1 & 5 \\ 3 & 2 & 9 \\ 5 & 4 & 17 \end{bmatrix}, \quad |A_2| = -2$$

$$A_3 = \begin{bmatrix} 3 & 3 & 1 \\ 3 & 5 & 2 \\ 5 & 9 & 4 \end{bmatrix}, \quad |A_3| = 2$$

The solution is

$$x_1 = \frac{|A_1|}{|A|} = \frac{0}{4} = 0$$

$$x_2 = \frac{|A_2|}{|A|} = -\frac{2}{4} = -\frac{1}{2}$$

$$x_3 = \frac{|A_3|}{|A|} = \frac{2}{4} = \frac{1}{2}.$$

33. The coefficient matrix is

$$A = \begin{bmatrix} -0.4 & 0.8 \\ 2 & -4 \end{bmatrix}.$$

Using a graphing utility or a computer software program, $|A| = 0$.

Cramer's rule does not apply because the coefficient matrix has a determinant of zero.

35. The coefficient matrix is

$$A = \begin{bmatrix} -\frac{1}{4} & \frac{3}{8} \\ \frac{3}{2} & \frac{3}{4} \end{bmatrix}.$$

Using a graphing utility or a computer software program, $|A| = -\frac{3}{4}$.

$$A_1 = \begin{bmatrix} -2 & \frac{3}{8} \\ -12 & \frac{3}{4} \end{bmatrix}$$

Using a graphing utility or a computer software program, $|A_1| = 3$.

So, $x_1 = |A_1| / |A| = 3 \div \left(-\frac{3}{4} \right) = -4.$

37. The coefficient matrix is

$$A = \begin{bmatrix} 4 & -1 & 1 \\ 2 & 2 & 3 \\ 5 & -2 & 6 \end{bmatrix}.$$

Using a graphing utility or a computer software program, $|A| = 55$.

$$A_1 = \begin{bmatrix} -5 & -1 & 1 \\ 10 & 2 & 3 \\ 1 & -2 & 6 \end{bmatrix}$$

Using a graphing utility or a computer software program, $|A_1| = -55$.

So, $x_1 = |A_1|/|A| = \frac{-55}{55} = -1$.

39. The coefficient matrix is

$$A = \begin{bmatrix} 3 & -2 & 1 \\ -4 & 1 & -3 \\ 1 & -5 & 1 \end{bmatrix}.$$

Using a graphing utility or a computer software program, $|A| = -25$.

$$A_1 = \begin{bmatrix} -29 & -2 & 1 \\ 37 & 1 & -3 \\ -24 & -5 & 1 \end{bmatrix}.$$

Using a graphing utility or a computer software program, $|A_1| = 175$.

So, $x_1 = |A_1|/|A| = \frac{175}{-25} = -7$.

41. The coefficient matrix is

$$A = \begin{bmatrix} 3 & -2 & 9 & 4 \\ -1 & 0 & -9 & -6 \\ 0 & 0 & 3 & 1 \\ 2 & 2 & 0 & 8 \end{bmatrix}.$$

Using a graphing utility or a computer software program, $|A| = 36$.

$$A_1 = \begin{bmatrix} 35 & -2 & 9 & 4 \\ -17 & 0 & -9 & -6 \\ 5 & 0 & 3 & 1 \\ -4 & 2 & 0 & 8 \end{bmatrix}$$

Using a graphing utility or a computer software program, $|A_1| = 180$.

So, $x_1 = |A_1|/|A| = \frac{180}{36} = 5$.

43. The coefficient matrix is

$$A = \begin{bmatrix} k & 1-k \\ 1-k & k \end{bmatrix}, \text{ and } |A| = k^2 - (1-k)^2 = 2k-1.$$

Replacing the ith column of A with the column of constants yields A_i.

$$A_1 = \begin{bmatrix} 1 & 1-k \\ 3 & k \end{bmatrix}, \quad |A_1| = 4k - 3$$

$$A_2 = \begin{bmatrix} k & 1 \\ 1-k & 3 \end{bmatrix}, \quad |A_2| = 4k - 1$$

The solution is

$$x = \frac{|A_1|}{|A|} = \frac{4k-3}{2k-1}$$

$$y = \frac{|A_2|}{|A|} = \frac{4k-1}{2k-1}.$$

Notice that when $k = \frac{1}{2}, |A| = 2k - 1 = 0$ and the system will be inconsistent.

45. Use the formula for area as follows.

$$\text{Area} = \pm \frac{1}{2} \begin{vmatrix} x_1 & y_1 & 1 \\ x_2 & y_2 & 1 \\ x_3 & y_3 & 1 \end{vmatrix}$$

Because

$$\begin{vmatrix} x_1 & y_1 & 1 \\ x_2 & y_2 & 1 \\ x_3 & y_3 & 1 \end{vmatrix} = \begin{vmatrix} 0 & 0 & 1 \\ 2 & 0 & 1 \\ 0 & 3 & 1 \end{vmatrix} = 6,$$

the area is $\frac{1}{2}(6) = 3$.

47. Use the formula for area as follows.

$$\text{Area} = \pm \frac{1}{2} \begin{vmatrix} x_1 & y_1 & 1 \\ x_2 & y_2 & 1 \\ x_3 & y_3 & 1 \end{vmatrix} = \pm \frac{1}{2} \begin{vmatrix} -1 & 2 & 1 \\ 2 & 2 & 1 \\ -2 & 4 & 1 \end{vmatrix} = \pm \frac{1}{2}(6) = 3$$

49. Use the fact that

$$\begin{vmatrix} x_1 & y_1 & 1 \\ x_2 & y_2 & 1 \\ x_3 & y_3 & 1 \end{vmatrix} = \begin{vmatrix} 1 & 2 & 1 \\ 3 & 4 & 1 \\ 5 & 6 & 1 \end{vmatrix} = 0$$

to determine that the three points are collinear.

51. Use the fact that

$$\begin{vmatrix} x_1 & y_1 & 1 \\ x_2 & y_2 & 1 \\ x_3 & y_3 & 1 \end{vmatrix} = \begin{vmatrix} -2 & 5 & 1 \\ 0 & -1 & 1 \\ 3 & -9 & 1 \end{vmatrix} = 2$$

to determine that the three points are not collinear.

53. Use the equation

$$\begin{vmatrix} x & y & 1 \\ x_1 & y_1 & 1 \\ x_2 & y_2 & 1 \end{vmatrix} = 0$$

to find the equation of the line. So,

$$\begin{vmatrix} x & y & 1 \\ 0 & 0 & 1 \\ 3 & 4 & 1 \end{vmatrix} = 3y - 4x = 0.$$

55. Find the equation as follows.

$$0 = \begin{vmatrix} x & y & 1 \\ x_1 & y_1 & 1 \\ x_2 & y_2 & 1 \end{vmatrix} = \begin{vmatrix} x & y & 1 \\ -2 & 3 & 1 \\ -2 & -4 & 1 \end{vmatrix} = 7x + 14$$

So, an equation for the line is $x = -2$.

59. Use the formula for volume as follows.

$$\text{Volume} = \pm\frac{1}{6}\begin{vmatrix} x_1 & y_1 & z_1 & 1 \\ x_2 & y_2 & z_2 & 1 \\ x_3 & y_3 & z_3 & 1 \\ x_4 & y_4 & z_4 & 1 \end{vmatrix} = \pm\frac{1}{6}\begin{vmatrix} 3 & -1 & 1 & 1 \\ 4 & -4 & 4 & 1 \\ 1 & 1 & 1 & 1 \\ 0 & 0 & 1 & 1 \end{vmatrix} = \pm\frac{1}{6}(-12) = 2$$

61. Use the fact that

$$\begin{vmatrix} x_1 & y_1 & z_1 & 1 \\ x_2 & y_2 & z_2 & 1 \\ x_3 & y_3 & z_3 & 1 \\ x_4 & y_4 & z_4 & 1 \end{vmatrix} = \begin{vmatrix} -4 & 1 & 0 & 1 \\ 0 & 1 & 2 & 1 \\ 4 & 3 & -1 & 1 \\ 0 & 0 & 1 & 1 \end{vmatrix} = 28$$

to determine that the four points are not coplanar.

63. Use the fact that

$$\begin{vmatrix} x_1 & y_1 & z_1 & 1 \\ x_2 & y_2 & z_2 & 1 \\ x_3 & y_3 & z_3 & 1 \\ x_4 & y_4 & z_4 & 1 \end{vmatrix} = \begin{vmatrix} 0 & 0 & -1 & 1 \\ 0 & -1 & 0 & 1 \\ 1 & 1 & 0 & 1 \\ 2 & 1 & 2 & 1 \end{vmatrix} = 0$$

to determine that the four points are coplanar.

65. Use the equation

$$\begin{vmatrix} x & y & z & 1 \\ x_1 & y_1 & z_1 & 1 \\ x_2 & y_2 & z_2 & 1 \\ x_3 & y_3 & z_3 & 1 \end{vmatrix} = 0$$

to find the equation of the plane. So,

$$\begin{vmatrix} x & y & z & 1 \\ 1 & -2 & 1 & 1 \\ -1 & -1 & 7 & 1 \\ 2 & -1 & 3 & 1 \end{vmatrix} = x\begin{vmatrix} -2 & 1 & 1 \\ -1 & 7 & 1 \\ -1 & 3 & 1 \end{vmatrix} - y\begin{vmatrix} 1 & 1 & 1 \\ -1 & 7 & 1 \\ 2 & 3 & 1 \end{vmatrix} + z\begin{vmatrix} 1 & -2 & 1 \\ -1 & -1 & 1 \\ 2 & -1 & 1 \end{vmatrix} - \begin{vmatrix} 1 & -2 & 1 \\ -1 & -1 & 7 \\ 2 & -1 & 3 \end{vmatrix} = 0, \text{ or } 4x - 10y + 3z = 27.$$

57. Use the formula for volume as follows.

$$\text{Volume} = \pm\frac{1}{6}\begin{vmatrix} x_1 & y_1 & z_1 & 1 \\ x_2 & y_2 & z_2 & 1 \\ x_3 & y_3 & z_3 & 1 \\ x_4 & y_4 & z_4 & 1 \end{vmatrix}$$

Because

$$\begin{vmatrix} x_1 & y_1 & z_1 & 1 \\ x_2 & y_2 & z_2 & 1 \\ x_3 & y_3 & z_3 & 1 \\ x_4 & y_4 & z_4 & 1 \end{vmatrix} = \begin{vmatrix} 1 & 0 & 0 & 1 \\ 0 & 1 & 0 & 1 \\ 0 & 0 & 1 & 1 \\ 1 & 1 & 1 & 1 \end{vmatrix} = -2,$$

the volume of the tetrahedron is $-\frac{1}{6}(-2) = \frac{1}{3}$.

67. Find the equation as follows.

$$0 = \begin{vmatrix} x & y & z & 1 \\ x_1 & y_1 & z_1 & 1 \\ x_2 & y_2 & z_2 & 1 \\ x_3 & y_3 & z_3 & 1 \end{vmatrix} = \begin{vmatrix} x & y & z & 1 \\ 0 & 0 & 0 & 1 \\ 1 & -1 & 0 & 1 \\ 0 & 1 & -1 & 1 \end{vmatrix} = x\begin{vmatrix} 0 & 0 & 1 \\ -1 & 0 & 1 \\ 1 & -1 & 1 \end{vmatrix} - y\begin{vmatrix} 0 & 0 & 1 \\ 1 & 0 & 1 \\ 0 & -1 & 1 \end{vmatrix} + z\begin{vmatrix} 0 & 0 & 1 \\ 1 & -1 & 1 \\ 0 & 1 & 1 \end{vmatrix} - \begin{vmatrix} 0 & 0 & 0 \\ 1 & -1 & 0 \\ 0 & 1 & -1 \end{vmatrix} = x + y + z = 0$$

69. The given use of Cramer's Rule to solve for y is not correct. The numerator and denominator have been reversed. The determinant of the coefficient matrix should be in the denominator.

71. Cramer's Rule was used correctly.

73. (a) $49a + 7b + c = 4380$

$64a + 8b + c = 4439$

$81a + 9b + c = 4524$

(b) The coefficient matrix is

$$A = \begin{bmatrix} 49 & 7 & 1 \\ 64 & 8 & 1 \\ 81 & 9 & 1 \end{bmatrix} \quad \text{and} \quad |A| = -2.$$

Also, $A_1 = \begin{bmatrix} 4380 & 7 & 1 \\ 4439 & 8 & 1 \\ 4524 & 9 & 1 \end{bmatrix}$ and $|A_1| = -26$,

$$A_2 = \begin{bmatrix} 49 & 4380 & 1 \\ 64 & 4439 & 1 \\ 81 & 4524 & 1 \end{bmatrix} \quad \text{and} \quad |A_2| = 272,$$

$$A_3 = \begin{bmatrix} 49 & 7 & 4380 \\ 64 & 8 & 4439 \\ 81 & 9 & 4524 \end{bmatrix} \quad \text{and} \quad |A_3| = -9390.$$

So, $a = \frac{-26}{-2} = 13, b = \frac{272}{-2} = -136,$ and $c = \frac{-9390}{-2} = 4695.$

(c)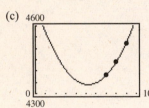

(d) The function fits the data exactly.

Review Exercises for Chapter 3

1. Using the formula for the determinant of a 2×2 matrix,

$$\begin{vmatrix} 4 & -1 \\ 2 & 2 \end{vmatrix} = 4(2) - 2(-1) = 10.$$

3. Using the formula for the determinant of a 2×2 matrix,

$$\begin{vmatrix} -3 & 1 \\ 6 & -2 \end{vmatrix} = (-3)(-2) - 6(1) = 0.$$

5. Expansion by cofactors along the first column produces

$$\begin{vmatrix} 1 & 4 & -2 \\ 0 & -3 & 1 \\ 1 & 1 & -1 \end{vmatrix} = 1\begin{vmatrix} -3 & 1 \\ 1 & -1 \end{vmatrix} - 0\begin{vmatrix} 4 & -2 \\ 1 & -1 \end{vmatrix} + 1\begin{vmatrix} 4 & -2 \\ -3 & 1 \end{vmatrix} = 1(2) + (-2) = 0.$$

7. The determinant of a diagonal matrix is the product of the entries along the main diagonal.

$$\begin{vmatrix} -2 & 0 & 0 \\ 0 & -3 & 0 \\ 0 & 0 & -1 \end{vmatrix} = (-2)(-3)(-1) = -6$$

9. Expansion by cofactors along the first column produces

$$\begin{vmatrix} -3 & 6 & 9 \\ 9 & 12 & -3 \\ 0 & 15 & -6 \end{vmatrix} = -3\begin{vmatrix} 12 & -3 \\ 15 & -6 \end{vmatrix} - 9\begin{vmatrix} 6 & 9 \\ 15 & -6 \end{vmatrix} + 0\begin{vmatrix} 6 & 9 \\ 12 & -3 \end{vmatrix} = 81 + 1539 + 0 = 1620.$$

11. Expansion by cofactors along the second column produces

$$\begin{vmatrix} 2 & 0 & -1 & 4 \\ -1 & 2 & 0 & 3 \\ 3 & 0 & 1 & 2 \\ -2 & 0 & 3 & 1 \end{vmatrix} = 2\begin{vmatrix} 2 & -1 & 4 \\ 3 & 1 & 2 \\ -2 & 3 & 1 \end{vmatrix} = 2\begin{vmatrix} 5 & 0 & 6 \\ 3 & 1 & 2 \\ -11 & 0 & -5 \end{vmatrix} = 2\begin{vmatrix} 5 & 6 \\ -11 & -5 \end{vmatrix} = 2(-25 + 66) = 82.$$

13.

$$\begin{vmatrix} -4 & 1 & 2 & 3 \\ 1 & -2 & 1 & 2 \\ 2 & -1 & 3 & 4 \\ 1 & 2 & 2 & -1 \end{vmatrix} = -\begin{vmatrix} 1 & 2 & 2 & -1 \\ 1 & -2 & 1 & 2 \\ 2 & -1 & 3 & 4 \\ -4 & 1 & 2 & 3 \end{vmatrix}$$

$$= -\begin{vmatrix} 1 & 2 & 2 & -1 \\ 0 & -4 & -1 & 3 \\ 0 & -5 & -1 & 6 \\ 0 & 9 & 10 & -1 \end{vmatrix}$$

$$= -\begin{vmatrix} 1 & 2 & 2 & -1 \\ 0 & 1 & 0 & -3 \\ 0 & -5 & -1 & 6 \\ 0 & 9 & 10 & -1 \end{vmatrix}$$

$$= -\begin{vmatrix} 1 & 2 & 2 & -1 \\ 0 & 1 & 0 & -3 \\ 0 & 0 & -1 & -9 \\ 0 & 0 & 10 & 26 \end{vmatrix}$$

$$= -\begin{vmatrix} 1 & 2 & 2 & -1 \\ 0 & 1 & 0 & -3 \\ 0 & 0 & -1 & -9 \\ 0 & 0 & 0 & -64 \end{vmatrix}$$

$$= -64$$

15.

$$\begin{vmatrix} -1 & 1 & -1 & 0 & 0 \\ 0 & 1 & -1 & 0 & 1 \\ 1 & 0 & 1 & -1 & 0 \\ 0 & -1 & 0 & 1 & -1 \\ 0 & 1 & 1 & -1 & 1 \end{vmatrix} = \begin{vmatrix} -1 & 1 & -1 & 0 & 0 \\ 0 & 1 & -1 & 0 & 1 \\ 0 & 1 & 0 & -1 & 0 \\ 0 & -1 & 0 & 1 & -1 \\ 0 & 1 & 1 & -1 & 1 \end{vmatrix}$$

$$= (-1)\begin{vmatrix} 1 & -1 & 0 & 1 \\ 1 & 0 & -1 & 0 \\ -1 & 0 & 1 & -1 \\ 1 & 1 & -1 & 1 \end{vmatrix}$$

$$= (-1)\begin{vmatrix} 1 & -1 & 0 & 1 \\ 1 & 0 & -1 & 0 \\ -1 & 0 & 1 & -1 \\ 2 & 0 & -1 & 2 \end{vmatrix}$$

$$= (-1)\begin{vmatrix} 1 & -1 & 0 \\ -1 & 1 & -1 \\ 2 & -1 & 2 \end{vmatrix}$$

$$= (-1)(1) = -1$$

17. The determinant of a diagonal matrix is the product of its main diagonal entries. So,

$$\begin{vmatrix} -1 & 0 & 0 & 0 & 0 \\ 0 & -1 & 0 & 0 & 0 \\ 0 & 0 & -1 & 0 & 0 \\ 0 & 0 & 0 & -1 & 0 \\ 0 & 0 & 0 & 0 & -1 \end{vmatrix} = (-1)^5 = -1.$$

19. Because the second row is a multiple of the first row, the determinant is zero.

21. Because -4 has been factored out of the second column, and 3 factored out of the third column, the first determinant is -12 times the second one.

23. (a) $|A| = \begin{vmatrix} -1 & 2 \\ 0 & 1 \end{vmatrix} = -1$

  (b) $|B| = \begin{vmatrix} 3 & 4 \\ 2 & 1 \end{vmatrix} = -5$

  (c) $AB = \begin{bmatrix} -1 & 2 \\ 0 & 1 \end{bmatrix}\begin{bmatrix} 3 & 4 \\ 2 & 1 \end{bmatrix} = \begin{bmatrix} 1 & -2 \\ 2 & 1 \end{bmatrix}$

  (d) $|AB| = \begin{vmatrix} 1 & -2 \\ 2 & 1 \end{vmatrix} = 5$

  Notice that $|A||B| = |AB| = 5$.

31. $A^{-1} = \begin{bmatrix} \dfrac{12}{5} & -\dfrac{3}{5} & -\dfrac{1}{10} \\ -\dfrac{4}{5} & \dfrac{1}{5} & \dfrac{1}{5} \\ -\dfrac{7}{5} & \dfrac{3}{5} & \dfrac{1}{10} \end{bmatrix}$

$$|A^{-1}| = \begin{vmatrix} \dfrac{12}{5} & -\dfrac{3}{5} & -\dfrac{1}{10} \\ -\dfrac{4}{5} & \dfrac{1}{5} & \dfrac{1}{5} \\ -\dfrac{7}{5} & \dfrac{3}{5} & \dfrac{1}{10} \end{vmatrix} = \begin{vmatrix} 1 & 0 & 0 \\ -\dfrac{4}{5} & \dfrac{1}{5} & \dfrac{1}{5} \\ -\dfrac{7}{5} & \dfrac{3}{5} & \dfrac{1}{10} \end{vmatrix} = -\dfrac{1}{10}$$

Notice that $|A| = \begin{vmatrix} 1 & 0 & 1 \\ 2 & -1 & 4 \\ 2 & 6 & 0 \end{vmatrix} = 1(0 - 24) + 1(12 + 2) = -10$, so $|A^{-1}| = \dfrac{1}{|A|} = -\dfrac{1}{10}$.

33. (a) $\begin{bmatrix} 3 & 3 & 5 & 1 \\ 3 & 5 & 9 & 2 \\ 5 & 9 & 17 & 4 \end{bmatrix} \Rightarrow \begin{bmatrix} 1 & 1 & \dfrac{5}{3} & \dfrac{1}{3} \\ 3 & 5 & 9 & 2 \\ 5 & 9 & 17 & 4 \end{bmatrix} \Rightarrow \begin{bmatrix} 1 & 1 & \dfrac{5}{3} & \dfrac{1}{3} \\ 0 & 2 & 4 & 1 \\ 0 & 4 & \dfrac{26}{3} & \dfrac{7}{3} \end{bmatrix} \Rightarrow \begin{bmatrix} 1 & 1 & \dfrac{5}{3} & \dfrac{1}{3} \\ 0 & 1 & 2 & \dfrac{1}{2} \\ 0 & 0 & 1 & \dfrac{1}{2} \end{bmatrix}$

So, $x_3 = \dfrac{1}{2}, x_2 = \dfrac{1}{2} - 2\left(\dfrac{1}{2}\right) = -\dfrac{1}{2}$, and $x_1 = \dfrac{1}{3} - \dfrac{5}{3}\left(\dfrac{1}{2}\right) - 1\left(-\dfrac{1}{2}\right) = 0.$

25. First find

$$|A| = \begin{vmatrix} -2 & 6 \\ 1 & 3 \end{vmatrix} = -12.$$

  (a) $|A^T| = |A| = -12$

  (b) $|A^3| = |A|^3 = (-12)^3 = -1728$

  (c) $|A^T A| = |A^T||A| = -12(-12) = 144$

  (d) $|5A| = 5^2|A| = 25(-12) = -300$

27. (a) $|A| = \begin{vmatrix} 1 & 0 & -4 \\ 0 & 3 & 2 \\ -2 & 7 & 6 \end{vmatrix} = \begin{vmatrix} 1 & 0 & -4 \\ 0 & 3 & 2 \\ 0 & 7 & -2 \end{vmatrix} = \begin{vmatrix} 3 & 2 \\ 7 & -2 \end{vmatrix} = -20$

  (b) $|A^{-1}| = \dfrac{1}{|A|} = -\dfrac{1}{20}$

29. $A^{-1} = \dfrac{1}{6}\begin{bmatrix} 4 & 1 \\ -2 & 1 \end{bmatrix} = \begin{bmatrix} \dfrac{2}{3} & \dfrac{1}{6} \\ -\dfrac{1}{3} & \dfrac{1}{6} \end{bmatrix}$

$$|A^{-1}| = \dfrac{2}{3}\left(\dfrac{1}{6}\right) - \left(-\dfrac{1}{3}\right)\left(\dfrac{1}{6}\right) = \dfrac{1}{9} + \dfrac{1}{18} = \dfrac{1}{6}$$

Notice that $|A| = 6$, so $|A^{-1}| = \dfrac{1}{|A|} = \dfrac{1}{6}$.

(b) $\begin{bmatrix} 1 & 1 & \frac{5}{3} & \frac{1}{3} \\ 0 & 1 & 2 & \frac{1}{2} \\ 0 & 0 & 1 & \frac{1}{2} \end{bmatrix} \Rightarrow \begin{bmatrix} 1 & 0 & -\frac{1}{3} & -\frac{1}{6} \\ 0 & 1 & 2 & \frac{1}{2} \\ 0 & 0 & 1 & \frac{1}{2} \end{bmatrix} \Rightarrow \begin{bmatrix} 1 & 0 & 0 & 0 \\ 0 & 1 & 2 & \frac{1}{2} \\ 0 & 0 & 1 & \frac{1}{2} \end{bmatrix} \Rightarrow \begin{bmatrix} 1 & 0 & 0 & 0 \\ 0 & 1 & 0 & -\frac{1}{2} \\ 0 & 0 & 1 & \frac{1}{2} \end{bmatrix}$

So, $x_1 = 0$, $x_2 = -\frac{1}{2}$, and $x_3 = \frac{1}{2}$.

(c) The coefficient matrix is

$A = \begin{bmatrix} 3 & 3 & 5 \\ 3 & 5 & 9 \\ 5 & 9 & 17 \end{bmatrix}$ and $|A| = 4$.

Also, $A_1 = \begin{bmatrix} 1 & 3 & 5 \\ 2 & 5 & 9 \\ 4 & 9 & 17 \end{bmatrix}$ and $|A_1| = 0$,

$A_2 = \begin{bmatrix} 3 & 1 & 5 \\ 3 & 2 & 9 \\ 5 & 4 & 17 \end{bmatrix}$ and $|A_2| = -2$,

$A_3 = \begin{bmatrix} 3 & 3 & 1 \\ 3 & 5 & 2 \\ 5 & 9 & 4 \end{bmatrix}$ and $|A_3| = 2$.

So, $x_1 = \frac{0}{4} = 0$, $x_2 = \frac{-2}{4} = -\frac{1}{2}$, and $x_3 = \frac{2}{4} = \frac{1}{2}$.

35. (a) $\begin{bmatrix} 1 & 2 & -1 & -7 \\ 2 & -2 & -2 & -8 \\ -1 & 3 & 4 & 8 \end{bmatrix} \Rightarrow \begin{bmatrix} 1 & 2 & -1 & -7 \\ 0 & -6 & 0 & 6 \\ 0 & 5 & 3 & 1 \end{bmatrix} \Rightarrow \begin{bmatrix} 1 & 2 & -1 & -7 \\ 0 & 1 & 0 & -1 \\ 0 & 5 & 3 & 1 \end{bmatrix} \Rightarrow \begin{bmatrix} 1 & 2 & -1 & -7 \\ 0 & 1 & 0 & -1 \\ 0 & 0 & 1 & 2 \end{bmatrix}$

So, $x_3 = 2$, $x_2 = -1$, and $x_1 = -7 + 1(2) - 2(-1) = -3$.

(b) $\begin{bmatrix} 1 & 2 & -1 & -7 \\ 0 & 1 & 0 & -1 \\ 0 & 0 & 1 & 2 \end{bmatrix} \Rightarrow \begin{bmatrix} 1 & 0 & -1 & -5 \\ 0 & 1 & 0 & -1 \\ 0 & 0 & 1 & 2 \end{bmatrix} \Rightarrow \begin{bmatrix} 1 & 0 & 0 & -3 \\ 0 & 1 & 0 & -1 \\ 0 & 0 & 1 & 2 \end{bmatrix}$

So, $x_1 = -3$, $x_2 = -1$, and $x_3 = 2$.

(c) The coefficient matrix is

$A = \begin{bmatrix} 1 & 2 & -1 \\ 2 & -2 & -2 \\ -1 & 3 & 4 \end{bmatrix}$ and $|A| = -18$.

Also, $A_1 = \begin{bmatrix} -7 & 2 & -1 \\ -8 & -2 & -2 \\ 8 & 3 & 4 \end{bmatrix}$ and $|A_1| = 54$,

$A_2 = \begin{bmatrix} 1 & -7 & -1 \\ 2 & -8 & -2 \\ -1 & 8 & 4 \end{bmatrix}$ and $|A_2| = 18$,

$A_3 = \begin{bmatrix} 1 & 2 & -7 \\ 2 & -2 & -8 \\ -1 & 3 & 8 \end{bmatrix}$ and $|A_3| = -36$.

So, $x_1 = \frac{54}{-18} = -3$, $x_2 = \frac{18}{-18} = -1$, and $x_3 = \frac{-36}{-18} = 2$.

37. Because the determinant of the coefficient matrix is

$$\begin{vmatrix} 5 & 4 \\ -1 & 1 \end{vmatrix} = 9 \neq 0,$$

the system has a unique solution.

39. Because the determinant of the coefficient matrix is

$$\begin{vmatrix} -1 & 1 & 2 \\ 2 & 3 & 1 \\ 5 & 4 & 2 \end{vmatrix} = -15 \neq 0,$$

the system has a unique solution.

41. Because the determinant of the coefficient matrix is

$$\begin{vmatrix} 1 & 2 & 6 \\ 2 & 5 & 15 \\ 3 & 1 & 3 \end{vmatrix} = 0,$$

the system does not have a unique solution.

43. (a) False. See "Definitions of Minors and Cofactors of a Matrix," page 124.

(b) False. See Theorem 3.3, part 1, page 134.

(c) False. See Theorem 3.9, page 148.

45. Using the fact that $|cA| = c^n |A|$, where A is an $n \times n$ matrix, $|4A| = 4^3 |A| = 64(2) = 128$.

47. Expand the determinant on the left along the third row

$$\begin{vmatrix} a_{11} & a_{12} & a_{13} \\ a_{21} & a_{22} & a_{23} \\ (a_{31} + c_{31}) & (a_{32} + c_{32}) & (a_{33} + c_{33}) \end{vmatrix}$$

$$= (a_{31} + c_{31}) \begin{vmatrix} a_{12} & a_{13} \\ a_{22} & a_{23} \end{vmatrix} - (a_{32} + c_{32}) \begin{vmatrix} a_{11} & a_{13} \\ a_{21} & a_{23} \end{vmatrix} + (a_{33} + c_{33}) \begin{vmatrix} a_{11} & a_{12} \\ a_{21} & a_{22} \end{vmatrix}$$

$$= a_{31} \begin{vmatrix} a_{12} & a_{13} \\ a_{22} & a_{23} \end{vmatrix} + c_{31} \begin{vmatrix} a_{12} & a_{13} \\ a_{22} & a_{23} \end{vmatrix} - a_{32} \begin{vmatrix} a_{11} & a_{13} \\ a_{21} & a_{23} \end{vmatrix} - c_{32} \begin{vmatrix} a_{11} & a_{13} \\ a_{21} & a_{23} \end{vmatrix} + a_{33} \begin{vmatrix} a_{11} & a_{12} \\ a_{21} & a_{22} \end{vmatrix} + c_{33} \begin{vmatrix} a_{11} & a_{12} \\ a_{21} & a_{22} \end{vmatrix}.$$

The first, third and fifth terms in this sum correspond to the determinant

$$\begin{vmatrix} a_{11} & a_{12} & a_{13} \\ a_{21} & a_{22} & a_{23} \\ a_{31} & a_{32} & a_{33} \end{vmatrix}$$

expanded along the third row. Similarly, the second, fourth and sixth terms of the sum correspond to the determinant

$$\begin{vmatrix} a_{11} & a_{12} & a_{13} \\ a_{21} & a_{22} & a_{23} \\ c_{31} & c_{32} & c_{33} \end{vmatrix}$$

expanded along the third row.

49. Each row consists of $n - 1$ ones and one element equal to $1 - n$. The sum of these elements is then

$$(n - 1)1 + (1 - n) = 0.$$

In Section 3.3, Exercise 53, you showed that a matrix whose rows each add up to zero has a determinant of zero. So, the determinant of this matrix is zero.

51. $|\lambda I - A| = \begin{vmatrix} \lambda + 3 & -10 \\ -5 & \lambda - 2 \end{vmatrix} = \lambda^2 + \lambda - 56 = (\lambda + 8)(\lambda - 7)$

$$\lambda = -8: \begin{bmatrix} -5 & -10 \\ -5 & -10 \end{bmatrix} \Rightarrow \begin{bmatrix} 1 & 2 \\ 0 & 0 \end{bmatrix} \text{ and } \mathbf{x}_1 = \begin{bmatrix} -2 \\ 1 \end{bmatrix}$$

$$\lambda = 7: \begin{bmatrix} 10 & -10 \\ -5 & 5 \end{bmatrix} \Rightarrow \begin{bmatrix} 1 & -1 \\ 0 & 0 \end{bmatrix} \text{ and } \mathbf{x}_2 = \begin{bmatrix} 1 \\ 1 \end{bmatrix}$$

53. $|\lambda I - A| = \begin{vmatrix} \lambda - 1 & 0 & 0 \\ 2 & \lambda - 3 & 0 \\ 0 & 0 & \lambda - 4 \end{vmatrix} = (\lambda - 1)(\lambda - 3)(\lambda - 4)$

$\lambda = 1: \begin{bmatrix} 0 & 0 & 0 \\ 2 & -2 & 0 \\ 0 & 0 & 3 \end{bmatrix} \Rightarrow \begin{bmatrix} 1 & -1 & 0 \\ 0 & 0 & 1 \\ 0 & 0 & 0 \end{bmatrix}$ and $\mathbf{x}_1 = \begin{bmatrix} 1 \\ 1 \\ 0 \end{bmatrix}$

$\lambda = 3: \begin{bmatrix} 2 & 0 & 0 \\ 2 & 0 & 0 \\ 0 & 0 & -1 \end{bmatrix} \Rightarrow \begin{bmatrix} 1 & 0 & 0 \\ 0 & 0 & 1 \\ 0 & 0 & 0 \end{bmatrix}$ and $\mathbf{x}_2 = \begin{bmatrix} 0 \\ 1 \\ 0 \end{bmatrix}$

$\lambda = 4: \begin{bmatrix} 3 & 0 & 0 \\ 2 & 1 & 0 \\ 0 & 0 & 0 \end{bmatrix} \Rightarrow \begin{bmatrix} 1 & 0 & 0 \\ 0 & 1 & 0 \\ 0 & 0 & 0 \end{bmatrix}$ and $\mathbf{x}_3 = \begin{bmatrix} 0 \\ 0 \\ 1 \end{bmatrix}$

55. By definition of the Jacobian,

$$J(u, v) = \begin{vmatrix} \dfrac{\partial x}{\partial u} & \dfrac{\partial x}{\partial v} \\ \dfrac{\partial y}{\partial u} & \dfrac{\partial y}{\partial v} \end{vmatrix} = \begin{vmatrix} -\dfrac{1}{2} & \dfrac{1}{2} \\ \dfrac{1}{2} & \dfrac{1}{2} \end{vmatrix} = -\dfrac{1}{4} - \dfrac{1}{4} = -\dfrac{1}{2}.$$

57. $J(u, v, w) = \begin{vmatrix} \dfrac{1}{2} & \dfrac{1}{2} & 0 \\ \dfrac{1}{2} & -\dfrac{1}{2} & 0 \\ 2vw & 2uw & 2uv \end{vmatrix} = 2uv\left(-\dfrac{1}{4} - \dfrac{1}{4}\right) = -uv$

59. Row reduction is generally preferred for matrices with few zeros. For a matrix with many zeros, it is often easier to expand along a row or column having many zeros.

61. The matrix of cofactors is given by $\begin{bmatrix} 1 & 2 \\ -1 & 0 \end{bmatrix}$.

So, the adjoint is adj $\begin{bmatrix} 0 & 1 \\ -2 & 1 \end{bmatrix} = \begin{bmatrix} 1 & -1 \\ 2 & 0 \end{bmatrix}$.

63. The determinant of the coefficient matrix is

$\begin{vmatrix} 0.2 & -0.1 \\ 0.4 & -0.5 \end{vmatrix} = -0.06 \neq 0.$

So, the system has a unique solution. Form the matrices A_1 and A_2 and find their determinants.

$A_1 = \begin{bmatrix} 0.07 & -0.1 \\ -0.01 & -0.5 \end{bmatrix}, \quad |A_1| = -0.036$

$A_2 = \begin{bmatrix} 0.2 & 0.07 \\ 0.4 & -0.01 \end{bmatrix}, \quad |A_2| = -0.03$

So,

$x = \dfrac{|A_1|}{|A|} = \dfrac{-0.036}{-0.06} = 0.6$

$y = \dfrac{|A_2|}{|A|} = \dfrac{-0.03}{-0.06} = 0.5.$

65. The determinant of the coefficient matrix is

$\begin{vmatrix} 2 & 3 & 3 \\ 6 & 6 & 12 \\ 12 & 9 & -1 \end{vmatrix} = 168 \neq 0.$

So, the system has a unique solution. Using Cramer's Rule

$A_1 = \begin{bmatrix} 3 & 3 & 3 \\ 13 & 6 & 12 \\ 2 & 9 & -1 \end{bmatrix}, \quad |A_1| = 84$

$A_2 = \begin{bmatrix} 2 & 3 & 3 \\ 6 & 13 & 12 \\ 12 & 2 & -1 \end{bmatrix}, \quad |A_2| = -56$

$A_3 = \begin{bmatrix} 2 & 3 & 3 \\ 6 & 6 & 13 \\ 12 & 9 & 2 \end{bmatrix}, \quad |A_3| = 168.$

So,

$x_1 = \dfrac{|A_1|}{|A|} = \dfrac{84}{168} = \dfrac{1}{2}$

$x_2 = \dfrac{|A_2|}{|A|} = \dfrac{-56}{168} = -\dfrac{1}{3}$

$x_3 = \dfrac{|A_3|}{|A|} = \dfrac{168}{168} = 1.$

67. (a) $100a + 10b + c = 308.9$

$400a + 20b + c = 335.8$

$900a + 30b + c = 363.6$

(b) The coefficient matrix is

$$A = \begin{bmatrix} 100 & 10 & 1 \\ 400 & 20 & 1 \\ 900 & 30 & 1 \end{bmatrix} \quad \text{and} \quad |A| = -2000.$$

Also, $A_1 = \begin{bmatrix} 308.9 & 10 & 1 \\ 335.8 & 20 & 1 \\ 363.6 & 30 & 1 \end{bmatrix}$ and $|A_1| = -9,$

$$A_2 = \begin{bmatrix} 100 & 308.9 & 1 \\ 400 & 335.8 & 1 \\ 900 & 363.6 & 1 \end{bmatrix} \quad \text{and} \quad |A_2| = -5110,$$

$$A_3 = \begin{bmatrix} 100 & 10 & 308.9 \\ 400 & 20 & 335.8 \\ 900 & 30 & 363.6 \end{bmatrix} \quad \text{and} \quad |A_3| = -565,800.$$

So, $a = \dfrac{-9}{-2000} = 0.0045, b = \dfrac{-5110}{-2000} = 2.555,$ and $c = \dfrac{-565,800}{-2000} = 282.9.$

(c)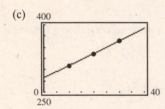

(d) The function fits the data exactly.

69. The formula for area yields

$$\text{Area} = \pm\frac{1}{2} \begin{vmatrix} x_1 & y_1 & 1 \\ x_2 & y_2 & 1 \\ x_3 & y_3 & 1 \end{vmatrix} = \pm\frac{1}{2} \begin{vmatrix} 1 & 0 & 1 \\ 5 & 0 & 1 \\ 5 & 8 & 1 \end{vmatrix}$$

$$= \pm\frac{1}{2}(-8)(1 - 5) = 16.$$

71. Use the equation

$$\begin{vmatrix} x & y & 1 \\ x_1 & y_1 & 1 \\ x_2 & y_2 & 1 \end{vmatrix} = 0$$

to find the equation of the line.

$$\begin{vmatrix} x & y & 1 \\ -4 & 0 & 1 \\ 4 & 4 & 1 \end{vmatrix} = -4x + 8y - 16 = 0, \text{ or } x - 2y = -4$$

73. The equation of the plane is given by the equation

$$\begin{vmatrix} x & y & z & 1 \\ 0 & 0 & 0 & 1 \\ 1 & 0 & 3 & 1 \\ 0 & 3 & 4 & 1 \end{vmatrix} = 0.$$

Expanding by cofactors along the second row yields

$$\begin{vmatrix} x & y & z \\ 1 & 0 & 3 \\ 0 & 3 & 4 \end{vmatrix} = 0$$

or, $9x + 4y - 3z = 0.$

75. (a) False. See Theorem 3.11, page 163.

(b) False. See "Test for Collinear Points in the *xy*-Plane," page 165.

Cumulative Test for Chapters 1–3

1. Interchange the first equation and the third equation.

$$x_1 + x_2 + x_3 = -3$$
$$2x_1 - 3x_2 + 2x_3 = 9$$
$$4x_1 + x_2 - 3x_3 = 11$$

Adding −2 times the first equation to the second equation produces a new second equation.

$$x_1 + x_2 + x_3 = -3$$
$$- 5x_2 = 15$$
$$4x_1 + x_2 - 3x_3 = 11$$

Adding −4 times the first equation to the third equation produces a new third equation.

$$x_1 + x_2 + x_3 = -3$$
$$- 5x_2 = 15$$
$$- 3x_2 - 7x_3 = 23$$

Dividing the second equation by −5 produces a new second equation.

$$x_1 + x_2 + x_3 = -3$$
$$x_2 = -3$$
$$- 3x_2 - 7x_3 = 23$$

Adding 3 times the second equation to the third equation produces a new third equation.

$$x_1 + x_2 + x_3 = -3$$
$$x_2 = -3$$
$$- 7x_3 = 14$$

Dividing the third equation by −7 produces a new third equation.

$$x_1 + x_2 + x_3 = -3$$
$$x_2 = -3$$
$$x_3 = -2$$

Using back-substitution, the answers are found to be $x_1 = 2$, $x_2 = -3$, and $x_3 = -2$.

2.
$$\begin{bmatrix} 0 & 1 & -1 & 0 & 2 \\ 1 & 0 & 2 & -1 & 0 \\ 1 & 2 & 0 & -1 & 4 \end{bmatrix} \Rightarrow \begin{bmatrix} 1 & 0 & 2 & -1 & 0 \\ 0 & 1 & -1 & 0 & 2 \\ 0 & 0 & 0 & 0 & 0 \end{bmatrix}$$

$$x_1 = s - 2t$$
$$x_2 = 2 + t$$
$$x_3 = t$$
$$x_4 = s$$

3.
$$\begin{bmatrix} 1 & 2 & 1 & -2 \\ 0 & 0 & 2 & -4 \\ -2 & -4 & 1 & -2 \end{bmatrix} \Rightarrow \begin{bmatrix} 1 & 2 & 0 & 0 \\ 0 & 0 & 1 & -2 \\ 0 & 0 & 0 & 0 \end{bmatrix}$$

$$x_1 = -2s$$
$$x_2 = s$$
$$x_3 = 2t$$
$$x_4 = t$$

4.
$$\begin{bmatrix} 1 & 2 & -1 & 3 \\ -1 & -1 & 1 & 2 \\ -1 & 1 & 1 & k \end{bmatrix} \Rightarrow \begin{bmatrix} 1 & 2 & -1 & 3 \\ 0 & 1 & 0 & 5 \\ 0 & 3 & 0 & 3+k \end{bmatrix} \Rightarrow \begin{bmatrix} 1 & 2 & -1 & 3 \\ 0 & 1 & 0 & 5 \\ 0 & 0 & 0 & -12+k \end{bmatrix}$$

$k = 12$ (for consistent system)

5. $BA = \begin{bmatrix} 12.50 & 9.00 & 21.50 \end{bmatrix} \begin{bmatrix} 200 & 300 \\ 600 & 350 \\ 250 & 400 \end{bmatrix} = \begin{bmatrix} 13,275.00 & 15,500.00 \end{bmatrix}$

This product represents the total value of the three products sent to the two warehouses.

6. $2A - B = \begin{bmatrix} -2 & 2 \\ 4 & 6 \end{bmatrix} - \begin{bmatrix} x & 2 \\ y & 5 \end{bmatrix} = \begin{bmatrix} 1 & 0 \\ 0 & 1 \end{bmatrix}$ $\Rightarrow$ $\begin{aligned} -2 - x &= 1 \\ 4 - y &= 0 \\ x &= -3 \\ y &= 4 \end{aligned}$

7. $A^T A = \begin{bmatrix} 17 & 22 & 27 \\ 22 & 29 & 36 \\ 27 & 36 & 45 \end{bmatrix}$

8. (a) $\begin{bmatrix} -2 & 3 \\ 4 & 6 \end{bmatrix}^{-1} = -\frac{1}{24}\begin{bmatrix} 6 & -3 \\ -4 & -2 \end{bmatrix}\begin{bmatrix} -\frac{1}{4} & \frac{1}{8} \\ \frac{1}{6} & \frac{1}{12} \end{bmatrix}$

(b) $\begin{bmatrix} -2 & 3 \\ 3 & 6 \end{bmatrix}^{-1} = -\frac{1}{21}\begin{bmatrix} 6 & -3 \\ -3 & -2 \end{bmatrix} = \begin{bmatrix} -\frac{2}{7} & \frac{1}{7} \\ \frac{1}{7} & \frac{2}{21} \end{bmatrix}$

9. $\begin{bmatrix} 1 & 1 & 0 \\ -3 & 6 & 5 \\ 0 & 1 & 0 \end{bmatrix}^{-1} = \begin{bmatrix} 1 & 0 & -1 \\ 0 & 0 & 1 \\ \frac{3}{5} & \frac{1}{5} & -\frac{9}{5} \end{bmatrix}$

10. $\begin{bmatrix} 2 & -4 \\ 1 & 0 \end{bmatrix} \Rightarrow \begin{bmatrix} 1 & 0 \\ 2 & -4 \end{bmatrix} \Rightarrow \begin{bmatrix} 1 & 0 \\ 0 & -4 \end{bmatrix} \Rightarrow \begin{bmatrix} 1 & 0 \\ 0 & 1 \end{bmatrix}$

$\begin{bmatrix} 1 & 0 \\ 0 & -\frac{1}{4} \end{bmatrix}\begin{bmatrix} 1 & 0 \\ -2 & 1 \end{bmatrix}\begin{bmatrix} 0 & 1 \\ 1 & 0 \end{bmatrix}\begin{bmatrix} 2 & -4 \\ 1 & 0 \end{bmatrix} = \begin{bmatrix} 1 & 0 \\ 0 & 1 \end{bmatrix}$

$A = \begin{bmatrix} 0 & 1 \\ 1 & 0 \end{bmatrix}\begin{bmatrix} 1 & 0 \\ 2 & 1 \end{bmatrix}\begin{bmatrix} 1 & 0 \\ 0 & -4 \end{bmatrix}$.

(The answer is not unique.)

11. Because the fourth row already has two zeros, choose it for cofactor expansion. An additional zero can be created by adding 4 times the first column to the fourth column.

$\begin{vmatrix} 5 & 1 & 2 & 24 \\ 1 & 0 & -2 & 1 \\ 1 & 1 & 6 & 5 \\ 1 & 0 & 0 & 0 \end{vmatrix} = 1(-1)^5\begin{vmatrix} 1 & 2 & 24 \\ 0 & -2 & 1 \\ 1 & 6 & 5 \end{vmatrix} = -\begin{vmatrix} 1 & 2 & 24 \\ 0 & -2 & 1 \\ 1 & 6 & 5 \end{vmatrix}$

Because the first column already has a zero, choose it for the next cofactor expansion. An additional zero can be created by adding -1 times the first row to the third row.

$-\begin{vmatrix} 1 & 2 & 24 \\ 0 & -2 & 1 \\ 0 & 4 & -19 \end{vmatrix} = -(1)(-1)^2\begin{vmatrix} -2 & 1 \\ 4 & -19 \end{vmatrix} = -(38 - 4) = -34$

12. (a) $|A| = 14$

(b) $|B| = -10$

(c) $|AB| = -140$

(d) $|A^{-1}| = \frac{1}{|A|} = \frac{1}{14}$

13. (a) $|3A| = 3^4 \cdot 7 = 567$

(b) $|A^T| = |A| = 7$

(c) $|A^{-1}| = \frac{1}{7}$

(d) $|A^3| = 7^3 = 343$

14. The matrix of cofactors is

$\begin{bmatrix} \begin{vmatrix} -2 & 1 \\ 0 & 2 \end{vmatrix} & -\begin{vmatrix} 0 & 1 \\ 1 & 2 \end{vmatrix} & \begin{vmatrix} 0 & -2 \\ 1 & 0 \end{vmatrix} \\ -\begin{vmatrix} -5 & -1 \\ 0 & 2 \end{vmatrix} & \begin{vmatrix} 1 & -1 \\ 1 & 2 \end{vmatrix} & -\begin{vmatrix} 1 & -5 \\ 1 & 0 \end{vmatrix} \\ \begin{vmatrix} -5 & -1 \\ -2 & 1 \end{vmatrix} & -\begin{vmatrix} 1 & -1 \\ 0 & 1 \end{vmatrix} & \begin{vmatrix} 1 & -5 \\ 0 & -2 \end{vmatrix} \end{bmatrix} = \begin{bmatrix} -4 & 1 & 2 \\ 10 & 3 & -5 \\ -7 & -1 & -2 \end{bmatrix}$

So, the adjoint of A is $\begin{bmatrix} -4 & 10 & -7 \\ 1 & 3 & -1 \\ 2 & -5 & -2 \end{bmatrix}$.

Because $|A| = -11$, the inverse of A is

$A^{-1} = \frac{1}{|A|}\text{adj}(A) = -\frac{1}{11}\begin{bmatrix} -4 & 10 & -7 \\ 1 & 3 & -1 \\ 2 & -5 & -2 \end{bmatrix} = \begin{bmatrix} \frac{4}{11} & -\frac{10}{11} & \frac{7}{11} \\ -\frac{1}{11} & -\frac{3}{11} & \frac{1}{11} \\ -\frac{2}{11} & \frac{5}{11} & \frac{2}{11} \end{bmatrix}$.

15. $a\begin{bmatrix} 1 \\ 0 \\ 1 \end{bmatrix} + b\begin{bmatrix} 1 \\ 1 \\ 0 \end{bmatrix} + c\begin{bmatrix} 0 \\ 1 \\ 1 \end{bmatrix} = \begin{bmatrix} 1 \\ 2 \\ 3 \end{bmatrix}$

The solution of this system is $a = 1$, $b = 0$, and $c = 2$.

(The answer is not unique.)

16. $\begin{aligned} a - b + c &= 2 \\ c &= 1 \\ 4a + 2b + c &= 6 \end{aligned}$

The solution of this system is $a = \frac{7}{6}$, $b = \frac{1}{6}$, and

$c = 1$, so $y = \frac{7}{6}x^2 + \frac{1}{6}x + 1$.

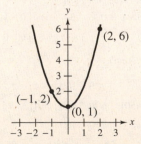

17. Find the equation using

$$0 = \begin{vmatrix} x & y & 1 \\ x_1 & y_1 & 1 \\ x_2 & y_2 & 1 \end{vmatrix} = \begin{vmatrix} x & y & 1 \\ 1 & 4 & 1 \\ 5 & -2 & 1 \end{vmatrix} = x(6) - y(-4) + 1(-22) = 6x + 4y - 22.$$

An equation of the line is $6x + 4y - 22 = 0$, or $3x + 2y = 11$.

18. Use the formula for area.

$$\text{Area} = \pm\frac{1}{2}\begin{vmatrix} x_1 & y_1 & 1 \\ x_2 & y_2 & 1 \\ x_3 & y_3 & 1 \end{vmatrix} = \pm\frac{1}{2}\begin{vmatrix} 3 & 1 & 1 \\ 7 & 1 & 1 \\ 7 & 9 & 1 \end{vmatrix} = \pm\frac{1}{2}\begin{vmatrix} 3 & 1 & 1 \\ 4 & 0 & 0 \\ 7 & 9 & 1 \end{vmatrix} = \frac{1}{2}(-4)(-8) = 16$$

19. $\left|\lambda I - A\right| = \begin{vmatrix} \lambda - 1 & -4 & -6 \\ -1 & \lambda - 2 & -2 \\ 1 & 2 & \lambda + 4 \end{vmatrix} = (\lambda - 1)(\lambda^2 + 2\lambda - 8 + 4) + 1(-4\lambda - 16 + 12) + 1(8 + 6\lambda - 12)$

$$= \lambda^3 + \lambda^2 - 6\lambda + 4 - 4\lambda - 4 + 6\lambda - 4$$
$$= \lambda^3 + \lambda^2 - 4\lambda - 4$$

Because $\lambda^3 + \lambda^2 - 4\lambda - 4 = 0$

$$\lambda^2(\lambda + 1) - 4(\lambda + 1) = 0$$
$$(\lambda^2 - 4)(\lambda + 1) = 0,$$

the eigenvalues are $\lambda_1 = -2, \lambda_2 = -1$, and $\lambda_3 = 2$.

For $\lambda_1 = -2$: $\begin{bmatrix} -3 & -4 & -6 \\ -1 & -4 & -2 \\ 1 & 2 & 2 \end{bmatrix} \Rightarrow \begin{bmatrix} 1 & 0 & 2 \\ 0 & 1 & 0 \\ 0 & 0 & 0 \end{bmatrix}$ and $\mathbf{x}_1 = \begin{bmatrix} -2 \\ 0 \\ 1 \end{bmatrix}$

For $\lambda_2 = -1$: $\begin{bmatrix} -2 & -4 & -6 \\ -1 & -3 & -2 \\ 1 & 2 & 3 \end{bmatrix} \Rightarrow \begin{bmatrix} 1 & 0 & 5 \\ 0 & 1 & -1 \\ 0 & 0 & 0 \end{bmatrix}$ and $\mathbf{x}_2 = \begin{bmatrix} -5 \\ 1 \\ 1 \end{bmatrix}$

For $\lambda_3 = 2$: $\begin{bmatrix} 1 & -4 & -6 \\ -1 & 0 & -2 \\ 1 & 2 & 6 \end{bmatrix} \Rightarrow \begin{bmatrix} 1 & 0 & 2 \\ 0 & 1 & 2 \\ 0 & 0 & 0 \end{bmatrix}$ and $\mathbf{x}_3 = \begin{bmatrix} -2 \\ -2 \\ 1 \end{bmatrix}$

20. No. C could be singular. $\underbrace{\begin{bmatrix} 0 & 0 \\ 0 & 1 \end{bmatrix}}_{A}\underbrace{\begin{bmatrix} 0 & 1 \\ 0 & 0 \end{bmatrix}}_{C} = \underbrace{\begin{bmatrix} 0 & 1 \\ 0 & 0 \end{bmatrix}}_{B}\underbrace{\begin{bmatrix} 0 & 1 \\ 0 & 0 \end{bmatrix}}_{C}$

21. $\left(B^T B\right)^T = B^T \left(B^T\right)^T = B^T B$

22. Let B, C be inverses of A. Then

$$B = (CA)B = C(AB) = C.$$

23. (a) A is row equivalent to B if there exist elementary matrices $E_1, \cdots, E_k$ such that $A = E_k \cdots E_1 B$.

(b) A row equivalent to $B \Rightarrow A = E_k \cdots E_1 B$ $(E_1, \cdots, E_k$ elementary$)$.

B row equivalent to $C \Rightarrow B = F_l \cdots F_1 C$ $(F_1, \cdots, F_l$ elementary$)$.

Then $A = E_k \cdots E_1(F_l \cdots F_1)C \Rightarrow A$ row equivalent to C.

C H A P T E R 4
Vector Spaces

CHAPTER 4
Vector Spaces

Section 4.1   Vectors in R^n

1. $\mathbf{v} = (4, 5)$

3.

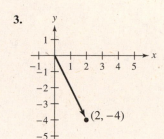

5.

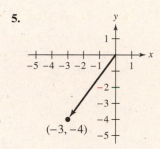

7. $\mathbf{u} + \mathbf{v} = (1, 3) + (2, -2)$

$= (1 + 2, 3 - 2)$

$= (3, 1)$

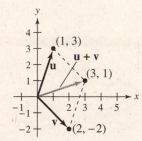

9. $\mathbf{u} + \mathbf{v} = (2, -3) + (-3, -1)$

$= (2 - 3, -3 - 1)$

$= (-1, -4)$

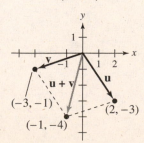

11. $\mathbf{v} = \frac{3}{2}\mathbf{u} = \frac{3}{2}(-2, 3) = \left(-3, \frac{9}{2}\right)$

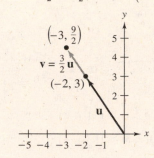

13. $\mathbf{v} = \mathbf{u} + 2\mathbf{w}$

$= (-2, 3) + 2(-3, -2)$

$= (-2, 3) + (-6, -4)$

$= (-2 - 6, 3 - 4)$

$= (-8, -1)$

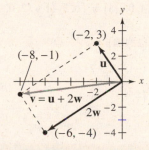

15. $\mathbf{v} = \frac{1}{2}(3\mathbf{u} + \mathbf{w})$

$= \frac{1}{2}(3(-2, 3) + (-3, -2))$

$= \frac{1}{2}((-6, 9) + (-3, -2))$

$= \frac{1}{2}(-9, 7) = \left(-\frac{9}{2}, \frac{7}{2}\right)$

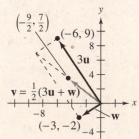

17. (a) $2\mathbf{v} = 2(2, 1) = \big(2(2), 2(1)\big) = (4, 2)$

(b) $-3\mathbf{v} = -3(2, 1) = \big(-3(2), -3(1)\big) = (-6, -3)$

(c) $\frac{1}{2}\mathbf{v} = \frac{1}{2}(2, 1) = \big(\frac{1}{2}(2), \frac{1}{2}(1)\big) = \big(1, \frac{1}{2}\big)$

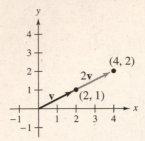 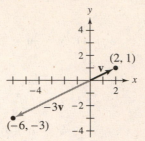 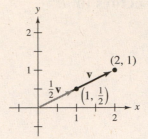

19. $\mathbf{u} - \mathbf{v} = (1, 2, 3) - (2, 2, -1) = (-1, 0, 4)$

$\mathbf{v} - \mathbf{u} = (2, 2, -1) - (1, 2, 3) = (1, 0, -4)$

21. $2\mathbf{u} + 4\mathbf{v} - \mathbf{w} = 2(1, 2, 3) + 4(2, 2, -1) - (4, 0, -4)$

$= (2, 4, 6) + (8, 8, -4) - (4, 0, -4)$

$= \big(2 + 8 - 4, 4 + 8 - 0, 6 + (-4) - (-4)\big) = (6, 12, 6)$

23. $2\mathbf{z} - 3\mathbf{u} = \mathbf{w}$ implies that $2\mathbf{z} = 3\mathbf{u} + \mathbf{w}$, or $\mathbf{z} = \frac{3}{2}\mathbf{u} + \frac{1}{2}\mathbf{w}$.

So, $\mathbf{z} = \frac{3}{2}(1, 2, 3) + \frac{1}{2}(4, 0, -4) = \big(\frac{3}{2}, 3, \frac{9}{2}\big) + (2, 0, -2) = \big(\frac{7}{2}, 3, \frac{5}{2}\big)$.

25. (a) $2\mathbf{v} = 2(1, 2, 2) = (2, 4, 4)$

(b) $-\mathbf{v} = -(1, 2, 2) = (-1, -2, -2)$

(c) $\frac{1}{2}\mathbf{v} = \frac{1}{2}(1, 2, 2) = \big(\frac{1}{2}, 1, 1\big)$

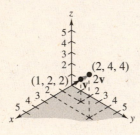 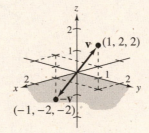 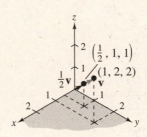

27. (a) Because $(-6, -4, 10) = -2(3, 2, -5)$, $\mathbf{u}$ is a scalar multiple of $\mathbf{z}$.

(b) Because $\big(2, \frac{4}{3}, -\frac{10}{3}\big) = \frac{2}{3}(3, 2, -5)$, $\mathbf{v}$ is a scalar multiple of $\mathbf{z}$.

(c) Because $(6, 4, 10) \neq c(3, 2, -5)$ for any c, $\mathbf{w}$ is *not* a scalar multiple of $\mathbf{z}$.

29. (a) $\mathbf{u} - \mathbf{v} = (4, 0, -3, 5) - (0, 2, 5, 4) = (4 - 0, 0 - 2, -3 - 5, 5 - 4) = (4, -2, -8, 1)$

(b) $2(\mathbf{u} + 3\mathbf{v}) = 2\big[(4, 0, -3, 5) + 3(0, 2, 5, 4)\big]$

$= 2\big[(4, 0, -3, 5) + (0, 6, 15, 12)\big]$

$= 2(4 + 0, 0 + 6, -3 + 15, 5 + 12) = 2(4, 6, 12, 17) = (8, 12, 24, 34)$

(c) $2\mathbf{v} - \mathbf{u} = 2(0, 2, 5, 4) - (4, 0, -3, 5) = (0, 4, 10, 8) - (4, 0, -3, 5) = (-4, 4, 13, 3)$

31. (a) $\mathbf{u} - \mathbf{v} = (-7, 0, 0, 0, 9) - (2, -3, -2, 3, 3)$

$$= (-9, 3, 2, -3, 6)$$

(b) $2(\mathbf{u} + 3\mathbf{v}) = 2\big[(-7, 0, 0, 0, 9) + 3(2, -3, -2, 3, 3)\big]$

$$= 2\big[(-7, 0, 0, 0, 9) + (6, -9, -6, 9, 9)\big]$$

$$= 2(-1, -9, -6, 9, 18)$$

$$= (-2, -18, -12, 18, 36)$$

(c) $2\mathbf{v} - \mathbf{u} = 2(2, -3, -2, 3, 3) - (-7, 0, 0, 0, 9)$

$$= (4, -6, -4, 6, 6) - (-7, 0, 0, 0, 9)$$

$$= (11, -6, -4, 6, -3)$$

33. Using a graphing utility with $\mathbf{u} = (1, 2, -3, 1)$,

$\mathbf{v} = (0, 2, -1, -2)$ and $\mathbf{w} = (2, -2, 1, 3)$ you have

(a) $\mathbf{u} + 2\mathbf{v} = (1, 6, -5, -3)$

(b) $\mathbf{w} - 3\mathbf{u} = (-1, -8, 10, 0)$

(c) $4\mathbf{v} + \frac{1}{2}\mathbf{u} - \mathbf{w} = (-1.5, 11, -6.5, -10.5)$

(d) $\frac{1}{4}(3\mathbf{u} + 2\mathbf{v} - \mathbf{w}) = (0.25, 3, -3, -1)$

35. $2\mathbf{w} = \mathbf{u} - 3\mathbf{v}$

$\mathbf{w} = \frac{1}{2}\mathbf{u} - \frac{3}{2}\mathbf{v}$

$$= \frac{1}{2}(1, -1, 0, 1) - \frac{3}{2}(0, 2, 3, -1)$$

$$= \left(\frac{1}{2}, -\frac{1}{2}, 0, \frac{1}{2}\right) - \left(0, 3, \frac{9}{2}, -\frac{3}{2}\right)$$

$$= \left(\frac{1}{2} - 0, -\frac{1}{2} - 3, 0 - \frac{9}{2}, \frac{1}{2} - \left(-\frac{3}{2}\right)\right)$$

$$= \left(\frac{1}{2}, -\frac{7}{2}, -\frac{9}{2}, 2\right)$$

37. $\frac{1}{2}\mathbf{w} = 2\mathbf{u} + 3\mathbf{v}$

$\mathbf{w} = 4\mathbf{u} + 6\mathbf{v}$

$$= 4(1, -1, 0, 1) + 6(0, 2, 3, -1)$$

$$= (4, -4, 0, 4) + (0, 12, 18, -6)$$

$$= (4, 8, 18, -2)$$

39. The equation

$$a\mathbf{u} + b\mathbf{w} = \mathbf{v}$$

$a(1, 2) + b(1, -1) = (2, 1)$

yields the system

$a + b = 2$

$2a - b = 1.$

Solving this system produces $a = 1$ and $b = 1$.

So, $\mathbf{v} = \mathbf{u} + \mathbf{w}.$

41. The equation

$$a\mathbf{u} + b\mathbf{w} = \mathbf{v}$$

$a(1, 2) + b(1, -1) = (3, 0)$

yields the system

$a + b = 3$

$2a - b = 0.$

Solving this system produces $a = 1$ and $b = 2$.

So, $\mathbf{v} = \mathbf{u} + 2\mathbf{w}.$

43. The equation

$$a\mathbf{u} + b\mathbf{w} = \mathbf{v}$$

$a(1, 2) + b(1, -1) = (-1, -2)$

yields the system

$a + b = -1$

$2a - b = -2.$

Solving this system produces $a = -1$ and $b = 0$.

So, $\mathbf{v} = -\mathbf{u}.$

45. $2\mathbf{u} + \mathbf{v} - 3\mathbf{w} = \mathbf{0}$

$\mathbf{w} = \frac{2}{3}\mathbf{u} + \frac{1}{3}\mathbf{v}$

$$= \frac{2}{3}(0, 2, 7, 5) + \frac{1}{3}(-3, 1, 4, -8)$$

$$= \left(0, \frac{4}{3}, \frac{14}{3}, \frac{10}{3}\right) + \left(-1, \frac{1}{3}, \frac{4}{3}, -\frac{8}{3}\right)$$

$$= \left(0 + (-1), \frac{4}{3} + \frac{1}{3}, \frac{14}{3} + \frac{4}{3}, \frac{10}{3} + \left(-\frac{8}{3}\right)\right)$$

$$= \left(-1, \frac{5}{3}, 6, \frac{2}{3}\right)$$

47. The equation

$$a\mathbf{u}_1 + b\mathbf{u}_2 + c\mathbf{u}_3 = \mathbf{v}$$

$a(2, 3, 5) + b(1, 2, 4) + c(-2, 2, 3) = (10, 1, 4)$

yields the system

$2a + b - 2c = 10$

$3a + 2b + 2c = 1$

$5a + 4b + 3c = 4.$

Solving this system produces $a = 1$, $b = 2$, and

$c = -3$. So, $\mathbf{v} = \mathbf{u}_1 + 2\mathbf{u}_2 - 3\mathbf{u}_3.$

49. The equation

$$a\mathbf{u}_1 + b\mathbf{u}_2 + c\mathbf{u}_3 = \mathbf{v}$$

$a(1, 1, 2, 2) + b(2, 3, 5, 6) + c(-3, 1, -4, 2) = (0, 5, 3, 0)$

yields the system

$a + 2b - 3c = 0$

$a + 3b + c = 5$

$2a + 5b - 4c = 3$

$2a + 6b + 2c = 0.$

The second and fourth equations cannot both be true. So, the system has no solution. It is not possible to write $\mathbf{v}$ as a linear combination of $\mathbf{u}_1$, $\mathbf{u}_2$, and $\mathbf{u}_3$.

51. Write a matrix using the given $\mathbf{u}_1, \mathbf{u}_2, ..., \mathbf{u}_5$ as columns and augment this matrix with $\mathbf{v}$ as a column.

$$A = \begin{bmatrix} 1 & 1 & 0 & 2 & 0 & 5 \\ 2 & 2 & 1 & 1 & 2 & 3 \\ -3 & 0 & 1 & -1 & 2 & -11 \\ 4 & 2 & 1 & 2 & -1 & 11 \\ -1 & 1 & -4 & 1 & -1 & 9 \end{bmatrix}$$

The reduced row-echelon form for A is

$$A = \begin{bmatrix} 1 & 0 & 0 & 0 & 0 & 2 \\ 0 & 1 & 0 & 0 & 0 & 1 \\ 0 & 0 & 1 & 0 & 0 & -2 \\ 0 & 0 & 0 & 1 & 0 & 1 \\ 0 & 0 & 0 & 0 & 1 & -1 \end{bmatrix}.$$

So, $\mathbf{v} = 2\mathbf{u}_1 + \mathbf{u}_2 - 2\mathbf{u}_3 + \mathbf{u}_4 - \mathbf{u}_5$. Verify the solution by showing that

$$2(1, 2, -3, 4, -1) + (1, 2, 0, 2, 1) - 2(0, 1, 1, 1, -4) + (2, 1, -1, 2, 1) - (0, 2, 2, -1, -1) \text{ equals } (5, 3, -11, 11, 9).$$

53. Write a matrix using the given $\mathbf{u}_1, \mathbf{u}_2, ..., \mathbf{u}_6$ as columns and augment this matrix with $\mathbf{v}$ as a column.

$$A = \begin{bmatrix} 1 & 1 & 0 & 1 & 1 & 3 & 10 \\ 2 & -2 & 2 & 0 & -2 & 2 & 30 \\ -3 & 1 & -1 & 3 & 1 & 1 & -13 \\ 4 & -1 & 2 & -4 & -1 & -2 & 14 \\ -1 & 2 & -1 & 1 & 2 & 3 & -7 \\ 2 & 1 & -1 & 2 & -3 & 0 & 27 \end{bmatrix}$$

The reduced row-echelon form for A is

$$A = \begin{bmatrix} 1 & 0 & 0 & 0 & 0 & 0 & 5 \\ 0 & 1 & 0 & 0 & 0 & 0 & -1 \\ 0 & 0 & 1 & 0 & 0 & 0 & 1 \\ 0 & 0 & 0 & 1 & 0 & 0 & 2 \\ 0 & 0 & 0 & 0 & 1 & 0 & -5 \\ 0 & 0 & 0 & 0 & 0 & 1 & 3 \end{bmatrix}.$$

So, $\mathbf{v} = 5\mathbf{u}_1 - \mathbf{u}_2 + \mathbf{u}_3 + 2\mathbf{u}_4 - 5\mathbf{u}_5 + 3\mathbf{u}_6$. Verify the solution by showing that

$$5(1, 2, -3, 4, -1, 2) - (1, -2, 1, -1, 2, 1) + (0, 2, -1, 2, -1, -1) + 2(1, 0, 3, -4, 1, 2) - 5(1, -2, 1, -1, 2, 3) + 3(3, 2, 1, -2, 3, 0).$$

equals $(10, 30, -13, 14, -7, 27)$.

55. (a) True. See the discussion before "Definition of Vector Addition and Scalar Multiplication in R^n," page 183.

(b) False. The vector $c\mathbf{v}$ is $|c|$ times as long as $\mathbf{v}$ and has the same direction as $\mathbf{v}$ if c is positive and the opposite direction as $\mathbf{v}$ if c is negative.

57. The equation

$$a\mathbf{v}_1 + b\mathbf{v}_2 + c\mathbf{v}_3 = \mathbf{0}$$
$$a(1, 0, 1) + b(-1, 1, 2) + c(0, 1, 4) = (0, 0, 0)$$

yields the homogeneous system

$$\begin{aligned} a - b & = 0 \\ b + c &= 0 \\ a + 2b + 4c &= 0. \end{aligned}$$

This system has only the trivial solution $a = b = c = 0$. So, you cannot find a nontrivial way of writing $\mathbf{0}$ as a combination of $\mathbf{v}_1, \mathbf{v}_2,$ and $\mathbf{v}_3$.

59. (1) $\mathbf{u} + \mathbf{v} = (2, -1, 3, 6) + (1, 4, 0, 1) = (3, 3, 3, 7)$ is a vector in R^4.

(2) $\mathbf{u} + \mathbf{v} = (2, -1, 3, 6) + (1, 4, 0, 1) = (3, 3, 3, 7)$ and

$\mathbf{v} + \mathbf{u} = (1, 4, 0, 1) + (2, -1, 3, 6) = (3, 3, 3, 7)$

So, $\mathbf{u} + \mathbf{v} = \mathbf{v} + \mathbf{u}$.

(3) $(\mathbf{u} + \mathbf{v}) + \mathbf{w} = [(2, -1, 3, 6) + (1, 4, 0, 1)] + (3, 0, 2, 0) = (3, 3, 3, 7) + (3, 0, 2, 0) = (6, 3, 5, 7)$

$\mathbf{u} + (\mathbf{v} + \mathbf{w}) = (2, -1, 3, 6) + [(1, 4, 0, 1) + (3, 0, 2, 0)] = (2, -1, 3, 6) + (4, 4, 2, 1) = (6, 3, 5, 7)$

So, $(\mathbf{u} + \mathbf{v}) + \mathbf{w} = \mathbf{u} + (\mathbf{v} + \mathbf{w})$.

(4) $\mathbf{u} + \mathbf{0} = (2, -1, 3, 6) + (0, 0, 0, 0) = (2, -1, 3, 6) = \mathbf{u}$

(5) $\mathbf{u} + (-\mathbf{u}) = (2, -1, 3, 6) + (-2, 1, -3, -6) = (0, 0, 0, 0) = \mathbf{0}$

(6) $c\mathbf{u} = 5(2, -1, 3, 6) = (10, -5, 15, 30)$ is a vector in R^4.

(7) $c(\mathbf{u} + \mathbf{v}) = 5[(2, -1, 3, 6) + (1, 4, 0, 1)] = 5(3, 3, 3, 7) = (15, 15, 15, 35)$

$c\mathbf{u} + c\mathbf{v} = 5(2, -1, 3, 6) + 5(1, 4, 0, 1) = (10, -5, 15, 30) + (5, 20, 0, 5) = (15, 15, 15, 35)$

So, $c(\mathbf{u} + \mathbf{v}) = c\mathbf{u} + c\mathbf{v}$.

(8) $(c + d)\mathbf{u} = (5 + (-2))(2, -1, 3, 6) = 3(2, -1, 3, 6) = (6, -3, 9, 18)$

$c\mathbf{u} + d\mathbf{u} = 5(2, -1, 3, 6) + (-2)(2, -1, 3, 6) = (10, -5, 15, 30) + (-4, 2, -6, -12) = (6, -3, 9, 18)$

So, $(c + d)\mathbf{u} = c\mathbf{u} + d\mathbf{u}$.

(9) $c(d\mathbf{u}) = 5((-2)(2, -1, 3, 6)) = 5(-4, 2, -6, -12) = (-20, 10, -30, -60)$

$(cd)\mathbf{u} = (5(-2))(2, -1, 3, 6) = -10(2, -1, 3, 6) = (-20, 10, -30, -60)$

So, $c(d\mathbf{u}) = (cd)\mathbf{u}$.

(10) $1(\mathbf{u}) = 1(2, -1, 3, 6) = (2, -1, 3, 6) = \mathbf{u}$

61. Prove the remaining eight properties.

(1) $\mathbf{u} + \mathbf{v} = (u_1, u_2) + (v_1, v_2) = (u_1 + v_1, u_2 + v_2)$ is a vector in the plane.

(2) $\mathbf{u} + \mathbf{v} = (u_1, u_2) + (v_1, v_2) = (u_1 + v_1, u_2 + v_2) = (v_1 + u_1, v_2 + u_2) = (v_1, v_2) + (u_1, u_2) = \mathbf{v} + \mathbf{u}$

(4) $\mathbf{u} + \mathbf{0} = (u_1, u_2) + (0, 0) = (u_1 + 0, u_2 + 0) = (u_1, u_2) = \mathbf{u}$

(5) $\mathbf{u} + (-\mathbf{u}) = (u_1, u_2) + (-u_1, -u_2) = (u_1 - u_1, u_2 - u_2) = (0, 0) = \mathbf{0}$

(6) $c\mathbf{u} = c(u_1, u_2) = (cu_1, cu_2)$ is a vector in the plane.

(7) $c(\mathbf{u} + \mathbf{v}) = c[(u_1, u_2) + (v_1, v_2)] = c(u_1 + v_1, u_2 + v_2)$

$= (c(u_1 + v_1), c(u_2 + v_2)) = (cu_1 + cv_1, cu_2 + cv_2)$

$= (cu_1, cu_2) + (cv_1, cv_2)$

$= c(u_1, u_2) + c(v_1, v_2) = c\mathbf{u} + c\mathbf{v}$

(9) $c(d\mathbf{u}) = c(d(u_1, u_2)) = c(du_1, du_2) = (cdu_1, cdu_2) = (cd)(u_1, u_2) = (cd)\mathbf{u}$

(10) $1(\mathbf{u}) = 1(u_1, u_2) = (u_1, u_2) = \mathbf{u}$

63. (a) Add $-\mathbf{v}$ to both sides

(b) Associative property and Additive identity

(c) Additive inverse

(d) Commutative property

(e) Additive identity

65. (a) Additive identity

(b) Distributive property

(c) Add $-c\mathbf{0}$ to both sides

(d) Additive inverse and Associative property

(e) Additive inverse

(f) Additive identity

67. (a) Additive inverse

(b) Transitive property

(c) Add **v** to both sides

(d) Associative property

(e) Additive inverse

(f) Additive identity

69. $\begin{bmatrix} 1 & 2 & 3 \\ 7 & 8 & 9 \\ 4 & 5 & 7 \end{bmatrix} \Rightarrow \begin{bmatrix} 1 & 0 & 0 \\ 0 & 1 & 0 \\ 0 & 0 & 1 \end{bmatrix}$

No.

Section 4.2 Vector Spaces

1. The additive identity of R^4 is the vector $(0, 0, 0, 0)$.

3. The additive identity of $M_{2,3}$ is the 2×3 zero matrix

$$\begin{bmatrix} 0 & 0 & 0 \\ 0 & 0 & 0 \end{bmatrix}.$$

5. P_3 is the set of all polynominals of degree less than or equal to 3. Its additive identity is

$$0x^3 + 0x^2 + 0x + 0 = 0.$$

7. In R^4, the additive inverse of (v_1, v_2, v_3, v_4) is

$$(-v_1, -v_2, -v_3, -v_4).$$

9. $M_{2,3}$ is the set of all 2×3 matrices. The additive inverse of

$$\begin{bmatrix} a_{11} & a_{12} & a_{13} \\ a_{21} & a_{22} & a_{23} \end{bmatrix}$$

is $-\begin{bmatrix} a_{11} & a_{12} & a_{13} \\ a_{21} & a_{22} & a_{23} \end{bmatrix} = \begin{bmatrix} -a_{11} & -a_{12} & -a_{13} \\ -a_{21} & -a_{22} & -a_{23} \end{bmatrix}.$

11. P_3 is the set of all polynominals of degree less than or equal to 3. The additive inverse of

$$a_3x^3 + a_2x^2 + a_1x + a_0 \text{ is}$$

$$-\left(a_3x^3 + a_2x^2 + a_1x + a_0\right) = -a_3x^3 - a_2x^2 - a_1x - a_0.$$

71. You can describe vector subtraction $\mathbf{u} - \mathbf{v}$ as follows

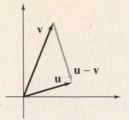

Or, write subtraction in terms of addition,

$$\mathbf{u} - \mathbf{v} = \mathbf{u} + (-1)\mathbf{v}.$$

13. $M_{4,6}$ with the standard operations is a vector space. All ten vector space axioms hold.

15. This set is *not* a vector space. Axiom 1 fails. The set is not closed under addition. For example,

$\left(-x^3 + 4x^2\right) + \left(x^3 + 2x\right) = 4x^2 + 2x$ is not a third-degree polynomial.

17. This set is *not* a vector space. Axiom 1 fails. For example, given $f(x) = x$ and $g(x) = -x$,

$$f(x) + g(x) = 0 \text{ is not of the form } ax + b \text{ where}$$

$a \neq 0$.

19. This set is *not* a vector space. The set is not closed under scalar multiplication. For example,

$(-1)(3, 2) = (-3, -2)$ is not in the set.

21. This set is a vector space. All ten vector space axioms hold.

23. This set is a vector space. All ten vector space axioms hold.

25. This set is *not* a vector space because it is not closed under addition. A counterexample is

$$\begin{bmatrix} 1 & 0 \\ 0 & 0 \end{bmatrix} + \begin{bmatrix} 0 & 0 \\ 0 & 1 \end{bmatrix} = \begin{bmatrix} 1 & 0 \\ 0 & 1 \end{bmatrix}.$$

Each matrix on the left is singular, while the sum is nonsingular.

27. This set is a vector space. All ten vector space axioms hold.

29. (a) Axiom 8 fails. For example,

$$(1 + 2)(1, 1) = 3(1, 1) = (3, 1) \quad \left(\text{Because } c(x, y) = (cx, y)\right)$$

$$1(1, 1) + 2(1, 1) = (1, 1) + (2, 1) = (3, 2).$$

So, R^2 is not a vector space with these operations.

(b) Axiom 2 fails. For example,

$$(1, 2) + (2, 1) = (1, 0)$$

$$(2, 1) + (1, 2) = (2, 0).$$

So, R^2 is not a vector space with these operations.

(c) Axiom 6 fails. For example, $(-1)(1, 1) = \left(\sqrt{-1}, \sqrt{-1}\right)$, which is not in R^2.

So, R^2 is not a vector space with these operations.

31. Verify the ten axioms in the definition of vector space.

(1) $\mathbf{u} + \mathbf{v} = \begin{bmatrix} u_1 & u_2 \\ u_3 & u_4 \end{bmatrix} + \begin{bmatrix} v_1 & v_2 \\ v_3 & v_4 \end{bmatrix} = \begin{bmatrix} u_1 + v_1 & u_2 + v_2 \\ u_3 + v_3 & u_4 + v_4 \end{bmatrix}$ is in $M_{2,2}$.

(2) $\mathbf{u} + \mathbf{v} = \begin{bmatrix} u_1 & u_2 \\ u_3 & u_4 \end{bmatrix} + \begin{bmatrix} v_1 & v_2 \\ v_3 & v_4 \end{bmatrix} = \begin{bmatrix} u_1 + v_1 & u_2 + v_2 \\ u_3 + v_3 & u_4 + v_4 \end{bmatrix}$

$$= \begin{bmatrix} v_1 + u_1 & v_2 + u_2 \\ v_3 + u_3 & v_4 + u_4 \end{bmatrix} = \begin{bmatrix} v_1 & v_2 \\ v_3 & v_4 \end{bmatrix} + \begin{bmatrix} u_1 & u_2 \\ u_3 & u_4 \end{bmatrix} = \mathbf{v} + \mathbf{u}$$

(3) $\mathbf{u} + (\mathbf{v} + \mathbf{w}) = \begin{bmatrix} u_1 & u_2 \\ u_3 & u_4 \end{bmatrix} + \left(\begin{bmatrix} v_1 & v_2 \\ v_3 & v_4 \end{bmatrix} + \begin{bmatrix} w_1 & w_2 \\ w_3 & w_4 \end{bmatrix} \right)$

$$= \begin{bmatrix} u_1 & u_2 \\ u_3 & u_4 \end{bmatrix} + \begin{bmatrix} v_1 + w_1 & v_2 + w_2 \\ v_3 + w_3 & v_4 + w_4 \end{bmatrix}$$

$$= \begin{bmatrix} u_1 + (v_1 + w_1) & u_2 + (v_2 + w_2) \\ u_3 + (v_3 + w_3) & u_4 + (v_4 + w_4) \end{bmatrix}$$

$$= \begin{bmatrix} (u_1 + v_1) + w_1 & (u_2 + v_2) + w_2 \\ (u_3 + v_3) + w_3 & (u_4 + v_4) + w_4 \end{bmatrix}$$

$$= \begin{bmatrix} u_1 + v_1 & u_2 + v_2 \\ u_3 + v_3 & u_4 + v_4 \end{bmatrix} + \begin{bmatrix} w_1 & w_2 \\ w_3 & w_4 \end{bmatrix}$$

$$= \left(\begin{bmatrix} u_1 & u_2 \\ u_3 & u_4 \end{bmatrix} + \begin{bmatrix} v_1 & v_2 \\ v_3 & v_4 \end{bmatrix} \right) + \begin{bmatrix} w_1 & w_2 \\ w_3 & w_4 \end{bmatrix} = (\mathbf{u} + \mathbf{v}) + \mathbf{w}$$

(4) The zero vector is

$$\mathbf{0} = \begin{bmatrix} 0 & 0 \\ 0 & 0 \end{bmatrix}. \text{ So,}$$

$$\mathbf{u} + \mathbf{0} = \begin{bmatrix} u_1 & u_2 \\ u_3 & u_4 \end{bmatrix} + \begin{bmatrix} 0 & 0 \\ 0 & 0 \end{bmatrix} = \begin{bmatrix} u_1 & u_2 \\ u_3 & u_4 \end{bmatrix} = \mathbf{u}.$$

(5) For every

$$\mathbf{u} = \begin{bmatrix} u_1 & u_2 \\ u_3 & u_4 \end{bmatrix}, \text{ you have } -\mathbf{u} = \begin{bmatrix} -u_1 & -u_2 \\ -u_3 & -u_4 \end{bmatrix}.$$

$$\mathbf{u} + (-\mathbf{u}) = \begin{bmatrix} u_1 & u_2 \\ u_3 & u_4 \end{bmatrix} + \begin{bmatrix} -u_1 & -u_2 \\ -u_3 & -u_4 \end{bmatrix}$$

$$= \begin{bmatrix} 0 & 0 \\ 0 & 0 \end{bmatrix}$$

$$= \mathbf{0}$$

(6) $c\mathbf{u} = c\begin{bmatrix} u_1 & u_2 \\ u_3 & u_4 \end{bmatrix} = \begin{bmatrix} cu_1 & cu_2 \\ cu_3 & cu_4 \end{bmatrix}$ is in $M_{2,2}$.

(7) $c(\mathbf{u} + \mathbf{v}) = c\left(\begin{bmatrix} u_1 & u_2 \\ u_3 & u_4 \end{bmatrix} + \begin{bmatrix} v_1 & v_2 \\ v_3 & v_4 \end{bmatrix}\right) = c\begin{bmatrix} u_1 + v_1 & u_2 + v_2 \\ u_3 + v_3 & u_4 + v_4 \end{bmatrix}$

$= \begin{bmatrix} c(u_1 + v_1) & c(u_2 + v_2) \\ c(u_3 + v_3) & c(u_4 + v_4) \end{bmatrix} = \begin{bmatrix} cu_1 + cv_1 & cu_2 + cv_2 \\ cu_3 + cv_3 & cu_4 + cv_4 \end{bmatrix}$

$= \begin{bmatrix} cu_1 & cu_2 \\ cu_3 & cu_4 \end{bmatrix} + \begin{bmatrix} cv_1 & cv_2 \\ cv_3 & cv_4 \end{bmatrix} = c\begin{bmatrix} u_1 & u_2 \\ u_3 & u_4 \end{bmatrix} + c\begin{bmatrix} v_1 & v_2 \\ v_3 & v_4 \end{bmatrix}$

$= c\mathbf{u} + c\mathbf{v}$

(8) $(c + d)\mathbf{u} = (c + d)\begin{bmatrix} u_1 & u_2 \\ u_3 & u_4 \end{bmatrix} = \begin{bmatrix} (c + d)u_1 & (c + d)u_2 \\ (c + d)u_3 & (c + d)u_4 \end{bmatrix}$

$= \begin{bmatrix} cu_1 + du_1 & cu_2 + du_2 \\ cu_3 + du_3 & cu_4 + du_4 \end{bmatrix} = \begin{bmatrix} cu_1 & cu_2 \\ cu_3 & cu_4 \end{bmatrix} + \begin{bmatrix} du_1 & du_2 \\ du_3 & du_4 \end{bmatrix}$

$= c\begin{bmatrix} u_1 & u_2 \\ u_3 & u_4 \end{bmatrix} + d\begin{bmatrix} u_1 & u_2 \\ u_3 & u_4 \end{bmatrix} = c\mathbf{u} + d\mathbf{u}$

(9) $c(d\mathbf{u}) = c\left(d\begin{bmatrix} u_1 & u_2 \\ u_3 & u_4 \end{bmatrix}\right) = c\begin{bmatrix} du_1 & du_2 \\ du_3 & du_4 \end{bmatrix} = \begin{bmatrix} c(du_1) & c(du_2) \\ c(du_3) & c(du_4) \end{bmatrix}$

$= \begin{bmatrix} (cd)u_1 & (cd)u_2 \\ (cd)u_3 & (cd)u_4 \end{bmatrix} = (cd)\begin{bmatrix} u_1 & u_2 \\ u_3 & u_4 \end{bmatrix} = (cd)\mathbf{u}$

(10) $1(\mathbf{u}) = 1\begin{bmatrix} u_1 & u_2 \\ u_3 & u_4 \end{bmatrix} = \begin{bmatrix} 1u_1 & 1u_2 \\ 1u_3 & 1u_4 \end{bmatrix} = \mathbf{u}$

33. This set is not a vector space because Axiom 5 fails. The additive identity is $(1, 1)$ and so $(0, 0)$ has no additive inverse. Axioms 7 and 8 also fail.

35. (a) True. See the first paragraph of the section, page 191.

(b) False. See Example 6, page 195.

(c) False. With standard operations on R^2, the additive inverse axiom is not satisfied.

37. (a) Add $-\mathbf{w}$ to both sides

(b) Associative property

(c) Additive inverse

(d) Additive identity.

39. $(-1)\mathbf{v} + 1(\mathbf{v}) = (-1 + 1)\mathbf{v} = 0\mathbf{v} = \mathbf{0}$. Also,

$-\mathbf{v} + \mathbf{v} = \mathbf{0}$. So, $(-1)\mathbf{v}$ and $-\mathbf{v}$ are both additive

inverses of $\mathbf{v}$. Because the additive inverse of a vector is

unique (Problem 41), $(-1)\mathbf{v} = -\mathbf{v}$.

41. Let $\mathbf{u}$ be an element of the vector space V. Then $-\mathbf{u}$ is the additive inverse of $\mathbf{u}$. Assume, to the contrary, that $\mathbf{v}$ is another additive inverse of $\mathbf{u}$. Then

$$\mathbf{u} + \mathbf{v} = \mathbf{0}$$
$$-\mathbf{u} + \mathbf{u} + \mathbf{v} = -\mathbf{u} + \mathbf{0}$$
$$\mathbf{0} + \mathbf{v} = -\mathbf{u} + \mathbf{0}$$
$$\mathbf{v} = -\mathbf{u}$$

Section 4.3 Subspaces of Vector Spaces

1. Because W is nonempty and $W \subset R^4$, you need only check that W is closed under addition and scalar multiplication. Given

$(x_1, x_2, x_3, 0) \in W$ and $(y_1, y_2, y_3, 0) \in W$,

it follows that

$(x_1, x_2, x_3, 0) + (y_1, y_2, y_3, 0) = (x_1 + y_1, x_2 + y_2, x_3 + y_3, 0) \in W$.

Furthermore, for any real number c and $(x_1, x_2, x_3, 0) \in W$, it follows that

$c(x_1, x_2, x_3, 0) = (cx_1, cx_2, cx_3, 0) \in W$.

3. Because W is nonempty and $W \subset M_{2,2}$, you need only check that W is closed under addition and scalar multiplication. Given

$$\begin{bmatrix} 0 & a_1 \\ b_1 & 0 \end{bmatrix} \in W \text{ and } \begin{bmatrix} 0 & a_2 \\ b_2 & 0 \end{bmatrix} \in W \text{ it follows that } \begin{bmatrix} 0 & a_1 \\ b_1 & 0 \end{bmatrix} + \begin{bmatrix} 0 & a_2 \\ b_2 & 0 \end{bmatrix} = \begin{bmatrix} 0 & a_1 + a_2 \\ b_1 + b_2 & 0 \end{bmatrix} \in W.$$

Furthermore, for any real number c and $\begin{bmatrix} 0 & a \\ b & o \end{bmatrix} \in W$ it follows that $c\begin{bmatrix} 0 & a \\ b & 0 \end{bmatrix} = \begin{bmatrix} 0 & ca \\ cb & 0 \end{bmatrix} \in W.$

5. Recall from calculus that continuity implies integrability, so $W \subset V$. Furthermore, because W is nonempty, you need only check that W is closed under addition and scalar multiplication. Given continuous functions $f, g \in W$, it follows that $f + g$ is continuous, so $f + g \in W$. Also, for any real number c and for a continuous function $f \in W$, cf is continuous. So, $cf \in W$.

7. The vectors in W are of the form $(a, b, -1)$. This set is not closed under addition or scalar multiplication. For example,

$$(0, 0, -1) + (0, 0, -1) = (0, 0, -2) \notin W$$

and

$$2(0, 0, -1) = (0, 0, -2) \notin W.$$

9. This set is not closed under scalar multiplication. For example,

$$\sqrt{2}, (1, 1) = \left(\sqrt{2}, \sqrt{2}\right) \notin W.$$

11. Consider $f(x) = e^x$, which is continuous and nonnegative. So, $f \in W$. The function $(-1)f = -f$ is negative. So, $-f \notin W$, and W is not closed under scalar multiplication.

13. This set is not closed under scalar multiplication. For example,

$$(-2)(1, 1, 1) = (-2, -2, -2) \notin W.$$

15. This set is not closed under addition. For example,

$$\begin{bmatrix} 1 & 0 \\ 0 & 0 \end{bmatrix} + \begin{bmatrix} 0 & 0 \\ 0 & 1 \end{bmatrix} = \begin{bmatrix} 1 & 0 \\ 0 & 1 \end{bmatrix} \notin W.$$

17. The vectors in W are of the form (a, a^3). This set is not closed under addition or scalar multiplication. For example,

$$(1, 1) + (2, 8) = (3, 9) \notin W$$

and

$$3(2, 8) = (6, 24) \notin W.$$

19. This set is *not* a subspace of $C(-\infty, \infty)$ because it is not closed under scalar multiplication.

21. This set is a subspace of $C(-\infty, \infty)$ because it is closed under addition and scalar multiplication.

23. This set is a subspace of $C(-\infty, \infty)$ because it is closed under addition and scalar multiplication.

25. This set is a subspace of $M_{m,n}$, because it is closed under addition and scalar multiplication.

27. This set is a subspace of $M_{m,n}$ because it is closed under addition and scalar multiplication.

29. This set is *not* a subspace because it is not closed under addition or scalar multiplication.

31. W is a subspace of R^3, because it is nonempty and closed under addition and scalar multiplication.

33. Note that $W \subset R^3$ and W is nonempty. If $(a_1, b_1, a_1 + 2b_1)$ and $(a_2, b_2, a_2 + 2b_2)$ are vectors in W, then their sum

$$(a_1, b_1, a_1 + 2b_1) + (a_2, b_2, a_2 + 2b_2) = (a_1 + a_2, b_1 + b_2, (a_1 + a_2) + 2(b_1 + b_2))$$

is also in W. Furthermore, for any real number c and $(a, b, a + 2b)$ in W,

$$c(a, b, a + 2b) = (ca, cb, ca + 2cb)$$

is in W. Because W is closed under addition and scalar multiplication, W is a subspace of R^3.

35. W is not a subspace of R^3 because it is not closed under addition or scalar multiplication. For example, $(1, 1, 1) \in W$, but

$$(1, 1, 1) + (1, 1, 1) = (2, 2, 2) \notin W.$$

Or,

$$2(1, 1, 1) = (2, 2, 2) \notin W.$$

37. (a) True. See "Remark," page 199.

(b) True. See Theorem 4.6, page 202.

(c) False. There may be elements of W which are not elements of U or vice-versa.

39. Let W be a nonempty subset of a vector space V. On the one hand, if W is a subspace of V, then for any scalars a, b, and any vectors $\mathbf{x}, \mathbf{y} \in W$, $a\mathbf{x} \in W$ and $b\mathbf{y} \in W$, and so, $a\mathbf{x} + b\mathbf{y} \in W$.

On the other hand, assume that $a\mathbf{x} + b\mathbf{y}$ is an element of W where a, b are scalars, and $\mathbf{x}, \mathbf{y} \in W$. To show that W is a subspace, you must verify the closure axioms. If $\mathbf{x}, \mathbf{y} \in W$, then $\mathbf{x} + \mathbf{y} \in W$ (by taking $a = b = 1$). Finally, if a is a scalar, $a\mathbf{x} \in W$ (by taking $b = 0$).

41. Assume A is a fixed 2×3 matrix. Assuming W is nonempty, let $\mathbf{x} \in W$. Then $A\mathbf{x} = \begin{bmatrix} 1 \\ 2 \end{bmatrix}$. Now, let c be a nonzero scalar such that $c \neq 1$. Then $c\mathbf{x} \in R^3$ and

$$A(c\mathbf{x}) = cA\mathbf{x} = c\begin{bmatrix} 1 \\ 2 \end{bmatrix} = \begin{bmatrix} c \\ 2c \end{bmatrix}.$$

So, $c\mathbf{x} \notin \mathbf{W}$. Therefore, W is not a subspace of R^3.

43. Let W be a subspace of the vector space V and let $\mathbf{0}_v$ be the zero vector of V and $\mathbf{0}_w$ be the zero vector of W. Because $\mathbf{0}_w \in W \subset V$,

$$\mathbf{0}_w = \mathbf{0}_w + (-\mathbf{0}_w) = \mathbf{0}_v$$

So, the zero vector in V is also the zero vector in W.

45. The set W is a nonempty subset of $M_{2,2}$. (For instance, $A \in W$). To show closure, let $X, Y \in W \Rightarrow AX = XA$ and $AY = YA$. Then, $(X + Y)A = XA + YA = AX + AY = A(X + Y) \Rightarrow X + Y \in W$. Similarly, if c is a scalar, then $(cX)A = c(XA) = c(AX) = A(cX) \Rightarrow cX \in W$.

47. $V + W$ is nonempty because $\mathbf{0} = \mathbf{0} + \mathbf{0} \in V + W$.

Let $\mathbf{u}_1, \mathbf{u}_2 \in V + W$. Then $\mathbf{u}_1 = \mathbf{v}_1 + \mathbf{w}_1, \mathbf{u}_2 = \mathbf{v}_2 + \mathbf{w}_2$, where $\mathbf{v}_i \in V$ and $\mathbf{w}_i \in W$. So,

$$\mathbf{u}_1 + \mathbf{u}_2 = (\mathbf{v}_1 + \mathbf{w}_1) + (\mathbf{v}_2 + \mathbf{w}_2) = (\mathbf{v}_1 + \mathbf{v}_2) + (\mathbf{w}_1 + \mathbf{w}_2) \in V + W.$$

For scalar c,

$$c\mathbf{u}_1 = c(\mathbf{v}_1 + \mathbf{w}_1) = c\mathbf{v}_1 + c\mathbf{w}_1 \in V + W.$$

If $V = \{(x, 0): x \text{ is a real number}\}$ and $W = \{(0, y): y \text{ is a real number}\}$, then $V + W = \mathbf{R}^2$.

Section 4.4 Spanning Sets and Linear Independence

1. (a) Solving the equation

$$c_1(2, -1, 3) + c_2(5, 0, 4) = (1, 1, -1)$$

for c_1 and c_2 yields the system

$$\begin{aligned} 2c_1 + 5c_2 &= 1 \\ -c_1 \quad\quad &= 1 \\ 3c_1 + 4c_2 &= -1. \end{aligned}$$

This system has no solution. So, $\mathbf{u}$ cannot be written as a linear combination of vectors in S.

(b) Proceed as in (a), substituting $\left(8, -\frac{1}{4}, \frac{27}{4}\right)$ for $(1, 1, -1)$, which yields the system

$$\begin{aligned} 2c_1 + 5c_2 &= 8 \\ -c_1 \quad\quad &= -\frac{1}{4} \\ 3c_1 + 4c_2 &= \frac{27}{4}. \end{aligned}$$

The solution to this system is $c_1 = \frac{1}{4}$ and $c_2 = \frac{3}{2}$. So, $\mathbf{v}$ can be written as a linear combination of vectors in S.

(c) Proceed as in (a), substituting $(1, -8, 12)$ for $(1, 1, -1)$, which yields the system

$$2c_1 + 5c_2 = 1$$
$$-c_1 \qquad = -8$$
$$3c_1 + 4c_2 = 12.$$

The solution to this system is $c_1 = 8$ and $c_2 = -3$. So, **w** can be written as a linear combination of the vectors in S.

(d) Proceed as in (a), substituting $(-1, -2, 2)$ for $(1, 1, -1)$, which yields the system

$$2c_1 + 5c_2 = -1$$
$$-c_1 \qquad = -2$$
$$3c_1 + 4c_2 = 2$$

The solution of this system is $c_1 = 2$ and $c_2 = -1$. So, **z** can be written as a linear combination of vectors in S.

3. (a) Solving the equation

$$c_1(2, 0, 7) + c_2(2, 4, 5) + c_3(2, -12, 13) = (-1, 5, -6)$$

for $c_1, c_2,$ and c_3, yields the system

$$2c_1 + 2c_2 + 2c_3 = -1$$
$$4c_2 - 12c_3 = 5$$
$$7c_1 + 5c_2 + 13c_3 = -6.$$

One solution is $c_1 = -\frac{7}{4}, c_2 = \frac{5}{4}$, and $c_3 = 0$. So, **u** can be written as a linear combination of vectors in S.

(b) Proceed as in (a), substituting $(-3, 15, 18)$ for $(-1, 5, -6)$, which yields the system

$$2c_1 + 2c_2 + 2c_3 = -3$$
$$4c_2 - 12c_3 = 15$$
$$7c_1 + 5c_2 + 13c_3 = 18.$$

This system has no solution. So, **v** cannot be written as a linear combination of vectors in S.

(c) Proceed as in (a), substituting $\left(\frac{1}{3}, \frac{4}{3}, \frac{1}{2}\right)$ for $(-1, 5, -6)$, which yields the system

$$2c_1 + 2c_2 + 2c_3 = \frac{1}{3}$$
$$4c_2 - 12c_3 = \frac{4}{3}$$
$$7c_1 + 5c_2 + 13c_3 = \frac{1}{2}.$$

One solution is $c_1 = -\frac{1}{6}, c_2 = \frac{1}{3}$, and $c_3 = 0$. So, **w** can be written as a linear combination of vectors in S.

(d) Proceed as in (a), substituting $(2, 20, -3)$ for $(-1, 5, -6)$, which yields the system

$$2c_1 + 2c_2 + 2c_3 = 2$$
$$4c_2 - 12c_3 = 20$$
$$7c_1 + 5c_2 + 13c_3 = -3.$$

One solution is $c_1 = -4, c_2 = 5$, and $c_3 = 0$. So, **z** can be written as a linear combination of vectors in S.

5. Let $\mathbf{u} = (u_1, u_2)$ be any vector in R^2. Solving the equation

$$c_1(2, 1) + c_2(-1, 2) = (u_1, u_2)$$

for c_1 and c_2 yields the system

$$2c_1 - c_2 = u_1$$
$$c_1 + 2c_2 = u_2.$$

This system has a unique solution because the determinant of the coefficient matrix is nonzero. So, S spans R^2.

7. Let $\mathbf{u} = (u_1, u_2)$ be any vector in R^2. Solving the equation

$$c_1(5, 0) + c_2(5, -4) = (u_1, u_2)$$

for c_1 and c_2 yields the system

$$5c_1 + 5c_2 = u_1$$
$${-4c_2} = u_2.$$

This system has a unique solution because the determinant of the coefficient matrix is nonzero. So, S spans R^2.

9. S does not span R^2 because only vectors of the form $t(-3, 5)$, are in span(S). For example, $(0, 1)$ is not in span(S). S spans a line in R^2.

11. S does not span R^2 because only vectors of the form $t(1, 3)$ are in span(S). For example, $(0, 1)$ is not in span(S). S spans a line in R^2.

13. S does not span R^2 because only vectors of the form $t(1, -2)$, are in span(S). For example, $(0, 1)$ is not in span(S). S spans a line in R^2.

15. S spans R^2. Let $\mathbf{u} = (u_1, u_2)$ be any vector in R^2. Solving the equation

$$c_1(-1, 4) + c_2(4, -1) + c_3(1, 1) = (u_1, u_2)$$

for c_1, c_2 and c_3 yields the system

$$-c_1 + 4c_2 + c_3 = u_1$$
$$4c_1 - c_2 + c_3 = u_2.$$

This system is equivalent to

$$c_1 - 4c_2 - c_3 = -u_1$$
$$15c_2 - 5c_3 = 4u_1 + u_2.$$

So, for any $\mathbf{u} = (u_1, u_2)$ in R^2, you can take

$$c_3 = 0, c_2 = (4u_1 + u_2)/15 \text{ and}$$
$$c_1 = 4c_2 - u_1 = (u_1 + 4u_2)/15.$$

17. Let $\mathbf{u} = (u_1, u_2, u_3)$ be any vector in R^3. Solving the equation

$$c_1(4, 7, 3) + c_2(-1, 2, 6) + c_3(2, -3, 5) = (u_1, u_2, u_3)$$

for c_1, c_2, and c_3 yields the system

$$4c_1 - c_2 + 2c_3 = u_1$$
$$7c_1 + 2c_2 - 3c_3 = u_2$$
$$3c_1 + 6c_2 + 5c_3 = u_3.$$

This system has a unique solution because the determinant of the coefficient matrix is nonzero. So, S spans R^3.

19. This set does not span R^3. S spans a plane in R^3.

21. Let $\mathbf{u} = (u_1, u_2, u_3)$ be any vector in R^3. Solving the equation

$$c_1(1, -2, 0) + c_2(0, 0, 1) + c_3(-1, 2, 0) = (u_1, u_2, u_3)$$

for c_1, c_2, and c_3 yields the system

$$c_1 - c_3 = u_1$$
$$-2c_1 + 2c_3 = u_2$$
$$ c_2 = u_3.$$

This system has an infinite number of solutions if $u_2 = -2u_1$, otherwise it has no solution.

For instance $(1, 1, 1)$ is not in the span of S. So, S does not span R^3. The subspace spanned by S is

$$\text{span}(S) = \{(a, -2a, b) : a \text{ and } b \text{ are any real numbers}\},$$

which is a plane in R^3.

23. Because $(-2, 2)$ is not a scalar multiple of $(3, 5)$, the set S is linearly independent.

25. This set is linearly dependent because

$$1(0, 0) + 0(1, -1) = (0, 0).$$

27. Because $(1, -4, 1)$ is not a scalar multiple of $(6, 3, 2)$, the set S is linearly independent.

29. Because these vectors are multiples of each other, the set S is linearly dependent.

31. From the vector equation

$$c_1(-4, -3, 4) + c_2(1, -2, 3) + c_3(6, 0, 0) = \mathbf{0}$$

you obtain the homogenous system

$$-4c_1 + c_2 + 6c_3 = 0$$
$$-3c_1 - 2c_2 = 0$$
$$4c_1 + 3c_2 = 0.$$

This system has only the trivial solution $c_1 = c_2 = c_3 = 0$. So, the set S is linearly independent.

33. From the vector equation

$$c_1(4, -3, 6, 2) + c_2(1, 8, 3, 1) + c_3(3, -2, -1, 0) = (0, 0, 0, 0)$$

you obtain the homogeneous system

$$
\begin{aligned}
4c_1 +\ c_2 + 3c_3 &= 0 \\
-3c_1 + 8c_2 - 2c_3 &= 0 \\
6c_1 + 3c_2 -\ c_3 &= 0 \\
2c_1 +\ c_2 \qquad &= 0.
\end{aligned}
$$

This system has only the trivial solution $c_1 = c_2 = c_3 = 0$. So, the set S is linearly independent.

35. One example of a nontrivial linear combination of vectors in S whose sum is the zero vector is

$$2(3, 4) - 8(-1, 1) - 7(2, 0) = (0, 0). \text{ Solving this equation for } (2, 0) \text{ yields } (2, 0) = \tfrac{2}{7}(3, 4) - \tfrac{8}{7}(-1, 1).$$

37. One example of a nontrivial linear combination of vectors in S whose sum is the zero vector is

$$(1, 1, 1) - (1, 1, 0) - 0(0, 1, 1) - (0, 0, 1) = (0, 0, 0). \text{ Solving this equation for } (1, 1, 1) \text{ yields}$$

$$(1, 1, 1) = (1, 1, 0) + (0, 0, 1) + 0(0, 1, 1).$$

39. (a) From the vector equation

$$c_1(t, 1, 1) + c_2(1, t, 1) + c_3(1, 1, t) = (0, 0, 0)$$

you obtain the homogeneous system

$$
\begin{aligned}
tc_1 +\ c_2 +\ c_3 &= 0 \\
c_1 + tc_2 +\ c_3 &= 0 \\
c_1 +\ c_2 + tc_3 &= 0.
\end{aligned}
$$

The coefficient matrix of this system will have a nonzero determinant if $t^3 - 3t + 2 \neq 0$. So, the vectors will be linearly independent for all values of t other than $t = -2$ or $t = 1$.

   (b) Proceeding as in (a), you obtain the homogeneous system

$$
\begin{aligned}
tc_1 +\ c_2 +\ c_3 &= 0 \\
c_1 \qquad +\ c_3 &= 0 \\
c_1 +\ c_2 + 3tc_3 &= 0.
\end{aligned}
$$

The coefficient matrix of this system will have a nonzero determinant if $2 - 4t \neq 0$.

So, the vectors will be linearly independent for all values of t other than $t = \tfrac{1}{2}$.

41. (a) From the vector equation

$$c_1 \begin{bmatrix} 2 & -3 \\ 4 & 1 \end{bmatrix} + c_2 \begin{bmatrix} 0 & 5 \\ 1 & -2 \end{bmatrix} = \begin{bmatrix} 6 & -19 \\ 10 & 7 \end{bmatrix}$$

you obtain the linear system

$$
\begin{aligned}
2c_1 \qquad &=\ 6 \\
-3c_1 + 5c_2 &= -19 \\
4c_1 +\ c_2 &=\ 10 \\
c_1 - 2c_2 &=\ 7.
\end{aligned}
$$

The solution to this system is $c_1 = 3$ and $c_2 = -2$.

So,

$$\begin{bmatrix} 6 & -19 \\ 10 & 7 \end{bmatrix} = 3\begin{bmatrix} 2 & -3 \\ 4 & 1 \end{bmatrix} - 2\begin{bmatrix} 0 & 5 \\ 1 & -2 \end{bmatrix} = 3A - 2B.$$

(b) Proceeding as in (a), you obtain the system

$$
\begin{aligned}
2c_1 \quad\quad &= 6 \\
-3c_1 + 5c_2 &= 2 \\
4c_1 + c_2 &= 9 \\
c_1 - 2c_2 &= 11.
\end{aligned}
$$

This system is inconsistent, and so the matrix is not a linear combination of A and B.

(c) Proceeding as in (a), you obtain

$$
\begin{bmatrix} -2 & 28 \\ 1 & -11 \end{bmatrix} = -\begin{bmatrix} 2 & -3 \\ 4 & 1 \end{bmatrix} + 5\begin{bmatrix} 0 & 5 \\ 1 & -2 \end{bmatrix} = -A + 5B
$$

and so the matrix is a linear combination of A and B.

(d) Proceeding as in (a), you obtain the trivial combination

$$
\begin{bmatrix} 0 & 0 \\ 0 & 0 \end{bmatrix} = 0\begin{bmatrix} 2 & -3 \\ 4 & 1 \end{bmatrix} + 0\begin{bmatrix} 0 & 5 \\ 1 & -2 \end{bmatrix} = 0A + 0B.
$$

43. From the vector equation $c_1(2 - x) + c_2(2x - x^2) + c_3(6 - 5x + x^2) = 0 + 0x + 0x^2$ you obtain the homogeneous system

$$
\begin{aligned}
2c_1 \quad\quad + 6c_3 &= 0 \\
-c_1 + 2c_2 - 5c_3 &= 0 \\
-c_2 + c_3 &= 0.
\end{aligned}
$$

This system has infinitely many solutions. For instance, $c_1 = -3, c_2 = 1, c_3 = 1$. So, S is linearly dependent.

45. From the vector equation $c_1(x^2 + 3x + 1) + c_2(2x^2 + x - 1) + c_3(4x) = 0 + 0x + 0x^2$ you obtain the homogeneous system

$$
\begin{aligned}
c_1 - c_2 \quad\quad &= 0 \\
3c_1 + c_2 + 4c_3 &= 0 \\
c_1 + 2c_2 \quad\quad &= 0.
\end{aligned}
$$

This system has only the trivial solution. So, S is linearly independent.

47. S does not span P_2 because only vectors of the form $s(x^2) + t(1)$ are in span(S). For example, $1 + x + x^2$ is not in span(S).

49. (a) Because $(-2, 4) = -2(1, -2)$, S is linearly dependent.

(b) Because $2(1, -6, 2) = (2, -12, 4)$, S is linearly dependent.

(c) Because $(0, 0) = 0(1, 0)$, S is linearly dependent.

51. Because the matrix $\begin{bmatrix} 1 & 2 & -1 \\ 0 & 1 & 1 \\ 2 & 5 & -1 \end{bmatrix}$ row reduces to $\begin{bmatrix} 1 & 0 & -3 \\ 0 & 1 & 1 \\ 0 & 0 & 0 \end{bmatrix}$ and $\begin{bmatrix} -2 & -6 & 0 \\ 1 & 1 & -2 \end{bmatrix}$ row reduces to $\begin{bmatrix} 1 & 0 & -3 \\ 0 & 1 & 1 \end{bmatrix}$, you see that S_1

and S_2 span the same subspace. You could also verify this by showing that each vector in S_1 is in the span of S_2, and conversely, each vector in S_2 is in the span of S_1. For example, $(1, 2, -1) = -\frac{1}{4}(-2, -6, 0) + \frac{1}{2}(1, 1, -2)$.

53. (a) False. See "Definition of Linear Dependences and Linear Independence," page 263.

(b) True. See corollary to Theorem 4.8, page 218.

55. The matrix $\begin{bmatrix} 1 & 1 & 1 \\ 1 & 1 & 0 \\ 1 & 0 & 0 \end{bmatrix}$ row reduces to $\begin{bmatrix} 1 & 0 & 0 \\ 0 & 1 & 0 \\ 0 & 0 & 1 \end{bmatrix}$, which

shows that the equation

$$c_1(1, 1, 1) + c_2(1, 1, 0) + c_3(1, 0, 0) = (0, 0, 0)$$

only has the trivial solution. So, the three vectors are linearly independent. Furthermore, the vectors span R^3 because the coefficient matrix of the linear system

$$\begin{bmatrix} 1 & 1 & 1 \\ 1 & 1 & 0 \\ 1 & 0 & 0 \end{bmatrix}\begin{bmatrix} c_1 \\ c_2 \\ c_3 \end{bmatrix} = \begin{bmatrix} u_1 \\ u_2 \\ u_3 \end{bmatrix}$$

is nonsingular.

57. Let S be a set of linearly independent vectors and $T \subset S$. If $T = \{v_1, \cdots, v_k\}$ and T were linearly dependent, then there would exist constants $c_1, \cdots, c_k$, not all zero, satisfying $c_1 v_1 + \cdots + c_k v_k = 0$. But, $v_i \in S$, and S is linearly independent, which is impossible. So, T is linearly independent.

59. If a set of vectors $\{v_1, v_2, \cdots\}$ contains the zero vector, then $0 = 0 v_1 + \ldots + 0 v_k + 1 \cdot 0$ which implies that the set is linearly dependent.

61. If the set $\{v_1, \cdots, v_{k-1}\}$ spanned v, then

$$v_k = c_1 v_1 + \cdots + c_{k-1} v_{k-1}$$

for some scalars $c_1, \cdots, c_{k-1}$. So,

$$cv_1 + \cdots + c_{k-1} v_{k-1} - v_k = 0$$

which is impossible because $\{v_1, \cdots, v_k\}$ is linearly independent.

63. Theorem 4.8 requires that only one of the vectors to be a linear combination of the others. In this case, $(-1, 0, 2) = 0(1, 2, 3) - (1, 0, -2)$, and so, there is no contradiction.

65. Consider the vector equation

$$c_1(u + v) + c_2(u - v) = 0.$$

Regrouping, you have

$$(c_1 + c_2)u + (c_1 - c_2)v = 0.$$

Because u and v are linearly independent, $c_1 + c_2 = c_1 - c_2 = 0$. So, $c_1 = c_2 = 0$, and the vectors $u + v$ and $u - v$ are linearly independent.

67. On $[0, 1]$, $f_2(x) = |x| = x = \frac{1}{3}(3x)$

$$= \frac{1}{3}f_1(x)$$

$$\Rightarrow \{f_1, f_2\} \text{ dependent.}$$

On $[-1, 1]$, f_1 and f_2 are not multiples of each other.

$f_2(x) \neq cf_1(x)$ for $-1 \leq x < 0$, that is

$f(x) = |x| \neq \frac{1}{3}(3x)$ for $-1 \leq x \leq 0$.

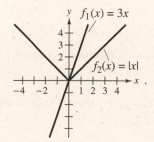

69. On the one hand, if u and v are linearly dependent, then there exist constants c_1 and c_2, not both zero, such that $c_1 u + c_2 v = 0$. Without loss of generality, you can assume $c_1 \neq 0$, and obtain $u = -\dfrac{c_2}{c_1}v$.

On the other hand, if one vector is a scalar multiple of another, $u = cv$, then $u - cv = 0$, which implies that u and v are linearly dependent.

Section 4.5   Basis and Dimension

1. There are six vectors in the standard basis for R^6.

$\{(1, 0, 0, 0, 0, 0), (0, 1, 0, 0, 0, 0), (0, 0, 1, 0, 0, 0),$
$(0, 0, 0, 1, 0, 0), (0, 0, 0, 0, 1, 0), (0, 0, 0, 0, 0, 1)\}$

3. There are eight vectors in the standard basis.

$$\left\{ \begin{bmatrix} 1 & 0 & 0 & 0 \\ 0 & 0 & 0 & 0 \end{bmatrix}, \begin{bmatrix} 0 & 1 & 0 & 0 \\ 0 & 0 & 0 & 0 \end{bmatrix}, \begin{bmatrix} 0 & 0 & 1 & 0 \\ 0 & 0 & 0 & 0 \end{bmatrix}, \begin{bmatrix} 0 & 0 & 0 & 1 \\ 0 & 0 & 0 & 0 \end{bmatrix}, \right.$$

$$\left. \begin{bmatrix} 0 & 0 & 0 & 0 \\ 1 & 0 & 0 & 0 \end{bmatrix}, \begin{bmatrix} 0 & 0 & 0 & 0 \\ 0 & 1 & 0 & 0 \end{bmatrix}, \begin{bmatrix} 0 & 0 & 0 & 0 \\ 0 & 0 & 1 & 0 \end{bmatrix}, \begin{bmatrix} 0 & 0 & 0 & 0 \\ 0 & 0 & 0 & 1 \end{bmatrix} \right\}$$

5. There are five vectors in the standard basis for P_4.

$$\{1, x, x^2, x^3, x^4\}$$

7. A basis for R^2 can only have two vectors. Because S has three vectors, it is not a basis for R^2.

9. S is linearly dependent $((0, 0) \in S)$ and does not span R^2. For instance, $(1, 1) \notin \text{span}(S)$.

11. S is linearly dependent and does not span R^2. For instance, $(1, 1) \notin \text{span}(S)$.

13. S does not span R^2, although it is linearly independent. For instance, $(1, 1) \notin \text{span}(S)$.

15. A basis for R^3 contains three linearly independent vectors. Because

$$-2(1, 3, 0) + (4, 1, 2) + (-2, 5, -2) = (0, 0, 0)$$

S is linearly dependent and is, therefore, not a basis for R^3.

17. S does not span R^3, although it is linearly independent. For instance, $(0, 1, 0) \notin \text{span}(S)$.

19. S is linearly dependent and does not span R^3. For instance, $(0, 0, 1) \notin \text{span}(S)$.

21. A basis for P_2 can have only three vectors. Because S has four vectors, it is not a basis for P_2.

23. S is not a basis because the vectors are linearly dependent.

$$-2(1 - x) + 3(1 - x^2) + (3x^2 - 2x - 1) = 0$$

25. A basis for $M_{2,2}$ must have four vectors. Because S has only two vectors, it is not a basis for $M_{2,2}$.

27. S is not a basis because the vectors are linearly dependent.

$$5\begin{bmatrix} 1 & 0 \\ 0 & 0 \end{bmatrix} - 4\begin{bmatrix} 0 & 1 \\ 1 & 0 \end{bmatrix} + 3\begin{bmatrix} 1 & 0 \\ 0 & 1 \end{bmatrix} - \begin{bmatrix} 8 & -4 \\ -4 & 3 \end{bmatrix} = \begin{bmatrix} 0 & 0 \\ 0 & 0 \end{bmatrix}$$

Also, S does not span $M_{2,2}$.

29. Because $\{\mathbf{v}_1, \mathbf{v}_2\}$ consists of exactly two linearly independent vectors, it is a basis for R^2.

31. Because $\mathbf{v}_1$ and $\mathbf{v}_2$ are multiplies of each other, they do not form a basis for R^2.

33. Because $\mathbf{v}_1$ and $\mathbf{v}_2$ are multiples of each other they do not form a basis for R^2.

35. Because the vectors in S are not scalar multiples of one another, they are linearly independent. Because S consists of exactly two linearly independent vectors, it is a basis for R^2.

37. To determine if the vectors in S are linearly independent, find the solution to

$$c_1(1, 5, 3) + c_2(0, 1, 2) + c_3(0, 0, 6) = (0, 0, 0)$$

which corresponds to the solution of

$$\begin{aligned} c_1 & & & = 0 \\ 5c_1 & + & c_2 & = 0 \\ 3c_1 & + & 2c_2 + 6c_3 & = 0. \end{aligned}$$

This system has only the trivial solution. So, S consists of exactly three linearly independent vectors, and is, therefore, a basis for R^3.

39. To determine if the vectors in S are linearly independent, find the solution to

$$c_1(0, 3, -2) + c_2(4, 0, 3) + c_3(-8, 15, -16) = (0, 0, 0)$$

which corresponds to the solution of

$$\begin{aligned} 4c_2 & - 8c_3 & = 0 \\ 3c_1 & + 15c_3 & = 0 \\ -2c_1 + 3c_2 & - 16c_3 & = 0. \end{aligned}$$

Because this system has nontrivial solutions (for instance, $c_1 = -5$, $c_2 = 2$ and $c_3 = 1$), the vectors are linearly dependent, and S is not a basis for R^3.

41. To determine if the vectors of S are linearly independent, find the solution to

$$c_1(-1, 2, 0, 0) + c_2(2, 0, -1, 0) + c_3(3, 0, 0, 4) + c_4(0, 0, 5, 0) = (0, 0, 0, 0)$$

which corresponds to the solution of

$$\begin{aligned} -c_1 & + 2c_2 + 3c_3 & & = 0 \\ 2c_1 & & & = 0 \\ & -c_2 & & + 5c_4 = 0 \\ & & 4c_3 & = 0. \end{aligned}$$

This system has only the trivial solution. So, S consists of exactly four linearly independent vectors, and is therefore a basis for R^4.

43. Form the equation

$$c_1 \begin{bmatrix} 2 & 0 \\ 0 & 3 \end{bmatrix} + c_2 \begin{bmatrix} 1 & 4 \\ 0 & 1 \end{bmatrix} + c_3 \begin{bmatrix} 0 & 1 \\ 3 & 2 \end{bmatrix} + c_4 \begin{bmatrix} 0 & 1 \\ 2 & 0 \end{bmatrix} = \begin{bmatrix} 0 & 0 \\ 0 & 0 \end{bmatrix}$$

which yields the homogeneous system

$$\begin{aligned} 2c_1 + c_2 &= 0 \\ 4c_2 + c_3 + c_4 &= 0 \\ 3c_3 + 2c_4 &= 0 \\ 3c_1 + c_2 + 2c_3 &= 0. \end{aligned}$$

This system has only the trivial solution. So, S consists of exactly four linearly independent vectors, and is therefore a basis for $M_{2,2}$.

45. Form the equation

$$c_1(t^3 - 2t^2 + 1) + c_2(t^2 - 4) + c_3(t^3 + 2t) + c_4(5t) = 0 + 0t + 0t^2 + 0t^3$$

which yields the homogeneous system

$$\begin{aligned} c_1 \qquad + c_3 \qquad &= 0 \\ -2c_1 + c_2 \qquad &= 0 \\ 2c_3 + 5c_4 &= 0 \\ c_1 - 4c_2 \qquad &= 0. \end{aligned}$$

This system has only the trivial solution. So, S consists of exactly four linearly independent vectors, and is, therefore, a basis for P_3.

47. Because a basis for P_3 can contain only four basis vectors and set S contains five vectors, you have that S is not a basis for P_3.

49. Form the equation
$c_1(4, 3, 2) + c_2(0, 3, 2) + c_3(0, 0, 2) = (0, 0, 0)$ which
yields the homogeneous system

$$\begin{aligned} 4c_1 \qquad &= 0 \\ 3c_1 + 3c_2 \qquad &= 0 \\ 2c_1 + 2c_2 + 2c_3 &= 0. \end{aligned}$$

This system has only the trivial solution, so S is a basis for R^3. Solving the system

$$\begin{aligned} 4c_1 \qquad &= 8 \\ 3c_1 + 3c_2 \qquad &= 3 \\ 2c_1 + 2c_2 + 2c_3 &= 8 \end{aligned}$$

yields $c_1 = 2, c_2 = -1,$ and $c_3 = 3.$ So,

$$\mathbf{u} = 2(4, 3, 2) - (0, 3, 2) + 3(0, 0, 2) = (8, 3, 8).$$

51. The set S contains the zero vector, and is, therefore, linearly dependent.

$$1(0, 0, 0) + 0(1, 3, 4) + 0(6, 1, -2) = (0, 0, 0).$$

So, S is not a basis for R^3.

53. Form the equation

$$c_1\left(\tfrac{2}{3}, \tfrac{5}{2}, 1\right) + c_2\left(1, \tfrac{3}{2}, 0\right) + c_3(2, 12, 6) = (0, 0, 0)$$

which yields the homogeneous system

$$\begin{aligned} \tfrac{2}{3}c_1 + c_2 + 2c_3 &= 0 \\ \tfrac{5}{2}c_1 + \tfrac{3}{2}c_2 + 12c_3 &= 0 \\ c_1 \qquad + 6c_3 &= 0. \end{aligned}$$

Because this system has nontrivial solutions (for instance, $c_1 = 6, c_2 = -2$ and $c_3 = -1$), the vectors are linearly dependent. So, S is not a basis for R^3.

55. Because a basis for R^6 has six linearly independent vectors, the dimension of R^6 is 6.

57. Because a basis for R has one linearly independent vector, the dimension of R is 1.

59. Because a basis for P_7 has eight linearly independent vectors, the dimension of P_7 is 8.

61. Because a basis for $M_{2,3}$ has six linearly independent vectors, the dimension of $M_{2,3}$ is 6.

63. One basis for $D_{3,3}$ is

$$\left\{ \begin{bmatrix} 1 & 0 & 0 \\ 0 & 0 & 0 \\ 0 & 0 & 0 \end{bmatrix}, \begin{bmatrix} 0 & 0 & 0 \\ 0 & 1 & 0 \\ 0 & 0 & 0 \end{bmatrix}, \begin{bmatrix} 0 & 0 & 0 \\ 0 & 0 & 0 \\ 0 & 0 & 1 \end{bmatrix} \right\}.$$

Because a basis for $D_{3,3}$ has 3 vectors,

$$\dim(D_{3,3}) = 3.$$

65. The following subsets of two vectors form a basis for R^2.

$\{(1, 0), (0, 1)\}, \{(1, 0), (1, 1)\}, \{(0, 1), (1, 1)\}.$

67. Add any vector that is not a multiple of $(1, 1)$. For instance, the set $\{(1, 1), (1, 0)\}$ is a basis for R^2.

69. (a) W is a line through the origin.

(b) A basis for W is $\{(2, 1)\}$.

(c) The dimension of W is 1.

71. (a) W is a line through the origin.

(b) A basis for W is $\{(2, 1, -1)\}$.

(c) The dimension of W is 1.

73. (a) A basis for W is $\{(2, 1, 0, 1), (-1, 0, 1, 0)\}$.

(b) The dimension of W is 2.

75. (a) A basis for W is $\{(0, 6, 1, -1)\}$.

(b) The dimension of W is 1.

77. (a) False. See paragraph before "Definition of Dimension of a Vector Space," page 227.

(b) True. Find a set of n basis vectors in V that will span V and add any other vector.

79. Because the set $S_1 = \{c\mathbf{v}_1, \cdots, c\mathbf{v}_n\}$ has n vectors, you only need to show that they are linearly independent. Consider the equation

$$a_1(c\mathbf{v}_1) + a_2(c\mathbf{v}_2) + \cdots + a_n(c\mathbf{v}_n) = \mathbf{0}$$
$$c(a_1\mathbf{v}_1 + a_2\mathbf{v}_2 + \cdots + a_n\mathbf{v}_n) = \mathbf{0}$$
$$a_1\mathbf{v}_1 + a_2\mathbf{v}_2 + \cdots + a_n\mathbf{v}_n = \mathbf{0}$$

Because $\{\mathbf{v}_1, \cdots, \mathbf{v}_n\}$ are linearly independent, the coefficients $a_1, \cdots, a_n$ must all be zero. So, S_1 is linearly independent.

81. Let $W \subset V$ and $\dim(V) = n$. Let $\mathbf{w}_1, \cdots, \mathbf{w}_k$ be a basis for W. Because $W \subset V$, the vectors $\mathbf{w}_1, \cdots, \mathbf{w}_k$ are linearly independent in V. If $\text{span}(\mathbf{w}_1, \cdots, \mathbf{w}_k) = V$, then $\dim(W) = \dim(V)$. If not, let $\mathbf{v} \in V, \mathbf{v} \notin W$. Then $\dim(W) < \dim(V)$.

83. (a) S_1–basis: $\{(1, 0, 0), (1, 1, 0)\}$ $\quad \dim(S_1) = 2$

S_2–basis: $\{(0, 0, 1), (0, 1, 0)\}$ $\quad \dim(S_2) = 2$

$S_1 \cap S_2$–basis: $\{(0, 1, 0)\}$ $\quad \dim(S_1 \cap S_2) = 1$

$S_1 + S_2$–basis: $\{(1, 0, 0), (0, 1, 0), (0, 0, 1)\}$ $\quad \dim(S_1 + S_2) = 3$

(Answers are not unique.)

(b) No, it is not possible, because two planes cannot intersect only at the origin.

85. If S spans V, you are done. If not, let $\mathbf{v}_1 \notin \text{span}(S)$, and consider the linearly independent set $S_1 = S \cup \{\mathbf{v}_1\}$. If S_1 spans V you are done. If not, let $\mathbf{v}_2 \notin \text{span}(S_1)$ and continue as before. Because the vector space is finite-dimensional, this process will ultimately produce a basis of V containing S.

Section 4.6   Rank of a Matrix and Systems of Linear Equations

1. (a) Because this matrix row reduces to

$$\begin{bmatrix} 1 & 0 \\ 0 & 1 \end{bmatrix}$$

the rank of the matrix is 2.

(b) A basis for the row space is $\{(1, 0), (0, 1)\}$.

(c) Row-reducing the transpose of the original matrix produces the identity matrix again, and so, a basis for the column space is $\left\{ \begin{bmatrix} 1 \\ 0 \end{bmatrix}, \begin{bmatrix} 0 \\ 1 \end{bmatrix} \right\}$.

3. (a) Because this matrix is row-reduced already, the rank is 1.

(b) A basis for the row space is $\{(1, 2, 3)\}$.

(c) A basis for the column space is $\{[1]\}$.

5. (a) Because this matrix row reduces to

$$\begin{bmatrix} 1 & 0 & \frac{1}{2} \\ 0 & 1 & -\frac{1}{2} \end{bmatrix}$$

the rank of the matrix is 2.

(b) A basis for the row space is $\left\{\left(1, 0, \frac{1}{2}\right), \left(0, 1, -\frac{1}{2}\right)\right\}$

(c) Row-reducing the transpose of the original matrix produces

$$\begin{bmatrix} 1 & 0 \\ 0 & 1 \\ 0 & 0 \end{bmatrix}.$$

So, a basis for the column space of the matrix is

$$\left\{\begin{bmatrix} 1 \\ 0 \end{bmatrix}, \begin{bmatrix} 0 \\ 1 \end{bmatrix}\right\}.$$

7. (a) Because this matrix row reduces to

$$\begin{bmatrix} 1 & 0 & \frac{1}{4} \\ 0 & 1 & \frac{3}{2} \\ 0 & 0 & 0 \end{bmatrix}$$

the rank of the matrix is 2.

(b) A basis for the row space is $\left\{\left(1, 0, \frac{1}{4}\right), \left(0, 1, \frac{3}{2}\right)\right\}.$

(c) Row-reducing the transpose of the original matrix produces

$$\begin{bmatrix} 1 & 0 & -\frac{2}{5} \\ 0 & 1 & \frac{3}{5} \\ 0 & 0 & 0 \end{bmatrix}.$$

So, a basis for the column space is $\left\{\begin{bmatrix} 1 \\ 0 \\ -\frac{2}{5} \end{bmatrix}, \begin{bmatrix} 0 \\ 1 \\ \frac{3}{5} \end{bmatrix}\right\}.$

Equivalently, a basis for the column space consists of columns 1 and 2 of the original matrix

$$\left\{\begin{bmatrix} 4 \\ 6 \\ 2 \end{bmatrix}, \begin{bmatrix} 20 \\ -5 \\ -11 \end{bmatrix}\right\}.$$

11. (a) Because this matrix row reduces to

$$\begin{bmatrix} 1 & 0 & 0 & 0 & 0 \\ 0 & 1 & 0 & 0 & 0 \\ 0 & 0 & 1 & 0 & 0 \\ 0 & 0 & 0 & 1 & 0 \\ 0 & 0 & 0 & 0 & 1 \end{bmatrix}$$

the rank of the matrix is 5.

(b) A basis for the row space is $\{(1, 0, 0, 0, 0), (0, 1, 0, 0, 0), (0, 0, 1, 0, 0), (0, 0, 0, 1, 0), (0, 0, 0, 0, 1)\}.$

9. (a) Because this matrix row reduces to

$$\begin{bmatrix} 1 & 2 & -2 & 0 \\ 0 & 0 & 0 & 1 \\ 0 & 0 & 0 & 0 \end{bmatrix}$$

the rank of the matrix is 2.

(b) A basis for the row space of the matrix is
$\{(1, 2, -2, 0), (0, 0, 0, 1)\}.$

(c) Row-reducing the transpose of the original matrix produces

$$\begin{bmatrix} 1 & 0 & \frac{19}{7} \\ 0 & 1 & \frac{8}{7} \\ 0 & 0 & 0 \\ 0 & 0 & 0 \end{bmatrix}.$$

So, a basis for the column space of the matrix is

$$\left\{\begin{bmatrix} 1 \\ 0 \\ \frac{19}{7} \end{bmatrix}, \begin{bmatrix} 0 \\ 1 \\ \frac{8}{7} \end{bmatrix}\right\}.$$ Equivalently, a basis for the column

space consists of columns 1 and 4 of the original

matrix $\left\{\begin{bmatrix} -2 \\ 3 \\ -2 \end{bmatrix}, \begin{bmatrix} 5 \\ -4 \\ 9 \end{bmatrix}\right\}$

(c) Row reducing the transpose of the original matrix produces

$$\begin{bmatrix} 1 & 0 & 0 & 0 & 0 \\ 0 & 1 & 0 & 0 & 0 \\ 0 & 0 & 1 & 0 & 0 \\ 0 & 0 & 0 & 1 & 0 \\ 0 & 0 & 0 & 0 & 1 \end{bmatrix}.$$

So, a basis for the column space is $\left\{ \begin{bmatrix} 1 \\ 0 \\ 0 \\ 0 \\ 0 \end{bmatrix}, \begin{bmatrix} 0 \\ 1 \\ 0 \\ 0 \\ 0 \end{bmatrix}, \begin{bmatrix} 0 \\ 0 \\ 1 \\ 0 \\ 0 \end{bmatrix}, \begin{bmatrix} 0 \\ 0 \\ 0 \\ 1 \\ 0 \end{bmatrix}, \begin{bmatrix} 0 \\ 0 \\ 0 \\ 0 \\ 1 \end{bmatrix} \right\}.$

13. Use $\mathbf{v}_1, \mathbf{v}_2,$ and $\mathbf{v}_3$ to form the rows of matrix A. Then write A in row-echelon form.

$$A = \begin{bmatrix} 1 & 2 & 4 \\ -1 & 3 & 4 \\ 2 & 3 & 1 \end{bmatrix} \begin{matrix} \mathbf{v}_1 \\ \mathbf{v}_2 \\ \mathbf{v}_3 \end{matrix} \rightarrow B = \begin{bmatrix} 1 & 0 & 0 \\ 0 & 1 & 0 \\ 0 & 0 & 1 \end{bmatrix} \begin{matrix} \mathbf{w}_1 \\ \mathbf{w}_2 \\ \mathbf{w}_3 \end{matrix}$$

So, the nonzero row vectors of B, $\mathbf{w}_1 = (1, 0, 0)$, $\mathbf{w}_2 = (0, 1, 0)$, and $\mathbf{w}_3 = (0, 0, 1)$, form a basis for the row space of A. That is, they form a basis for the subspace spanned by S.

15. Use $\mathbf{v}_1, \mathbf{v}_2,$ and $\mathbf{v}_3$ to form the rows of matrix A. Then write A in row-echelon form.

$$A = \begin{bmatrix} 4 & 4 & 8 \\ 1 & 1 & 2 \\ 1 & 1 & 1 \end{bmatrix} \begin{matrix} \mathbf{v}_1 \\ \mathbf{v}_2 \\ \mathbf{v}_3 \end{matrix} \rightarrow B = \begin{bmatrix} 1 & 1 & 0 \\ 0 & 0 & 1 \\ 0 & 0 & 0 \end{bmatrix} \begin{matrix} \mathbf{w}_1 \\ \mathbf{w}_2 \\ \end{matrix}$$

So, the nonzero row vectors of B, $\mathbf{w}_1 = (1, 1, 0)$ and $\mathbf{w}_2 = (0, 0, 1)$, form a basis for the row space of A. That is, they form a basis for the subspace spanned by S.

17. Begin by forming the matrix whose rows are vectors in S.

$$\begin{bmatrix} 2 & 9 & -2 & 53 \\ -3 & 2 & 3 & -2 \\ 8 & -3 & -8 & 17 \\ 0 & -3 & 0 & 15 \end{bmatrix}$$

This matrix reduces to

$$\begin{bmatrix} 1 & 0 & -1 & 0 \\ 0 & 1 & 0 & 0 \\ 0 & 0 & 0 & 1 \\ 0 & 0 & 0 & 0 \end{bmatrix}.$$

So, a basis for span(S) is
$$\{(1, 0, -1, 0), (0, 1, 0, 0), (0, 0, 0, 1)\}.$$

19. Form the matrix whose rows are the vectors in S, and then row-reduce.

$$\begin{bmatrix} -3 & 2 & 5 & 28 \\ -6 & 1 & -8 & -1 \\ 14 & -10 & 12 & -10 \\ 0 & 5 & 12 & 50 \end{bmatrix} \Rightarrow \begin{bmatrix} 1 & 0 & 0 & 0 \\ 0 & 1 & 0 & 0 \\ 0 & 0 & 1 & 0 \\ 0 & 0 & 0 & 1 \end{bmatrix}$$

So, a basis for span(S) is
$$\{(1, 0, 0, 0), (0, 1, 0, 0), (0, 0, 1, 0), (0, 0, 0, 1)\}.$$

21. Solving the system $A\mathbf{x} = \mathbf{0}$ yields only the trivial solution $\mathbf{x} = (0, 0)$. So, the dimension of the solution space is 0. The solution space consists of the zero vector itself.

23. Solving the system $A\mathbf{x} = \mathbf{0}$ yields solutions of the form $(-2s - 3t, s, t)$, where s and t are any real numbers. The dimension of the solution space is 2, and a basis is $\{(-2, 1, 0), (-3, 0, 1)\}$.

25. Solving the system $A\mathbf{x} = \mathbf{0}$ yields solutions of the form $(-3t, 0, t)$, where t is any real number. The dimension of the solution space is 1, and a basis is $\{(-3, 0, 1)\}$.

27. Solving the system $A\mathbf{x} = \mathbf{0}$ yields solutions of the form $(-t, 2t, t)$, where t is any real number. The dimension of the solution space is 1, and a basis for the solution space is $\{(-1, 2, 1)\}$.

29. Solving the system $A\mathbf{x} = \mathbf{0}$ yields solutions of the form $(-s + 2t, s - 2t, s, t)$, where s and t are any real numbers. The dimension of the solution space is 2, and a basis is $\{(2, -2, 0, 1), (-1, 1, 1, 0)\}$.

31. The only solution to the system $A\mathbf{x} = \mathbf{0}$ is the trivial solution. So, the solution space is $\{(0, 0, 0, 0)\}$ whose dimension is 0.

33. (a) This system yields solutions of the form $(-t, -3t, 2t)$, where t is any real number and a basis is $\{(-1, -3, 2)\}$.

(b) The dimension of the solution space is 1.

35. (a) This system yields solutions of the form $(2s - 3t, s, t)$, where s and t are any real numbers and a basis for the solution space is $\{(2, 1, 0), (-3, 0, 1)\}$.

(b) The dimension of the solution space is 2.

37. (a) This system yields solutions of the form $\left(-4s - 3t, -s - \frac{2}{3}t, s, t\right)$, where s and t is any real numbers and a basis is $\left\{(-4, -1, 1, 0), \left(-3, -\frac{2}{3}, 0, 1\right)\right\}$.

(b) The dimension of the solution space is 2.

39. (a) This system yields solutions of the form $\left(\frac{4}{3}t, -\frac{3}{2}t, -t, t\right)$, where t is any real number and a basis is $\left\{\left(\frac{4}{3}, -\frac{3}{2}, -1, 1\right)\right\}$ or $\{(8, -9, -6, 6)\}$.

(b) The dimension of the solution space is 1.

41. (a) This system $A\mathbf{x} = \mathbf{b}$ is consistent because its augmented matrix reduces to

$$\begin{bmatrix} 1 & 0 & -2 & 3 \\ 0 & 1 & 4 & 5 \\ 0 & 0 & 0 & 0 \\ 0 & 0 & 0 & 0 \end{bmatrix}.$$

(b) The solutions of $A\mathbf{x} = \mathbf{b}$ are of the form $(3 + 2t, 5 - 4t, t)$ where t is any real number. That is,

$$\mathbf{x} = t\begin{bmatrix} 2 \\ -4 \\ 1 \end{bmatrix} + \begin{bmatrix} 3 \\ 5 \\ 0 \end{bmatrix},$$

where

$$\mathbf{x}_h = t\begin{bmatrix} 2 \\ -4 \\ 1 \end{bmatrix} \quad \text{and} \quad \mathbf{x}_p = \begin{bmatrix} 3 \\ 5 \\ 0 \end{bmatrix}.$$

43. (a) This system $A\mathbf{x} = \mathbf{b}$ is inconsistent because its augmented matrix reduces to

$$\begin{bmatrix} 1 & 0 & 4 & 2 & 0 \\ 0 & 1 & -2 & 4 & 0 \\ 0 & 0 & 0 & 0 & 1 \end{bmatrix}.$$

45. (a) This system $A\mathbf{x} = \mathbf{b}$ is consistent because its augmented matrix reduces to

$$\begin{bmatrix} 1 & 2 & 0 & 0 & -5 & 1 \\ 0 & 0 & 1 & 0 & 6 & 2 \\ 0 & 0 & 0 & 1 & 4 & -3 \\ 0 & 0 & 0 & 0 & 0 & 0 \end{bmatrix}.$$

(b) The solutions of the system are of the form $(1 - 2s + 5t, s, 2 - 6t, -3 - 4t, t)$. where s and t are any real numbers. That is,

$$\mathbf{x} = s\begin{bmatrix} -2 \\ 1 \\ 0 \\ 0 \\ 0 \end{bmatrix} + t\begin{bmatrix} 5 \\ 0 \\ -6 \\ -4 \\ 1 \end{bmatrix} + \begin{bmatrix} 1 \\ 0 \\ 2 \\ -3 \\ 0 \end{bmatrix},$$

where

$$\mathbf{x}_h = s\begin{bmatrix} -2 \\ 1 \\ 0 \\ 0 \\ 0 \end{bmatrix} + t\begin{bmatrix} 5 \\ 0 \\ -6 \\ -4 \\ 1 \end{bmatrix} \quad \text{and} \quad \mathbf{x}_p = \begin{bmatrix} 1 \\ 0 \\ 2 \\ -3 \\ 0 \end{bmatrix}.$$

47. The vector $\mathbf{b}$ is in the column space of A if the equation $A\mathbf{x} = \mathbf{b}$ is consistent. Because $A\mathbf{x} = \mathbf{b}$ has the solution

$$\mathbf{x} = \begin{bmatrix} 1 \\ 2 \end{bmatrix},$$

$\mathbf{b}$ is in the column space of A. Furthermore,

$$\mathbf{b} = 1\begin{bmatrix} -1 \\ 4 \end{bmatrix} + 2\begin{bmatrix} 2 \\ 0 \end{bmatrix} = \begin{bmatrix} 3 \\ 4 \end{bmatrix}.$$

49. The vector $\mathbf{b}$ is in the column space of A if the equation $A\mathbf{x} = \mathbf{b}$ is consistent. Because $A\mathbf{x} = \mathbf{b}$ has the solution

$$\mathbf{x} = \begin{bmatrix} -\frac{5}{4} \\ \frac{3}{4} \\ -\frac{1}{2} \end{bmatrix},$$

$\mathbf{b}$ is in the column space of A. Furthermore,

$$\mathbf{b} = -\frac{5}{4}\begin{bmatrix} 1 \\ -1 \\ 2 \end{bmatrix} + \frac{3}{4}\begin{bmatrix} 3 \\ 1 \\ 0 \end{bmatrix} - \frac{1}{2}\begin{bmatrix} 0 \\ 0 \\ 1 \end{bmatrix} = \begin{bmatrix} 1 \\ 2 \\ -3 \end{bmatrix}.$$

51. The rank of the matrix is at most 3. So, the dimension of the row space is at most 3, and any four vectors in the row space must form a linearly dependent set.

53. Assume that A is an $m \times n$ matrix where $n > m$. Then the set of n column vectors of A are vectors in R^m and must be linearly dependent. Similarly, if $m > n$, then the set of m row vectors of A are vectors in R^n, and must be linearly dependent.

55. (a) Let

$$A = \begin{bmatrix} 1 & 0 \\ 0 & 1 \end{bmatrix} \quad \text{and} \quad B = \begin{bmatrix} 0 & 1 \\ 1 & 0 \end{bmatrix}. \qquad \text{Then } A + B = \begin{bmatrix} 1 & 1 \\ 1 & 1 \end{bmatrix}.$$

Note that $\text{rank}(A) = \text{rank}(B) = 2$, and $\text{rank}(A + B) = 1$.

(b) Let

$$A = \begin{bmatrix} 1 & 0 \\ 0 & 0 \end{bmatrix} \quad \text{and} \quad B = \begin{bmatrix} 0 & 1 \\ 0 & 0 \end{bmatrix}. \qquad \text{Then } A + B = \begin{bmatrix} 1 & 1 \\ 0 & 0 \end{bmatrix}.$$

Note that $\text{rank}(A) = \text{rank}(B) = 1$, and $\text{rank}(A + B) = 1$.

(c) Let

$$A = \begin{bmatrix} 1 & 0 \\ 0 & 0 \end{bmatrix} \quad \text{and} \quad B = \begin{bmatrix} 0 & 0 \\ 0 & 1 \end{bmatrix}. \qquad \text{Then } A + B = \begin{bmatrix} 1 & 0 \\ 0 & 1 \end{bmatrix}.$$

Note that $\text{rank}(A) = \text{rank}(B) = 1$, and $\text{rank}(A + B) = 2$.

57. (a) Because the row (or column) space has dimension no larger than the smaller of m and n, $r \le m$ (because $m < n$).

(b) There are r vectors in a basis for the row space of A.

(c) There are r vectors in a basis for the column space of A.

(d) The row space of A is a subspace of R^n.

(e) The column space of A is a subspace of R^m.

59. Consider the first row of the product AB.

$$\begin{bmatrix} a_{11} & \cdots & a_{1n} \\ \vdots & & \vdots \\ a_{m1} & \cdots & a_{mn} \end{bmatrix} \begin{bmatrix} b_{11} & \cdots & b_{1k} \\ \vdots & & \vdots \\ b_{n1} & \cdots & b_{nk} \end{bmatrix} = \begin{bmatrix} (a_{11}b_{11} + \cdots + a_{1n}b_{n1}) & \cdots & (a_{11}b_{1k} + \cdots + a_{1n}b_{nk}) \\ \vdots & & \vdots \\ (a_{m1}b_{11} + \cdots + a_{mn}b_{n1}) & \cdots & (a_{m1}b_{1k} + \cdots + a_{mn}b_{nk}) \end{bmatrix}.$$

$$\qquad A \qquad\qquad\qquad B \qquad\qquad\qquad\qquad\qquad\qquad AB$$

First row of $AB = \left[(a_{11}b_{11} + \cdots + a_{1n}b_{n1}), \cdots, (a_{11}b_{1k} + \cdots + a_{1n}b_{nk}) \right]$.

You can express this first row as $a_{11}(b_{11}, b_{12}, \cdots b_{1k}) + a_{12}(b_{21}, \cdots, b_{2k}) + \cdots + a_{1n}(b_{n1}, b_{n2}, \cdots, b_{nk})$,

which means that the first row of AB is in the row space of B. The same argument applies to the other rows of A. A similar argument can be used to show that the column vectors of AB are in the column space of A. The first column of AB is

$$\begin{bmatrix} a_{11}b_{11} & + & \cdots & + & a_{1n}b_{n1} \\ & & \vdots & & \\ a_{m1}b_{11} & + & \cdots & + & a_{mn}b_{n1} \end{bmatrix} = b_{11}\begin{bmatrix} a_{11} \\ \vdots \\ a_{m1} \end{bmatrix} + \cdots + b_{n1}\begin{bmatrix} a_{1n} \\ \vdots \\ a_{mn} \end{bmatrix}.$$

61. Let $A\mathbf{x} = \mathbf{b}$ be a system of linear equations in n variables.

(a) If $\text{rank}(A) = \text{rank}([A \vdots \mathbf{b}]) = n$, then $\mathbf{b}$ is in the column space of A, and so $A\mathbf{x} = \mathbf{b}$ has a unique solution.

(b) If $\text{rank}(A) = \text{rank}([A \vdots \mathbf{b}]) < n$, then $\mathbf{b}$ is in the column space of A and $\text{rank}(A) < n$, which implies that $A\mathbf{x} = \mathbf{b}$ has an infinite number of solutions.

(c) If $\text{rank}(A) < \text{rank}([A \vdots \mathbf{b}])$, then $\mathbf{b}$ is *not* in the column space of A, and the system is inconsistent.

63. (a) True. See Theorem 4.13 on page 233.

(b) False. The dimension of the solution space of $A\mathbf{x} = \mathbf{0}$ for $m \times n$ matrix of rank r is $n - r$. See Theorem 4.17 on page 241.

65. (a) True. The columns of A become the rows of the transpose, A^T, so the columns of A span the same space as the rows of A^T.

(b) False. The elementary row operations on A do *not* change linear dependency relationships of the columns of A but may change the column space of A.

67. (a) $\text{rank}(A) = \text{rank}(B) = 3$.

nullity$(A) = n - r = 5 - 3 = 2$.

(b) Choosing $x_3 = s$ and $x_5 = t$ as the free variables, you have

$x_1 = -s - t$

$x_2 = 2s - 3t$

$x_3 = s$

$x_4 = 5t$

$x_5 = t$

A basis for nullspace is

$\{(-1, 2, 1, 0, 0), (-1, -3, 0, 5, 1)\}$.

(c) A basis for the row space of A (which is equal to the row space of B) is

$\{(1, 0, 1, 0, 1), (0, 1, -2, 0, 3), (0, 0, 0, 1, -5)\}$.

(d) A basis for the column space A (which is *not* the same as the column space of B) is

$\{(-2, 1, 3, 1), (-5, 3, 11, 7), (0, 1, 7, 5)\}$.

(e) Linearly dependent

(f) (i) and (iii) are linearly independent, while (ii) is linearly dependent.

69. (a) $A\mathbf{x} = B\mathbf{x} \Rightarrow (A - B)(\mathbf{x}) = \mathbf{0}$ for all

$\mathbf{x} \in R^n \Rightarrow \text{nullity}(A - B) = n$ and

$\text{rank}(A - B) = 0$

(b) So, $A - B = O \Rightarrow A = B$.

71. Let A and B be $2m \times n$ row equivalent matrices. The dependency relationships among the columns of A can be expressed in the form $A\mathbf{x} = \mathbf{0}$, while those of B in the form $B\mathbf{x} = \mathbf{0}$. Because A and B are row equivalent, $A\mathbf{x} = \mathbf{0}$ and $B\mathbf{x} = \mathbf{0}$ have the same solution sets, and therefore the same dependency relationships.

Section 4.7   Coordinates and Change of Basis

1. Because $[\mathbf{x}]_B = \begin{bmatrix} 4 \\ 1 \end{bmatrix}$, you can write

$\mathbf{x} = 4(2, -1) + 1(0, 1) = (8, -3)$.

Moreover, because $(8, -3) = 8(1, 0) - 3(0, 1)$, it follows that the coordinates of $\mathbf{x}$ relative to S are

$[\mathbf{x}]_S = \begin{bmatrix} 8 \\ -3 \end{bmatrix}$.

3. Because $[\mathbf{x}]_B = \begin{bmatrix} 2 \\ 3 \\ 1 \end{bmatrix}$, you can write

$\mathbf{x} = 2(1, 0, 1) + 3(1, 1, 0) + 1(0, 1, 1) = (5, 4, 3)$.

Moreover, because

$(5, 4, 3) = 5(1, 0, 0) + 4(0, 1, 0) + 3(0, 0, 1)$, it follows that the coordinates of $\mathbf{x}$ relative to S are

$[\mathbf{x}]_S = \begin{bmatrix} 5 \\ 4 \\ 3 \end{bmatrix}$.

5. Because $[\mathbf{x}]_B = \begin{bmatrix} 1 \\ -2 \\ 3 \\ -1 \end{bmatrix}$, you can write

$\mathbf{x} = 1(0, 0, 0, 1) - 2(0, 0, 1, 1) + 3(0, 1, 1, 1) - 1(1, 1, 1, 1)$

$= (-1, 2, 0, 1),$

which implies that the coordinates of $\mathbf{x}$ relative to the standard basis S are

$[\mathbf{x}]_S = \begin{bmatrix} -1 \\ 2 \\ 0 \\ 1 \end{bmatrix}$.

7. Begin by writing $\mathbf{x}$ as a linear combination of the vectors in B.

$\mathbf{x} = (12, 6) = c_1(4, 0) + c_2(0, 3)$

Equating corresponding components yields the following system of linear equations.

$4c_1 \qquad = 12$

$\qquad 3c_2 = 6$

The solution of this system is $c_1 = 3$ and $c_2 = 2$. So,

$\mathbf{x} = 3(4, 0) + 2(0, 3)$, and the coordinate vector of $\mathbf{x}$

relative to B is $[\mathbf{x}]_B = \begin{bmatrix} 3 \\ 2 \end{bmatrix}$.

9. Begin by writing **x** as a linear combination of the vector in B.

$$\mathbf{x} = (3, 19, 2) = c_1(8, 11, 0) + c_2(7, 0, 10) + c_3(1, 4, 6)$$

Equating corresponding components yields the following system of linear equations.

$$8c_1 + 7c_2 + c_3 = 3$$
$$11c_1 \qquad + 4c_3 = 19$$
$$10c_2 + 6c_3 = 2$$

The solution of this system is $c_1 = 1$, $c_2 = -1$, and $c_3 = 2$. So, $\mathbf{x} = 1(8, 11, 0) + (-1)(7, 0, 10) + 2(1, 4, 6)$, and the coordinate vector of **x** relative to B is

$$[\mathbf{x}]_B = \begin{bmatrix} 1 \\ -1 \\ 2 \end{bmatrix}.$$

11. Begin by writing **x** as a linear combination of the vector in B.

$$\mathbf{x} = (11, 18, -7) = c_1(4, 3, 3) + c_2(-11, 0, 11) + c_3(0, 9, 2)$$

Equating corresponding components yields the following system of linear equations.

$$4c_1 - 11c_2 \qquad = 11$$
$$3c_1 \qquad + 9c_3 = 18$$
$$3c_1 + 11c_2 + 2c_3 = -7$$

The solution to this system is $c_1 = 0$, $c_2 = -1$, and $c_3 = 2$. So,

$$\mathbf{x} = (11, 18, -7) = 0(4, 3, 3) - 1(-11, 0, 11) + 2(0, 9, 2)$$

and

$$[\mathbf{x}]_B = \begin{bmatrix} 0 \\ -1 \\ 2 \end{bmatrix}.$$

13. Begin by forming the matrix

$$[B' : B] = \begin{bmatrix} 2 & 1 & : & 1 & 0 \\ 4 & 3 & : & 0 & 1 \end{bmatrix}$$

and then use Gauss-Jordan elimination to produce

$$[I_2 : P^{-1}] = \begin{bmatrix} 1 & 0 & : & \frac{3}{2} & -\frac{1}{2} \\ 0 & 1 & : & -2 & 1 \end{bmatrix}.$$

So, the transition matrix from B to B' is

$$P^{-1} = \begin{bmatrix} \frac{3}{2} & -\frac{1}{2} \\ -2 & 1 \end{bmatrix}.$$

15. Begin by forming the matrix

$$[B' : B] = \begin{bmatrix} 1 & 0 & : & 2 & -1 \\ 0 & 1 & : & 4 & 3 \end{bmatrix}.$$

Because this matrix is already in the form $\left[I_2 : P^{-1} \right]$, you see that the transition matrix from B to B' is

$$P^{-1} = \begin{bmatrix} 2 & -1 \\ 4 & 3 \end{bmatrix}.$$

17. Begin by forming the matrix

$$[B' : B] = \begin{bmatrix} 1 & 0 & 6 & : & 1 & 0 & 0 \\ 0 & 2 & 0 & : & 0 & 1 & 0 \\ 0 & 8 & 12 & : & 0 & 0 & 1 \end{bmatrix}$$

and then use the Gauss-Jordan elimination to produce

$$[I_3 : P^{-1}] = \begin{bmatrix} 1 & 0 & 0 & : & 1 & 2 & -\frac{1}{2} \\ 0 & 1 & 0 & : & 0 & \frac{1}{2} & 0 \\ 0 & 0 & 1 & : & 0 & -\frac{1}{3} & \frac{1}{12} \end{bmatrix}.$$

So, the transition matrix from B to B' is

$$P^{-1} = \begin{bmatrix} 1 & 2 & -\frac{1}{2} \\ 0 & \frac{1}{2} & 0 \\ 0 & -\frac{1}{3} & \frac{1}{12} \end{bmatrix}.$$

19. Begin by forming the matrix

$$[B' : B] = \begin{bmatrix} 2 & -1 & : & 2 & 1 \\ 1 & 2 & : & 5 & 2 \end{bmatrix}$$

and then use Gauss-Jordan elimination to produce

$$[I_2 : P^{-1}] = \begin{bmatrix} 1 & 0 & : & \frac{9}{5} & \frac{4}{5} \\ 0 & 1 & : & \frac{8}{5} & \frac{3}{5} \end{bmatrix}.$$

So, the transition matrix from B to B' is

$$P^{-1} = \begin{bmatrix} \frac{9}{5} & \frac{4}{5} \\ \frac{8}{5} & \frac{3}{5} \end{bmatrix}.$$

21. Begin by forming the matrix

$$[B' : B] = \begin{bmatrix} 1 & 0 & 0 & : & 1 & 1 & 1 \\ 0 & 1 & 0 & : & 3 & 5 & 4 \\ 0 & 0 & 1 & : & 3 & 6 & 5 \end{bmatrix}$$

Because this matrix is already in the form $\left[I_3 : P^{-1} \right]$, you see that the transition matrix from B to B' is

$$P^{-1} = \begin{bmatrix} 1 & 1 & 1 \\ 3 & 5 & 4 \\ 3 & 6 & 5 \end{bmatrix}.$$

23. Begin by forming the matrix

$$[B' : B] = \begin{bmatrix} 0 & -2 & 1 & : & 1 & -1 & 2 \\ 2 & 1 & 1 & : & 2 & 2 & 4 \\ 1 & 0 & 1 & : & 4 & 0 & 0 \end{bmatrix}$$

and then use Gauss-Jordan elimination to produce

$$\left[I_3 : P^{-1}\right] = \begin{bmatrix} 1 & 0 & 0 & : & -7 & 3 & 10 \\ 0 & 1 & 0 & : & 5 & -1 & -6 \\ 0 & 0 & 1 & : & 11 & -3 & -10 \end{bmatrix}.$$

So, the transition matrix from B to B' is

$$P^{-1} = \begin{bmatrix} -7 & 3 & 10 \\ 5 & -1 & -6 \\ 11 & -3 & -10 \end{bmatrix}.$$

25. Begin by forming the matrix

$$[B' : B] = \begin{bmatrix} 1 & -2 & -1 & -2 & : & 1 & 0 & 0 & 0 \\ 3 & -5 & -2 & -3 & : & 0 & 1 & 0 & 0 \\ 2 & -5 & -2 & -5 & : & 0 & 0 & 1 & 0 \\ -1 & 4 & 4 & 11 & : & 0 & 0 & 0 & 1 \end{bmatrix}$$

and then use Gauss-Jordan elimination to produce

$$\left[I_4 : P^{-1}\right] = \begin{bmatrix} 1 & 0 & 0 & 0 & : & -24 & 7 & 1 & -2 \\ 0 & 1 & 0 & 0 & : & -10 & 3 & 0 & -1 \\ 0 & 0 & 1 & 0 & : & -29 & 7 & 3 & -2 \\ 0 & 0 & 0 & 1 & : & 12 & -3 & -1 & 1 \end{bmatrix}.$$

So, the transition matrix from B to B' is

$$P^{-1} = \begin{bmatrix} -24 & 7 & 1 & -2 \\ -10 & 3 & 0 & -1 \\ -29 & 7 & 3 & -2 \\ 12 & -3 & -1 & 1 \end{bmatrix}.$$

27. Begin by forming the matrix

$$[B' : B] = \begin{bmatrix} 1 & -2 & 0 & 0 & 1 & : & 1 & 0 & 0 & 0 & 0 \\ 2 & -3 & 1 & 1 & -1 & : & 0 & 1 & 0 & 0 & 0 \\ 4 & 4 & 2 & 2 & 0 & : & 0 & 0 & 1 & 0 & 0 \\ -1 & 2 & -2 & 2 & 1 & : & 0 & 0 & 0 & 1 & 0 \\ 2 & 1 & 1 & 1 & 2 & : & 0 & 0 & 0 & 0 & 1 \end{bmatrix}$$

and then use Gauss-Jordan elimination to produce

$$\left[I_5 : P^{-1}\right] = \begin{bmatrix} 1 & 0 & 0 & 0 & 0 & : & 1 & -\frac{3}{11} & \frac{5}{11} & 0 & -\frac{7}{11} \\ 0 & 1 & 0 & 0 & 0 & : & 0 & -\frac{2}{11} & \frac{3}{22} & 0 & -\frac{1}{11} \\ 0 & 0 & 1 & 0 & 0 & : & -\frac{5}{4} & \frac{9}{22} & -\frac{19}{44} & -\frac{1}{4} & \frac{21}{22} \\ 0 & 0 & 0 & 1 & 0 & : & -\frac{3}{4} & \frac{1}{2} & -\frac{1}{4} & \frac{1}{4} & \frac{1}{2} \\ 0 & 0 & 0 & 0 & 1 & : & 0 & -\frac{1}{11} & -\frac{2}{11} & 0 & \frac{5}{11} \end{bmatrix}.$$

So, the transition matrix from B to B' is

$$P^{-1} = \begin{bmatrix} 1 & -\frac{3}{11} & \frac{5}{11} & 0 & \frac{7}{11} \\ 0 & -\frac{2}{11} & \frac{3}{22} & 0 & -\frac{1}{11} \\ -\frac{5}{4} & \frac{9}{22} & -\frac{19}{44} & -\frac{1}{4} & \frac{21}{22} \\ -\frac{3}{4} & \frac{1}{2} & -\frac{1}{4} & \frac{1}{4} & \frac{1}{2} \\ 0 & -\frac{1}{11} & -\frac{2}{11} & 0 & \frac{5}{11} \end{bmatrix}.$$

29. (a) $[B' : B] = \begin{bmatrix} -12 & -4 & : & 1 & -2 \\ 0 & 4 & : & 3 & -2 \end{bmatrix} \Rightarrow \begin{bmatrix} 1 & 0 & : & -\frac{1}{3} & \frac{1}{3} \\ 0 & 1 & : & \frac{3}{4} & -\frac{1}{2} \end{bmatrix} = \left[I : P^{-1}\right]$

(b) $[B : B'] = \begin{bmatrix} 1 & -2 & : & -12 & -4 \\ 3 & -2 & : & 0 & 4 \end{bmatrix} \Rightarrow \begin{bmatrix} 1 & 0 & : & 6 & 4 \\ 0 & 1 & : & 9 & 4 \end{bmatrix} = [I : P]$

(c) $P^{-1}P = \begin{bmatrix} -\frac{1}{3} & \frac{1}{3} \\ \frac{3}{4} & -\frac{1}{2} \end{bmatrix}\begin{bmatrix} 6 & 4 \\ 9 & 4 \end{bmatrix} = \begin{bmatrix} 1 & 0 \\ 0 & 1 \end{bmatrix}$

(d) $[\mathbf{x}]_B = P[\mathbf{x}]_{B'} = \begin{bmatrix} 6 & 4 \\ 9 & 4 \end{bmatrix}\begin{bmatrix} -1 \\ 3 \end{bmatrix} = \begin{bmatrix} 6 \\ 3 \end{bmatrix}$

31. (a) $[B' \vdots B] = \begin{bmatrix} 2 & 1 & 0 & \vdots & 1 & 0 & 1 \\ 1 & 0 & 2 & \vdots & 0 & 1 & 1 \\ 1 & 0 & 1 & \vdots & 2 & 3 & 1 \end{bmatrix} \Rightarrow \begin{bmatrix} 1 & 0 & 0 & \vdots & 4 & 5 & 1 \\ 0 & 1 & 0 & \vdots & -7 & -10 & -1 \\ 0 & 0 & 1 & \vdots & -2 & -2 & 0 \end{bmatrix} = [I \vdots P^{-1}]$

(b) $[B \vdots B'] = \begin{bmatrix} 1 & 0 & 1 & \vdots & 2 & 1 & 0 \\ 0 & 1 & 1 & \vdots & 1 & 0 & 2 \\ 2 & 3 & 1 & \vdots & 1 & 0 & 1 \end{bmatrix} \Rightarrow \begin{bmatrix} 1 & 0 & 0 & \vdots & \frac{1}{2} & \frac{1}{2} & -\frac{5}{4} \\ 0 & 1 & 0 & \vdots & -\frac{1}{2} & -\frac{1}{2} & \frac{3}{4} \\ 0 & 0 & 1 & \vdots & \frac{3}{2} & \frac{1}{2} & \frac{5}{4} \end{bmatrix} = [I \vdots P]$

(c) $P^{-1}P = \begin{bmatrix} 4 & 5 & 1 \\ -7 & -10 & -1 \\ -2 & -2 & 0 \end{bmatrix} \begin{bmatrix} \frac{1}{2} & \frac{1}{2} & -\frac{5}{4} \\ -\frac{1}{2} & -\frac{1}{2} & \frac{3}{4} \\ \frac{3}{2} & \frac{1}{2} & \frac{5}{4} \end{bmatrix} = \begin{bmatrix} 1 & 0 & 0 \\ 0 & 1 & 0 \\ 0 & 0 & 1 \end{bmatrix}$

(d) $[\mathbf{x}]_B = P[\mathbf{x}]_{B'} = \begin{bmatrix} \frac{1}{2} & \frac{1}{2} & -\frac{5}{4} \\ -\frac{1}{2} & -\frac{1}{2} & \frac{3}{4} \\ \frac{3}{2} & \frac{1}{2} & \frac{5}{4} \end{bmatrix} \begin{bmatrix} 1 \\ 2 \\ -1 \end{bmatrix} = \begin{bmatrix} \frac{11}{4} \\ -\frac{9}{4} \\ \frac{5}{4} \end{bmatrix}$

33. (a) $[B' \vdots B] = \begin{bmatrix} 1 & 4 & 2 & \vdots & 4 & 6 & 2 \\ 0 & 2 & 5 & \vdots & 2 & -5 & -1 \\ 4 & 8 & -2 & \vdots & -4 & -6 & 8 \end{bmatrix}$

$[I \vdots P^{-1}] = \begin{bmatrix} 1 & 0 & 0 & \vdots & -\frac{48}{5} & -24 & \frac{4}{5} \\ 0 & 1 & 0 & \vdots & 4 & 10 & \frac{1}{2} \\ 0 & 0 & 1 & \vdots & -\frac{6}{5} & -5 & -\frac{2}{5} \end{bmatrix}$

So, the transition matrix from B to B' is

$P^{-1} = \begin{bmatrix} -\frac{48}{5} & -24 & \frac{4}{5} \\ 4 & 10 & \frac{1}{2} \\ -\frac{6}{5} & -5 & -\frac{2}{5} \end{bmatrix}.$

(b) $[B \vdots B'] = \begin{bmatrix} 4 & 6 & 2 & \vdots & 1 & 4 & 2 \\ 2 & -5 & -1 & \vdots & 0 & 2 & 5 \\ -4 & -6 & 8 & \vdots & 4 & 8 & -2 \end{bmatrix}$

$[I \vdots P] = \begin{bmatrix} 1 & 0 & 0 & \vdots & \frac{3}{32} & \frac{17}{20} & \frac{5}{4} \\ 0 & 1 & 0 & \vdots & \frac{1}{16} & -\frac{3}{10} & -\frac{1}{2} \\ 0 & 0 & 1 & \vdots & \frac{1}{2} & \frac{6}{5} & 0 \end{bmatrix}$

So, the transition matrix from B' to B is

$P = \begin{bmatrix} \frac{3}{32} & \frac{17}{20} & \frac{5}{4} \\ \frac{1}{16} & -\frac{3}{10} & -\frac{1}{2} \\ \frac{1}{2} & \frac{6}{5} & 0 \end{bmatrix}.$

(c) Using a graphing utility you have $PP^{-1} = I$.

(d) $[\mathbf{x}]_B = P[\mathbf{x}]_{B'} = P\begin{bmatrix} 1 \\ -1 \\ 2 \end{bmatrix} = \begin{bmatrix} \frac{279}{160} \\ -\frac{61}{80} \\ -\frac{7}{10} \end{bmatrix}$

35. The standard basis in P_2 is $S = \{1, x, x^2\}$ and because

$$p = 4(1) + 11(x) + 1(x^2),$$

it follows that

$$[p]_S = \begin{bmatrix} 4 \\ 11 \\ 1 \end{bmatrix}.$$

37. The standard basis in P_2 is $S = \{1, x, x^2\}$ and because

$$p = 1(1) + 5(x) - 2(x^2),$$

it follows that

$$[p]_S = \begin{bmatrix} 1 \\ 5 \\ -2 \end{bmatrix}.$$

39. The standard basis in $M_{3,1}$ is

$$S = \left\{ \begin{bmatrix} 1 \\ 0 \\ 0 \end{bmatrix}, \begin{bmatrix} 0 \\ 1 \\ 0 \end{bmatrix}, \begin{bmatrix} 0 \\ 0 \\ 1 \end{bmatrix} \right\}$$

and because

$$X = 0\begin{bmatrix} 1 \\ 0 \\ 0 \end{bmatrix} + 3\begin{bmatrix} 0 \\ 1 \\ 0 \end{bmatrix} + 2\begin{bmatrix} 0 \\ 0 \\ 1 \end{bmatrix},$$

it follows that

$$[X]_S = \begin{bmatrix} 0 \\ 3 \\ 2 \end{bmatrix}.$$

41. The standard basis in $M_{3,1}$ is

$$S = \left\{ \begin{bmatrix} 1 \\ 0 \\ 0 \end{bmatrix}, \begin{bmatrix} 0 \\ 1 \\ 0 \end{bmatrix}, \begin{bmatrix} 0 \\ 0 \\ 1 \end{bmatrix} \right\}$$

and because

$$X = 1\begin{bmatrix} 1 \\ 0 \\ 0 \end{bmatrix} + 2\begin{bmatrix} 0 \\ 1 \\ 0 \end{bmatrix} - 1\begin{bmatrix} 0 \\ 0 \\ 1 \end{bmatrix},$$

it follows that

$$[X]_S = \begin{bmatrix} 1 \\ 2 \\ -1 \end{bmatrix}.$$

43. (a) False. See Theorem 4.20, page 253.

(b) True. See the discussion following Example 4, page 257.

45. If P is the transition matrix from B'', to B', then $P[\mathbf{x}]_{B''} = [\mathbf{x}]_{B'}$. If Q is the transition matrix from B' to B, then $Q[\mathbf{x}]_{B'} = [\mathbf{x}]_B$. So,

$$[\mathbf{x}]_B = Q[\mathbf{x}]_{B'} = QP[\mathbf{x}]_{B''}$$

which means that QP, is the transition matrix from B'' to B.

47. If B is the standard basis, then

$$[B' \vdots B] = [B' \vdots I] \Rightarrow \left[I \vdots (B')^{-1} \right]$$

shows that P^{-1}, the transition matrix from B to B', is $(B')^{-1}$.

If B' is the standard basis, then

$$[B' \vdots B] = [I \vdots B]$$

shows that P^{-1}, the transition matrix from B to B', is B.

Section 4.8   Applications of Vector Spaces

1. (a) If $y = e^x$, then $y'' = e^x$ and $y'' + y = 2e^x \neq 0$. So, e^x is not a solution to the equation.

(b) If $y = \sin x$, then $y'' = -\sin x$ and $y'' + y = 0$. So, $\sin x$ is a solution to the equation.

(c) If $y = \cos x$, then $y'' = -\cos x$ and $y'' + y = 0$. So, $\cos x$ is a solution to the equation.

(d) If $y = \sin x - \cos x$, then $y'' = -\sin x + \cos x$ and $y'' + y = 0$. So, $\sin x - \cos x$ is a solution to the equation.

3. (a) If $y = e^{-2x}$, then $y' = -2e^{-2x}$ and $y'' = 4e^{-2x}$. So,

$$y'' + 4y' + 4y = 4e^{-2x} + 4(-2e^{-2x}) + 4(e^{-2x}) = 0, \text{ and } e^{-2x} \text{ is a solution.}$$

(b) If $y = xe^{-2x}$, then $y' = (1 - 2x)e^{-2x}$ and $y'' = (4x - 4)e^{-2x}$. So,

$$y'' + 4y' + 4y = (4x - 4)e^{-2x} + 4(1 - 2x)e^{-2x} + 4xe^{-2x} = 0, \text{ and } xe^{-2x} \text{ is a solution.}$$

(c) If $y = x^2 e^{-2x}$, then $y' = (2x - 2x^2)e^{-2x}$ and $y'' = (4x^2 - 8x + 2)e^{-2x}$. So,

$$y'' + 4y' + 4y = (4x^2 - 8x + 2)e^{-2x} + 4(2x - 2x^2)e^{-2x} + 4(x^2 e^{-2x}) \neq 0, \text{ and } x^2 e^{-2x} \text{ is not a solution.}$$

(d) If $y = (x + 2)e^{-2x}$, then $y' = (-3 - 2x)e^{-2x}$ and $y'' = (4 + 4x)e^{-2x}$. So,

$$y'' + 4y' + 4y = (4 + 4x)e^{-2x} + 4(-3 - 2x)e^{-2x} + 4(x + 2)e^{-2x} = 0, \text{ and } (x + 2)e^{-2x} \text{ is a solution.}$$

5. (a) If $y = \dfrac{1}{x^2}$, then $y'' = \dfrac{6}{x^4}$. So, $x^2 y'' - 2y = x^2\left(\dfrac{6}{x^4}\right) - 2\left(\dfrac{1}{x^2}\right) \neq 0$, and $y = \dfrac{1}{x^2}$ is not a solution.

(b) If $y = x^2$, then $y'' = 2$. So, $x^2 y'' - 2y = x^2(2) - 2x^2 = 0$, and $y = x^2$ is a solution.

(c) If $y = e^{x^2}$, then $y'' = 4x^2 e^{x^2} + 2e^{x^2}$. So, $x^2 y'' - 2y = x^2\left(4x^2 e^{x^2} + 2e^{x^2}\right) - 2\left(e^{x^2}\right) \neq 0$, and $y = e^{x^2}$ is not a solution.

(d) If $y = e^{-x^2}$, then $y'' = 4x^2 e^{-x^2} - 2e^{-x^2}$. So, $x^2 y'' - 2y = x^2\left(4x^2 e^{-x^2} - 2e^{-x^2}\right) - 2\left(e^{-x^2}\right) \neq 0$, and $y = e^{-x^2}$ is not a solution.

7. (a) If $y = xe^{2x}$, then $y' = 2xe^{2x} + e^{2x}$ and $y'' = 4xe^{2x} + 4e^{2x}$. So,

$$y'' - y' - 2y = 4xe^{2x} + 4e^{2x} - \left(2xe^{2x} + e^{2x}\right) - 2\left(xe^{2x}\right) \neq 0, \text{ and } y = xe^{2x} \text{ is not a solution.}$$

(b) If $y = 2e^{2x}$, then $y' = 4e^{2x}$ and $y'' = 8e^{2x}$. So,

$$y'' - y' - 2y = 8e^{2x} - 4e^{2x} - 2\left(2e^{2x}\right) = 0, \text{ and } y = 2e^{2x} \text{ is a solution.}$$

(c) If $y = 2e^{-2x}$, then $y' = -4e^{-2x}$ and $y'' = 8e^{-2x}$. So,

$$y'' - y' - 2y = 8e^{-2x} - \left(-4e^{-2x}\right) - 2\left(2e^{-2x}\right) \neq 0, \text{ and } y = 2e^{-2x} \text{ is not a solution.}$$

(d) If $y = xe^{-x}$, then $y' = e^{-x} - xe^{-x}$ and $y'' = xe^{-x} - 2e^{-x}$. So,

$$y'' - y' - 2y = xe^{-x} - 2e^{-x} - \left(e^{-x} - xe^{-x}\right) - 2\left(xe^{-x}\right) \neq 0, \text{ and } y = xe^{-x} \text{ is not a solution.}$$

9. $W\left(e^x, e^{-x}\right) = \begin{vmatrix} e^x & e^{-x} \\ \dfrac{d}{dx}\left(e^x\right) & \dfrac{d}{dx}\left(e^{-x}\right) \end{vmatrix} = \begin{vmatrix} e^x & e^{-x} \\ e^x & -e^{-x} \end{vmatrix} = -2$

11. $W(x, \sin x, \cos x) = \begin{vmatrix} x & \sin x & \cos x \\ 1 & \cos x & -\sin x \\ 0 & -\sin x & -\cos x \end{vmatrix} = x\left(-\cos^2 x - \sin^2 x\right) - 1(0) = -x$

13. $W\left(e^{-x}, xe^{-x}, (x+3)e^{-x}\right) = \begin{vmatrix} e^{-x} & xe^{-x} & (x+3)e^{-x} \\ -e^{-x} & (1-x)e^{-x} & (-x-2)e^{-x} \\ e^{-x} & (x-2)e^{-x} & (x+1)e^{-x} \end{vmatrix} = e^{-3x}\begin{vmatrix} 1 & x & x+3 \\ -1 & 1-x & -x-2 \\ 1 & x-2 & x+1 \end{vmatrix} = e^{-3x}\begin{vmatrix} 1 & x & x+3 \\ 0 & 1 & 1 \\ 0 & -2 & -2 \end{vmatrix} = 0$

15. $W\left(1, e^x, e^{2x}\right) = \begin{vmatrix} 1 & e^x & e^{2x} \\ 0 & e^x & 2e^{2x} \\ 0 & e^x & 4e^{2x} \end{vmatrix} = 4e^{3x} - 2e^{3x} = 2e^{3x}$

17. Because $W(\sin x, \cos x) = \begin{vmatrix} \sin x & \cos x \\ \cos x & -\sin x \end{vmatrix} = -\sin^2 x - \cos^2 x = -1 \neq 0$, the set is linearly independent.

19. $W\left(e^{-2x}, xe^{-2x}, (2x+1)e^{-2x}\right) = \begin{vmatrix} e^{-2x} & xe^{-2x} & (2x+3)e^{-2x} \\ -2e^{-2x} & (1-2x)e^{-2x} & -4xe^{-2x} \\ 4e^{-2x} & (4x-4)e^{-2x} & (8x-4)e^{-2x} \end{vmatrix}$

$$= e^{-6x}\begin{vmatrix} 1 & x & 2x+1 \\ -2 & 1-2x & -4x \\ 4 & 4x-4 & 8x-4 \end{vmatrix}$$

$$= e^{-6x}\begin{vmatrix} 1 & x & 2x+1 \\ 0 & 1 & 2 \\ 0 & -4 & -8 \end{vmatrix}$$

$$= 0,$$

the set is linearly dependent.

21. Because

$$W(2, -1 + 2\sin x, 1 + \sin x) = \begin{vmatrix} 2 & -1 + 2\sin x & 1 + \sin x \\ 0 & 2\cos x & \cos x \\ 0 & -2\sin x & -\sin x \end{vmatrix} = -4\cos x \sin x + 4\cos x \sin x = 0,$$

the set is linearly independent.

23. Note that $e^{-x} + xe^{-x}$ is the sum of the first two expressions in the set. So, the set is linearly dependent.

25. From Exercise 17 you have a set of two linearly independent solutions. Because $y'' + y = 0$ is second-degree, it has a general solution of the form $C_1 \sin x + C_2 \cos x$.

31. First calculate the Wronskian of the two functions

$$W\left(e^{ax}, xe^{ax}\right) = \begin{vmatrix} e^{ax} & xe^{ax} \\ ae^{ax} & (ax+1)e^{ax} \end{vmatrix} = (ax+1)e^{2ax} - axe^{2ax} = e^{2ax}.$$

Because $W\left(e^{ax}, xe^{ax}\right) \neq 0$ and the functions are solutions to $y'' - 2ay' + a^2y = 0$, they are linearly independent.

33. No, this is not true. For instance, consider the nonhomogeneous differential equation $y'' = 1$. Two solutions are $y_1 = x^2/2$ and $y_2 = x^2/2 + 1$, but $y_1 + y_2$ is not a solution.

35. The graph of this equation is a parabola $x = -y^2$ with the vertex at the origin. The parabola opens to the left.

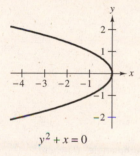

$$y^2 + x = 0$$

37. First rewrite the equation.

$$\frac{x^2}{16} + \frac{y^2}{4} = 1$$

You see that this is the equation of an ellipse centered at the origin with major axis falling along the x-axis.

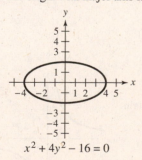

$$x^2 + 4y^2 - 16 = 0$$

27. From Exercise 20 you have a set of three linearly independent solutions. Because $y''' + y' = 0$ is third order, it has a general solution of the form $C_1 + C_2 \sin x + C_3 \cos x$.

29. Clearly $\cos ax$ and $\sin ax$ satisfy the differential equation $y'' + a^2y = 0$. Because $W(\cos ax, \sin ax) = a \neq 0$, they are linearly independent. So, the general solution is $y = C_1 \cos ax + C_2 \sin ax$.

39. First rewrite the equation.

$$\frac{x^2}{9} - \frac{y^2}{16} = 1$$

The graph of this equation is a hyperbola centered at the origin with transverse axis along the x-axis.

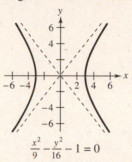

$$\frac{x^2}{9} - \frac{y^2}{16} - 1 = 0$$

41. First complete the square to find the standard form.

$$(x-1)^2 = 4(-2)(y+2)$$

You see that this is the equation of a parabola with vertex at $(1, -2)$ and opening downward.

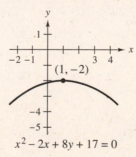

$$x^2 - 2x + 8y + 17 = 0$$

43. First complete the square to find the standard form.

$$(3x - 6)^2 + (5y - 5)^2 = 0$$

The graph of this equation is the single point $(2, 1)$.

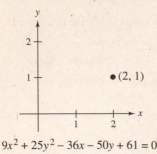

$$9x^2 + 25y^2 - 36x - 50y + 61 = 0$$

45. First complete the square to find the standard form.

$$\frac{(x + 3)^2}{\left(\frac{1}{3}\right)^2} - \frac{(y - 5)^2}{1} = 1$$

You see that this is the equation of a hyperbola centered at $(-3, 5)$ with transverse axis parallel to the x-axis.

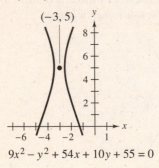

$$9x^2 - y^2 + 54x + 10y + 55 = 0$$

47. First complete the square to find the standard form.

$$\frac{(x + 2)^2}{2^2} + \frac{(y + 4)^2}{1^2} = 1$$

You see that this is the equation of an ellipse centered at $(-2, -4)$ with major axis parallel to the x-axis.

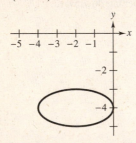

$$x^2 + 4y^2 + 4x + 32y + 64 = 0$$

49. First complete the square to find the standard form.

$$\frac{(y - 5)^2}{1} - \frac{(x + 1)^2}{\frac{1}{2}} = 1$$

You see that this is the equation of a hyperbola centered at $(-1, 5)$ with transverse axis parallel to the y-axis.

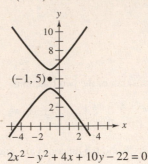

$$2x^2 - y^2 + 4x + 10y - 22 = 0$$

51. First complete the square to find the standard form.

$$(x + 2)^2 = -6(y - 1)$$

This is the equation of a parabola with vertex at $(-2, 1)$ and opening downward.

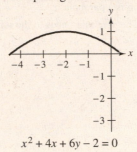

$$x^2 + 4x + 6y - 2 = 0$$

53. Begin by finding the rotation angle, θ, where

$$\cot 2\theta = \frac{a - c}{b} = \frac{0 - 0}{1} = 0, \text{ implying that } \theta = \frac{\pi}{4}.$$

So, $\sin \theta = 1/\sqrt{2}$ and $\cos \theta = 1/\sqrt{2}$. By substituting

$$x = x' \cos \theta - y' \sin \theta = \frac{1}{\sqrt{2}}(x' - y')$$

and

$$y = x' \sin \theta + y' \cos \theta = \frac{1}{\sqrt{2}}(x' + y')$$

into

$$xy + 1 = 0$$

and simplifying, you obtain

$$(x')^2 - (y')^2 + 2 = 0.$$

In standard form

$$\frac{(y')^2}{2} - \frac{(x')^2}{2} = 1.$$

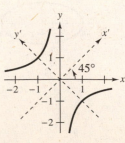

This is the equation of a hyperbola with a transverse axis along the y'-axis.

55. Begin by finding the rotation angle, θ, where

$$\cot 2\theta = \frac{a-c}{b} = \frac{4-4}{2} = 0 \Rightarrow \theta = \frac{\pi}{4}.$$

So, $\sin \theta = \dfrac{1}{\sqrt{2}}$ and $\cos \theta = \dfrac{1}{\sqrt{2}}$. By substituting

$$x = x' \cos \theta - y' \sin \theta = \frac{1}{\sqrt{2}}(x' - y')$$

and

$$y = x' \sin \theta + y' \cos \theta = \frac{1}{\sqrt{2}}(x' + y')$$

into $4x^2 + 2xy + 4y^2 - 15 = 0$ and simplifying, you obtain

$\dfrac{(x')^2}{3} + \dfrac{(y')^2}{5} = 1$, which is an ellipse with major axis along the y'-axis.

57. Begin by finding the rotation angle, θ, where

$$\cot 2\theta = \frac{5-5}{-2} = 0, \text{ implying that } \theta = \frac{\pi}{4}.$$

So, $\sin \theta = 1/\sqrt{2}$ and $\cos \theta = 1/\sqrt{2}$. By substituting

$$x = x' \cos \theta - y' \sin \theta = \frac{1}{\sqrt{2}}(x' - y')$$

and

$$y = x' \sin \theta + y' \cos \theta = \frac{1}{\sqrt{2}}(x' + y')$$

into

$5x^2 - 2xy + 5y^2 - 24 = 0$ and simplifying, you obtain

$4(x')^2 + 6(y')^2 - 24 = 0$.

In standard form $\dfrac{(x')^2}{6} + \dfrac{(y')^2}{4} = 1$.

This is the equation of an ellipse with major axis along the x'-axis.

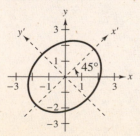

59. Begin by finding the rotation angle, θ, where

$$\cot 2\theta = \frac{a-c}{b} = \frac{13-7}{6\sqrt{3}} = \frac{1}{\sqrt{3}} \Rightarrow 2\theta = \frac{\pi}{3} \Rightarrow \theta = \frac{\pi}{6}.$$

So, $\sin \theta = \dfrac{1}{2}$ and $\cos \theta = \dfrac{\sqrt{3}}{2}$. By substituting

$$x = x' \cos \theta - y' \sin \theta = \frac{\sqrt{3}}{2}x' - \frac{1}{2}y'$$

and

$$y = x' \sin \theta + y' \cos \theta = \frac{1}{2}x' + \frac{\sqrt{3}}{2}y'$$

into $13x^2 + 6\sqrt{3}xy + 7y^2 - 16 = 0$ and simplifying,

you obtain $(x')^2 + \dfrac{(y')^2}{4} = 1$, which is an ellipse with

major axis along the y'-axis.

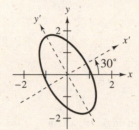

61. Begin by finding the rotation angle, θ, where

$$\cot 2\theta = \frac{1-3}{2\sqrt{3}} = -\frac{1}{\sqrt{3}}, \text{ implying that } \theta = \frac{\pi}{3}.$$

So, $\sin \theta = \sqrt{3}/2$ and $\cos \theta = 1/2$. By substituting

$$x = x' \cos \theta - y' \sin \theta = \frac{1}{2}\left(x' - \sqrt{3}y'\right)$$

and

$$y = x' \sin \theta + y' \cos \theta = \frac{1}{2}\left(\sqrt{3}x' + y'\right)$$

into $x^2 + 2\sqrt{3}xy + 3y^2 - 2\sqrt{3}x + 2y + 16 = 0$

and simplifying, you obtain

$4(x')^2 + 4y' + 16 = 0$.

In standard form $y' + 4 = -(x')^2$.

This is the equation of a parabola with axis on the y'-axis.

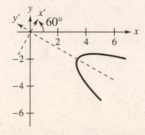

63. Begin by finding the rotation angle, θ, where

$$\cot 2\theta = \frac{1-1}{-2} = 0, \text{ implying that } \theta = \frac{\pi}{4}.$$

So, $\sin \theta = 1/\sqrt{2}$ and $\cos \theta = 1/\sqrt{2}$. By substituting

$$x = x' \cos \theta - y' \sin \theta = \frac{1}{\sqrt{2}}(x' - y')$$

and

$$y = x' \sin \theta + y' \cos \theta = \frac{1}{\sqrt{2}}(x' + y')$$

into $x^2 - 2xy + y^2 = 0$ and simplifying, you obtain

$$2(y')^2 = 0.$$

The graph of this equation is the line $y' = 0$.

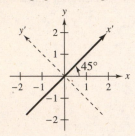

65. Begin by finding the rotation angle, θ, where

67. If $\theta = \frac{\pi}{4}$, then $\sin \theta = \frac{1}{\sqrt{2}}$ and $\cos \theta = \frac{1}{\sqrt{2}}$. So,

$$x = x' \cos \theta - y' \sin \theta = \frac{1}{\sqrt{2}}(x' - y')$$

and

$$y = x' \sin \theta - y' \cos \theta = \frac{1}{\sqrt{2}}(x' + y').$$

Substituting these expressions for x and y into $ax^2 + bxy + ay^2 + dx + ey + f = 0$, you obtain,

$$a\frac{1}{2}(x' - y')^2 + b\frac{1}{2}(x' - y')(x' + y') + a\frac{1}{2}(x' - y')^2 + d\frac{1}{\sqrt{2}}(x' - y') + e\frac{1}{\sqrt{2}}(x' + y') + f = 0.$$

Expanding out the first three terms, you see that $x'y'$-term has been eliminated.

69. Let $A = \begin{bmatrix} a & \dfrac{b}{2} \\ \dfrac{b}{2} & c \end{bmatrix}$ and assume $|A| = ac - \dfrac{b^2}{4} \neq 0$. If $a = 0$, then

$$ax^2 + bxy + cy^2 = bxy + cy^2 = y(cy + bx) = 0, \text{ which implies that } y = 0 \text{ or } y = \frac{-bx}{c},$$

the equations of two intersecting lines.

On the other hand, if $a \neq 0$, then you can divide $ax^2 + bxy + cy^2 = 0$ through by a to obtain

$$x^2 + \frac{b}{a}xy + \frac{c}{a}y^2 = x^2 + \frac{b}{a}xy + \left(\frac{b}{2a}\right)^2 y^2 + \frac{c}{a}y^2 - \left(\frac{b}{2a}\right)^2 y^2 = 0 \Rightarrow \left(x + \frac{b}{2a}y\right)^2 = \left(\left(\frac{b}{2a}\right)^2 - \frac{c}{a}\right)y^2.$$

Because $4ac \neq b^2$, you can see that this last equation represents two intersecting lines.

$$\cot 2\theta = \frac{1-1}{2} = 0, \text{ implying that } \theta = \frac{\pi}{4}.$$

So, $\sin \theta = 1/\sqrt{2}$ and $\cos \theta = 1/\sqrt{2}$. By substituting

$$x = x' \cos \theta - y' \sin \theta = \frac{1}{\sqrt{2}}(x' - y')$$

and

$$y = x' \sin \theta - y' \cos \theta = \frac{1}{\sqrt{2}}(x' + y')$$

into

$$x^2 + 2xy + y^2 - 1 = 0 \text{ and simplifying, you obtain}$$

$$2(x')^2 - 1 = 0.$$

The graph of this equation is two lines $x' = \pm\frac{\sqrt{2}}{2}$.

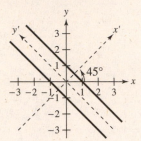

Review Exercises for Chapter 4

1. (a) $\mathbf{u} + \mathbf{v} = (-1, 2, 3) + (1, 0, 2) = (-1 + 1, 2 + 0, 3 + 2) = (0, 2, 5)$

(b) $2\mathbf{v} = 2(1, 0, 2) = (2, 0, 4)$

(c) $\mathbf{u} - \mathbf{v} = (-1, 2, 3) - (1, 0, 2) = (-1 - 1, 2 - 0, 3 - 2) = (-2, 2, 1)$

(d) $3\mathbf{u} - 2\mathbf{v} = 3(-1, 2, 3) - 2(1, 0, 2) = (-3, 6, 9) - (2, 0, 4)$

$\qquad = (-3 - 2, 6 - 0, 9 - 4) = (-5, 6, 5)$

3. (a) $\mathbf{u} + \mathbf{v} = (3, -1, 2, 3) + (0, 2, 2, 1) = (3, 1, 4, 4)$

(b) $2\mathbf{v} = 2(0, 2, 2, 1) = (0, 4, 4, 2)$

(c) $\mathbf{u} - \mathbf{v} = (3, -1, 2, 3) - (0, 2, 2, 1) = (3, -3, 0, 2)$

(d) $3\mathbf{u} - 2\mathbf{v} = 3(3, -1, 2, 3) - 2(0, 2, 2, 1)$

$\qquad = (9, -3, 6, 9) - (0, 4, 4, 2) = (9, -7, 2, 7)$

5. $\mathbf{x} = \frac{1}{2}\mathbf{u} - \frac{3}{2}\mathbf{v} - \frac{1}{2}\mathbf{w}$

$\qquad = \frac{1}{2}(1, -1, 2) - \frac{3}{2}(0, 2, 3) - \frac{1}{2}(0, 1, 1)$

$\qquad = \left(\frac{1}{2}, -4, -4\right)$

7. $5\mathbf{u} - 2\mathbf{x} = 3\mathbf{v} + \mathbf{w}$

$\qquad -2\mathbf{x} = -5\mathbf{u} + 3\mathbf{v} + \mathbf{w}$

$\qquad\quad \mathbf{x} = \frac{5}{2}\mathbf{u} - \frac{3}{2}\mathbf{v} - \frac{1}{2}\mathbf{w}$

$\qquad\quad = \frac{5}{2}(1, -1, 2) - \frac{3}{2}(0, 2, 3) - \frac{1}{2}(0, 1, 1)$

$\qquad\quad = \left(\frac{5}{2}, -\frac{5}{2}, 5\right) - \left(0, 3, \frac{9}{2}\right) - \left(0, \frac{1}{2}, \frac{1}{2}\right)$

$\qquad\quad = \left(\frac{5}{2} - 0 - 0, -\frac{5}{2} - 3 - \frac{1}{2}, 5 - \frac{9}{2} - \frac{1}{2}\right)$

$\qquad\quad = \left(\frac{5}{2}, -6, 0\right)$

9. To write $\mathbf{v}$ as a linear combination of $\mathbf{u}_1$, $\mathbf{u}_2$, and $\mathbf{u}_3$, solve the equation

$c_1\mathbf{u}_1 + c_2\mathbf{u}_2 + c_3\mathbf{u}_3 = \mathbf{v}$

For c_1, c_2, and c_3. This vector equation corresponds to the system of linear equations

$c_1 + 2c_2 + \;\; c_3 = \;\;\; 3$

$-c_1 + 4c_2 + 2c_3 = \;\;\; 0$

$2c_1 - 2c_2 - 4c_3 = -6.$

The solution of this system is $c_1 = 2$, $c_2 = -1$, and $c_3 = 3$. So, $\mathbf{v} = 2\mathbf{u}_1 - \mathbf{u}_2 + 3\mathbf{u}_3$.

11. To write $\mathbf{v}$ as a linear combination of $\mathbf{u}_1$, $\mathbf{u}_2$ and $\mathbf{u}_3$, solve the equation

$c_1\mathbf{u}_1 + c_2\mathbf{u}_2 + c_3\mathbf{u}_3 = \mathbf{v}$

for c_1, c_2, and c_3. This vector equation corresponds to the system of linear equations

$c_1 - \;\; c_2 \qquad\quad = 1$

$2c_1 - 2c_2 \qquad\quad = 2$

$3c_1 - 3c_2 + c_3 = 3$

$4c_1 + 4c_2 + c_3 = 5.$

The solution of this system is $c_1 = \frac{9}{8}$, $c_2 = \frac{1}{8}$, and $c_3 = 0$. So, $\mathbf{v} = \frac{9}{8}\mathbf{u}_1 + \frac{1}{8}\mathbf{u}_2$.

13. The zero vector is $\begin{bmatrix} 0 & 0 & 0 & 0 \\ 0 & 0 & 0 & 0 \\ 0 & 0 & 0 & 0 \end{bmatrix}$.

The additive inverse of $\begin{bmatrix} a_{11} & a_{12} & a_{13} & a_{14} \\ a_{21} & a_{22} & a_{23} & a_{24} \\ a_{31} & a_{32} & a_{33} & a_{34} \end{bmatrix}$ is

$\begin{bmatrix} -a_{11} & -a_{12} & -a_{13} & -a_{14} \\ -a_{21} & -a_{22} & -a_{23} & -a_{24} \\ -a_{31} & -a_{32} & -a_{33} & -a_{34} \end{bmatrix}$.

15. The zero vector is $(0, 0, 0)$. The additive inverse of a vector in R^3 is $(-a_1, -a_2, -a_3)$.

17. Because $W = \{(x, y) : x = 2y\}$ is nonempty and
$W \subset R^2$, you need only check that W is closed under
addition and scalar multiplication. Because
$$(2x_1, x_1) + (2x_2, x_2) = (2(x_1 + x_2), x_1 + x_2) \in W$$
and
$$c(2x_1, x_1) = (2cx_1, cx_1) \in W$$
Conclude that W is a subspace of R^2.

19. W is not a space of R^2.

Because $W = \{(x, y) : y = ax, a \text{ is an integer}\}$ is
nonempty and $W \subset R^2$, you only need to check that W
is closed under addition and scalar multiplication.
Because
$$(x_1, ax_1) + (x_2, ax_2) = (x_1 + x_2, a(x_1 + x_2)) \in W$$
and
$$c(x_1, ax_1) = (cx_1, acx_1) \text{ is not in } W, \text{ because } ac \text{ is not}$$
necessarily an integer. So, W is not closed under scalar
multiplication.

21. Because $W = \{(x, 2x, 3x) : x \text{ is a real number}\}$ is nonempty and $W \subset R^2$, you need only check that W is closed under
addition and scalar multiplication. Because
$$(x_1, 2x_1, 3x_1) + (x_2, 2x_2, 3x_2) = (x_1 + x_2, 2(x_1 + x_2), 3(x_1 + x_2)) \in W$$
and
$$c(x_1, 2x_1, 3x_1) = (cx_2, 2(cx_1), 3(cx_1)) \in W,$$
conclude that W is a subspace of R^3.

23. W is not a subspace of $C[-1, 1]$. For instance, $f(x) = x - 1$ and $g(x) = -1$ are in W, but their sum $(f + g)(x) = x - 2$ is
not in W, because $(f + g)(0) = -2 \neq -1$. So, W is closed under addition (nor scalar multiplication).

25. (a) The only vector in W is the zero vector. So, W is nonempty and $W \subset R^3$. Furthermore, because W is closed under
addition and scalar multiplication, it is a subspace of R^3.

(b) W is not closed under addition or scalar multiplication, so it is not a subspace of R^3. For example, $(1, 0, 0) \in W$, and yet
$$2(1, 0, 0) = (2, 0, 0) \notin W.$$

27. (a) To find out whether S spans R^3, form the vector equation
$$c_1(1, -5, 4) + c_2(11, 6, -1) + c_3(2, 3, 5) = (u_1, u_2, u_3).$$
This yields the system of linear equations
$$\begin{aligned} c_1 + 11c_2 + 2c_3 &= u_1 \\ -5c_1 + 6c_2 + 3c_3 &= u_2 \\ 4c_1 - c_2 + 5c_3 &= u_3. \end{aligned}$$
This system has a unique solution for every (u_1, u_2, u_3) because the determinant of the coefficient matrix is not zero. So, S
spans R^3.

(b) Solving the same system in (a) with $(u_1, u_2, u_3) = (0, 0, 0)$ yields the trivial solution. So, S is linearly independent.

(c) Because S is linearly independent and S spans R^3, it is a basis for R^3.

29. (a) To find out whether S spans R^3, form the vector equation

$$c_1\left(-\tfrac{1}{2}, \tfrac{3}{4}, -1\right) + c_2(5, 2, 3) + c_3(-4, 6, -8) = (u_1, u_2, u_3).$$

This yields the system

$$-\tfrac{1}{2}c_1 + 5c_2 - 4c_3 = u_1$$
$$\tfrac{3}{4}c_1 + 2c_2 + 6c_3 = u_2$$
$$-c_1 + 3c_2 - 8c_3 = u_3.$$

which is equivalent to the system

$$c_1 - 10c_2 + 8c_3 = -2u_1$$
$$7c_2 = 2u_2 - u_3$$
$$0 = -34u_1 + 28u_2 + 38u_3.$$

So, there are vectors (u_1, u_2, u_3) not spanned by S. For instance, $(0, 0, 1) \notin \text{span}(S)$.

(b) Solving the same system in (a) for $(u_1, u_2, u_3) = (0, 0, 0)$ yields nontrivial solutions. For instance, $c_1 = -8$, $c_2 = 0$ and $c_3 = 1$.

So, $-8\left(-\tfrac{1}{2}, \tfrac{3}{4}, -1\right) + 0(5, 2, 3) + 1(-4, 6, -8) = (0, 0, 0)$ and S is linearly dependent.

(c) S is not a basis because it does not span R^3 nor is it linearly independent.

31. (a) S span R^3 because the first three vectors in the set form the standard basis of R^3.

(b) S is linearly dependent because the fourth vector is a linear combination of the first three
$$(-1, 2, -3) = -1(1, 0, 0) + 2(0, 1, 0) - 3(0, 0, 1).$$

(c) S is not a basis because it is not linearly independent.

33. S has four vectors, so you need only check that S is linearly independent. Form the vector equation
$$c_1(1 - t) + c_2(2t + 3t^2) + c_3(t^2 - 2t^3) + c_4(2 + t^3) = 0 + 0t + 0t^2 + 0t^3$$

which yields the homogenous system of linear equations

$$c_1 \qquad\qquad + 2c_4 = 0$$
$$-c_1 + 2c_2 \qquad\qquad = 0$$
$$3c_2 + c_3 \qquad = 0$$
$$-2c_3 + c_4 = 0.$$

This system has only the trivial solution. So, S is linearly independent and S is a basis for P_3.

35. S has four vectors, so you need only check that S is linearly independent. Form the vector equation

$$c_1\begin{bmatrix} 1 & 0 \\ 2 & 3 \end{bmatrix} + c_2\begin{bmatrix} -2 & 1 \\ -1 & 0 \end{bmatrix} + c_3\begin{bmatrix} 3 & 4 \\ 2 & 3 \end{bmatrix} + c_4\begin{bmatrix} -3 & -3 \\ 1 & 3 \end{bmatrix} = \begin{bmatrix} 0 & 0 \\ 0 & 0 \end{bmatrix},$$

which yields the homogeneous system of linear equations

$$c_1 - 2c_2 + 3c_3 - 3c_4 = 0$$
$$c_2 + 4c_3 - 3c_4 = 0$$
$$2c_1 - c_2 + 2c_3 + c_4 = 0$$
$$3c_1 + 3c_3 + 3c_4 = 0.$$

Because this system has nontrivial solutions, S is not a basis. For example, one solution is $c_1 = 2$, $c_2 = 1$, $c_3 = -1$, $c_4 = -1$.

$$2\begin{bmatrix} 1 & 0 \\ 2 & 3 \end{bmatrix} + \begin{bmatrix} -2 & 1 \\ -1 & 0 \end{bmatrix} - \begin{bmatrix} 3 & 4 \\ 2 & 3 \end{bmatrix} - \begin{bmatrix} -3 & -3 \\ 1 & 3 \end{bmatrix} = \begin{bmatrix} 0 & 0 \\ 0 & 0 \end{bmatrix}$$

37. (a) This system has solutions of the form
$(-2s - 3t, s, 4t, t)$, where s and t are any real
numbers. A basis for the solution space is
$\{(-2, 1, 0, 0), (-3, 0, 4, 1)\}$.

(b) The dimension of the solution space is 2—the
number of vectors in a basis for the solution space.

39. (a) This system has solutions of the form
$\left(\frac{2}{7}s - t, \frac{3}{7}s, s, t\right)$, where s and t are any real
numbers. A basis is $\{(2, 3, 7, 0), (-1, 0, 0, 1)\}$.

(b) The dimension of the solution space is 2—the
number of vectors in a basis for the solution space.

41. The system given by $A\mathbf{x} = \mathbf{0}$ has solutions of the form
$(8t, 5t)$, where t is any real number. So, a basis for the
solution space is $\{(8, 5)\}$. The rank of A is 1 (the number
of nonzero row vectors in the reduced row-echelon
matrix) and the nullity is 1.
Note that $\operatorname{rank}(A) + \operatorname{nullity}(A) = 1 + 1 = 2 = n$.

43. The system given by $A\mathbf{x} = \mathbf{0}$ has solutions of the form
$(3s - t, -2t, s, t)$, where s and t are any real numbers.
So, a basis for the solution space of $A\mathbf{x} = \mathbf{0}$ is
$\{(3, 0, 1, 0), (-1, -2, 0, 1)\}$. The rank of A is 2 (the number
of nonzero row vectors in the reduced row-echelon
matrix) and the nullity of A is 2. Note that
$\operatorname{rank}(A) + \operatorname{nullity}(A) = 2 + 2 = 4 = n$.

45. The system given by $A\mathbf{x} = \mathbf{0}$ has solutions of the form
$(4t, -2t, t)$, where t is any real number. So, a basis for
the solution space is $\{(4, -2, 1)\}$. The rank of A is 2 (the
number of nonzero row vectors in the reduced row-
echelon matrix) and the nullity is 1. Note that
$\operatorname{rank}(A) + \operatorname{nullity}(A) = 2 + 1 = 3 = n$.

47. (a) Using Gauss-Jordan elimination, the matrix reduces to
$\begin{bmatrix} 1 & 0 \\ 0 & 1 \\ 0 & 0 \end{bmatrix}$. So, the rank is 2.

(b) A basis for the row space is $\{(1, 0), (0, 1)\}$.

49. (a) Because the matrix is already row-reduced, its rank is 1.

(b) A basis for the row space is $\{(1, -4, 0, 4)\}$.

51. (a) Using Gauss-Jordan elimination, the matrix reduces to
$\begin{bmatrix} 1 & 0 & 0 \\ 0 & 1 & 0 \\ 0 & 0 & 1 \end{bmatrix}$. So, the rank is 3.

(b) A basis for the row space is
$\{(1, 0, 0), (0, 1, 0), (0, 0, 1)\}$.

53. Because $[\mathbf{x}]_B = \begin{bmatrix} 3 \\ 5 \end{bmatrix}$, write $\mathbf{x}$ as

$\mathbf{x} = 3(1, 1) + 5(-1, 1) = (-2, 8)$.

Because $(-2, 8) = -2(1, 0) + 8(0, 1)$, the coordinate
vector of $\mathbf{x}$ relative to the standard basis is

$[\mathbf{x}]_S = \begin{bmatrix} -2 \\ 8 \end{bmatrix}$.

55. Because $[\mathbf{x}]_B = \begin{bmatrix} \frac{1}{2} \\ \frac{1}{2} \end{bmatrix}$, write $\mathbf{x}$ as

$\mathbf{x} = \frac{1}{2}\left(\frac{1}{2}, \frac{1}{2}\right) + \frac{1}{2}(1, 0) = \left(\frac{3}{4}, \frac{1}{4}\right)$.

Because $\left(\frac{3}{4}, \frac{1}{4}\right) = \frac{3}{4}(1, 0) + \frac{1}{4}(0, 1)$, the coordinate vector
of $\mathbf{x}$ relative to the standard basis is

$[\mathbf{x}]_S = \begin{bmatrix} \frac{3}{4} \\ \frac{1}{4} \end{bmatrix}$.

57. Because $[\mathbf{x}]_B = \begin{bmatrix} 2 \\ 0 \\ -1 \end{bmatrix}$, write $\mathbf{x}$ as

$\mathbf{x} = 2(1, 0, 0) + 0(1, 1, 0) - 1(0, 1, 1) = (2, -1, -1)$.

Because
$(-2, -1, -1) = 2(1, 0, 0) - 1(0, 1, 0) - 1(0, 0, 1)$, the
coordinate vector of $\mathbf{x}$ relative to the standard basis is

$[\mathbf{x}]_S = \begin{bmatrix} 2 \\ -1 \\ -1 \end{bmatrix}$.

59. To find $[\mathbf{x}]_{B'} = \begin{bmatrix} c_1 \\ c_2 \end{bmatrix}$, solve the equation

$c_1(5, 0) + c_2(0, -8) = (2, 2)$.

The resulting system of linear equations is
$5c_1 \quad\ = 2$
$\quad -8c_2 = 2$.

So, $c_1 = \frac{2}{5}, c_2 = -\frac{1}{4}$, and $[\mathbf{x}]_{B'} = \begin{bmatrix} \frac{2}{5} \\ -\frac{1}{4} \end{bmatrix}$.

61. To find $[\mathbf{x}]_{B'} = \begin{bmatrix} c_1 \\ c_2 \\ c_3 \end{bmatrix}$ solve the equation

$$c_1(1, 2, 3) + c_2(1, 2, 0) + c_3(0, -6, 2) = (3, -3, 0).$$

The resulting system of linear equations is

$$
\begin{aligned}
c_1 + c_2 &= 3 \\
2c_1 + 2c_2 - 6c_3 &= -3 \\
3c_1 \quad\quad + 2c_3 &= 0.
\end{aligned}
$$

The solution to this system is $c_1 = -1$, $c_2 = 4$, $c_3 = \frac{3}{2}$, and

$$[\mathbf{x}]_{B'} = \begin{bmatrix} -1 \\ 4 \\ \frac{3}{2} \end{bmatrix}.$$

63. To find $[\mathbf{x}]_{B'} = \begin{bmatrix} c_1 \\ c_2 \\ c_3 \\ c_4 \end{bmatrix}$, solve the equation

$$c_1(9, -3, 15, 4) + c_2(-3, 0, 0, -1) + c_3(0, -5, 6, 8) + c_4(-3, 4, -2, 3) = (21, -5, 43, 14).$$

Forming the corresponding linear system, you find its solution to be

$c_1 = 3$, $c_2 = 1$, $c_3 = 0$ and $c_4 = 1$. So,

$$[\mathbf{x}]_{B'} = \begin{bmatrix} 3 \\ 1 \\ 0 \\ 1 \end{bmatrix}.$$

65. Begin by finding $\mathbf{x}$ relative to the standard basis

$$\mathbf{x} = 3(1, 1) + (-3)(-1, 1) = (6, 0).$$

Then solve for $[\mathbf{x}]_{B'} = \begin{bmatrix} c_1 \\ c_2 \end{bmatrix}$

by forming the equation $c_1(0, 1) + c_2(1, 2) = (6, 0)$. The resulting system of linear equations is

$$
\begin{aligned}
c_2 &= 6 \\
c_1 + 2c_2 &= 0.
\end{aligned}
$$

The solution to this system is $c_1 = -12$ and $c_2 = 6$. So,

$$[\mathbf{x}]_{B'} = \begin{bmatrix} -12 \\ 6 \end{bmatrix}.$$

67. Begin by finding $\mathbf{x}$ relative to the standard basis

$$\mathbf{x} = -(1, 0, 0) + 2(1, 1, 0) - 3(1, 1, 1) = (-2, -1, -3).$$

Then solve for $[\mathbf{x}]_{B'} = \begin{bmatrix} c_1 \\ c_2 \\ c_3 \end{bmatrix}$

by forming the equation

$c_1(0, 0, 1) + c_2(0, 1, 1) + c_3(1, 1, 1) = (-2, -1, -3)$. The resulting system of linear equations is

$$
\begin{aligned}
c_3 &= -2 \\
c_2 + c_3 &= -1 \\
c_1 + c_2 + c_3 &= -3.
\end{aligned}
$$

The solution to this system is $c_1 = -2$, $c_2 = 1$, and $c_3 = -2$.

So, you have

$$[\mathbf{x}]_{B'} = \begin{bmatrix} -2 \\ 1 \\ -2 \end{bmatrix}.$$

69. Begin by forming

$$[B' \vdots B] = \begin{bmatrix} 1 & 0 & \vdots & 1 & 3 \\ 0 & 1 & \vdots & -1 & 1 \end{bmatrix}.$$

Because this matrix is already in the form $[I_2 \vdots P^{-1}]$, you have

$$P^{-1} = \begin{bmatrix} 1 & 3 \\ -1 & 1 \end{bmatrix}.$$

71. Begin by forming.

$$[B' \vdots B] = \begin{bmatrix} 0 & 0 & 1 & \vdots & 1 & 0 & 0 \\ 0 & 1 & 0 & \vdots & 0 & 1 & 0 \\ 1 & 0 & 0 & \vdots & 0 & 0 & 1 \end{bmatrix}.$$

Then use Gauss-Jordan elimination to obtain

$$[I_3 \vdots P^{-1}] = \begin{bmatrix} 1 & 0 & 0 & \vdots & 0 & 0 & 1 \\ 0 & 1 & 0 & \vdots & 0 & 1 & 0 \\ 0 & 0 & 1 & \vdots & 1 & 0 & 0 \end{bmatrix}.$$

So, we have

$$P^{-1} = \begin{bmatrix} 0 & 0 & 1 \\ 0 & 1 & 0 \\ 1 & 0 & 0 \end{bmatrix}.$$

73. Begin by finding a basis for W. The polynomials in W must have x as a factor. Consequently, a polynomial in W is of the form

$$p = x(c_1 + c_2 x + c_3 x^2) = c_1 x + c_2 x^2 + c_3 x^3.$$

A basis for W is $\{x, x^2, x^3\}$. Similarly, the polynomials in U must have $(x - 1)$ as a factor. A polynomial in U is of the form

$$p = (x - 1)(c_1 + c_2 x + c_3 x^2)$$
$$= c_1(x - 1) + c_2(x^2 - x) + c_3(x^3 - x^2).$$

So, a basis for U is $\{x - 1, x^2 - x, x^3 - x^2\}$. The intersection of W and U contains polynomials with x and $(x - 1)$ as a factor. A polynomial in $W \cap M$ is of the form

$$p = x(x - 1)(c_1 + c_2 x) = c_1(x^2 - x) + c_2(x^3 - x^2).$$

So, a basis for $W \cap U$ is $\{x^2 - x, x^3 - x^2\}$.

75. No. For example, the set $\{x^2 + x, x^2 - x, 1\}$ is a basis for P_2.

77. Because W is a nonempty subset of V, you need only show that W is closed under addition and scalar multiplication. If $(x^3 + x)p(x)$ and $(x^3 + x)q(x)$ are in W, then $(x^3 + x)p(x) + (x^3 + x)q(x) = (x^3 + x)(p(x) + q(x)) \in W$. Finally, $c(x^3 + x)p(x) = (x^3 + x)(cp(x)) \in W$. So, W is a subspace of $P_5 = V$.

79. The row vectors of A are linearly dependent if and only if the rank of A is less than n, which is equivalent to the column vectors of A being linearly dependent.

81. (a) Consider the equation $c_1 f + c_2 g = c_1 x + c_2|x| = 0$. If $x = \frac{1}{2}$, then $\frac{1}{2}c_1 + \frac{1}{2}c_2 = 0$, while if $x = -\frac{1}{2}$, you obtain $-\frac{1}{2}c_1 + \frac{1}{2}c_2 = 0$. This implies that $c_1 = c_2 = 0$, and f and g are linearly independent.

(b) On the interval $[0, 1]$, $f = g = x$, and so they are linearly dependent.

83. (a) True. See discussion above "Definition of Vector Addition and Scalar Multiplication in R^n," page 183.

(b) False. See Theorem 4.3, part 2, page 186.

(c) True. See "Definition of Vector Space" and the discussion following, page 191.

85. (a) True. See discussion under "Vectors in R^n," page 183.

(b) False. See "Definition of Vector Space," part 4, page 191.

(c) True. See discussion following "Summary of Important Vector Spaces," page 194.

87. (a) Because $y' = 3e^{3x}$ and $y'' = 9e^{3x}$, you have

$$y'' - y' - 6y = 9e^{3x} - 3e^{3x} - 6(e^{3x}) = 0.$$

Therefore, e^{3x} is a solution.

(b) Because $y' = 2e^{2x}$ and $y'' = 4e^{2x}$, you have

$$y'' - y' - 6y = 4e^{2x} - 2e^{2x} - 6(e^{2x}) = -4e^{2x} \neq 0.$$

Therefore, e^{2x} is *not* a solution.

(c) Because $y' = -3e^{-3x}$ and $y'' = 9e^{-3x}$, you have

$$y'' - y' - 6y = 9e^{-3x} - \left(-3e^{-3x}\right) - 6\left(e^{-3x}\right) = 6e^{-3x} \neq 0.$$

Therefore, e^{-3x} is *not* a solution.

(d) Because $y' = -2e^{-2x}$ and $y'' = 4e^{-2x}$, you have

$$y'' - y' - 6y = 4e^{-2x} - \left(-2e^{-2x}\right) - 6\left(e^{-2x}\right) = 0.$$

Therefore, e^{-2x} is a solution.

89. (a) Because $y' = -2e^{-2x}$, you have $y' + 2y = -2e^{-2x} + 2e^{-2x} = 0$.

Therefore, e^{-2x} is a solution.

(b) Because $y' = e^{-2x} - 2xe^{-2x}$, you have $y' + 2y = e^{-2x} - 2xe^{-2x} + 2xe^{-2x} = e^{-2x} \neq 0$.

Therefore, xe^{-2x} is *not* a solution.

(c) Because $y' = 2xe^{-x} - x^2e^{-x}$, you have $y' + 2y = 2xe^{-x} - x^2e^{-x} + 2x^2e^{-x} \neq 0$.

Therefore, x^2e^{-x} is *not* a solution.

(d) Because $y' = 2e^{-2x} - 4xe^{-2x}$, you have $y' + 2y = 2e^{-2x} - 4xe^{-2x} + 2\left(2xe^{-2x}\right) = 2e^{-2x} \neq 0$.

Therefore, $2xe^{-2x}$ is *not* a solution.

91. $W\left(1, x, e^x\right) = \begin{vmatrix} 1 & x & e^x \\ 0 & 1 & e^x \\ 0 & 0 & e^x \end{vmatrix} = e^x$

93. $W\left(1, \sin 2x, \cos 2x\right) = \begin{vmatrix} 1 & \sin 2x & \cos 2x \\ 0 & 2\cos 2x & -2\sin 2x \\ 0 & -4\sin 2x & -4\cos 2x \end{vmatrix} = -8.$

95. The Wronskian of this set is

$$W\left(e^{-3x}, xe^{-3x}\right) = \begin{vmatrix} e^{-3x} & xe^{-3x} \\ -3e^{-3x} & (1 - 3x)e^{-3x} \end{vmatrix}$$

$$= (1 - 3x)e^{-6x} + 3xe^{-6x}$$

$$= e^{-6x}.$$

Because $W\left(e^{-3x}, xe^{-3x}\right) = e^{-6x} \neq 0$, the set is linearly independent.

97. The Wronskian of this set is

$$W\left(e^x, e^{2x}, e^x - e^{2x}\right) = \begin{vmatrix} e^x & e^{2x} & e^x - e^{2x} \\ e^x & 2e^{2x} & e^x - 2e^{2x} \\ e^x & 4e^{2x} & e^x - 4e^{2x} \end{vmatrix} = 0.$$

Because the third column is the difference of the first two columns, the set is linearly dependent.

99. Begin by completing the square.

$$\left(x^2 - 4x + 4\right) + \left(y^2 + 2y + 1\right) = 4 + 4 + 1$$

$$(x - 2)^2 + (y + 1)^2 = 9$$

This is the equation of a circle of radius

$\sqrt{9} = 3$, centered at $(2, -1)$.

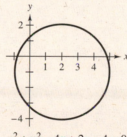

$$x^2 + y^2 - 4x + 2y - 4 = 0$$

101. Begin by completing the square.

$$x^2 - y^2 + 2x - 3 = 0$$

$$\left(x^2 + 2x + 1\right) - y^2 = 3 + 1$$

$$(x + 1)^2 - y^2 = 4$$

$$\frac{(x + 1)^2}{2^2} - \frac{y^2}{2^2} = 1$$

This is the equation of a hyperbola with center $(-1, 0)$.

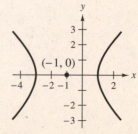

$$x^2 - y^2 + 2x - 3 = 0$$

103. Begin by completing the square.

$$2x^2 - 20x - y + 46 = 0$$

$$2(x^2 - 10x + 25) = y - 46 + 50$$

$$2(x - 5)^2 = y + 4$$

This is the equation of a parabola with vertex $(5, -4)$.

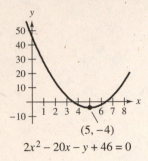

$$2x^2 - 20x - y + 46 = 0$$

105. Begin by completing the square.

$$4x^2 + y^2 + 32x + 4y + 63 = 0$$

$$4(x^2 + 8x + 16) + (y^2 + 4y + 4) = -63 + 64 + 4$$

$$4(x + 4)^2 + (y + 2)^2 = 5$$

This is the equation of an ellipse with center $(-4, -2)$.

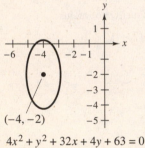

$$4x^2 + y^2 + 32x + 4y + 63 = 0$$

107. From the equation.

$$\cot 2\theta = \frac{a - c}{b} = \frac{0 - 0}{1} = 0$$

you find that the angle of rotation is $\theta = \frac{\pi}{4}$. Therefore, $\sin \theta = 1/\sqrt{2}$ and $\cos \theta = 1/\sqrt{2}$. By substituting

$$x = x' \cos \theta - y' \sin \theta = \frac{1}{\sqrt{2}} = (x' - y')$$

and

$$y = x' \sin \theta + y' \cos \theta = \frac{1}{\sqrt{2}} = (x' + y')$$

into $xy = 3$, you obtain $\frac{1}{2}(x')^2 - \frac{1}{2}(y')^2 = 3$.
In standard form,

$$\frac{(x')^2}{6} - \frac{(y')^2}{6} = 1$$

you can recognize this to be the equation of a hyperbola whose transverse axis is the x'-axix.

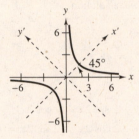

109. From the equation

$$\cos 2\theta = \frac{a - c}{b} = \frac{16 - 9}{-24} = -\frac{7}{24}$$

you find that the angle of rotation is $\theta \approx -36.87°$.
Therefore, $\sin \theta \approx -0.6$ and $\cos \theta \approx 0.8$.

By substituting

$$x = x' \cos \theta - y' \sin \theta = 0.8x' + 0.6y'$$

and

$$y = x' \sin \theta + y' \cos \theta = -0.6x' + 0.8y'$$

into $16x^2 - 24xy + 9y^2 - 60x - 80y + 100 = 0$, you obtain $25(x')^2 - 100y' = -100$. In standard form,

$$(x')^2 = 4(y' - 1)$$

you can recognize this to be the equation of a parabola with vertex at $(x', y') = (0, 1)$.

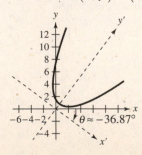

C H A P T E R 5
Inner Product Spaces

CHAPTER 5
Inner Product Spaces

Section 5.1   Length and Dot Product in R^n

1. $\|\mathbf{v}\| = \sqrt{4^2 + 3^2} = \sqrt{25} = 5$

3. $\|\mathbf{v}\| = \sqrt{1^2 + 2^2 + 2^2} = \sqrt{9} = 3$

5. $\|\mathbf{v}\| = \sqrt{2^2 + 0^2 + (-5)^2 + 5^2} = \sqrt{54} = 3\sqrt{6}$

7. (a) $\|\mathbf{u}\| = \sqrt{(-1)^2 + \left(\dfrac{1}{4}\right)^2} = \sqrt{\dfrac{17}{16}} = \dfrac{\sqrt{17}}{4}$

(b) $\|\mathbf{v}\| = \sqrt{4^2 + \left(-\dfrac{1}{8}\right)^2} = \sqrt{\dfrac{1025}{64}} = \dfrac{5\sqrt{41}}{8}$

(c) $\|\mathbf{u} + \mathbf{v}\| = \left\|\left(3, \dfrac{1}{8}\right)\right\|$

$= \sqrt{3^2 + \left(\dfrac{1}{8}\right)^2} = \sqrt{\dfrac{577}{64}} = \dfrac{\sqrt{577}}{8}$

9. (a) $\|\mathbf{u}\| = \sqrt{0^2 + 4^2 + 3^2} = \sqrt{25} = 5$

(b) $\|\mathbf{v}\| = \sqrt{1^2 + (-2)^2 + 1^2} = \sqrt{6}$

(c) $\|\mathbf{u} + \mathbf{v}\| = \|(1, 2, 4)\| + \sqrt{1^2 + 2^2 + 4^2} = \sqrt{21}$

11. (a) $\|\mathbf{u}\| = \sqrt{0^2 + 1^2 + (-1)^2 + 2^2} = \sqrt{6}$

(b) $\|\mathbf{v}\| = \sqrt{1^2 + 1^2 + 3^2 + 0^2} = \sqrt{11}$

(c) $\|\mathbf{u} + \mathbf{v}\| = \|(1, 2, 2, 2)\|$

$= \sqrt{1^2 + 2^2 + 2^2 + 2^2} = \sqrt{13}$

13. (a) A unit vector $\mathbf{v}$ in the direction of $\mathbf{u}$ is given by

$\mathbf{v} = \dfrac{\mathbf{u}}{\|\mathbf{u}\|} = \dfrac{1}{\sqrt{(-5)^2 + 12^2}}(-5, 12)$

$= \dfrac{1}{13}(-5, 12) = \left(-\dfrac{5}{13}, \dfrac{12}{13}\right)$

(b) A unit vector in the direction opposite that of $\mathbf{u}$ is given by

$-\mathbf{v} = -\left(-\dfrac{5}{13}, \dfrac{12}{13}\right) = \left(\dfrac{5}{13}, -\dfrac{12}{13}\right)$

15. (a) A unit vector $\mathbf{v}$ in the direction of $\mathbf{u}$ is given by

$\mathbf{v} = \dfrac{\mathbf{u}}{\|\mathbf{u}\|} = \dfrac{1}{\sqrt{3^2 + 2^2 + (-5)^2}}(3, 2, -5)$

$= \dfrac{1}{\sqrt{38}}(3, 2, -5) = \left(\dfrac{3}{\sqrt{38}}, \dfrac{2}{\sqrt{38}}, -\dfrac{5}{\sqrt{38}}\right)$

(b) A unit vector in the direction opposite that of $\mathbf{u}$ is given by

$-\mathbf{v} = -\left(\dfrac{3}{\sqrt{38}}, \dfrac{2}{\sqrt{38}}, -\dfrac{5}{\sqrt{38}}\right)$

$= \left(-\dfrac{3}{\sqrt{38}}, -\dfrac{2}{\sqrt{38}}, \dfrac{5}{\sqrt{38}}\right)$

17. (a) A unit vector $\mathbf{v}$ in the direction of $\mathbf{u}$ is given by

$\mathbf{v} = \dfrac{\mathbf{u}}{\|\mathbf{u}\|} = \dfrac{1}{\sqrt{1^2 + 0^2 + 2^2 + 2^2}}(1, 0, 2, 2)$

$= \dfrac{1}{3}(1, 0, 2, 2) = \left(\dfrac{1}{3}, 0, \dfrac{2}{3}, \dfrac{2}{3}\right)$

(b) A unit vector in the direction opposite that of $\mathbf{u}$ is given by

$-\mathbf{v} = -\left(\dfrac{1}{3}, 0, \dfrac{2}{3}, \dfrac{2}{3}\right) = \left(-\dfrac{1}{3}, 0, -\dfrac{2}{3}, -\dfrac{2}{3}\right)$

19. Solve the equation for c as follows.

$\|c(1, 2, 3)\| = 1$

$|c|\|(1, 2, 3)\| = 1$

$|c| = \dfrac{1}{\|(1, 2, 3)\|} = \dfrac{1}{\sqrt{14}}$

$c = \pm\dfrac{1}{\sqrt{14}}$

21. First find a unit vector in the direction of $\mathbf{u}$.

$\dfrac{\mathbf{u}}{\|\mathbf{u}\|} = \dfrac{1}{\sqrt{1^2 + 1^2}}(1, 1) = \left(\dfrac{1}{\sqrt{2}}, \dfrac{1}{\sqrt{2}}\right)$

Then $\mathbf{v}$ is four times this vector.

$\mathbf{v} = 4\dfrac{\mathbf{u}}{\|\mathbf{u}\|} = 4\left(\dfrac{1}{\sqrt{2}}, \dfrac{1}{\sqrt{2}}\right) = \left(2\sqrt{2}, 2\sqrt{2}\right)$

23. First find a unit vector in the direction of **u**.

$$\frac{\mathbf{u}}{\|\mathbf{u}\|} = \frac{1}{\sqrt{3+9+0}}(\sqrt{3}, 3, 0)$$

$$= \frac{1}{2\sqrt{3}}(\sqrt{3}, 3, 0) = \left(\frac{1}{2}, \frac{\sqrt{3}}{2}, 0\right)$$

Then **v** is twice this vector.

$$\mathbf{v} = 2\left(\frac{1}{2}, \frac{\sqrt{3}}{2}, 0\right) = (1, \sqrt{3}, 0)$$

25. First find a unit vector in the direction of **u**.

$$\frac{\mathbf{u}}{\|\mathbf{u}\|} = \frac{1}{\sqrt{0+4+1+1}}(0, 2, 1, -1) = \frac{1}{\sqrt{6}}(0, 2, 1, -1)$$

Then **v** is three times this vector.

$$\mathbf{v} = 3\frac{1}{\sqrt{6}}(0, 2, 1, -1) = \left(0, \frac{6}{\sqrt{6}}, \frac{3}{\sqrt{6}}, -\frac{3}{\sqrt{6}}\right)$$

27. (a) Because $\mathbf{v}/\|\mathbf{v}\|$ is a unit vector in the direction of **v**,

$$\mathbf{u} = \frac{\|\mathbf{v}\|}{2}\frac{\mathbf{v}}{\|\mathbf{v}\|} = \frac{1}{2}\mathbf{v} = \frac{1}{2}(8, 8, 6) = (4, 4, 3).$$

(b) Because $-\mathbf{v}/\|\mathbf{v}\|$ is a unit vector with direction opposite that of **v**,

$$\mathbf{u} = \frac{\|\mathbf{v}\|}{4}\left(-\frac{\mathbf{v}}{\|\mathbf{v}\|}\right)$$

$$= \frac{-1}{4}\mathbf{v} = -\frac{1}{4}(8, 8, 6) = \left(-2, -2, -\frac{3}{2}\right).$$

(c) Because $-\mathbf{v}/\|\mathbf{v}\|$ is a unit vector with direction opposite that of **v**,

$$\mathbf{u} = 2\|\mathbf{v}\|\left(-\frac{\mathbf{v}}{\|\mathbf{v}\|}\right)$$

$$= -2\mathbf{v} = -2(8, 8, 6) = (-16, -16, -12).$$

29. $d(\mathbf{u}, \mathbf{v}) = \|\mathbf{u} - \mathbf{v}\| = \|(2, -2)\| = \sqrt{4+4} = 2\sqrt{2}$

31. $d(\mathbf{u}, \mathbf{v}) = \|\mathbf{u} - \mathbf{v}\| = \|(2, -2, 2)\|$

$$= \sqrt{2^2 + (-2)^2 + 2^2} = 2\sqrt{3}$$

33. $d(\mathbf{u}, \mathbf{v}) = \|\mathbf{u} - \mathbf{v}\| = \|(-1, 1, -2, 4)\|$

$$= \sqrt{1+1+4+16} = \sqrt{22}$$

35. (a) $\mathbf{u} \cdot \mathbf{v} = 3(2) + 4(-3) = 6 - 12 = -6$

(b) $\mathbf{u} \cdot \mathbf{u} = 3(3) + 4(4) = 9 + 16 = 25$

(c) $\|\mathbf{u}\|^2 = \mathbf{u} \cdot \mathbf{u} = 25$

(d) $(\mathbf{u} \cdot \mathbf{v})\mathbf{v} = -6(2, -3) = (-12, 18)$

(e) $\mathbf{u} \cdot (5\mathbf{v}) = 5(\mathbf{u} \cdot \mathbf{v}) = 5(-6) = -30$

37. (a) $\mathbf{u} \cdot \mathbf{v} = (-1)(1) + 1(-3) + (-2)(-2) = 0$

(b) $\mathbf{u} \cdot \mathbf{u} = (-1)(-1) + 1(1) + (-2)(-2) = 6$

(c) $\|\mathbf{u}\|^2 = \mathbf{u} \cdot \mathbf{u} = 6$

(d) $(\mathbf{u} \cdot \mathbf{v})\mathbf{v} = 0(1, -3, -2) = (0, 0, 0) = \mathbf{0}$

(e) $\mathbf{u} \cdot (5\mathbf{v}) = 5(\mathbf{u} \cdot \mathbf{v}) = 5 \cdot 0 = 0$

39. (a) $\mathbf{u} \cdot \mathbf{v} = 4(0) + 0(2) + (-3)5 + 5(4) = 5$

(b) $\mathbf{u} \cdot \mathbf{u} = 4(4) + 0(0) + (-3)(-3) + 5(5) = 50$

(c) $\|\mathbf{u}\|^2 = \mathbf{u} \cdot \mathbf{u} = 50$

(d) $(\mathbf{u} \cdot \mathbf{v})\mathbf{v} = 5(0, 2, 5, 4) = (0, 10, 25, 20)$

(e) $\mathbf{u} \cdot (5\mathbf{v}) = 5(\mathbf{u} \cdot \mathbf{v}) = 5 \cdot 5 = 25$

41. $(\mathbf{u} + \mathbf{v}) \cdot (2\mathbf{u} - \mathbf{v}) = \mathbf{u} \cdot (2\mathbf{u} - \mathbf{v}) + \mathbf{v} \cdot (2\mathbf{u} - \mathbf{v})$

$$= 2\mathbf{u} \cdot \mathbf{u} - \mathbf{u} \cdot \mathbf{v} + 2\mathbf{v} \cdot \mathbf{u} - \mathbf{v} \cdot \mathbf{v}$$

$$= 2(\mathbf{u} \cdot \mathbf{u}) + \mathbf{u} \cdot \mathbf{v} - \mathbf{v} \cdot \mathbf{v}$$

$$= 2(4) + (-5) - 10 = -7$$

43. (a) $\|\mathbf{u}\| = \sqrt{5^2 + (-12)^2} = \sqrt{169} = 13,$

$$\|\mathbf{v}\| = \sqrt{(-8)^2 + (-15)^2} = \sqrt{289} = 17$$

(b) $\dfrac{\mathbf{v}}{\|\mathbf{v}\|} = \left(-\dfrac{8}{17}, -\dfrac{15}{17}\right)$

(c) $-\dfrac{\mathbf{u}}{\|\mathbf{u}\|} = \left(-\dfrac{5}{13}, \dfrac{12}{13}\right)$

(d) $\mathbf{u} \cdot \mathbf{v} = (5)(-8) + (-12)(-15)$

$$= -40 + 180 = 140$$

(e) $\mathbf{u} \cdot \mathbf{u} = (5)(5) + (-12)(-12) = 169$

(f) $\mathbf{v} \cdot \mathbf{v} = (-8)(-8) + (-15)(-15) = 64 + 225 = 289$

45. (a) $\|\mathbf{u}\| = \sqrt{5^2 + 12^2} = \sqrt{169} = 13,$

$$\|\mathbf{v}\| = \sqrt{(-12)^2 + 5^2} = \sqrt{169} = 13$$

(b) $\dfrac{\mathbf{v}}{\|\mathbf{v}\|} = \left(-\dfrac{12}{13}, \dfrac{5}{13}\right)$

(c) $-\dfrac{\mathbf{u}}{\|\mathbf{u}\|} = \left(-\dfrac{5}{13}, -\dfrac{12}{13}\right)$

(d) $\mathbf{u} \cdot \mathbf{v} = (5)(-12) + (12)(5) = -60 + 60 = 0$

(e) $\mathbf{u} \cdot \mathbf{u} = (5)(5) + (12)(12) = 169$

(f) $\mathbf{v} \cdot \mathbf{v} = (-12)(-12) + (5)(5) = 144 + 25 = 169$

47. (a) $\|\mathbf{u}\| = \sqrt{10^2 + (-24)^2} = \sqrt{676} = 26,$

$\|\mathbf{v}\| = \sqrt{(-5)^2 + (-12)^2} = \sqrt{169} = 13$

(b) $\dfrac{\mathbf{v}}{\|\mathbf{v}\|} = \left(-\dfrac{5}{13}, -\dfrac{12}{13}\right)$

(c) $-\dfrac{\mathbf{u}}{\|\mathbf{u}\|} = \left(-\dfrac{10}{26}, \dfrac{24}{26}\right) = \left(-\dfrac{5}{13}, \dfrac{12}{13}\right)$

(d) $\mathbf{u} \cdot \mathbf{v} = (10)(-5) + (-24)(-12)$

$= -50 + 288 = 238$

(e) $\mathbf{u} \cdot \mathbf{u} = (10)(10) + (-24)(-24) = 676$

(f) $\mathbf{v} \cdot \mathbf{v} = (-5)(-5) + (-12)(-12)$

$= 25 + 144 = 169$

49. (a) $\|\mathbf{u}\| = \sqrt{0^2 + 5^2 + 12^2} = \sqrt{169} = 13,$

$\|\mathbf{v}\| = \sqrt{0^2 + (-5)^2 + (-12)^2} = \sqrt{169} = 13$

(b) $\dfrac{\mathbf{v}}{\|\mathbf{v}\|} = \left(0, -\dfrac{5}{13}, -\dfrac{12}{13}\right)$

(c) $-\dfrac{\mathbf{u}}{\|\mathbf{u}\|} = \left(0, -\dfrac{5}{13}, -\dfrac{12}{13}\right)$

(d) $\mathbf{u} \cdot \mathbf{v} = (0)(0) + (5)(-5) + (12)(-12)$

$= 0 - 25 - 144 = -169$

(e) $\mathbf{u} \cdot \mathbf{u} = (0)(0) + (5)(5) + (12)(12) = 169$

(f) $\mathbf{v} \cdot \mathbf{v} = (0)(0) + (-5)(-5) + (-12)(-12)$

$= 0 + 25 + 144 = 169$

55. (a) $\|\mathbf{u}\| = 3.7417, \ \|\mathbf{v}\| = 3.7417$

(b) $\dfrac{1}{\|\mathbf{v}\|}\mathbf{v} = (0.5345, 0, 0.2673, 0.2673, 0.5345, -0.5345)$

(c) $-\dfrac{1}{\|\mathbf{u}\|}\mathbf{u} = (0, -0.5345, -0.5345, 0.2673, -0.2673, 0.5345)$

(d) $\mathbf{u} \cdot \mathbf{v} = 7$

(e) $\mathbf{u} \cdot \mathbf{u} = 14$

(f) $\mathbf{v} \cdot \mathbf{v} = 14$

57. (a) $\|\mathbf{u}\| = 3.7417, \ \|\mathbf{u}\| = 4$

(b) $\dfrac{1}{\|\mathbf{v}\|}\mathbf{v} = \left(-\dfrac{1}{4}, 0, \dfrac{1}{4}, \dfrac{1}{2}, -\dfrac{1}{2}, \dfrac{1}{4}, \dfrac{1}{4}, -\dfrac{1}{2}\right)$

(c) $-\dfrac{1}{\|\mathbf{u}\|}\mathbf{u} = (0.2673, -0.2673, -0.5345, 0.2673, -0.2673, -0.2673, 0.5345, -0.2673)$

(d) $\mathbf{u} \cdot \mathbf{v} = -4$

(e) $\mathbf{u} \cdot \mathbf{u} = 14$

(f) $\mathbf{v} \cdot \mathbf{v} = 16$

51. (a) $\|\mathbf{u}\| = 1.0843, \ \|\mathbf{v}\| = 0.3202$

(b) $\dfrac{\mathbf{v}}{\|\mathbf{v}\|} = (0, 0.7809, 0.6247)$

(c) $-\dfrac{\mathbf{u}}{\|\mathbf{u}\|} = (-0.9223, -0.1153, -0.3689)$

(d) $\mathbf{u} \cdot \mathbf{v} = 0.1113$

(e) $\mathbf{u} \cdot \mathbf{u} = 1.1756$

(f) $\mathbf{v} \cdot \mathbf{v} = 0.1025$

53. (a) $\|\mathbf{u}\| = 1.7321, \ \|\mathbf{v}\| = 2$

(b) $\dfrac{\mathbf{v}}{\|\mathbf{v}\|} = (-0.5, 0.7071, -0.5)$

(c) $-\dfrac{\mathbf{u}}{\|\mathbf{u}\|} = (0, -0.5774, -0.8165)$

(d) $\mathbf{u} \cdot \mathbf{v} = 0$

(e) $\mathbf{u} \cdot \mathbf{u} = 3$

(f) $\mathbf{v} \cdot \mathbf{v} = 4$

59. You have

$$\mathbf{u} \cdot \mathbf{v} = 3(2) + 4(-3) = -6,$$

$$\|\mathbf{u}\| = \sqrt{3^2 + 4^2} = \sqrt{25} = 5, \text{ and}$$

$$\|\mathbf{v}\| = \sqrt{2^2 + (-3)^2} = \sqrt{13}. \text{ So,}$$

$$|\mathbf{u} \cdot \mathbf{v}| \le \|\mathbf{u}\| \|\mathbf{v}\|$$

$$|-6| \le 5(\sqrt{13})$$

$$6 \le 5\sqrt{3} \approx 8.66.$$

61. You have

$$\mathbf{u} \cdot \mathbf{v} = (1) + 1(-3) + (-2)(-2) = 2,$$

$$\|\mathbf{u}\| = \sqrt{1^2 + 1^2 + (-2)^2} = \sqrt{6}, \text{ and}$$

$$\|\mathbf{v}\| = \sqrt{1^2 + (-3)^2 + (-2)^2} = \sqrt{14}. \text{ So,}$$

$$|\mathbf{u} \cdot \mathbf{v}| \le \|\mathbf{u}\| \|\mathbf{v}\|$$

$$|2| \le \sqrt{6}\sqrt{14}$$

$$2 \le 2\sqrt{21} \approx 9.17.$$

63. The cosine of the angle θ between $\mathbf{u}$ and $\mathbf{v}$ is given by

$$\cos \theta = \frac{\mathbf{u} \cdot \mathbf{v}}{\|\mathbf{u}\| \|\mathbf{v}\|} = \frac{3(-2) + 1(4)}{\sqrt{3^2 + 1^2}\sqrt{(-2)^2 + 4^2}}$$

$$= -\frac{2}{10\sqrt{2}} = -\frac{\sqrt{2}}{10}.$$

So, $\theta = \cos^{-1}\left(-\frac{\sqrt{2}}{10}\right) \approx 1.713$ radians $(98.13°)$.

65. The cosine of the angle θ between $\mathbf{u}$ and $\mathbf{v}$ is given by

$$\cos \theta = \frac{\mathbf{u} \cdot \mathbf{v}}{\|\mathbf{u}\| \|\mathbf{v}\|}$$

$$= \frac{\cos\frac{\pi}{6}\cos\frac{3\pi}{4} + \sin\frac{\pi}{6}\sin\frac{3\pi}{4}}{\sqrt{\cos^2\frac{\pi}{6} + \sin^2\frac{\pi}{6}} \cdot \sqrt{\cos^2\frac{3\pi}{4} + \sin^2\frac{3\pi}{4}}}$$

$$= \frac{\cos\left(\frac{\pi}{6} - \frac{3\pi}{4}\right)}{1 \cdot 1}$$

$$= \cos\left(-\frac{7\pi}{12}\right)$$

$$= \cos\left(\frac{7\pi}{12}\right).$$

So, $\theta = \frac{7\pi}{12}$ radians $(105°)$.

67. The cosine of the angle θ between $\mathbf{u}$ and $\mathbf{v}$ is given by

$$\cos \theta = \frac{\mathbf{u} \cdot \mathbf{v}}{\|\mathbf{u}\| \|\mathbf{v}\|} = \frac{1(2) + 1(1) + 1(-1)}{\sqrt{1^2 + 1^2 + 1^2}\sqrt{2^2 + 1^2 + (-1)^2}}$$

$$= \frac{2}{3\sqrt{2}} = \frac{\sqrt{2}}{3}.$$

So, $\theta = \cos^{-1}\left(\frac{\sqrt{2}}{3}\right) \approx 1.080$ radians $(61.87°)$.

69. The cosine of the angle θ between $\mathbf{u}$ and $\mathbf{v}$ is given by

$$\cos \theta = \frac{\mathbf{u} \cdot \mathbf{v}}{\|\mathbf{u}\| \|\mathbf{v}\|}$$

$$= \frac{0(3) + 1(3) + 0(3) + 1(3)}{\sqrt{0^2 + 1^2 + 0^2 + 1^2}\sqrt{3^2 + 3^2 + 3^2 + 3^2}}$$

$$= \frac{6}{6\sqrt{2}} = \frac{\sqrt{2}}{2}.$$

So, $\theta = \cos^{-1}\left(\frac{\sqrt{2}}{2}\right) = \frac{\pi}{4}.$

71. The cosine of the angle θ between $\mathbf{u}$ and $\mathbf{v}$ is given by

$$\cos \theta = \frac{\mathbf{u} \cdot \mathbf{v}}{\|\mathbf{u}\| \|\mathbf{v}\|}$$

$$= \frac{1(-1) + 3(4) + (-1)(5) + 2(-3) + 0(2)}{\sqrt{1^2 + 3^2 + (-1)^2 + 2^2 + 0^2}\sqrt{(-1)^2 + 4^2 + 5^2 + (-3)^2 + 2^2}}$$

$$= \frac{0}{\sqrt{15}\sqrt{55}} = 0.$$

So, $\theta = \cos^{-1}(0) = \frac{\pi}{2}.$

73.
$$\mathbf{u} \cdot \mathbf{v} = 0$$
$$(0, 5) \cdot (v_1, v_2) = 0$$
$$0v_1 + 5v_2 = 0$$
$$v_2 = 0$$

So, $\mathbf{v} = (t, 0)$, where t is any real number.

75.
$$\mathbf{u} \cdot \mathbf{v} = 0$$
$$(-3, 2) \cdot (v_1, v_2) = 0$$
$$-3v_1 + 2v_2 = 0$$

So, $\mathbf{v} = (2t, 3t)$, where t is any real number.

77.
$$\mathbf{u} \cdot \mathbf{v} = 0$$
$$(4, -1, 0) \cdot (v_1, v_2, v_3) = 0$$
$$4v_1 + (-1)v_2 + 0v_3 = 0$$
$$4v_1 - v_2 = 0$$

So, $\mathbf{v} = (t, 4t, s)$, where s and t are any real numbers.

79.
$$\mathbf{u} \cdot \mathbf{v} = 0$$
$$(0, 1, 0, 0, 0) \cdot (v_1, v_2, v_3, v_4, v_5) = 0$$
$$v_2 = 0$$

So, $\mathbf{v} = (r, 0, s, t, w)$, where r, s, t and w are any real numbers.

81. Because $\mathbf{u} \cdot \mathbf{v} = 2\left(\frac{3}{2}\right) + 18\left(-\frac{1}{6}\right) = 0$, the vectors $\mathbf{u}$ and $\mathbf{v}$ are orthogonal.

83. Because $\mathbf{v} = -6\mathbf{u}$, the vectors are parallel.

85. Because $\mathbf{u} \cdot \mathbf{v} = 0(1) + 1(-2) + 0(0) = -2 \neq 0$, the vectors $\mathbf{u}$ and $\mathbf{v}$ are not orthogonal. Moreover, because one is not a scalar multiple of the other, they are not parallel.

87. Because
$$\mathbf{u} \cdot \mathbf{v} = -2\left(\frac{1}{4}\right) + 5\left(-\frac{5}{4}\right) + 1(0) + 0(1) = -\frac{27}{4} \neq 0, \text{ the}$$
vectors $\mathbf{u}$ and $\mathbf{v}$ are not orthogonal. Moreover, because one is not a scalar multiple of the other, they are not parallel.

89. $\mathbf{u} = \left(-2, \frac{1}{2}, -1, 3\right)$ and $\mathbf{v} = \left(\frac{3}{2}, 1, -\frac{5}{2}, 0\right)$.

Using a graphing utility or a computer software program you have $\mathbf{u} \cdot \mathbf{v} = 0$. Because $\mathbf{u} \cdot \mathbf{v} = 0$, the vectors are orthogonal.

91. $\mathbf{u} = \left(-\frac{3}{4}, \frac{3}{2}, -\frac{9}{2}, -6\right)$ and $\mathbf{v} = \left(\frac{3}{8}, -\frac{3}{4}, \frac{9}{8}, 3\right)$.

Using a graphing utility or a computer software program you have $\mathbf{u} \cdot \mathbf{v} = -24.46875 \neq 0$. Because $\mathbf{u} \cdot \mathbf{v} \neq 0$, the vectors are not orthogonal. Because one is not a scalar multiple of the other, they are not parallel.

93. Because $\mathbf{u} \cdot \mathbf{v} = \cos\theta \sin\theta + \sin\theta(-\cos\theta) - 1(0)$
$$= \cos\theta \sin\theta - \cos\theta \sin\theta = 0,$$
the vectors $\mathbf{u}$ and $\mathbf{v}$ are orthogonal.

95. (a) False. See "Definition of Length of a Vector in R^n," page 278.

(b) False. See "Definition of Dot Product in R^n," page 282.

97. (a) $\|\mathbf{u} \cdot \mathbf{v}\|$ is meaningless because $\mathbf{u} \cdot \mathbf{v}$ is a scalar.

(b) $\mathbf{u} + (\mathbf{u} \cdot \mathbf{v})$ is meaningless because $\mathbf{u}$ is a vector and $\mathbf{u} \cdot \mathbf{v}$ is a scalar.

99. Because $\mathbf{u} + \mathbf{v} = (4, 0) + (1, 1) = (5, 1)$, you have
$$\|\mathbf{u} + \mathbf{v}\| \le \|\mathbf{u}\| + \|\mathbf{v}\|$$
$$\|(5, 1)\| \le \|(4, 0)\| + \|(1, 1)\|$$
$$\sqrt{26} \le 4 + \sqrt{2}.$$

101. Because $\mathbf{u} + \mathbf{v} = (-1, 1) + (2, 0) = (1, 1)$, you have
$$\|\mathbf{u} + \mathbf{v}\| \le \|\mathbf{u}\| + \|\mathbf{v}\|$$
$$\|(1, 1)\| \le \|(-1, 1)\| + \|(2, 0)\|$$
$$\sqrt{2} \le \sqrt{2} + 2.$$

103. First note that $\mathbf{u}$ and $\mathbf{v}$ are orthogonal, because $\mathbf{u} \cdot \mathbf{v} = (1, -1) \cdot (1, 1) = 0$. Then note
$$\|\mathbf{u} + \mathbf{v}\|^2 = \|\mathbf{u}\|^2 + \|\mathbf{v}\|^2$$
$$\|(2, 0)\|^2 = \|(1, -1)\|^2 + \|(1, 1)\|^2$$
$$4 = 2 + 2.$$

105. First note that $\mathbf{u}$ and $\mathbf{v}$ are orthogonal, because $\mathbf{u} \cdot \mathbf{v} = (3, 4, -2) \cdot (4, -3, 0) = 0$. Then note
$$\|\mathbf{u} + \mathbf{v}\|^2 = \|\mathbf{u}\|^2 + \|\mathbf{v}\|^2$$
$$\|(7, 1, -2)\|^2 = \|(3, 4, -2)\|^2 + \|(4, -3, 0)\|^2$$
$$54 = 29 + 25.$$

107. (a) If $\mathbf{u} \cdot \mathbf{v} = 0$, then $\dfrac{\mathbf{u} \cdot \mathbf{v}}{\|\mathbf{u}\|\|\mathbf{v}\|} = \dfrac{0}{\|\mathbf{u}\|\|\mathbf{v}\|}$. So, $\cos\theta = 0$ and $\theta = \dfrac{\pi}{2}$, provided $\mathbf{u} \neq \mathbf{0}$ and $\mathbf{v} \neq \mathbf{0}$.

(b) If $\mathbf{u} \cdot \mathbf{v} > 0$, then $\dfrac{\mathbf{u} \cdot \mathbf{v}}{\|\mathbf{u}\|\|\mathbf{v}\|} > 0$.

So, $\cos\theta > 0$ and $0 \le \theta < \dfrac{\pi}{2}$.

(c) If $\mathbf{u} \cdot \mathbf{v} < 0$, then $\dfrac{\mathbf{u} \cdot \mathbf{v}}{\|\mathbf{u}\|\|\mathbf{v}\|} < 0$.

So, $\cos\theta < 0$, and $\dfrac{\pi}{2} < \theta \le \pi$.

109. $\mathbf{v} = (v_1, v_2) = (8, 15), (v_2, -v_1) = (15, -8)$

$(8, 15) \cdot (15, -8) = 8(15) + 15(-8) = 120 - 120 = 0$

So, $(v_2, -v_1)$ is orthogonal to $\mathbf{v}$.

Two unit vectors orthogonal to $\mathbf{v}$:

$-1(15, -8) = (-15, 8)$: $(8, 15) \cdot (-15, 8) = 8(-15) + 15(8)$
$$= -120 + 120$$
$$= 0$$

$3(15, -8) = (45, -24)$: $(8, 15) \cdot (45, -24) = 8(45) + (15)(-24)$
$$= 360 - 360$$
$$= 0$$

(Answer is not unique.)

111. Let $\mathbf{v} = (t, t, t)$ be the diagonal of the cube, and $\mathbf{u} = (t, t, 0)$ the diagonal of one of its sides. Then,

$$\cos \theta = \frac{\mathbf{u} \cdot \mathbf{v}}{\|\mathbf{u}\|\|\mathbf{v}\|} = \frac{2t^2}{(\sqrt{2}\,t)(\sqrt{3}\,t)} = \frac{2}{\sqrt{6}} = \frac{\sqrt{6}}{3}$$

and $\theta = \cos^{-1}\left(\dfrac{\sqrt{6}}{3}\right) \approx 35.26°$.

113. Given $\mathbf{u} \cdot \mathbf{v} = 0$ and $\mathbf{u} \cdot \mathbf{w} = 0$,

$\mathbf{u} \cdot (c\mathbf{v} + d\mathbf{w}) = \mathbf{u} \cdot (c\mathbf{v}) + \mathbf{u} \cdot (d\mathbf{w})$
$$= c(\mathbf{u} \cdot \mathbf{v}) + d(\mathbf{u} \cdot \mathbf{w})$$
$$= c(0) + d(0)$$
$$= 0.$$

So, $\mathbf{u}$ is orthogonal to $c\mathbf{v} + d\mathbf{w}$.

115. $\|\mathbf{u} + \mathbf{v}\|^2 + \|\mathbf{u} - \mathbf{v}\|^2 = (\mathbf{u} + \mathbf{v}) \cdot (\mathbf{u} + \mathbf{v}) + (\mathbf{u} - \mathbf{v}) \cdot (\mathbf{u} - \mathbf{v})$
$$= (\mathbf{u} \cdot \mathbf{u} + \mathbf{v} \cdot \mathbf{v} + 2\mathbf{u} \cdot \mathbf{v}) + (\mathbf{u} \cdot \mathbf{u} + \mathbf{v} \cdot \mathbf{v} - 2\mathbf{u} \cdot \mathbf{v})$$
$$= 2\|\mathbf{u}\|^2 + 2\|\mathbf{v}\|^2$$

117. If $\mathbf{u}$ and $\mathbf{v}$ have the same direction, then $\mathbf{u} = c\mathbf{v}, c > 0$, and

$\|\mathbf{u} + \mathbf{v}\| = \|c\mathbf{v} + \mathbf{v}\| = (c + 1)\|\mathbf{v}\|$
$$= c\|\mathbf{v}\| + \|\mathbf{v}\| = \|c\mathbf{v}\| + \|\mathbf{v}\|$$
$$= \|\mathbf{u}\| + \|\mathbf{v}\|.$$

On the other hand, if

$$\|\mathbf{u} + \mathbf{v}\| = \|\mathbf{u}\| + \|\mathbf{v}\|, \text{ then}$$
$$\|\mathbf{u} + \mathbf{v}\|^2 = (\|\mathbf{u}\| + \|\mathbf{v}\|)^2$$
$$(\mathbf{u} + \mathbf{v}) \cdot (\mathbf{u} + \mathbf{v}) = \|\mathbf{u}\|^2 + \|\mathbf{v}\|^2 + 2\|\mathbf{u}\|\|\mathbf{v}\|$$
$$\|\mathbf{u}\|^2 + \|\mathbf{v}\|^2 + 2\mathbf{u} \cdot \mathbf{v} = \|\mathbf{u}\|^2 + \|\mathbf{v}\|^2 + 2\|\mathbf{u}\|\|\mathbf{v}\|$$
$$2\mathbf{u} \cdot \mathbf{v} = 2\|\mathbf{u}\|\|\mathbf{v}\|$$

$\Rightarrow \cos \theta = \dfrac{\mathbf{u} \cdot \mathbf{v}}{\|\mathbf{u}\|\|\mathbf{v}\|} = 1 \quad \Rightarrow \quad \theta = 0 \quad \Rightarrow \quad \mathbf{u}$ and $\mathbf{v}$ have the same direction.

119. $A\mathbf{x} = \mathbf{0}$ means that the dot product of each row of A with the column vector $\mathbf{x}$ is zero. So, $\mathbf{x}$ is orthogonal to the row vectors of A.

121. Property 1: $\mathbf{u} \cdot \mathbf{v} = \mathbf{u}^T\mathbf{v} = \left(\mathbf{u}^T\mathbf{v}\right)^T = \mathbf{v}^T\mathbf{u} = \mathbf{v} \cdot \mathbf{u}$

Property 2: $\mathbf{u} \cdot (\mathbf{v} + \mathbf{w}) = \mathbf{u}^T(\mathbf{v} + \mathbf{w}) = \mathbf{u}^T\mathbf{v} + \mathbf{u}^T\mathbf{w} = \mathbf{u} \cdot \mathbf{v} + \mathbf{u} \cdot \mathbf{w}$

Property 3: $c(\mathbf{u} \cdot \mathbf{v}) = c\left(\mathbf{u}^T\mathbf{v}\right) = (c\mathbf{u})^T\mathbf{v} = (c\mathbf{u}) \cdot \mathbf{v}$ and $c(\mathbf{u} \cdot \mathbf{v}) = c\left(\mathbf{u}^T\mathbf{v}\right) = \mathbf{u}^T(c\mathbf{v}) = \mathbf{u} \cdot (c\mathbf{v})$

Section 5.2   Inner Product Spaces

1. (a) $\langle \mathbf{u}, \mathbf{v} \rangle = \mathbf{u} \cdot \mathbf{v} = 3(5) + 4(-12) = -33$

(b) $\|\mathbf{u}\| = \sqrt{\langle \mathbf{u}, \mathbf{u} \rangle} = \sqrt{\mathbf{u} \cdot \mathbf{u}} = \sqrt{3(3) + 4(4)} = 5$

(c) $\|\mathbf{v}\| = \sqrt{\langle \mathbf{v}, \mathbf{v} \rangle} = \sqrt{\mathbf{v} \cdot \mathbf{v}} = \sqrt{5(5) + (-12)(-12)} = \sqrt{169} = 13$

(d) $d(\mathbf{u}, \mathbf{v}) = \|\mathbf{u} - \mathbf{v}\| = \sqrt{\langle \mathbf{u} - \mathbf{v}, \mathbf{u} - \mathbf{v} \rangle} = \sqrt{(\mathbf{u} - \mathbf{v}) \cdot (\mathbf{u} - \mathbf{v})} = \sqrt{(-2)(-2) + 16(16)} = 2\sqrt{65}$

3. (a) $\langle \mathbf{u}, \mathbf{v} \rangle = 3u_1v_1 + u_2v_2 = 3(-4)(0) + 3(5) = 15$

(b) $\|\mathbf{u}\| = \sqrt{\langle \mathbf{u}, \mathbf{u} \rangle} = \sqrt{3(-4)^2 + 3^2} = \sqrt{57}$

(c) $\|\mathbf{v}\| = \sqrt{\langle \mathbf{v}, \mathbf{v} \rangle} = \sqrt{3 \cdot 0^2 + 5^2} = 5$

(d) $d(\mathbf{u}, \mathbf{v}) = \|\mathbf{u} - \mathbf{v}\| = \sqrt{\langle \mathbf{u} - \mathbf{v}, \mathbf{u} - \mathbf{v} \rangle} = \sqrt{3(-4)^2 + (-2)^2} = 2\sqrt{13}$

5. (a) $\langle \mathbf{u}, \mathbf{v} \rangle = \mathbf{u} \cdot \mathbf{v} = 0(9) + 9(-2) + 4(-4) = -34$

(b) $\|\mathbf{u}\| = \sqrt{\langle \mathbf{u}, \mathbf{u} \rangle} = \sqrt{\mathbf{u} \cdot \mathbf{u}} = \sqrt{0 + 9^2 + 4^2} = \sqrt{97}$

(c) $\|\mathbf{v}\| = \sqrt{\langle \mathbf{v}, \mathbf{v} \rangle} = \sqrt{\mathbf{v} \cdot \mathbf{v}} = \sqrt{9^2 + (-2)^2 + (-4)^2} = \sqrt{101}$

(d) $d(\mathbf{u}, \mathbf{v}) = \|\mathbf{u} - \mathbf{v}\| = \|(-9, 11, 8)\| = \sqrt{9^2 + 11^2 + 8^2} = \sqrt{266}$

7. (a) $\langle \mathbf{u}, \mathbf{v} \rangle = 2u_1v_1 + 3u_2v_2 + u_3v_3 = 2 \cdot 8 \cdot 8 + 3 \cdot 0 \cdot 3 + (-8) \cdot 16 = 0.$

(b) $\|\mathbf{u}\| = \sqrt{2 \cdot 8 \cdot 8 + 3 \cdot 0 \cdot 0 + (-8)^2} = 8\sqrt{3}.$

(c) $\|\mathbf{v}\| = \sqrt{\langle \mathbf{v}, \mathbf{v} \rangle} = \sqrt{2 \cdot 8^2 + 3 \cdot 3^2 + 16^2} = \sqrt{411}$

(d) $d(\mathbf{u}, \mathbf{v}) = \|\mathbf{u} - \mathbf{v}\| = \|(0, -3, -24)\| = 3\sqrt{67}.$

9. (a) $\langle \mathbf{u}, \mathbf{v} \rangle = \mathbf{u} \cdot \mathbf{v} = 2(2) + 0(2) + 1(0) + (-1)(1) = 3$

(b) $\|\mathbf{u}\| = \sqrt{\langle \mathbf{u}, \mathbf{u} \rangle} = \sqrt{\mathbf{u} \cdot \mathbf{u}} = \sqrt{2^2 + 0^2 + 1^2 + (-1)^2} = \sqrt{6}$

(c) $\|\mathbf{v}\| = \sqrt{\langle \mathbf{v}, \mathbf{v} \rangle} = \sqrt{\mathbf{v} \cdot \mathbf{v}} = \sqrt{2^2 + 2^2 + 0^2 + 1^2} = 3$

(d) $d(\mathbf{u}, \mathbf{v}) = \|\mathbf{u} - \mathbf{v}\| = \|(0, -2, 1, -2)\| = \sqrt{0^2 + (-2)^2 + 1^2 + (-2)^2} = 3$

11. (a) $\langle f, g \rangle = \int_{-1}^{1} f(x)g(x)\,dx = \int_{-1}^{1} x^2(x^2 + 1)\,dx = \int_{-1}^{1} (x^4 + x^2)\,dx = \left[\dfrac{x^5}{5} + \dfrac{x^3}{3} \right]_{-1}^{1} = \dfrac{16}{15}$

(b) $\|f\|^2 = \langle f, f \rangle = \int_{-1}^{1} (x^2)^2\,dx = \left. \dfrac{x^5}{5} \right]_{-1}^{1} = \dfrac{2}{5}$

$\|f\| = \sqrt{\dfrac{2}{5}} = \dfrac{\sqrt{10}}{5}$

(c) $\|g\|^2 = \langle g, g \rangle = \int_{-1}^{1} (x^2 + 1)^2\,dx = \int_{-1}^{1} (x^4 + 2x^2 + 1)\,dx = \left[\dfrac{x^5}{5} + \dfrac{2x^3}{3} + x \right]_{-1}^{1} = \dfrac{56}{15}$

$\|g\| = \sqrt{\dfrac{56}{15}} = \dfrac{2\sqrt{210}}{15}$

(d) Use the fact that $d(f, g) = \|f - g\|$. Because $f - g = x^2 - (x^2 + 1) = -1$, you have

$\langle f - g, f - g \rangle = \int_{-1}^{1} (-1)(-1)\,dx = \left. x \right]_{-1}^{1} = 2.$

So, $d(f, g) = \sqrt{\langle f - g, f - g \rangle} = \sqrt{2}.$

13. (a) $\langle f, g \rangle = \int_{-1}^{1} xe^x\,dx = \left. (x - 1)e^x \right]_{-1}^{1} = \dfrac{2}{e}$

(b) $\|f\|^2 = \langle f, f \rangle = \int_{-1}^{1} x^2\,dx = \left. \dfrac{x^3}{3} \right]_{-1}^{1} = \dfrac{2}{3}$

$\|f\| = \dfrac{\sqrt{6}}{3}.$

(c) $\|g\|^2 = \langle g, g \rangle = \int_{-1}^{1} e^{2x}\,dx = \left. \dfrac{e^{2x}}{2} \right]_{-1}^{1} = \dfrac{1}{2}(e^2 - e^{-2})$

$\|g\| = \sqrt{\dfrac{1}{2}(e^2 - e^{-2})}$

(d) Use the fact that $d(f, g) = \|f - g\|$. Because $f - g = x - e^x$, you have

$\langle f - g, f - g \rangle = \int_{-1}^{1} (x - e^x)^2\,dx$

$= \int_{-1}^{1} (x^2 - 2xe^x + e^{2x})\,dx$

$= \left[\dfrac{x^3}{3} - 2(x - 1)e^x + \dfrac{e^{2x}}{2} \right]_{-1}^{1}$

$= \dfrac{2}{3} - \dfrac{4}{e} + \dfrac{e^2 - e^{-2}}{2}$

$= \dfrac{2}{3} - \dfrac{4}{e} + \dfrac{e^2}{2} - \dfrac{1}{2e^2}.$

So, $d(f, g) = \sqrt{\langle f - g, f - g \rangle} = \sqrt{\dfrac{2}{3} - \dfrac{4}{e} + \dfrac{e^2}{2} - \dfrac{1}{2e^2}}.$

15. (a) $\langle f, g \rangle = \int_{-1}^{1} 1(3x^2 - 1)dx = \left[x^3 - x \right]_{-1}^{1} = (1 - 1) - (-1 + 1) = 0$

(b) $\|f\|^2 = \langle f, f \rangle = \int_{-1}^{1} 1 dx = x \Big]_{-1}^{1} = 2 \Rightarrow \|f\| = \sqrt{2}$

(c) $\|g\|^2 = \langle g, g \rangle = \int_{-1}^{1} (3x^2 - 1)^2 dx = \int_{-1}^{1} (9x^4 - 6x^2 + 1)dx = \left[\frac{9x^5}{5} - 2x^3 + x \right]_{-1}^{1} = \frac{8}{5}$

So, $\|g\| = \sqrt{\frac{8}{5}} = \frac{2\sqrt{10}}{5}$.

(d) Use the fact that $d(f, g) = \|f - g\|$. Because $f - g = 1 - (3x^2 - 1) = 2 - 3x^2$, you have

$\langle f - g, f - g \rangle = \int_{-1}^{1} (2 - 3x^2)dx = \int_{-1}^{1} (9x^4 - 12x^2 + 4)dx = \left[\frac{9x^5}{5} - 4x^3 + 4x \right]_{-1}^{1} = \frac{18}{5}$.

So, $d(f, g) = \sqrt{\langle f - g, f - g \rangle} = \sqrt{\frac{18}{5}} = \frac{3\sqrt{10}}{5}$.

17. (a) $\langle A, B \rangle = 2(-1)(0) + 3(-2) + 4(1) + 2(-2)(1) = -6$

(b) $\|A\|^2 = \langle A, A \rangle = 2(-1)^2 + 3^2 + 4^2 + 2(-2)^2 = 35$

$\|A\| = \sqrt{\langle A, A \rangle} = \sqrt{35}$

(c) $\|B\|^2 = \langle B, B \rangle = 2 \cdot 0^2 + (-2)^2 + 1^2 + 2 \cdot 1^2 = 7$

$\|B\| = \sqrt{\langle B, B \rangle} = \sqrt{7}$

(d) Use the fact that $d(A, B) = \|A - B\|$.

$\langle A - B, A - B \rangle = 2(-1)^2 + 5^2 + 3^2 + 2(-3)^2 = 54$

$d(A, B) = \sqrt{\langle A - B, A - B \rangle} = 3\sqrt{6}$

19. (a) $\langle A, B \rangle = 2(1)(0) + (-1)(1) + (2)(-2) + 2(4)(0) = -5$

(b) $\|A\|^2 = \langle A, A \rangle = 2(1)^2 + (-1)^2 + (2)^2 + 2(4)^2 = 39$

$\|A\| = \sqrt{\langle A, A \rangle} = \sqrt{39}$

(c) $\|B\|^2 = \langle B, B \rangle = 2(0)^2 + 1^2 + (-2)^2 + 0^2 = 5$

$\|B\| = \sqrt{\langle B, B \rangle} = \sqrt{5}$

(d) Use the fact that $d(A, B) = \|A - B\|$.

$\langle A - B, A - B \rangle = 2(1)^2 + (-2)^2 + 4^2 + 2(4)^2 = 54$

$d(A, B) = \sqrt{\langle A - B, A - B \rangle} = \sqrt{54} = 3\sqrt{6}$

21. (a) $\langle p, q \rangle = 1(0) + (-1)(1) + 3(-1) = -4$

(b) $\|p\| = \sqrt{\langle p, p \rangle} = \sqrt{1^2 + (-1)^2 + 3^2} = \sqrt{11}$

(c) $\|q\| = \sqrt{\langle q, q \rangle} = \sqrt{0^2 + 1^2 + (-1)^2} = \sqrt{2}$

(d) $d(p, q) = \|p - q\| = \sqrt{\langle p - q, p - q \rangle} = \sqrt{1^2 + (-2)^2 + 4^2} = \sqrt{21}$

23. (a) $\langle p, q \rangle = 1(1) + 0(0) + 1(-1) = 0$

(b) $\|p\| = \sqrt{\langle p, p \rangle} = \sqrt{1^2 + 0^2 + 1^2} = \sqrt{2}$

(c) $\|q\| = \sqrt{\langle q, q \rangle} = \sqrt{1^2 + 0^2 + (-1)^2} = \sqrt{2}$

(d) $d(p, q) = \|p - q\| = \sqrt{\langle p - q, p - q \rangle} = \sqrt{0^2 + 0^2 + 2^2} = 2$

25. Verify that the function $\langle \mathbf{u}, \mathbf{v} \rangle = 3u_1v_1 + u_2v_2$ satisfies the four parts of the definition.

1. $\langle \mathbf{u}, \mathbf{v} \rangle = 3u_1v_1 + u_2v_2 = 3v_1u_1 + v_2u_2 = \langle \mathbf{v}, \mathbf{u} \rangle$

2. $\langle \mathbf{u}, \mathbf{v} + \mathbf{w} \rangle = 3u_1(v_1 + w_1) + u_2(v_2 + w_2) = 3u_1v_1 + u_2v_2 + 3u_1w_1 + u_2w_2$
$= \langle \mathbf{u}, \mathbf{v} \rangle + \langle \mathbf{u}, \mathbf{w} \rangle$

3. $c\langle \mathbf{u}, \mathbf{v} \rangle = c(3u_1v_1 + u_2v_2) = 3(cu_1)v_1 + (cu_2)v_2 = \langle c\mathbf{u}, \mathbf{v} \rangle$

4. $\langle \mathbf{v}, \mathbf{v} \rangle = 3v_1^2 + v_2^2 \geq 0$ and $\langle \mathbf{v}, \mathbf{v} \rangle = 0$ if and only if $\mathbf{v} = (0, 0)$.

27. Verify that the function $\langle A, B \rangle = 2a_{11}b_{11} + a_{12}b_{12} + a_{21}b_{21} + 2a_{22}b_{22}$ satisfies the four parts of the definition.

1. $\langle A, B \rangle = 2a_{11}b_{11} + a_{12}b_{12} + a_{21}b_{21} + 2a_{22}b_{22}$
$= 2b_{11}a_{11} + b_{12}a_{12} + b_{21}a_{21} + 2b_{22}a_{22} = \langle B, A \rangle$

2. $\langle A, B + C \rangle = 2a_{11}(b_{11} + c_{11}) + a_{12}(b_{12} + c_{12}) + a_{21}(b_{21} + c_{21}) + 2a_{22}(b_{22} + c_{22})$
$= 2a_{11}b_{11} + a_{12}b_{12} + a_{21}b_{21} + 2a_{22}b_{22} + 2a_{11}c_{11} + a_{12}c_{12} + a_{21}c_{21} + 2a_{22}c_{22}$
$= \langle A, B \rangle + \langle A, C \rangle$

3. $c\langle A, B \rangle = c(2a_{11}b_{11} + a_{12}b_{12} + a_{21}b_{21} + 2a_{22}b_{22})$
$= 2(ca_{11})b_{11} + (ca_{12})b_{12} + (ca_{21})b_{21} + 2(ca_{22})b_{22} = \langle cA, B \rangle$

4. $\langle A, A \rangle = 2a_{11}^2 + a_{12}^2 + a_{21}^2 + 2a_{22}^2 \geq 0$, and

$\langle A, A \rangle = 0$ if and only if $A = \begin{bmatrix} 0 & 0 \\ 0 & 0 \end{bmatrix}$.

29. The product $\langle \mathbf{u}, \mathbf{v} \rangle$ is not an inner product because nonzero vectors can have a norm of zero. For example, if $\mathbf{v} = (0, 1)$, then $\langle \mathbf{v}, \mathbf{v} \rangle = 0^2 = 0$.

31. The product $\langle \mathbf{u}, \mathbf{v} \rangle$ is not an inner product because nonzero vectors can have a norm of zero. For example, if $\mathbf{v} = (1, 1)$, then $\langle (1, 1), (1, 1) \rangle = 0$.

33. The product $\langle \mathbf{u}, \mathbf{v} \rangle$ is not an inner product because it is not distributive over addition.

For example, if $\mathbf{u} = (1, 0)$, $\mathbf{v} = (1, 0)$, and $\mathbf{w} = (1, 0)$, then $\langle \mathbf{u}, \mathbf{v} + \mathbf{w} \rangle = 1^2(2)^2 + 0^2(0)^2 = 4$ and

$\langle \mathbf{u}, \mathbf{v} \rangle + \langle \mathbf{u}, \mathbf{w} \rangle = 1^2(1)^2 + 0^2(0)^2 + 1^2(1)^2 + 0^2(0)^2 = 2$.

So, $\langle \mathbf{u}, \mathbf{v} + \mathbf{w} \rangle \neq \langle \mathbf{u}, \mathbf{v} \rangle + \langle \mathbf{u}, \mathbf{w} \rangle$.

35. The product $\langle \mathbf{u}, \mathbf{v} \rangle$ is not an inner product because it is not commutative. For example, if $\mathbf{u} = (1, 2)$, and $\mathbf{v} = (2, 3)$, then $\langle \mathbf{u}, \mathbf{v} \rangle = 3(1)(3) - 2(2) = 5$ while $\langle \mathbf{v}, \mathbf{u} \rangle = 3(2)(2) - 3(1) = 9$.

37. Because

$$\frac{\langle \mathbf{u}, \mathbf{v} \rangle}{\|\mathbf{u}\|\|\mathbf{v}\|} = \frac{3(5) + 4(-12)}{\sqrt{3^2 + 4^2}\sqrt{5^2 + (-12)^2}} = \frac{-33}{5 \cdot 13} = \frac{-33}{65},$$

the angle between $\mathbf{u}$ and $\mathbf{v}$ is

$$\cos^{-1}\left(\frac{-33}{65}\right) \approx 2.103 \text{ radians } (120.51°).$$

39. Because

$$\frac{\langle \mathbf{u}, \mathbf{v} \rangle}{\|\mathbf{u}\|\|\mathbf{v}\|} = \frac{3(-4)(0) + (3)(5)}{\sqrt{3(-4)^2 + 3^2}\sqrt{3(0)^2 + 5^2}}$$

$$= \frac{15}{\sqrt{57} \cdot 5} = \frac{3}{\sqrt{57}},$$

the angle between $\mathbf{u}$ and $\mathbf{v}$ is

$$\cos^{-1}\left(\frac{3}{\sqrt{57}}\right) \approx 1.16 \text{ radians } (66.59°).$$

41. Because

$$\langle \mathbf{u}, \mathbf{v} \rangle = 1(2) + 2(1)(-2) + 1(2) = 0,$$

the angle between $\mathbf{u}$ and $\mathbf{v}$ is $\cos^{-1}(0) = \dfrac{\pi}{2}$.

43. Because

$$\frac{\langle p, q \rangle}{\|p\|\|q\|} = \frac{1 - 1 + 1}{\sqrt{3}\sqrt{3}} = \frac{1}{3},$$

the angle between p and q is

$$\cos^{-1}\left(\frac{1}{3}\right) \approx 1.23 \text{ radians } (70.53°).$$

45. Because

$$\langle f, g \rangle = \int_{-1}^{1} x^3 \, dx = \frac{x^4}{4}\bigg]_{-1}^{1} = 0,$$

the angle between f and g is $\cos^{-1}(0) = \dfrac{\pi}{2}$.

47. (a) To verify the Cauchy-Schwarz Inequality, observe

$$|\langle \mathbf{u}, \mathbf{v} \rangle| \leq \|\mathbf{u}\|\|\mathbf{v}\|$$

$$|(5, 12) \cdot (3, 4)| \leq \|(5, 12)\|\|(3, 4)\|$$

$$63 \leq (13)(5) = 65.$$

(b) To verify the Triangle Inequality, observe

$$\|\mathbf{u} + \mathbf{v}\| \leq \|\mathbf{u}\| + \|\mathbf{v}\|$$

$$\|(8, 16)\| \leq \|(5, 12)\| + \|(3, 4)\|$$

$$\sqrt{320} \leq 13 + 5$$

$$8\sqrt{5} \leq 18.$$

49. (a) To verify the Cauchy-Schwarz Inequality, observe

$$|\langle \mathbf{u}, \mathbf{v} \rangle| \leq \|\mathbf{u}\|\|\mathbf{v}\|$$

$$|(1, 0, 4) \cdot (-5, 4, 1)| \leq \|(1, 0, 4)\|\|(-5, 4, 1)\|$$

$$1 \leq \sqrt{17}\sqrt{42}$$

$$1 \leq \sqrt{714}.$$

(b) To verify the Triangle Inequality, observe

$$\|\mathbf{u} + \mathbf{v}\| \leq \|\mathbf{u}\| + \|\mathbf{v}\|$$

$$\|(-4, 4, 5)\| \leq \|(1, 0, 4)\| + \|(-5, 4, 1)\|$$

$$\sqrt{57} \leq \sqrt{17} + \sqrt{42}$$

$$7.5498 \leq 10.6038.$$

51. (a) To verify the Cauchy-Schwarz Inequality, observe

$$|\langle p, q \rangle| \leq \|p\|\|q\|$$

$$|0(1) + 2(0) + 0(3)| \leq (2)\sqrt{10}$$

$$0 \leq 2\sqrt{10}.$$

(b) To verify the Triangle Inequality, observe

$$\|p + q\| \leq \|p\| + \|q\|$$

$$\|1 + 2x + 3x^2\| \leq 2 + \sqrt{10}$$

$$\sqrt{14} \leq 2 + \sqrt{10}$$

$$3.742 \leq 5.162.$$

53. (a) To verify the Cauchy-Schwarz Inequality, observe

$$|\langle A, B \rangle| \leq \|A\|\|B\|$$

$$|0(-3) + 3(1) + 2(4) + 1(3)| \leq \sqrt{14}\sqrt{35}$$

$$14 \leq \sqrt{14}\sqrt{35}.$$

(b) To verify the Triangle Inequality, observe

$$\|A + B\| \leq \|A\| + \|B\|$$

$$\left\|\begin{bmatrix} -3 & 4 \\ 6 & 4 \end{bmatrix}\right\| \leq \sqrt{14} + \sqrt{35}$$

$$\sqrt{77} \leq \sqrt{14} + \sqrt{35}$$

$$8.775 \leq 9.658.$$

55. (a) To verify the Cauchy-Schwarz Inequality, compute

$$\langle f, g \rangle = \langle \sin x, \cos x \rangle = \int_{-\pi}^{\pi} \sin x \cos x \, dx = \frac{\sin^2 x}{2} \Big]_{-\pi}^{\pi} = 0$$

$$\|f\|^2 = \langle \sin x, \sin x \rangle = \int_{-\pi}^{\pi} \sin^2 x \, dx = \int_{-\pi}^{\pi} \frac{1 - \cos 2x}{2} dx = \left[\frac{1}{2} x - \frac{\sin 2x}{4} \right]_{-\pi}^{\pi} = \pi \Rightarrow \|f\| = \sqrt{\pi}$$

$$\|g\|^2 = \langle \cos x, \cos x \rangle = \int_{-\pi}^{\pi} \cos^2 x \, dx = \int_{-\pi}^{\pi} \frac{1 + \cos 2x}{2} dx = \left[\frac{1}{2} x + \frac{\sin 2x}{4} \right]_{-\pi}^{\pi} = \pi \Rightarrow \|g\| = \sqrt{\pi}$$

and observe that

$$|\langle f, g \rangle| \le \|f\| \|g\|$$
$$0 \le \sqrt{\pi}\sqrt{\pi}.$$

(b) To verify the Triangle Inequality, compute

$$\|f + g\|^2 = \langle \sin x + \cos x, \sin x + \cos x \rangle = \int_{-\pi}^{\pi} (\sin x + \cos x)^2 \, dx$$

$$= \int_{-\pi}^{\pi} \sin^2 x \, dx + \int_{-\pi}^{\pi} \cos^2 x \, dx + 2\int_{-\pi}^{\pi} \sin x \cos x \, dx = \pi + \pi + 0 \Rightarrow \|f + g\| = \sqrt{2\pi}$$

and observe that

$$\|f + g\| \le \|f\| + \|g\|$$
$$\sqrt{2\pi} \le \sqrt{\pi} + \sqrt{\pi}.$$

57. (a) To verify the Cauchy-Schwarz Inequality, compute

$$\langle f, g \rangle = \langle x, e^x \rangle = \int_{0}^{1} x e^x \, dx = \left[e^x (x - 1) \right]_{0}^{1} = 1$$

$$\|f\|^2 = \langle x, x \rangle = \int_{0}^{1} x^2 \, dx = \frac{x^3}{3} \Big]_{0}^{1} = \frac{1}{3} \Rightarrow \|f\| = \frac{\sqrt{3}}{3}$$

$$\|g\|^2 = \langle e^x, e^x \rangle = \int_{0}^{1} e^{2x} \, dx = \frac{e^{2x}}{2} \Big]_{0}^{1} = \frac{e^2}{2} - \frac{1}{2} \Rightarrow \|g\| = \sqrt{\frac{e^2}{2} - \frac{1}{2}}$$

and observe that

$$|\langle f, g \rangle| \le \|f\| \|g\|$$
$$1 \le \left(\sqrt{\frac{3}{3}} \right)\left(\sqrt{\frac{e^2}{2} - \frac{1}{2}} \right)$$
$$1 \le 1.032.$$

(b) To verify the Triangle Inequality, compute

$$\|f + g\|^2 = \langle x + e^x, x + e^x \rangle = \int_{0}^{1} (x + e^x)^2 \, dx = \left[\frac{e^{2x}}{2} + 2e^x (x - 1) + \frac{x^3}{3} \right]_{0}^{1}$$

$$= \left[\frac{e^2}{2} + 2e(0) + \frac{1}{3} \right] - \left[\frac{1}{2} + 2(1)(-1) + 0 \right]$$

$$= \frac{e^2}{2} + \frac{11}{6} \Rightarrow \|f + g\| = \sqrt{\frac{e^2}{2} + \frac{11}{6}}$$

and observe that

$$\|f + g\| \le \|f\| + \|g\|$$
$$\sqrt{\frac{e^2}{2} + \frac{11}{6}} \le \frac{\sqrt{3}}{3} + \sqrt{\frac{e^2}{2} - \frac{1}{2}}$$
$$2.351 \le 2.364.$$

59. Because

$$\langle f, g \rangle = \int_{-\pi}^{\pi} \cos x \sin x \, dx = \frac{1}{2} \sin^2 x \bigg]_{-\pi}^{\pi} = 0,$$

f and g are orthogonal.

61. Because

$$\langle f, g \rangle = \int_{-1}^{1} x \frac{1}{2} \left(5x^3 - 3x \right) dx = \frac{1}{2} \int_{-1}^{1} \left(5x^4 - 3x^2 \right) dx = \frac{1}{2} \left(x^5 - x^3 \right) \bigg]_{-1}^{1} = 0,$$

f and g are orthogonal.

63. (a) $\operatorname{proj}_{\mathbf{v}} \mathbf{u} = \dfrac{\langle \mathbf{u}, \mathbf{v} \rangle}{\langle \mathbf{v}, \mathbf{v} \rangle} \mathbf{v} = \dfrac{1(2) + 2(1)}{2^2 + 1^2}(2, 1) = \dfrac{4}{5}(2, 1) = \left(\dfrac{8}{5}, \dfrac{4}{5} \right)$

(b) $\operatorname{proj}_{\mathbf{u}} \mathbf{v} = \dfrac{\langle \mathbf{v}, \mathbf{u} \rangle}{\langle \mathbf{u}, \mathbf{u} \rangle} \mathbf{u} = \dfrac{2(1) + 1(2)}{1^2 + 2^2}(1, 2) = \dfrac{4}{5}(1, 2) = \left(\dfrac{4}{5}, \dfrac{8}{5} \right)$

(c)

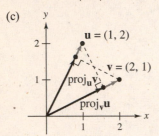

65. (a) $\operatorname{proj}_{\mathbf{v}} \mathbf{u} = \dfrac{\langle \mathbf{u}, \mathbf{v} \rangle}{\langle \mathbf{v}, \mathbf{v} \rangle} \mathbf{v} = \dfrac{(-1)(4) + 3(4)}{4(4) + 4(4)}(4, 4) = \dfrac{1}{4}(4, 4) = (1, 1)$

(b) $\operatorname{proj}_{\mathbf{u}} \mathbf{v} = \dfrac{\langle \mathbf{v}, \mathbf{u} \rangle}{\langle \mathbf{u}, \mathbf{u} \rangle} \mathbf{u} = \dfrac{4(-1) + 4(3)}{(-1)(-1) + 3(3)}(-1, 3) = \dfrac{4}{5}(-1, 3) = \left(-\dfrac{4}{5}, \dfrac{12}{5} \right)$

(c)

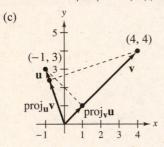

67. (a) $\operatorname{proj}_{\mathbf{v}} \mathbf{u} = \dfrac{\langle \mathbf{u}, \mathbf{v} \rangle}{\langle \mathbf{v}, \mathbf{v} \rangle} \mathbf{v} = \dfrac{1(0) + 3(-1) + (-2)(1)}{0^2 + (-1)^2 + 1^2}(0, -1, 1) = \dfrac{-5}{2}(0, -1, 1) = \left(0, \dfrac{5}{2}, -\dfrac{5}{2} \right)$

(b) $\operatorname{proj}_{\mathbf{u}} \mathbf{v} = \dfrac{\langle \mathbf{v}, \mathbf{u} \rangle}{\langle \mathbf{u}, \mathbf{u} \rangle} \mathbf{u} = \dfrac{0(1) + (-1)(3) + 1(-2)}{1^2 + 3^2 + (-2)^2}(1, 3, -2) = \dfrac{-5}{14}(1, 3, -2) = \left(-\dfrac{5}{14}, -\dfrac{15}{14}, \dfrac{5}{7} \right)$

69. (a) $\operatorname{proj}_{\mathbf{v}} \mathbf{u} = \dfrac{\langle \mathbf{u}, \mathbf{v} \rangle}{\langle \mathbf{v}, \mathbf{v} \rangle} \mathbf{v} = \dfrac{0(-1) + 1(1) + 3(2) + (-6)(2)}{(-1)(-1) + 1(1) + 2(2) + 2(2)} \mathbf{v} = -\dfrac{5}{10}(-1, 1, 2, 2) = \left(\dfrac{1}{2}, -\dfrac{1}{2}, -1, -1 \right)$

(b) $\operatorname{proj}_{\mathbf{u}} \mathbf{v} = \dfrac{\langle \mathbf{v}, \mathbf{u} \rangle}{\langle \mathbf{u}, \mathbf{u} \rangle} \mathbf{u} = \dfrac{(-1)(0) + 1(1) + 2(3) + 2(-6)}{0 + 1(1) + 3(3) + (-6)(-6)} \mathbf{u} = -\dfrac{5}{46}(0, 1, 3, -6) = \left(0, -\dfrac{5}{46}, -\dfrac{15}{46}, \dfrac{15}{23} \right)$

71. The inner products $\langle f, g \rangle$ and $\langle g, g \rangle$ are as follows.

$$\langle f, g \rangle = \int_{-1}^{1} x \, dx = \frac{x^2}{2} \Big]_{-1}^{1} = 0$$

$$\langle g, g \rangle = \int_{-1}^{1} dx = x \Big]_{-1}^{1} = 2$$

So, the projection of f onto g is

$$\text{proj}_g f = \frac{\langle f, g \rangle}{\langle g, g \rangle} g = \frac{0}{2}(1) = 0.$$

73. The inner products $\langle f, g \rangle$ and $\langle g, g \rangle$ are as follows.

$$\langle f, g \rangle = \int_{0}^{1} xe^x \, dx = \left[(x-1)e^x\right]_{0}^{1} = 0 + 1 = 1$$

$$\langle g, g \rangle = \int_{0}^{1} e^{2x} \, dx = \frac{1}{2}e^{2x} \Big]_{0}^{1} = \frac{e^2 - 1}{2}$$

So, the projection of f onto g is

$$\text{proj}_g f = \frac{\langle f, g \rangle}{\langle g, g \rangle} g = \frac{1}{(e^2 - 1)/2}e^x = \frac{2e^x}{e^2 - 1}.$$

75. The inner product $\langle f, g \rangle$ is

$$\langle f, g \rangle = \int_{-\pi}^{\pi} \sin x \cos x \, dx = \frac{\sin^2 x}{2} \Big]_{-\pi}^{\pi} = 0.$$

So, the projection of f onto g is

$$\text{proj}_g f = \frac{\langle f, g \rangle}{\langle g, g \rangle} g = \frac{0}{\langle g, g \rangle} g = 0.$$

77. The inner products $\langle f, g \rangle$ and $\langle g, g \rangle$ are as follows.

$$\langle f, g \rangle = \int_{-\pi}^{\pi} x \sin 2x \, dx = \left[\frac{\sin 2x}{4} - \frac{x \cos 2x}{2}\right]_{-\pi}^{\pi} = -\pi$$

$$\langle g, g \rangle = \int_{-\pi}^{\pi} (\sin 2x)^2 \, dx = \left[\frac{x}{2} - \frac{\sin 4x}{8}\right]_{-\pi}^{\pi} = \pi$$

So, the projection of f onto g is

$$\text{proj}_g f = \frac{\langle f, g \rangle}{\langle g, g \rangle} g = \frac{-\pi}{\pi}(\sin 2x) = -\sin 2x.$$

79. (a) False. See the introduction to this section, page 292.

(b) True. See "Remark," page 301.

81. (a) $\langle \mathbf{u}, \mathbf{v} \rangle = 4(2) + 2(2)(-2) = 0 \Rightarrow \mathbf{u}$ and $\mathbf{v}$ are orthogonal.

(b) The vectors are not orthogonal in the Euclidean sense.

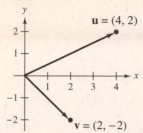

83. Verify the four parts of the definition for the function

$$\langle \mathbf{u}, \mathbf{v} \rangle = c_1 u_1 v_1 + \cdots + c_n u_n v_n = \sum_{i=1}^{n} c_i u_i v_i.$$

1. $\langle \mathbf{u}, \mathbf{v} \rangle = \sum_{i=1}^{n} c_i u_i v_i = \sum_{i=1}^{n} c_i v_i u_i = \langle \mathbf{v}, \mathbf{u} \rangle$

2. $\langle \mathbf{u}, \mathbf{v} + \mathbf{w} \rangle = \sum_{i=1}^{n} c_i u_i (v_i + w_i) = \sum_{i=1}^{n} c_i u_i v_i + \sum_{i=1}^{n} c_i u_i w_i$

$$= \langle \mathbf{u}, \mathbf{v} \rangle + \langle \mathbf{u}, \mathbf{w} \rangle$$

3. $d\langle \mathbf{u}, \mathbf{v} \rangle = d\sum_{i=1}^{n} c_i u_i v_i = \sum_{i=1}^{n} c_i (du_i) v_i = \langle d\mathbf{u}, \mathbf{v} \rangle$

4. $\langle \mathbf{v}, \mathbf{v} \rangle = \sum_{i=1}^{n} c_i v_i^2 \geq 0$, and $\langle \mathbf{v}, \mathbf{v} \rangle = 0$ if and only if $\mathbf{v} = \mathbf{0}$.

85. From the definition of inner product,

$$\langle \mathbf{u} + \mathbf{v}, \mathbf{w} \rangle = \langle \mathbf{w}, \mathbf{u} + \mathbf{v} \rangle$$

$$= \langle \mathbf{w}, \mathbf{u} \rangle + \langle \mathbf{w}, \mathbf{v} \rangle = \langle \mathbf{u}, \mathbf{w} \rangle + \langle \mathbf{v}, \mathbf{w} \rangle.$$

87. (i) $W^{\perp} = \{\mathbf{v} \in V : \langle \mathbf{v}, \mathbf{w} \rangle = 0 \text{ for all } \mathbf{w} \in W\}$ is nonempty because $\mathbf{0} \in W^{\perp}$.

(ii) Let $\mathbf{v}_1, \mathbf{v}_2 \in W^{\perp}$. Then $\langle \mathbf{v}_1, \mathbf{w} \rangle = \langle \mathbf{v}_2, \mathbf{w} \rangle = 0$ for all $\mathbf{w} \in W$.

So,

$$\langle \mathbf{v}_1 + \mathbf{v}_2, \mathbf{w} \rangle = \langle \mathbf{v}_1, \mathbf{w} \rangle + \langle \mathbf{v}_2, \mathbf{w} \rangle = 0 + 0 = 0$$

for all $\mathbf{w} \in W \Rightarrow \mathbf{v}_1 + \mathbf{v}_2 \in W^{\perp}$.

(iii) Let $\mathbf{v} \in W^{\perp}$ and $c \in R$. Then, $\langle \mathbf{v}, \mathbf{w} \rangle = 0$ for all $\mathbf{w} \in W$, and $\langle c\mathbf{v}, \mathbf{w} \rangle = c\langle \mathbf{v}, \mathbf{w} \rangle = c0 = 0$ for all $\mathbf{w} \in W \Rightarrow c\mathbf{v} \in W^{\perp}$.

89. (a) Let $\langle \mathbf{u}, \mathbf{v} \rangle$ be the Euclidean inner product on R^n. Because $\langle \mathbf{u}, \mathbf{v} \rangle = \mathbf{u}^T \mathbf{v}$, it follows that

$$\langle A^T \mathbf{u}, \mathbf{v} \rangle = \left(A^T \mathbf{u}\right)^T \mathbf{v} = \mathbf{u}^T \left(A^T\right)^T \mathbf{v} = \mathbf{u}^T A \mathbf{v} = \langle \mathbf{u}, A\mathbf{v} \rangle.$$

(b) $\langle A^T A \mathbf{u}, \mathbf{u} \rangle = \left(A^T A \mathbf{u}\right)^T \mathbf{u} = \mathbf{u}^T A^T \left(A^T\right)^T \mathbf{u} = \mathbf{u}^T A^T A \mathbf{u} = (A\mathbf{u})^T A\mathbf{u} = \langle A\mathbf{u}, A\mathbf{u} \rangle = \|A\mathbf{u}\|^2$

Section 5.3 Orthonormal Bases: Gram-Schmidt Process

1. The set is orthogonal because

$$(2, -4) \cdot (2, 1) = 2(2) - 4(1) = 0.$$

However, the set is not orthonormal because

$$\|(2, -4)\| = \sqrt{2^2 + (-4)^2} = \sqrt{20} \neq 1.$$

3. The set is *not* orthogonal because

$$(-4, 6) \cdot (5, 0) = -4(5) + 6(0) = -20 \neq 0.$$

5. The set is orthogonal because

$$\left(\tfrac{3}{5}, \tfrac{4}{5}\right) \cdot \left(-\tfrac{4}{5}, \tfrac{3}{5}\right) = \tfrac{3}{5}\left(-\tfrac{4}{5}\right) + \tfrac{4}{5}\left(\tfrac{3}{5}\right) = -\tfrac{12}{25} + \tfrac{12}{25} = 0.$$

Furthermore, the set is orthonormal because

$$\left\|\left(\tfrac{3}{5}, \tfrac{4}{5}\right)\right\| = \sqrt{\left(\tfrac{3}{5}\right)^2 + \left(\tfrac{4}{5}\right)^2} = 1$$

$$\left\|-\tfrac{4}{5}, \tfrac{3}{5}\right\| = \sqrt{\left(-\tfrac{4}{5}\right)^2 + \left(\tfrac{3}{5}\right)^2} = 1.$$

7. The set is orthogonal because

$$(4, -1, 1) \cdot (-1, 0, 4) = -4 + 4 = 0$$

$$(4, -1, 1) \cdot (-4, -17, -1) = -16 + 17 - 1 = 0$$

$$(-1, 0, 4) \cdot (-4, -17, -1) = 4 - 4 = 0.$$

However, the set is *not* orthonormal because

$$\|(4, -1, 1)\| = \sqrt{(4)^2 + (-1)^2 + 1^2} = \sqrt{18} \neq 1.$$

9. The set is orthogonal because

$$\left(\frac{\sqrt{2}}{2}, 0, \frac{\sqrt{2}}{2}\right) \cdot \left(-\frac{\sqrt{6}}{6}, \frac{\sqrt{6}}{3}, \frac{\sqrt{6}}{6}\right) = 0$$

$$\left(\frac{\sqrt{2}}{2}, 0, \frac{\sqrt{2}}{2}\right) \cdot \left(\frac{\sqrt{3}}{3}, \frac{\sqrt{3}}{3}, -\frac{\sqrt{3}}{3}\right) = 0$$

$$\left(-\frac{\sqrt{6}}{6}, \frac{\sqrt{6}}{3}, \frac{\sqrt{6}}{6}\right) \cdot \left(\frac{\sqrt{3}}{3}, \frac{\sqrt{3}}{3}, -\frac{\sqrt{3}}{3}\right) = 0.$$

Furthermore, the set is orthonormal because

$$\left\|\left(\frac{\sqrt{2}}{2}, 0, \frac{\sqrt{2}}{2}\right)\right\| = \sqrt{\frac{1}{2} + 0 + \frac{1}{2}} = 1$$

$$\left\|\left(-\frac{\sqrt{6}}{6}, \frac{\sqrt{6}}{3}, \frac{\sqrt{6}}{6}\right)\right\| = \sqrt{\frac{1}{6} + \frac{2}{3} + \frac{1}{6}} = 1$$

$$\left\|\left(\frac{\sqrt{3}}{3}, \frac{\sqrt{3}}{3}, -\frac{\sqrt{3}}{3}\right)\right\| = \sqrt{\frac{1}{3} + \frac{1}{3} + \frac{1}{3}} = 1.$$

11. The set is orthogonal because

$$(2, -5, -3) \cdot (4, -2, 6) = 8 + 10 - 18 = 0.$$

However, the set is not orthonormal because

$$\|(2, -5, -3)\| = \sqrt{2^2 + (-5)^2 + (-3)^2} = \sqrt{38} \neq 1.$$

13. The set is orthogonal because

$$\left(\frac{\sqrt{2}}{2}, 0, 0, \frac{\sqrt{2}}{2}\right) \cdot \left(0, \frac{\sqrt{2}}{2}, \frac{\sqrt{2}}{2}, 0\right) = 0$$

$$\left(\frac{\sqrt{2}}{2}, 0, 0, \frac{\sqrt{2}}{2}\right) \cdot \left(-\frac{1}{2}, \frac{1}{2}, -\frac{1}{2}, \frac{1}{2}\right) = 0$$

$$\left(0, \frac{\sqrt{2}}{2}, \frac{\sqrt{2}}{2}, 0\right) \cdot \left(-\frac{1}{2}, \frac{1}{2}, -\frac{1}{2}, \frac{1}{2}\right) = 0.$$

Furthermore, the set is orthonormal because

$$\left\|\left(\frac{\sqrt{2}}{2}, 0, 0, \frac{\sqrt{2}}{2}\right)\right\| = \sqrt{\frac{1}{2} + 0 + 0 + \frac{1}{2}} = 1$$

$$\left\|\left(0, \frac{\sqrt{2}}{2}, \frac{\sqrt{2}}{2}, 0\right)\right\| = \sqrt{0 + \frac{1}{2} + \frac{1}{2} + 0} = 1$$

$$\left\|\left(-\frac{1}{2}, \frac{1}{2}, -\frac{1}{2}, \frac{1}{2}\right)\right\| = \sqrt{\frac{1}{4} + \frac{1}{4} + \frac{1}{4} + \frac{1}{4}} = 1.$$

15. The set is orthogonal because

$$(-1, 4) \cdot (8, 2) = -8 + 8 = 0.$$

However, the set is not orthonormal because

$$\|(-1, 4)\| = \sqrt{(-1)^2 + 4^2} = \sqrt{17} \neq 1.$$

So, normalize the set to produce an orthonormal set.

$$\mathbf{u}_1 = \frac{\mathbf{v}_1}{\|\mathbf{v}_1\|} = \frac{1}{\sqrt{17}}(-1, 4) = \left(-\frac{\sqrt{17}}{17}, \frac{4\sqrt{17}}{17}\right)$$

$$\mathbf{u}_2 = \frac{\mathbf{v}_2}{\|\mathbf{v}_2\|} = \frac{1}{2\sqrt{17}}(8, 2) = \left(\frac{4\sqrt{17}}{17}, \frac{\sqrt{17}}{17}\right)$$

17. The set is orthogonal because

$$\left(\sqrt{3}, \sqrt{3}, \sqrt{3}\right) \cdot \left(-\sqrt{2}, 0, \sqrt{2}\right) = -\sqrt{6} + 0 + \sqrt{6} = 0.$$

However, the set is not orthonormal because

$$\left\|\left(\sqrt{3}, \sqrt{3}, \sqrt{3}\right)\right\| = \sqrt{\left(\sqrt{3}\right)^2 + \left(\sqrt{3}\right)^2 + \left(\sqrt{3}\right)^2}$$

$$= \sqrt{9} = 3 \neq 1.$$

So, normalize the set to produce an orthonormal set.

$$\mathbf{u}_1 = \frac{\mathbf{v}_1}{\|\mathbf{v}_1\|} = \frac{1}{3}\left(\sqrt{3}, \sqrt{3}, \sqrt{3}\right) = \left(\frac{\sqrt{3}}{3}, \frac{\sqrt{3}}{3}, \frac{\sqrt{3}}{3}\right)$$

$$\mathbf{u}_2 = \frac{\mathbf{v}_2}{\|\mathbf{v}_2\|} = \frac{1}{2}\left(-\sqrt{2}, 0, \sqrt{2}\right) = \left(-\frac{\sqrt{2}}{2}, 0, \frac{\sqrt{2}}{2}\right)$$

19. The set $\{1, x, x^2, x^3\}$ is orthogonal because

$$\langle 1, x\rangle = 0, \langle 1, x^2\rangle = 0, \langle 1, x^3\rangle = 0,$$

$$\langle x, x^2\rangle = 0, \langle x, x^3\rangle = 0, \text{ and } \langle x^2, x^3\rangle = 0.$$

Furthermore, the set is orthonormal because

$$\|1\| = 1, \|x\| = 1, \|x^2\| = 1 \text{ and } \|x^3\| = 1.$$

So, $\{1, x, x^2, x^3\}$ is an orthonormal basis for P_3.

21. Use Theorem 5.11 to find the coordinates of $\mathbf{x} = (1, 2)$ relative to B.

$$(1, 2) \cdot \left(-\frac{2\sqrt{13}}{13}, \frac{3\sqrt{13}}{13}\right) = -\frac{2\sqrt{13}}{13} + \frac{6\sqrt{13}}{13} = \frac{4\sqrt{13}}{13}$$

$$(1, 2) \cdot \left(\frac{3\sqrt{13}}{13}, \frac{2\sqrt{13}}{13}\right) = \frac{3\sqrt{13}}{13} + \frac{4\sqrt{13}}{13} = \frac{7\sqrt{13}}{13}$$

So, $[\mathbf{x}]_B = \begin{bmatrix} \dfrac{4\sqrt{13}}{13} \\ \dfrac{7\sqrt{13}}{13} \end{bmatrix}$.

23. Use Theorem 5.11 to find the coordinates of $\mathbf{x} = (2, -2, 1)$ relative to B.

$$(2, -2, 1) \cdot \left(\frac{\sqrt{10}}{10}, 0, \frac{3\sqrt{10}}{10}\right) = \frac{2\sqrt{10}}{10} + \frac{3\sqrt{10}}{10} = \frac{\sqrt{10}}{2}$$

$$(2, -2, 1) \cdot (0, 1, 0) = -2$$

$$(2, -2, 1) \cdot \left(-\frac{3\sqrt{10}}{10}, 0, \frac{\sqrt{10}}{10}\right) = -\frac{6\sqrt{10}}{10} + \frac{\sqrt{10}}{10} = -\frac{\sqrt{10}}{2}$$

So, $[\mathbf{x}]_B = \begin{bmatrix} \dfrac{\sqrt{10}}{2} \\ -2 \\ -\dfrac{\sqrt{10}}{2} \end{bmatrix}$.

25. Use Theorem 5.11 to find the coordinates of $\mathbf{x} = (5, 10, 15)$ relative to B.

$$(5, 10, 15) \cdot \left(\tfrac{3}{5}, \tfrac{4}{5}, 0\right) = 3 + 8 = 11$$

$$(5, 10, 15) \cdot \left(-\tfrac{4}{5}, \tfrac{3}{5}, 0\right) = -4 + 6 = 2$$

$$(5, 10, 15) \cdot (0, 0, 1) = 15$$

So, $[\mathbf{x}]_B = \begin{bmatrix} 11 \\ 2 \\ 15 \end{bmatrix}$.

27. First, orthogonalize each vector in B.

$$\mathbf{w}_1 = \mathbf{v}_1 = (3, 4)$$

$$\mathbf{w}_2 = \mathbf{v}_2 - \frac{\langle \mathbf{v}_2, \mathbf{w}_1\rangle}{\langle \mathbf{w}_1, \mathbf{w}_1\rangle}\mathbf{w}_1 = (1, 0) - \frac{1(3) + 0(4)}{3^2 + 4^2}(3, 4) = (1, 0) - \frac{3}{25}(3, 4) = \left(\frac{16}{25}, -\frac{12}{25}\right)$$

Then, normalize the vectors.

$$\mathbf{u}_1 = \frac{\mathbf{w}_1}{\|\mathbf{w}_1\|} = \frac{1}{\sqrt{3^2 + 4^2}}(3, 4) = \left(\frac{3}{5}, \frac{4}{5}\right)$$

$$\mathbf{u}_2 = \frac{\mathbf{w}_2}{\|\mathbf{w}_2\|} = \frac{1}{\sqrt{\left(\frac{16}{25}\right)^2 + \left(-\frac{12}{25}\right)^2}}\left(\frac{16}{25}, -\frac{12}{25}\right) = \left(\frac{4}{5}, -\frac{3}{5}\right)$$

So, the orthonormal basis is $\left\{\left(\frac{3}{5}, \frac{4}{5}\right), \left(\frac{4}{5}, -\frac{3}{5}\right)\right\}$.

29. First, orthogonalize each vector in B.

$$\mathbf{w}_1 = \mathbf{v}_1 = (0, 1)$$

$$\mathbf{w}_2 = \mathbf{v}_2 - \frac{\langle \mathbf{v}_2, \mathbf{w}_1 \rangle}{\langle \mathbf{w}_1, \mathbf{w}_1 \rangle} \mathbf{w}_1 = (2, 5) - \frac{2(0) + 5(1)}{0^2 + 1^2}(0, 1) = (2, 5) - 5(0, 1) = (2, 0)$$

Then, normalize the vectors.

$$\mathbf{u}_1 = \frac{\mathbf{w}_1}{\|\mathbf{w}_1\|} = \mathbf{w}_1 = (0, 1)$$

$$\mathbf{u}_2 = \frac{\mathbf{w}_2}{\|\mathbf{w}_2\|} = \frac{1}{2}(2, 0) = (1, 0)$$

So, the orthonormal basis is $\{(0, 1), (1, 0)\}$.

31. Because $\mathbf{v}_i \cdot \mathbf{v}_j = 0$ for $i \neq j$, the given vectors are orthogonal. Normalize the vectors.

$$\mathbf{u}_1 = \frac{\mathbf{v}_1}{\|\mathbf{v}_1\|} = \frac{1}{3}(1, -2, 2) = \left(\frac{1}{3}, -\frac{2}{3}, \frac{2}{3}\right)$$

$$\mathbf{u}_2 = \frac{\mathbf{v}_2}{\|\mathbf{v}_2\|} = \frac{1}{3}(2, 2, 1) = \left(\frac{2}{3}, \frac{2}{3}, \frac{1}{3}\right)$$

$$\mathbf{u}_3 = \frac{\mathbf{v}_3}{\|\mathbf{v}_3\|} = \frac{1}{3}(2, -1, -2) = \left(\frac{2}{3}, -\frac{1}{3}, -\frac{2}{3}\right)$$

So, the orthonormal basis is $\left\{\left(\frac{1}{3}, -\frac{2}{3}, \frac{2}{3}\right), \left(\frac{2}{3}, \frac{2}{3}, \frac{1}{3}\right), \left(\frac{2}{3}, -\frac{1}{3}, -\frac{2}{3}\right)\right\}$.

33. First, orthogonalize each vector in B.

$$\mathbf{w}_1 = \mathbf{v}_1 = (4, -3, 0)$$

$$\mathbf{w}_2 = \mathbf{v}_2 - \frac{\langle \mathbf{v}_2, \mathbf{w}_1 \rangle}{\langle \mathbf{w}_1, \mathbf{w}_1 \rangle} \mathbf{w}_1 = (1, 2, 0) - \frac{-2}{25}(4, -3, 0) = \left(\frac{33}{25}, \frac{44}{25}, 0\right)$$

$$\mathbf{w}_3 = \mathbf{v}_3 - \frac{\langle \mathbf{v}_3, \mathbf{w}_1 \rangle}{\langle \mathbf{w}_1, \mathbf{w}_1 \rangle} \mathbf{w}_1 - \frac{\langle \mathbf{v}_3, \mathbf{w}_2 \rangle}{\langle \mathbf{w}_2, \mathbf{w}_2 \rangle} \mathbf{w}_2 = (0, 0, 4) - 0(4, -3, 0) - 0\left(\frac{33}{25}, \frac{44}{25}, 0\right) = (0, 0, 4)$$

Then, normalize the vectors.

$$\mathbf{u}_1 = \frac{\mathbf{w}_1}{\|\mathbf{w}_1\|} = \frac{1}{5}(4, -3, 0) = \left(\frac{4}{5}, -\frac{3}{5}, 0\right)$$

$$\mathbf{u}_2 = \frac{\mathbf{w}_2}{\|\mathbf{w}_2\|} = \frac{5}{11}\left(\frac{33}{25}, \frac{44}{25}, 0\right) = \left(\frac{3}{5}, \frac{4}{5}, 0\right)$$

$$\mathbf{u}_3 = \frac{\mathbf{w}_3}{\|\mathbf{w}_3\|} = \frac{1}{4}(0, 0, 4) = (0, 0, 1)$$

So, the orthonormal basis is $\left\{\left(\frac{4}{5}, -\frac{3}{5}, 0\right), \left(\frac{3}{5}, \frac{4}{5}, 0\right), (0, 0, 1)\right\}$.

35. First, orthogonalize each vector in B.

$$\mathbf{w}_1 = \mathbf{v}_1 = (0, 1, 1)$$

$$\mathbf{w}_2 = \mathbf{v}_2 - \frac{\langle \mathbf{v}_2, \mathbf{w}_1 \rangle}{\langle \mathbf{w}_1, \mathbf{w}_1 \rangle}\mathbf{w}_1 = (1, 1, 0) - \frac{1(0) + 1(1) + 0(1)}{0^2 + 1^2 + 1^2}(0, 1, 1) = (1, 1, 0) - \frac{1}{2}(0, 1, 1) = \left(1, \frac{1}{2}, -\frac{1}{2}\right)$$

$$\mathbf{w}_3 = \mathbf{v}_3 - \frac{\langle \mathbf{v}_3, \mathbf{w}_2 \rangle}{\langle \mathbf{w}_2, \mathbf{w}_2 \rangle}\mathbf{w}_2 - \frac{\langle \mathbf{v}_3, \mathbf{w}_1 \rangle}{\langle \mathbf{w}_1, \mathbf{w}_1 \rangle}\mathbf{w}_1$$

$$= (1, 0, 1) - \frac{1(1) + 0\left(\frac{1}{2}\right) + 1\left(-\frac{1}{2}\right)}{1^2 + \left(\frac{1}{2}\right)^2 + \left(-\frac{1}{2}\right)^2}\left(1, \frac{1}{2}, -\frac{1}{2}\right) - \frac{1(0) + 0(1) + 1(1)}{0^2 + 1^2 + 1^2}(0, 1, 1)$$

$$= (1, 0, 1) - \frac{1}{3}\left(1, \frac{1}{2}, -\frac{1}{2}\right) - \frac{1}{2}(0, 1, 1) = \left(\frac{2}{3}, -\frac{2}{3}, \frac{2}{3}\right)$$

Then, normalize the vectors.

$$\mathbf{u}_1 = \frac{\mathbf{w}_1}{\|\mathbf{w}_1\|} = \frac{1}{\sqrt{0^2 + 1^2 + 1^2}}(0, 1, 1) = \left(0, \frac{\sqrt{2}}{2}, \frac{\sqrt{2}}{2}\right)$$

$$\mathbf{u}_2 = \frac{\mathbf{w}_2}{\|\mathbf{w}_2\|} = \frac{1}{\sqrt{1^2 + \left(\frac{1}{2}\right)^2 + \left(-\frac{1}{2}\right)^2}}\left(1, \frac{1}{2}, -\frac{1}{2}\right) = \left(\frac{\sqrt{6}}{3}, \frac{\sqrt{6}}{6}, -\frac{\sqrt{6}}{6}\right)$$

$$\mathbf{u}_3 = \frac{\mathbf{w}_3}{\|\mathbf{w}_3\|} = \frac{1}{\sqrt{\left(\frac{2}{3}\right)^2 + \left(-\frac{2}{3}\right)^2 + \left(\frac{2}{3}\right)^2}}\left(\frac{2}{3}, -\frac{2}{3}, \frac{2}{3}\right) = \left(\frac{\sqrt{3}}{3}, -\frac{\sqrt{3}}{3}, \frac{\sqrt{3}}{3}\right)$$

So, the orthonormal basis is $\left\{\left(0, \frac{\sqrt{2}}{2}, \frac{\sqrt{2}}{2}\right), \left(\frac{\sqrt{6}}{3}, \frac{\sqrt{6}}{6}, -\frac{\sqrt{6}}{6}\right), \left(\frac{\sqrt{3}}{3}, -\frac{\sqrt{3}}{3}, \frac{\sqrt{3}}{3}\right)\right\}$.

37. Because there is just one vector, you simply need to normalize it.

$$\mathbf{u}_1 = \frac{1}{\sqrt{(-8)^2 + 3^2 + 5^2}}(-8, 3, 5) = \left(-\frac{4\sqrt{2}}{7}, \frac{3\sqrt{2}}{14}, \frac{5\sqrt{2}}{14}\right)$$

So, the orthonormal basis is $\left\{\left(-\frac{4\sqrt{2}}{7}, \frac{3\sqrt{2}}{14}, \frac{5\sqrt{2}}{14}\right)\right\}$.

39. First orthogonalize each vector in B.

$$\mathbf{w}_1 = \mathbf{v}_1 = (3, 4, 0)$$

$$\mathbf{w}_2 = \mathbf{v}_2 - \frac{\langle \mathbf{v}_2, \mathbf{w}_1 \rangle}{\langle \mathbf{w}_1, \mathbf{w}_1 \rangle}\mathbf{w}_1 = (1, 0, 0) - \frac{3}{25}(3, 4, 0) = \left(\frac{16}{25}, -\frac{12}{25}, 0\right)$$

Then, normalize the vectors.

$$\mathbf{u}_1 = \frac{\mathbf{w}_1}{\|\mathbf{w}_1\|} = \frac{1}{\sqrt{3^2 + 4^2 + 0^2}}(3, 4, 0) = \left(\frac{3}{5}, \frac{4}{5}, 0\right)$$

$$\mathbf{u}_2 = \frac{\mathbf{w}_2}{\|\mathbf{w}_2\|} = \frac{1}{\sqrt{\left(\frac{16}{25}\right)^2 + \left(-\frac{12}{25}\right)^2 + 0^2}}\left(\frac{16}{25}, -\frac{12}{25}, 0\right) = \left(\frac{4}{5}, -\frac{3}{5}, 0\right)$$

So, the orthonormal basis is $\left\{\left(\frac{3}{5}, \frac{4}{5}, 0\right), \left(\frac{4}{5}, -\frac{3}{5}, 0\right)\right\}$.

41. First, orthogonalize each vector in B.

$$\mathbf{w}_1 = \mathbf{v}_1 = (1, 2, -1, 0)$$

$$\mathbf{w}_2 = \mathbf{v}_2 - \frac{\langle \mathbf{v}_2, \mathbf{w}_1 \rangle}{\langle \mathbf{w}_1, \mathbf{w}_1 \rangle}\mathbf{w}_1 = (2, 2, 0, 1) - \frac{2(1) + 2(2) + 0(-1) + 1(0)}{1^2 + 2^2 + (-1)^2 + 0^2}(1, 2, -1, 0) = (2, 2, 0, 1) - (1, 2, -1, 0) = (1, 0, 1, 1)$$

$$\mathbf{w}_3 = \mathbf{v}_3 - \frac{\langle \mathbf{v}_3, \mathbf{w}_2 \rangle}{\langle \mathbf{w}_2, \mathbf{w}_2 \rangle}\mathbf{w}_2 - \frac{\langle \mathbf{v}_3, \mathbf{w}_1 \rangle}{\langle \mathbf{w}_1, \mathbf{w}_1 \rangle}\mathbf{w}_1$$

$$= (1, 1, -1, 0) - \frac{1(1) + 1(0) - 1(1) + 0(1)}{1^2 + 0^2 + 1^2 + 1^2}(1, 0, 1, 1) - \frac{1(1) + 1(2) - 1(-1) + 0(0)}{1^2 + 2^2 + (-1)^2 + 0^2}(1, 2, -1, 0)$$

$$= (1, 1, -1, 0) - \frac{0}{3}(1, 0, 1, 1) - \frac{2}{3}(1, 2, -1, 0) = \left(\frac{1}{3}, -\frac{1}{3}, -\frac{1}{3}, 0\right)$$

Then, normalize the vectors.

$$\mathbf{u}_1 = \frac{\mathbf{w}_1}{\|\mathbf{w}_1\|} = \frac{1}{\sqrt{1^2 + 2^2 + (-1)^2 + 0^2}}(1, 2, -1, 0) = \left(\frac{\sqrt{6}}{6}, \frac{\sqrt{6}}{3}, -\frac{\sqrt{6}}{6}, 0\right)$$

$$\mathbf{u}_2 = \frac{\mathbf{w}_2}{\|\mathbf{w}_2\|} = \frac{1}{\sqrt{1^2 + 0^2 + 1^2 + 1^2}}(1, 0, 1, 1) = \left(\frac{\sqrt{3}}{3}, 0, \frac{\sqrt{3}}{3}, \frac{\sqrt{3}}{3}\right)$$

$$\mathbf{u}_3 = \frac{\mathbf{w}_3}{\|\mathbf{w}_3\|} = \frac{1}{\sqrt{\left(\frac{1}{3}\right)^2 + \left(-\frac{1}{3}\right)^2 + \left(-\frac{1}{3}\right)^2 + 0^2}}\left(\frac{1}{3}, -\frac{1}{3}, -\frac{1}{3}, 0\right) = \left(\frac{\sqrt{3}}{3}, -\frac{\sqrt{3}}{3}, -\frac{\sqrt{3}}{3}, 0\right)$$

So, the orthonormal basis is $\left\{\left(\frac{\sqrt{6}}{6}, \frac{\sqrt{6}}{3}, -\frac{\sqrt{6}}{6}, 0\right), \left(\frac{\sqrt{3}}{3}, 0, \frac{\sqrt{3}}{3}, \frac{\sqrt{3}}{3}\right), \left(\frac{\sqrt{3}}{3}, -\frac{\sqrt{3}}{3}, -\frac{\sqrt{3}}{3}, 0\right)\right\}$.

43. $\langle x, 1 \rangle = \displaystyle\int_{-1}^{1} x \, dx = \frac{x^2}{2}\Big]_{-1}^{1} = 0$

45. $\langle x^2, 1 \rangle = \displaystyle\int_{-1}^{1} x^2 \, dx = \frac{x^3}{3}\Big]_{-1}^{1} = \frac{2}{3}$

47. (a) True. See "Definition of Orthogonal and Orthornormal Sets," page 306.

   (b) True. See Corollary to Theorem 5.10, page 310.

   (c) False. See "Remark," page 316.

49. The solutions of the homogeneous system are of the form $(3s, -2t, s, t)$, where s and t are any real numbers. So, a basis for the solution space is $\{(3, 0, 1, 0), (0, -2, 0, 1)\}$.

Orthogonalize this basis as follows.

$$\mathbf{w}_1 = \mathbf{v}_1 = (3, 0, 1, 0)$$

$$\mathbf{w}_2 = \mathbf{v}_2 - \frac{\langle \mathbf{v}_2, \mathbf{w}_1 \rangle}{\langle \mathbf{w}_1, \mathbf{w}_1 \rangle}\mathbf{w}_1 = (0, -2, 0, 1) - \frac{0(3) + (-2)(0) + 0(1) + 1(0)}{3^2 + 0^2 + 1^2 + 0^2}(3, 0, 1, 0) = (0, -2, 0, 1)$$

Then, normalize these vectors.

$$\mathbf{u}_1 = \frac{\mathbf{w}_1}{\|\mathbf{w}_1\|} = \frac{1}{\sqrt{3^2 + 0^2 + 1^2 + 0^2}}(3, 0, 1, 0) = \left(\frac{3\sqrt{10}}{10}, 0, \frac{\sqrt{10}}{10}, 0\right)$$

$$\mathbf{u}_2 = \frac{\mathbf{w}_2}{\|\mathbf{w}_2\|} = \frac{1}{\sqrt{0^2 + (-2)^2 + 0^2 + 1^2}}(0, -2, 0, 1) = \left(0, -\frac{2\sqrt{5}}{5}, 0, \frac{\sqrt{5}}{5}\right)$$

So, the orthonormal basis for the solution set is $\left\{\left(\frac{3\sqrt{10}}{10}, 0, \frac{\sqrt{10}}{10}, 0\right), \left(0, -\frac{2\sqrt{5}}{5}, 0, \frac{\sqrt{5}}{5}\right)\right\}$.

51. The solutions of the homogeneous system are of the form $(s + t, 0, s, t)$, where s and t are any real numbers. So, a basis for the solution space is $\{(1, 0, 1, 0), (1, 0, 0, 1)\}$.

Orthogonalize this basis as follows.

$$\mathbf{w}_1 = \mathbf{v}_1 = (1, 0, 1, 0)$$

$$\mathbf{w}_2 = \mathbf{v}_2 - \frac{\langle \mathbf{v}_2, \mathbf{w}_1 \rangle}{\langle \mathbf{w}_1, \mathbf{w}_1 \rangle}\mathbf{w}_1 = (1, 0, 0, 1) - \frac{1}{2}(1, 0, 1, 0) = \left(\frac{1}{2}, 0, -\frac{1}{2}, 1\right)$$

Then, normalize these vectors.

$$\mathbf{u}_1 = \frac{\mathbf{w}_1}{\|\mathbf{w}_1\|} = \frac{1}{\sqrt{2}}(1, 0, 1, 0) = \left(\frac{\sqrt{2}}{2}, 0, \frac{\sqrt{2}}{2}, 0\right)$$

$$\mathbf{u}_2 = \frac{\mathbf{w}_2}{\|\mathbf{w}_2\|} = \frac{1}{\sqrt{3/2}}\left(\frac{1}{2}, 0, -\frac{1}{2}, 1\right) = \left(\frac{\sqrt{6}}{6}, 0, -\frac{\sqrt{6}}{6}, \frac{\sqrt{6}}{3}\right)$$

So, an orthonormal basis for the solution space is $\left\{\left(\frac{\sqrt{2}}{2}, 0, \frac{\sqrt{2}}{2}, 0\right), \left(\frac{\sqrt{6}}{6}, 0, -\frac{\sqrt{6}}{6}, \frac{\sqrt{6}}{3}\right)\right\}$.

53. The solutions of the homogeneous system are of the form $(-3s + 3t, s, t)$, where s and t are any real numbers. So, a basis for the solution space is $\{(-3, 1, 0), (3, 0, 1)\}$.

Orthogonalize this basis as follows.

$$\mathbf{w}_1 = \mathbf{v}_1 = (-3, 1, 0)$$

$$\mathbf{w}_2 = \mathbf{v}_2 - \frac{\langle \mathbf{v}_2, \mathbf{w}_1 \rangle}{\langle \mathbf{w}_1, \mathbf{w}_1 \rangle}\mathbf{w}_1 = (3, 0, 1) + \frac{9}{10}(-3, 1, 0) = \left(\frac{3}{10}, \frac{9}{10}, 1\right)$$

Then, normalize these vectors.

$$\mathbf{u}_1 = \frac{\mathbf{w}_1}{\|\mathbf{w}_1\|} = \frac{1}{\sqrt{10}}(-3, 1, 0) = \left(-\frac{3\sqrt{10}}{10}, \frac{\sqrt{10}}{10}, 0\right)$$

$$\mathbf{u}_2 = \frac{\mathbf{w}_2}{\|\mathbf{w}_2\|} = \frac{1}{\sqrt{19/10}}\left(\frac{3}{10}, \frac{9}{10}, 1\right) = \left(\frac{3\sqrt{190}}{190}, \frac{9\sqrt{190}}{190}, \frac{\sqrt{190}}{19}\right)$$

So, an orthonormal basis for the solution space is

$$\left\{\left(-\frac{3\sqrt{10}}{10}, \frac{\sqrt{10}}{10}, 0\right), \left(\frac{3\sqrt{190}}{190}, \frac{9\sqrt{190}}{190}, \frac{\sqrt{190}}{19}\right)\right\}.$$

55. Let

$$p(x) = \frac{x^2 + 1}{\sqrt{2}} \quad \text{and} \quad q(x) = \frac{x^2 + x - 1}{\sqrt{3}}.$$

Then

$$\langle p, q \rangle = \frac{1}{\sqrt{2}}\left(-\frac{1}{\sqrt{3}}\right) + 0\left(\frac{1}{\sqrt{3}}\right) + \frac{1}{\sqrt{2}}\left(\frac{1}{\sqrt{3}}\right) = 0. \text{ Furthermore,}$$

$$\|p\| = \sqrt{\left(\frac{1}{\sqrt{2}}\right)^2 + 0^2 + \left(\frac{1}{\sqrt{2}}\right)^2} = 1$$

$$\|q\| = \sqrt{\left(-\frac{1}{\sqrt{3}}\right)^2 + \left(\frac{1}{\sqrt{3}}\right)^2 + \left(\frac{1}{\sqrt{3}}\right)^2} = 1.$$

So, $\{p, q\}$ is an orthonormal set.

57. Let $p_1(x) = x^2$, $p_2(x) = x^2 + 2x$, and $p_3(x) = x^2 + 2x + 1$.

Then, because $\langle p_1, p_2 \rangle = 0(0) + 0(2) + 1(1) = 1 \neq 0$, the set is not orthogonal. Orthogonalize the set as follows.

$$\mathbf{w}_1 = p_1 = x^2$$

$$\mathbf{w}_2 = p_2 - \frac{\langle p_2, \mathbf{w}_1 \rangle}{\langle \mathbf{w}_1, \mathbf{w}_1 \rangle}\mathbf{w}_1 = x^2 + 2x - \frac{0(0) + 2(0) + 1(1)}{0^2 + 0^2 + 1^2}x^2 = 2x$$

$$\mathbf{w}_3 = p_3 - \frac{\langle p_3, \mathbf{w}_2 \rangle}{\langle \mathbf{w}_2, \mathbf{w}_2 \rangle}\mathbf{w}_2 - \frac{\langle p_3, \mathbf{w}_1 \rangle}{\langle \mathbf{w}_1, \mathbf{w}_1 \rangle}\mathbf{w}_1$$

$$= x^2 + 2x + 1 - \frac{1(0) + 2(2) + 1(0)}{0^2 + 2^2 + 0^2}(2x) - \frac{1(0) + 2(0) + 1(1)}{0^2 + 0^2 + 1^2}x^2$$

$$= x^2 + 2x + 1 - 2x - x^2 = 1$$

Then, normalize the vectors.

$$\mathbf{u}_1 = \frac{\mathbf{w}_1}{\|\mathbf{w}_1\|} = \frac{1}{\sqrt{0^2 + 0^2 + 1^2}}x^2 = x^2$$

$$\mathbf{u}_2 = \frac{\mathbf{w}_2}{\|\mathbf{w}_2\|} = \frac{1}{\sqrt{0^2 + 2^2 + 0^2}}(2x) = x$$

$$\mathbf{u}_3 = \frac{\mathbf{w}_3}{\|\mathbf{w}_3\|} = \frac{1}{\sqrt{1^2 + 0^2 + 0^2}}(1) = 1$$

So, the orthonormal set is $\{x^2, x, 1\}$.

59. Let $p(x) = x^2 - 1$ and $q(x) = x - 1$. Then, because $\langle p, q \rangle = 1 \neq 0$, the set is not orthogonal. Orthogonalize the set as follows.

$$\mathbf{w}_1 = p = x^2 - 1$$

$$\mathbf{w}_2 = q - \frac{\langle q, \mathbf{w}_1 \rangle}{\langle \mathbf{w}_1, \mathbf{w}_1 \rangle}\mathbf{w}_1 = (x - 1) - \frac{1}{2}(x^2 - 1) = -\frac{1}{2}x^2 + x - \frac{1}{2}$$

Then, normalize the vectors.

$$\mathbf{u}_1 = \frac{\mathbf{w}_1}{\|\mathbf{w}_1\|} = \frac{1}{\sqrt{2}}(x^2 - 1) = \frac{\sqrt{2}}{2}(x^2 - 1)$$

$$\mathbf{u}_2 = \frac{\mathbf{w}_2}{\|\mathbf{w}_2\|} = \frac{1}{\sqrt{3/2}}\left(-\frac{1}{2}x^2 + x - \frac{1}{2}\right) = \frac{\sqrt{6}}{3}\left(-\frac{1}{2}x^2 + x - \frac{1}{2}\right) = \frac{-\sqrt{6}}{6}(x^2 - 2x + 1)$$

So, the orthonormal set is $\left\{\frac{\sqrt{2}}{2}(x^2 - 1), -\frac{\sqrt{6}}{6}(x^2 - 2x + 1)\right\}$.

61. Begin by orthogonalizing the set.

$$\mathbf{w}_1 = \mathbf{v}_1 = (2, -1)$$

$$\mathbf{w}_2 = \mathbf{v}_2 - \frac{\langle \mathbf{v}_2, \mathbf{w}_1 \rangle}{\langle \mathbf{w}_1, \mathbf{w}_1 \rangle}\mathbf{w}_1 = (-2, 10) - \frac{2(-2)(2) + 10(-1)}{2(2)^2 + (-1)^2}(2, -1) = (-2, 10) + 2(2, -1) = (2, 8)$$

Then, normalize each vector.

$$\mathbf{u}_1 = \frac{\mathbf{w}_1}{\|\mathbf{w}_1\|} = \frac{1}{\sqrt{2(2)^2 + (-1)^2}}(2, -1) = \left(\frac{2}{3}, -\frac{1}{3}\right)$$

$$\mathbf{u}_2 = \frac{\mathbf{w}_2}{\|\mathbf{w}_2\|} = \frac{1}{\sqrt{2(2)^2 + 8^2}}(2, 8) = \left(\frac{\sqrt{2}}{6}, \frac{2\sqrt{2}}{3}\right)$$

So, an orthonormal basis, using the given inner product is $\left\{\left(\frac{2}{3}, -\frac{1}{3}\right), \left(\frac{\sqrt{2}}{6}, \frac{2\sqrt{2}}{3}\right)\right\}$.

63. For $\{u_1, u_2, ..., u_n\}$ an orthonormal basis for R^n and v any vector in R^n,

$$v = \langle v, u_1\rangle u_1 + \langle v, u_2\rangle u_2 + ... + \langle v, u_n\rangle u_n$$

$$\|v\|^2 = \|\langle v, u_1\rangle u_1 + \langle v, u_2\rangle u_2 + ... + \langle v, u_n\rangle u_n\|^2.$$

Because $u_i \cdot u_j = 0$ for $i \neq j$, it follows that

$$\|v\|^2 = (\langle v, u_1\rangle)^2\|u_1\|^2 + (\langle v, u_2\rangle)^2\|u_2\|^2 + ... + (\langle v, u_n\rangle)^2\|u_n\|^2$$

$$\|v\|^2 = (\langle v, u_1\rangle)^2(1) + (\langle v, u_2\rangle)^2(1) + ... + (\langle v, u_n\rangle)^2(1)$$

$$\|v\|^2 = |v \cdot u_1|^2 + |v \cdot u_2|^2 + ... + |v \cdot u_n|^2.$$

65. First prove that condition (a) implies (b). If
$P^{-1} = P^T$, consider p_i the ith row vector of P. Because
$P P^T = I_n$, you have $p_i \cdot p_i = 1$ and $p_i \cdot p_j = 0$, for
$i \neq j$. So the row vectors of P form an orthonormal
basis for R^n.

(b) implies (c) if the row vectors of P form an
orthonormal basis, then $P P^T = I_n \Rightarrow P^T P = I_n$,
which implies that the column vectors of P form an
orthonormal basis.

(c) implies (a) because the column vectors of P form an
orthonormal basis, you have $P^T P = I_n$, which implies
that $P^{-1} = P^T$.

67. Note that v_1 and v_2 are orthogonal unit vectors.
Furthermore, a vector (c_1, c_2, c_3, c_4) orthogonal to v_1 and
v_2 satisfies the homogeneous system of linear equations

$$\frac{1}{\sqrt{2}}c_1 \qquad + \frac{1}{\sqrt{2}}c_3 \qquad = 0$$

$$-\frac{1}{\sqrt{2}}c_2 \qquad + \frac{1}{\sqrt{2}}c_4 = 0,$$

which has solutions of the form $(-s, t, s, t)$, where s and
t are any real numbers. A basis for the solution set is
$\{(1, 0, -1, 0), (0, 1, 0, 1)\}$. Because $(1, 0, -1, 0)$ and
$(0, 1, 0, 1)$ are already orthogonal, you simply normalize

them to yield $\left(\frac{1}{\sqrt{2}}, 0, -\frac{1}{\sqrt{2}}, 0\right)$ and $\left(0, \frac{1}{\sqrt{2}}, 0, \frac{1}{\sqrt{2}}\right)$.

So,

$$\left\{\left(\frac{1}{\sqrt{2}}, 0, \frac{1}{\sqrt{2}}, 0\right), \left(0, -\frac{1}{\sqrt{2}}, 0, \frac{1}{\sqrt{2}}\right),\right.$$

$$\left.\left(\frac{1}{\sqrt{2}}, 0, -\frac{1}{\sqrt{2}}, 0\right), \left(0, \frac{1}{\sqrt{2}}, 0, \frac{1}{\sqrt{2}}\right)\right\}$$

is an orthonormal basis.

69. $A = \begin{bmatrix} 1 & 1 & -1 \\ 0 & 2 & 1 \\ 1 & 3 & 0 \end{bmatrix} \Rightarrow \begin{bmatrix} 2 & 0 & -3 \\ 0 & 2 & 1 \\ 0 & 0 & 0 \end{bmatrix}$

$A^T = \begin{bmatrix} 1 & 0 & 1 \\ 1 & 2 & 3 \\ -1 & 1 & 0 \end{bmatrix} \Rightarrow \begin{bmatrix} 1 & 0 & 1 \\ 0 & 1 & 1 \\ 0 & 0 & 0 \end{bmatrix}$

$N(A)$-basis: $\left\{ \begin{bmatrix} 3 \\ -1 \\ 2 \end{bmatrix} \right\}$

$N(A^T)$-basis: $\left\{ \begin{bmatrix} -1 \\ -1 \\ 1 \end{bmatrix} \right\}$

$R(A)$-basis: $\left\{ \begin{bmatrix} 1 \\ 0 \\ 1 \end{bmatrix}, \begin{bmatrix} 1 \\ 2 \\ 3 \end{bmatrix} \right\}$

$R(A^T)$-basis: $\left\{ \begin{bmatrix} 1 \\ 1 \\ -1 \end{bmatrix}, \begin{bmatrix} 0 \\ 2 \\ 1 \end{bmatrix} \right\}$

$N(A) = R(A^T)^{\perp}$ and $N(A^T) = R(A)^{\perp}$

71. $A = \begin{bmatrix} 1 & 0 & 1 \\ 1 & 1 & 1 \end{bmatrix} \Rightarrow \begin{bmatrix} 1 & 0 & 1 \\ 0 & 1 & 0 \end{bmatrix}$

$A^T = \begin{bmatrix} 1 & 1 \\ 0 & 1 \\ 1 & 1 \end{bmatrix} \Rightarrow \begin{bmatrix} 1 & 0 \\ 0 & 1 \\ 0 & 0 \end{bmatrix}$

$N(A)$-basis: $\left\{ \begin{bmatrix} 1 \\ 0 \\ -1 \end{bmatrix} \right\}$

$N(A^T) = \left\{ \begin{bmatrix} 0 \\ 0 \end{bmatrix} \right\}$

$R(A)$-basis: $\left\{ \begin{bmatrix} 1 \\ 1 \end{bmatrix}, \begin{bmatrix} 0 \\ 1 \end{bmatrix} \right\}$ $\left(R(A) = R^2 \right)$

$R(A^T)$-basis: $\left\{ \begin{bmatrix} 1 \\ 0 \\ 1 \end{bmatrix}, \begin{bmatrix} 1 \\ 1 \\ 1 \end{bmatrix} \right\}$

$N(A) = R(A^T)^{\perp}$ and $N(A^T) = R(A)^{\perp}$

73. (a) The row space of A is the column space of A^T, $R(A^T)$.

   (b) Let $\mathbf{x} \in N(A) \Rightarrow A\mathbf{x} = \mathbf{0} \Rightarrow \mathbf{x}$ is orthogonal to all the rows of $A \Rightarrow \mathbf{x}$ is orthogonal to all the columns of $A^T \Rightarrow \mathbf{x} \in R(A^T)^{\perp}$.

   (c) Let $\mathbf{x} \in R(A^T)^{\perp} \Rightarrow \mathbf{x}$ is orthogonal to each column vector of
   $A^T \Rightarrow A\mathbf{x} = \mathbf{0} \Rightarrow \mathbf{x} \in N(A)$. Combining this with part (b), $N(A) = R(A^T)^{\perp}$.

   (d) Substitute A^T for A in part (c).

Section 5.4 Mathematical Models and Least Squares Analysis

1. Not orthogonal: $\begin{bmatrix} 0 \\ 1 \\ 1 \end{bmatrix} \cdot \begin{bmatrix} -1 \\ 2 \\ 0 \end{bmatrix} = 2 \neq 0$

3. Orthogonal: $\begin{bmatrix} 1 \\ 1 \\ 1 \\ 1 \end{bmatrix} \cdot \begin{bmatrix} -1 \\ 1 \\ -1 \\ 1 \end{bmatrix} = \begin{bmatrix} 1 \\ 1 \\ 1 \\ 1 \end{bmatrix} \cdot \begin{bmatrix} 0 \\ 2 \\ -2 \\ 0 \end{bmatrix} = 0$

5. $S = \text{span}\left(\begin{bmatrix} 1 \\ 0 \\ 0 \end{bmatrix}, \begin{bmatrix} 0 \\ 0 \\ 1 \end{bmatrix} \right) \Rightarrow S^{\perp} = \text{span}\left\{ \begin{bmatrix} 0 \\ 1 \\ 0 \end{bmatrix} \right\}$ (The y-axis)

7. $A^T = \begin{bmatrix} 1 & 2 & 0 & 0 \\ 0 & 1 & 0 & 1 \end{bmatrix} \Rightarrow \begin{bmatrix} 1 & 0 & 0 & -2 \\ 0 & 1 & 0 & 1 \end{bmatrix} \Rightarrow S^{\perp} = \text{span}\left\{ \begin{bmatrix} 0 \\ 0 \\ 1 \\ 0 \end{bmatrix}, \begin{bmatrix} 2 \\ -1 \\ 0 \\ 1 \end{bmatrix} \right\}$

9. The orthogonal complement of span $\left\{ \begin{bmatrix} 0 \\ 0 \\ 1 \\ 0 \end{bmatrix}, \begin{bmatrix} 2 \\ -1 \\ 0 \\ 1 \end{bmatrix} \right\} = S^{\perp}$

is $\left(S^{\perp} \right)^{\perp} = S = \text{span}\left(\begin{bmatrix} 1 \\ 2 \\ 0 \\ 0 \end{bmatrix}, \begin{bmatrix} 0 \\ 1 \\ 0 \\ 1 \end{bmatrix} \right)$.

11. An orthonormal basis for S is $\left\{ \begin{bmatrix} 0 \\ 0 \\ -\dfrac{1}{\sqrt{2}} \\ \dfrac{1}{\sqrt{2}} \end{bmatrix}, \begin{bmatrix} 0 \\ \dfrac{1}{\sqrt{3}} \\ \dfrac{1}{\sqrt{3}} \\ \dfrac{1}{\sqrt{3}} \end{bmatrix} \right\}.$

$$\text{proj}_S \mathbf{v} = (\mathbf{v} \cdot \mathbf{u}_1)\mathbf{u}_1 + (\mathbf{v} \cdot \mathbf{u}_2)\mathbf{u}_2 = 0\mathbf{u}_1 + \frac{2}{\sqrt{3}}\begin{bmatrix} 0 \\ \dfrac{1}{\sqrt{3}} \\ \dfrac{1}{\sqrt{3}} \\ \dfrac{1}{\sqrt{3}} \end{bmatrix} = \begin{bmatrix} 0 \\ \dfrac{2}{3} \\ \dfrac{2}{3} \\ \dfrac{2}{3} \end{bmatrix}$$

13. Use Gram-Schmidt to construct an orthonormal basis for S.

$$\begin{bmatrix} 0 \\ 1 \\ 1 \end{bmatrix} - \frac{1}{2}\begin{bmatrix} 1 \\ 0 \\ 1 \end{bmatrix} = \begin{bmatrix} -\dfrac{1}{2} \\ 1 \\ \dfrac{1}{2} \end{bmatrix}$$

orthonormal basis: $\left\{ \begin{bmatrix} \dfrac{1}{\sqrt{2}} \\ 0 \\ \dfrac{1}{\sqrt{2}} \end{bmatrix}, \begin{bmatrix} -\dfrac{1}{\sqrt{6}} \\ \dfrac{2}{\sqrt{6}} \\ \dfrac{1}{\sqrt{6}} \end{bmatrix} \right\}$

$$\text{proj}_S \mathbf{v} = (\mathbf{u}_1 \cdot \mathbf{v})\mathbf{u}_1 + (\mathbf{u}_2 \cdot \mathbf{v})\mathbf{u}_2 = \frac{6}{\sqrt{2}}\begin{bmatrix} \dfrac{1}{\sqrt{2}} \\ 0 \\ \dfrac{1}{\sqrt{2}} \end{bmatrix} + \frac{8}{\sqrt{6}}\begin{bmatrix} -\dfrac{1}{\sqrt{6}} \\ \dfrac{2}{\sqrt{6}} \\ \dfrac{1}{\sqrt{6}} \end{bmatrix} = \begin{bmatrix} 3 \\ 0 \\ 3 \end{bmatrix} + \begin{bmatrix} -\dfrac{4}{3} \\ \dfrac{8}{3} \\ \dfrac{4}{3} \end{bmatrix} = \begin{bmatrix} \dfrac{5}{3} \\ \dfrac{8}{3} \\ \dfrac{13}{3} \end{bmatrix}$$

15. Use Gram-Schmidt to construct an orthonormal basis for the column space of A.

$$\begin{bmatrix} 2 \\ 1 \\ 1 \end{bmatrix} - \frac{3}{2}\begin{bmatrix} 1 \\ 0 \\ 1 \end{bmatrix} = \begin{bmatrix} \dfrac{1}{2} \\ 1 \\ -\dfrac{1}{2} \end{bmatrix}$$

orthonormal basis: $\left\{ \begin{bmatrix} \dfrac{1}{\sqrt{2}} \\ 0 \\ \dfrac{1}{\sqrt{2}} \end{bmatrix}, \begin{bmatrix} \dfrac{1}{\sqrt{6}} \\ \dfrac{2}{\sqrt{6}} \\ -\dfrac{1}{\sqrt{6}} \end{bmatrix} \right\}$

$$\text{proj}_S \mathbf{b} = (\mathbf{u}_1 \cdot \mathbf{b})\mathbf{u}_1 + (\mathbf{u}_2 \cdot \mathbf{b})\mathbf{u}_2 = \frac{3}{\sqrt{2}}\begin{bmatrix} \dfrac{1}{\sqrt{2}} \\ 0 \\ \dfrac{1}{\sqrt{2}} \end{bmatrix} + \frac{-3}{\sqrt{6}}\begin{bmatrix} \dfrac{1}{\sqrt{6}} \\ \dfrac{2}{\sqrt{6}} \\ -\dfrac{1}{\sqrt{6}} \end{bmatrix} = \begin{bmatrix} \dfrac{3}{2} \\ 0 \\ \dfrac{3}{2} \end{bmatrix} + \begin{bmatrix} -\dfrac{1}{2} \\ -1 \\ \dfrac{1}{2} \end{bmatrix} = \begin{bmatrix} 1 \\ -1 \\ 2 \end{bmatrix}$$

17. $A = \begin{bmatrix} 1 & 2 & 3 \\ 0 & 1 & 0 \end{bmatrix} \Rightarrow \begin{bmatrix} 1 & 0 & 3 \\ 0 & 1 & 0 \end{bmatrix}$

$A^T = \begin{bmatrix} 1 & 0 \\ 2 & 1 \\ 3 & 0 \end{bmatrix} \Rightarrow \begin{bmatrix} 1 & 0 \\ 0 & 1 \\ 0 & 0 \end{bmatrix}$

$N(A)$-basis: $\left\{ \begin{bmatrix} -3 \\ 0 \\ 1 \end{bmatrix} \right\}$

$N(A^T) = \left\{ \begin{bmatrix} 0 \\ 0 \end{bmatrix} \right\}$

$R(A)$-basis: $\left\{ \begin{bmatrix} 1 \\ 0 \end{bmatrix}, \begin{bmatrix} 2 \\ 1 \end{bmatrix} \right\}$ $\left(R(A) = R^2 \right)$

$R(A^T)$-basis: $\left\{ \begin{bmatrix} 1 \\ 2 \\ 3 \end{bmatrix}, \begin{bmatrix} 0 \\ 1 \\ 0 \end{bmatrix} \right\}$

19. $A = \begin{bmatrix} 1 & 0 & 0 & 1 \\ 0 & 1 & 1 & 1 \\ 1 & 1 & 1 & 2 \\ 1 & 2 & 2 & 3 \end{bmatrix} \Rightarrow \begin{bmatrix} 1 & 0 & 0 & 1 \\ 0 & 1 & 1 & 1 \\ 0 & 0 & 0 & 0 \\ 0 & 0 & 0 & 0 \end{bmatrix}$

$A^T = \begin{bmatrix} 1 & 0 & 1 & 1 \\ 0 & 1 & 1 & 2 \\ 0 & 1 & 1 & 2 \\ 1 & 1 & 2 & 3 \end{bmatrix} \Rightarrow \begin{bmatrix} 1 & 0 & 1 & 1 \\ 0 & 1 & 1 & 2 \\ 0 & 0 & 0 & 0 \\ 0 & 0 & 0 & 0 \end{bmatrix}$

$N(A)$-basis: $\left\{ \begin{bmatrix} -1 \\ -1 \\ 0 \\ 1 \end{bmatrix}, \begin{bmatrix} 0 \\ -1 \\ 1 \\ 0 \end{bmatrix} \right\}$

$N(A^T)$-basis: $\left\{ \begin{bmatrix} -1 \\ -1 \\ 1 \\ 0 \end{bmatrix}, \begin{bmatrix} -1 \\ -2 \\ 0 \\ 1 \end{bmatrix} \right\}$

$R(A)$-basis: $\left\{ \begin{bmatrix} 1 \\ 0 \\ 1 \\ 1 \end{bmatrix}, \begin{bmatrix} 0 \\ 1 \\ 1 \\ 2 \end{bmatrix} \right\}$

$R(A^T)$-basis: $\left\{ \begin{bmatrix} 1 \\ 0 \\ 0 \\ 1 \end{bmatrix}, \begin{bmatrix} 0 \\ 1 \\ 1 \\ 1 \end{bmatrix} \right\}$

21. $A^T A = \begin{bmatrix} 2 & 1 & 1 \\ 1 & 2 & 1 \end{bmatrix} \begin{bmatrix} 2 & 1 \\ 1 & 2 \\ 1 & 1 \end{bmatrix} = \begin{bmatrix} 6 & 5 \\ 5 & 6 \end{bmatrix}$

$A^T \mathbf{b} = \begin{bmatrix} 2 & 1 & 1 \\ 1 & 2 & 1 \end{bmatrix} \begin{bmatrix} 2 \\ 0 \\ -3 \end{bmatrix} = \begin{bmatrix} 1 \\ -1 \end{bmatrix}$

$\begin{bmatrix} 6 & 5 & 1 \\ 5 & 6 & -1 \end{bmatrix} \Rightarrow \begin{bmatrix} 1 & 0 & 1 \\ 0 & 1 & -1 \end{bmatrix} \Rightarrow \mathbf{x} = \begin{bmatrix} 1 \\ -1 \end{bmatrix}$

23. $A^T A = \begin{bmatrix} 1 & 1 & 0 & 1 \\ 0 & 1 & 1 & 1 \\ 1 & 1 & 1 & 0 \end{bmatrix} \begin{bmatrix} 1 & 0 & 1 \\ 1 & 1 & 1 \\ 0 & 1 & 1 \\ 1 & 1 & 0 \end{bmatrix} = \begin{bmatrix} 3 & 2 & 2 \\ 2 & 3 & 2 \\ 2 & 2 & 3 \end{bmatrix}$

$A^T \mathbf{b} = \begin{bmatrix} 1 & 1 & 0 & 1 \\ 0 & 1 & 1 & 1 \\ 1 & 1 & 1 & 0 \end{bmatrix} \begin{bmatrix} 4 \\ -1 \\ 0 \\ 1 \end{bmatrix} = \begin{bmatrix} 4 \\ 0 \\ 3 \end{bmatrix}$

$\begin{bmatrix} 3 & 2 & 2 & 4 \\ 2 & 3 & 2 & 0 \\ 2 & 2 & 3 & 3 \end{bmatrix} \Rightarrow \begin{bmatrix} 1 & 0 & 0 & 2 \\ 0 & 1 & 0 & -2 \\ 0 & 0 & 1 & 1 \end{bmatrix} \Rightarrow \mathbf{x} = \begin{bmatrix} 2 \\ -2 \\ 1 \end{bmatrix}$

25. $A^T A = \begin{bmatrix} 3 & 2 & 0 \\ 2 & 6 & 1 \\ 0 & 1 & 3 \end{bmatrix}, A^T \mathbf{b} = \begin{bmatrix} 1 \\ 1 \\ 1 \end{bmatrix}$

$\begin{bmatrix} 3 & 2 & 0 & 1 \\ 2 & 6 & 1 & 1 \\ 0 & 1 & 3 & 1 \end{bmatrix} \Rightarrow \begin{bmatrix} 1 & 0 & 0 & \frac{1}{3} \\ 0 & 1 & 0 & 0 \\ 0 & 0 & 1 & \frac{1}{3} \end{bmatrix} \Rightarrow \mathbf{x} = \begin{bmatrix} \frac{1}{3} \\ 0 \\ \frac{1}{3} \end{bmatrix}$

27. $A^T A = \begin{bmatrix} 1 & 1 & 1 \\ -1 & 1 & 3 \end{bmatrix} \begin{bmatrix} 1 & -1 \\ 1 & 1 \\ 1 & 3 \end{bmatrix} = \begin{bmatrix} 3 & 3 \\ 3 & 11 \end{bmatrix}$

$A^T \mathbf{b} = \begin{bmatrix} 1 & 1 & 1 \\ -1 & 1 & 3 \end{bmatrix} \begin{bmatrix} 1 \\ 0 \\ -3 \end{bmatrix} = \begin{bmatrix} -2 \\ -10 \end{bmatrix}$

$\begin{bmatrix} 3 & 3 & -2 \\ 3 & 11 & -10 \end{bmatrix} \Rightarrow \begin{bmatrix} 1 & 0 & \frac{1}{3} \\ 0 & 1 & -1 \end{bmatrix} \Rightarrow \mathbf{x} = \begin{bmatrix} \frac{1}{3} \\ -1 \end{bmatrix}$

line: $y = \frac{1}{3} - x$

29. $A^T A = \begin{bmatrix} 1 & 1 & 1 & 1 \\ -2 & -1 & 1 & 2 \end{bmatrix} \begin{bmatrix} 1 & -2 \\ 1 & -1 \\ 1 & 1 \\ 1 & 2 \end{bmatrix} = \begin{bmatrix} 4 & 0 \\ 0 & 10 \end{bmatrix}$

$A^T \mathbf{b} = \begin{bmatrix} 1 & 1 & 1 & 1 \\ -2 & -1 & 1 & 2 \end{bmatrix} \begin{bmatrix} -1 \\ 0 \\ 0 \\ 2 \end{bmatrix} = \begin{bmatrix} 1 \\ 6 \end{bmatrix}$

$\begin{bmatrix} 4 & 0 & 1 \\ 0 & 10 & 6 \end{bmatrix} \Rightarrow \begin{bmatrix} 1 & 0 & \frac{1}{4} \\ 0 & 1 & \frac{3}{5} \end{bmatrix} \Rightarrow \mathbf{x} = \begin{bmatrix} \frac{1}{4} \\ \frac{3}{5} \end{bmatrix}$

line: $y = \frac{1}{4} + \frac{3}{5}x$

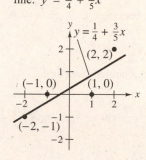

31. $A^T A = \begin{bmatrix} 1 & 1 & 1 & 1 & 1 \\ -2 & -1 & 0 & 1 & 2 \end{bmatrix} \begin{bmatrix} 1 & -2 \\ 1 & -1 \\ 1 & 0 \\ 1 & 1 \\ 1 & 2 \end{bmatrix} = \begin{bmatrix} 5 & 0 \\ 0 & 10 \end{bmatrix}$

$A^T \mathbf{b} = \begin{bmatrix} 1 & 1 & 1 & 1 & 1 \\ -2 & -1 & 0 & 1 & 2 \end{bmatrix} \begin{bmatrix} 1 \\ 2 \\ 1 \\ 2 \\ 1 \end{bmatrix} = \begin{bmatrix} 7 \\ 0 \end{bmatrix}$

$\begin{bmatrix} 5 & 0 & 7 \\ 0 & 10 & 0 \end{bmatrix} \Rightarrow \begin{bmatrix} 1 & 0 & \frac{7}{5} \\ 0 & 1 & 0 \end{bmatrix} \Rightarrow \mathbf{x} = \begin{bmatrix} \frac{7}{5} \\ 0 \end{bmatrix}$

line: $y = \frac{7}{5}$

33. $A^T A = \begin{bmatrix} 1 & 1 & 1 & 1 \\ 0 & 2 & 3 & 4 \\ 0 & 4 & 9 & 16 \end{bmatrix} \begin{bmatrix} 1 & 0 & 0 \\ 1 & 2 & 4 \\ 1 & 3 & 9 \\ 1 & 4 & 16 \end{bmatrix} = \begin{bmatrix} 4 & 9 & 29 \\ 9 & 29 & 99 \\ 29 & 99 & 353 \end{bmatrix}$

$A^T \mathbf{b} = \begin{bmatrix} 1 & 1 & 1 & 1 \\ 0 & 2 & 3 & 4 \\ 0 & 4 & 9 & 16 \end{bmatrix} \begin{bmatrix} 0 \\ 2 \\ 6 \\ 12 \end{bmatrix} = \begin{bmatrix} 20 \\ 70 \\ 254 \end{bmatrix}$

$\begin{bmatrix} 4 & 9 & 29 & 20 \\ 9 & 29 & 99 & 70 \\ 29 & 99 & 353 & 254 \end{bmatrix} \Rightarrow \begin{bmatrix} 1 & 0 & 0 & 0 \\ 0 & 1 & 0 & -1 \\ 0 & 0 & 1 & 1 \end{bmatrix}$

$\Rightarrow \mathbf{x} = \begin{bmatrix} 0 \\ -1 \\ 1 \end{bmatrix}$

Quadratic Polynomial: $y = x^2 - x$

35. $A^T A = \begin{bmatrix} 1 & 1 & 1 & 1 & 1 \\ -2 & -1 & 0 & 1 & 2 \\ 4 & 1 & 0 & 1 & 4 \end{bmatrix} \begin{bmatrix} 1 & -2 & 4 \\ 1 & -1 & 1 \\ 1 & 0 & 0 \\ 1 & 1 & 1 \\ 1 & 2 & 4 \end{bmatrix} = \begin{bmatrix} 5 & 0 & 10 \\ 0 & 10 & 0 \\ 10 & 0 & 34 \end{bmatrix}$

$A^T \mathbf{b} = \begin{bmatrix} 1 & 1 & 1 & 1 & 1 \\ -2 & -1 & 0 & 1 & 2 \\ 4 & 1 & 0 & 1 & 4 \end{bmatrix} \begin{bmatrix} 0 \\ 0 \\ 1 \\ 2 \\ 5 \end{bmatrix} = \begin{bmatrix} 8 \\ 12 \\ 22 \end{bmatrix}$

$\begin{bmatrix} 5 & 0 & 10 & 8 \\ 0 & 10 & 0 & 12 \\ 10 & 0 & 34 & 22 \end{bmatrix} \Rightarrow \begin{bmatrix} 1 & 0 & 0 & \frac{26}{35} \\ 0 & 1 & 0 & \frac{6}{5} \\ 0 & 0 & 1 & \frac{3}{7} \end{bmatrix} \Rightarrow \mathbf{x} = \begin{bmatrix} \frac{26}{35} \\ \frac{6}{5} \\ \frac{3}{7} \end{bmatrix}$

Quadratic Polynomial: $y = \frac{26}{35} + \frac{6}{5}x + \frac{3}{7}x^2$

37. Using a graphing utility, you find that the least squares cubic polynomial is the best fit for both companies.

Advanced Auto Parts:

$S = 2.859t^3 - 32.81t^2 + 492.0t + 2234$

For 2010, $t = 10$ so, $S \approx \$6732$ million.

Auto Zone:

$S = 8.444t^3 - 105.48t^2 + 578.6t + 4444$

For 2010, $t = 10$ so, $S \approx \$8126$ million.

39. Substitute the data points $(-1, 6325)$, $(0, 6505)$, $(1, 6578)$, $(2, 6668)$, $(3, 6999)$, and $(4, 7376)$ into the equation $y = c_0 + c_1 t + c_2 t^2$ to obtain the following system.

$$c_0 + c_1(-1) + c_2(-1)^2 = 6325$$
$$c_0 + c_1(0) + c_2(0)^2 = 6505$$
$$c_0 + c_1(1) + c_2(1)^2 = 6578$$
$$c_0 + c_1(2) + c_2(2)^2 = 6668$$
$$c_0 + c_1(3) + c_2(3)^2 = 6999$$
$$c_0 + c_1(4) + c_2(4)^2 = 7376$$

This produces the least squares problem

$$A\mathbf{x} = \mathbf{b}$$

$$\begin{bmatrix} 1 & -1 & 1 \\ 1 & 0 & 0 \\ 1 & 1 & 1 \\ 1 & 2 & 4 \\ 1 & 3 & 9 \\ 1 & 4 & 16 \end{bmatrix} \begin{bmatrix} c_0 \\ c_1 \\ c_2 \end{bmatrix} = \begin{bmatrix} 6325 \\ 6505 \\ 6578 \\ 6668 \\ 6999 \\ 7376 \end{bmatrix}.$$

The normal equations are

$$A^T A\mathbf{x} = A^T\mathbf{b}$$

$$\begin{bmatrix} 6 & 9 & 31 \\ 9 & 31 & 99 \\ 31 & 99 & 355 \end{bmatrix} \begin{bmatrix} c_0 \\ c_1 \\ c_2 \end{bmatrix} = \begin{bmatrix} 40,451 \\ 64,090 \\ 220,582 \end{bmatrix}$$

and their solution is

$$\mathbf{x} = \begin{bmatrix} c_0 \\ c_1 \\ c_2 \end{bmatrix} \approx \begin{bmatrix} 6425 \\ 87.0 \\ 36.02 \end{bmatrix}.$$

So, the least squares regression quadratic polynomial is

$$y = 6425 + 87.0t + 36.02t^2.$$

41. Use a graphing utility.

Least squares regression line:

$$y = 4946.7t + 28{,}231 \; (r^2 \approx 0.9869)$$

Least squares cubic regression polynomial:

$$y = 2.416t^3 - 36.74t^2 + 4989.3t + 28{,}549 \, (r^2 \approx 0.9871)$$

The cubic model is a better fit for the data.

43. (a) False. See discussion after Example 2, page 322.

(b) True. See "Definition of Direct Sum," page 323.

(c) True. See discussion preceding Example 7, page 328.

45. Let $\mathbf{v} \in S_1 \cap S_2$. Because $\mathbf{v} \in S_1$, $\mathbf{v} \cdot \mathbf{x}_2 = 0$ for all $\mathbf{x}_2 \in S_2 \Rightarrow \mathbf{v} \cdot \mathbf{v} = 0$, because $\mathbf{v} \in S_2 \Rightarrow \mathbf{v} = \mathbf{0}$.

47. Let $\mathbf{v} \in R^n$, $\mathbf{v} = \mathbf{v}_1 + \mathbf{v}_2$, $\mathbf{v}_1 \in S$, $\mathbf{v}_2 \in S^\perp$.

Let $\{\mathbf{u}_1, \dots, \mathbf{u}_t\}$ be an orthonormal basis for S.

Then

$$\mathbf{v} = \mathbf{v}_1 + \mathbf{v}_2 = c_1\mathbf{u}_1 + \cdots + c_t\mathbf{u}_t + \mathbf{v}_2, c_i \in R$$

and

$$\mathbf{v} \cdot \mathbf{u}_i = (c_1\mathbf{u}_1 + \cdots + c_t\mathbf{u}_t + \mathbf{v}_2) \cdot \mathbf{u}_i$$
$$= c_i(\mathbf{u}_i \cdot \mathbf{u}_i)$$
$$= c_i$$

which shows that

$$\mathbf{v}_1 = \text{proj}_S \mathbf{v} = (\mathbf{v} \cdot \mathbf{u}_1)\mathbf{u}_1 + \cdots + (\mathbf{v} \cdot \mathbf{u}_t)\mathbf{u}_t.$$

49. If A has orthonormal columns, then $A^T A = I$ and the normal equations become

$$A^T A\mathbf{x} = A^T\mathbf{b}$$
$$\mathbf{x} = A^T\mathbf{b}.$$

Section 5.5   Applications of Inner Product Spaces

1. $\mathbf{j} \times \mathbf{i} = \begin{vmatrix} \mathbf{i} & \mathbf{j} & \mathbf{k} \\ 0 & 1 & 0 \\ 1 & 0 & 0 \end{vmatrix}$

$$= \begin{vmatrix} 1 & 0 \\ 0 & 0 \end{vmatrix}\mathbf{i} - \begin{vmatrix} 0 & 0 \\ 1 & 0 \end{vmatrix}\mathbf{j} + \begin{vmatrix} 0 & 1 \\ 1 & 0 \end{vmatrix}\mathbf{k}$$

$$= 0\mathbf{i} - 0\mathbf{j} - \mathbf{k} = -\mathbf{k}$$

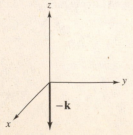

3. $\mathbf{j} \times \mathbf{k} = \begin{vmatrix} \mathbf{i} & \mathbf{j} & \mathbf{k} \\ 0 & 1 & 0 \\ 0 & 0 & 1 \end{vmatrix}$

$$= \begin{vmatrix} 1 & 0 \\ 0 & 1 \end{vmatrix}\mathbf{i} - \begin{vmatrix} 0 & 0 \\ 0 & 1 \end{vmatrix}\mathbf{j} + \begin{vmatrix} 0 & 1 \\ 0 & 0 \end{vmatrix}\mathbf{k}$$

$$= \mathbf{i} - 0\mathbf{j} + 0\mathbf{k} = \mathbf{i}$$

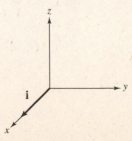

5. $\mathbf{i} \times \mathbf{k} = \begin{vmatrix} \mathbf{i} & \mathbf{j} & \mathbf{k} \\ 1 & 0 & 0 \\ 0 & 0 & 1 \end{vmatrix}$

$= \begin{vmatrix} 0 & 0 \\ 0 & 1 \end{vmatrix} \mathbf{i} - \begin{vmatrix} 1 & 0 \\ 0 & 1 \end{vmatrix} \mathbf{j} + \begin{vmatrix} 1 & 0 \\ 0 & 0 \end{vmatrix} \mathbf{k}$

$= 0\mathbf{i} - \mathbf{j} + 0\mathbf{k} = -\mathbf{j}$

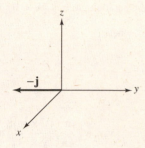

7. $\mathbf{u} \times \mathbf{v} = \begin{vmatrix} \mathbf{i} & \mathbf{j} & \mathbf{k} \\ 0 & 1 & -2 \\ 1 & -1 & 0 \end{vmatrix} = -2\mathbf{i} - 2\mathbf{j} - \mathbf{k} = (-2, -2, -1)$

Furthermore, $\mathbf{u} \times \mathbf{v} = (-2, -2, -1)$ is orthogonal to both $(0, 1, -2)$ and $(1, -1, 0)$ because

$(-2, -2, -1) \cdot (0, 1, -2) = 0$ and

$(-2, -2, -1) \cdot (1, -1, 0) = 0.$

9. $\mathbf{u} \times \mathbf{v} = \begin{vmatrix} \mathbf{i} & \mathbf{j} & \mathbf{k} \\ 12 & -3 & 1 \\ -2 & 5 & 1 \end{vmatrix} = -8\mathbf{i} - 14\mathbf{j} + 54\mathbf{k} = (-8, -14, 54)$

Furthermore, $\mathbf{u} \times \mathbf{v} = (-8, -14, 54)$ is orthogonal to both $(12, -3, 1)$ and $(-2, 5, 1)$ because

$(-8, -14, 54) \cdot (12, -3, 1) = 0$ and

$(-8, -14, 54) \cdot (-2, 5, 1) = 0.$

11. $\mathbf{u} \times \mathbf{v} = \begin{vmatrix} \mathbf{i} & \mathbf{j} & \mathbf{k} \\ 2 & -3 & 1 \\ 1 & -2 & 1 \end{vmatrix} = -\mathbf{i} - \mathbf{j} - \mathbf{k} = (-1, -1, -1)$

Furthermore, $\mathbf{u} \times \mathbf{v} = (-1, -1, -1)$ is orthogonal to both $(2, -3, 1)$ and $(1, -2, 1)$ because

$(-1, -1, -1) \cdot (2, -3, 1) = 0$ and

$(-1, -1, -1) \cdot (1, -2, 1) = 0.$

13. $\mathbf{u} \times \mathbf{v} = \begin{vmatrix} \mathbf{i} & \mathbf{j} & \mathbf{k} \\ 0 & 1 & 6 \\ 2 & 0 & -1 \end{vmatrix} = -\mathbf{i} + 12\mathbf{j} - 2\mathbf{k} = (-1, 12, -2)$

Furthermore, $\mathbf{u} \times \mathbf{v} = (-1, 12, -2)$ is orthogonal to both $(0, 1, 6)$ and $(2, 0, -1)$ because

$(-1, 12, -2) \cdot (0, 1, 6) = 0$ and

$(-1, 12, -2) \cdot (2, 0, -1) = 0.$

15. $\mathbf{u} \times \mathbf{v} = \begin{vmatrix} \mathbf{i} & \mathbf{j} & \mathbf{k} \\ 1 & 1 & 1 \\ 2 & 1 & -1 \end{vmatrix} = -2\mathbf{i} + 3\mathbf{j} - \mathbf{k} = (-2, 3, -1).$

Furthermore, $\mathbf{u} \times \mathbf{v} = (-2, 3, -1)$ is orthogonal to both $(1, 1, 1)$ and $(2, 1, -1)$ because

$(-2, 3, -1) \cdot (1, 1, 1) = 0$ and

$(-2, 3, -1) \cdot (2, 1, -1) = 0.$

17. Using a graphing utility:

$\mathbf{w} = \mathbf{u} \times \mathbf{v} = (5, -4, -3)$

Check if $\mathbf{w}$ is orthogonal to both $\mathbf{u}$ and $\mathbf{v}$:

$\mathbf{w} \cdot \mathbf{u} = (5, -4, -3) \cdot (1, 2, -1) = 5 - 8 + 3 = 0$

$\mathbf{w} \cdot \mathbf{v} = (5, -4, -3) \cdot (2, 1, 2) = 10 - 4 - 6 = 0$

19. Using a graphing utility:

$\mathbf{w} = \mathbf{u} \times \mathbf{v} = (2, -1, -1)$

Check if $\mathbf{w}$ is orthogonal to both $\mathbf{u}$ and $\mathbf{v}$:

$\mathbf{w} \cdot \mathbf{u} = (2, -1, -1) \cdot (0, 1, -1) = -1 + 1 = 0$

$\mathbf{w} \cdot \mathbf{v} = (2, -1, -1) \cdot (1, 2, 0) = 2 - 2 = 0$

21. Using a graphing utility:

$\mathbf{w} = \mathbf{u} \times \mathbf{v} = (1, -1, -3)$

Check if $\mathbf{w}$ is orthogonal to both $\mathbf{u}$ and $\mathbf{v}$:

$\mathbf{w} \cdot \mathbf{u} = (1, -1, -3) \cdot (2, -1, 1) = 2 + 1 - 3 = 0$

$\mathbf{w} \cdot \mathbf{v} = (1, -1, -3) \cdot (1, -2, 1) = 1 + 2 - 3 = 0$

23. Using a graphing utility:

$\mathbf{w} = \mathbf{u} \times \mathbf{v} = (1, -5, -3)$

Check if $\mathbf{w}$ is orthogonal to both $\mathbf{u}$ and $\mathbf{v}$:

$\mathbf{w} \cdot \mathbf{u} = (1, -5, -3) \cdot (2, 1, -1) = 2 - 5 + 3 = 0$

$\mathbf{w} \cdot \mathbf{v} = (1, -5, -3) \cdot (1, -1, 2) = 1 + 5 - 6 = 0$

25. Because

$$\mathbf{u} \times \mathbf{v} = \begin{vmatrix} \mathbf{i} & \mathbf{j} & \mathbf{k} \\ 0 & 1 & 0 \\ 0 & 1 & 1 \end{vmatrix} = \mathbf{i} = (1, 0, 0)$$

the area of the parallelogram is

$$\|\mathbf{u} \times \mathbf{v}\| = \|\mathbf{i}\| = 1.$$

27. Because

$$\mathbf{u} \times \mathbf{v} = \begin{vmatrix} \mathbf{i} & \mathbf{j} & \mathbf{k} \\ 3 & 2 & -1 \\ 1 & 2 & 3 \end{vmatrix} = 8\mathbf{i} - 10\mathbf{j} + 4\mathbf{k} = (8, -10, 4),$$

the area of the parallelogram is

$$\|(8, -10, 4)\| = \sqrt{8^2 + (-10)^2 + 4^2} = \sqrt{180} = 6\sqrt{5}.$$

29. $(2, 3, 4) - (1, 1, 1) = (1, 2, 3)$

$(7, 7, 5) - (6, 5, 2) = (1, 2, 3)$

$(7, 7, 5) - (2, 3, 4) = (5, 4, 1)$

$(6, 5, 2) - (1, 1, 1) = (5, 4, 1)$

$\mathbf{u} = (1, 2, 3)$ and $\mathbf{v} = (5, 4, 1)$

Because

$$\mathbf{u} \times \mathbf{v} = \begin{vmatrix} \mathbf{i} & \mathbf{j} & \mathbf{k} \\ 1 & 2 & 3 \\ 5 & 4 & 1 \end{vmatrix} = -10\mathbf{i} + 14\mathbf{j} - 6\mathbf{k} = (-10, 14, -6),$$

the area of the parallelogram is

$$\|\mathbf{u} \times \mathbf{v}\| = \sqrt{(-10)^2 + 14^2 + (-6)^2} = \sqrt{332} = 2\sqrt{83}.$$

31. Because

$$\mathbf{v} \times \mathbf{w} = \begin{vmatrix} \mathbf{i} & \mathbf{j} & \mathbf{k} \\ 0 & 1 & 0 \\ 0 & 0 & 1 \end{vmatrix} = \mathbf{i} = (1, 0, 0),$$

the triple scalar product of $\mathbf{u}$, $\mathbf{v}$, and $\mathbf{w}$ is

$$\mathbf{u} \cdot (\mathbf{v} \times \mathbf{w}) = (1, 0, 0) \cdot (1, 0, 0) = 1.$$

33. Because

$$\mathbf{v} \times \mathbf{w} = \begin{vmatrix} \mathbf{i} & \mathbf{j} & \mathbf{k} \\ 2 & 1 & 0 \\ 0 & 0 & 1 \end{vmatrix} = \mathbf{i} - 2\mathbf{j} = (1, -2, 0),$$

the triple scalar product of $\mathbf{u}$, $\mathbf{v}$, and $\mathbf{w}$ is

$$\mathbf{u} \cdot (\mathbf{v} \times \mathbf{w}) = (1, 1, 1) \cdot (1, -2, 0) = -1.$$

35. The area of the base of the parallelogram is $\|\mathbf{v} \times \mathbf{w}\|$.

The height is $|\cos \theta| \|\mathbf{u}\|$, where

$$|\cos \theta| = \frac{|\mathbf{u} \cdot (\mathbf{v} \times \mathbf{w})|}{\|\mathbf{u}\| \|\mathbf{v} \times \mathbf{w}\|}.$$

So,

$$\text{volume} = \text{base} \times \text{height} = \|\mathbf{v} \times \mathbf{w}\| \frac{|\mathbf{u} \cdot (\mathbf{v} \times \mathbf{w})|}{\|\mathbf{u}\| \|\mathbf{v} \times \mathbf{w}\|} \|\mathbf{u}\|$$

$$= |\mathbf{u} \cdot (\mathbf{v} \times \mathbf{w})|.$$

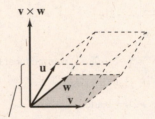

37. $(3, 3, 0) - (1, 3, 5) = (2, 0, -5)$

$(3, 3, 0) - (-2, 0, 5) = (5, 3, -5)$

Because

$$\mathbf{u} \times \mathbf{v} = \begin{vmatrix} \mathbf{i} & \mathbf{j} & \mathbf{k} \\ 2 & 0 & -5 \\ 5 & 3 & -5 \end{vmatrix} = 15\mathbf{i} - 15\mathbf{j} + 6\mathbf{k} = (15, -15, 6),$$

the area of the triangle is

$$A = \frac{1}{2}\|\mathbf{u} \times \mathbf{v}\|$$

$$= \frac{1}{2}\sqrt{15^2 + (-15)^2 + 6^2} = \frac{1}{2}\sqrt{486} = \frac{9\sqrt{6}}{2}.$$

39. Because

$$\mathbf{v} \times \mathbf{w} = \begin{vmatrix} \mathbf{i} & \mathbf{j} & \mathbf{k} \\ 0 & 1 & 1 \\ 1 & 0 & 1 \end{vmatrix} = \mathbf{i} + \mathbf{j} - \mathbf{k} = (1, 1, -1),$$

the volume is given by

$$|\mathbf{u} \cdot (\mathbf{v} \times \mathbf{w})| = |(1, 1, 0) \cdot (1, 1, -1)| = 2.$$

41. $\mathbf{u} \times (\mathbf{v} + \mathbf{w})$

$$= \begin{vmatrix} \mathbf{i} & \mathbf{j} & \mathbf{k} \\ u_1 & u_2 & u_3 \\ v_1 + w_1 & v_2 + w_2 & v_3 + w_3 \end{vmatrix}$$

$$= \big[u_2(v_3 + w_3) - u_3(v_2 + w_2) \big]\mathbf{i} - \big[u_1(v_3 + w_3) - u_3(v_1 + w_1) \big]\mathbf{j} + \big[u_1(v_2 + w_2) - u_2(v_1 + w_1) \big]\mathbf{k}$$

$$= (u_2v_3 - v_2u_3)\mathbf{i} - (u_1v_3 - u_3v_1)\mathbf{j} + (u_1v_2 - u_2v_1)\mathbf{k} + (u_2w_3 - u_3w_2)\mathbf{i} - (u_1w_3 - u_3w_1)\mathbf{j} + (u_1w_2 - u_2w_1)\mathbf{k}$$

$$= \begin{vmatrix} \mathbf{i} & \mathbf{j} & \mathbf{k} \\ u_1 & u_2 & u_3 \\ v_1 & v_2 & v_3 \end{vmatrix} + \begin{vmatrix} \mathbf{i} & \mathbf{j} & \mathbf{k} \\ u_1 & u_2 & u_3 \\ w_1 & w_2 & w_3 \end{vmatrix}$$

$$= (\mathbf{u} \times \mathbf{v}) + (\mathbf{u} \times \mathbf{w})$$

43. $\mathbf{u} \times \mathbf{u} = \begin{vmatrix} \mathbf{i} & \mathbf{j} & \mathbf{k} \\ u_1 & u_2 & u_3 \\ u_1 & u_2 & u_3 \end{vmatrix} = 0$, because two rows are the same.

45. Because $\mathbf{u} \times \mathbf{v} = (u_2v_3 - v_2u_3)\mathbf{i} - (u_1v_3 - u_3v_1)\mathbf{j} + (u_1v_2 - v_1u_2)\mathbf{k}$

you see that

$$\mathbf{u} \cdot (\mathbf{u} \times \mathbf{v}) = (u_1, u_2, u_3) \cdot (u_2v_3 - v_2u_3, -u_1v_3 + u_3v_1, u_1v_2 - v_1u_2)$$

$$= (u_1u_2v_3 - u_1v_2u_3 - u_2u_1v_3 + u_2u_3v_1 + u_3u_1v_2 - u_3v_1u_2) = 0,$$

which shows that $\mathbf{u}$ is orthogonal to $\mathbf{u} \times \mathbf{v}$. A similar computation shows that $\mathbf{v} \cdot (\mathbf{u} \times \mathbf{v}) = 0$. [Note that $\mathbf{v} \cdot (\mathbf{u} \times \mathbf{v}) = -\mathbf{v} \cdot (\mathbf{v} \times \mathbf{u}) = 0$ by the above with the roles of $\mathbf{u}$ and $\mathbf{v}$ reversed.]

47. You have the following equivalences.

$$\mathbf{u} \times \mathbf{v} = \mathbf{0} \Leftrightarrow \|\mathbf{u} \times \mathbf{v}\| = 0$$

$$\Leftrightarrow \|\mathbf{u}\|\|\mathbf{v}\|\sin\theta = 0 \ \big(\text{Theorem 5.18 (2)}\big)$$

$$\Leftrightarrow \sin\theta = 0$$

$$\Leftrightarrow \theta = 0$$

$$\Leftrightarrow \mathbf{u} \text{ and } \mathbf{v} \text{ are parallel.}$$

49. (a) $\mathbf{u} \times (\mathbf{v} \times \mathbf{w})$

$$= \mathbf{u} \times \begin{vmatrix} \mathbf{i} & \mathbf{j} & \mathbf{k} \\ v_1 & v_2 & v_3 \\ w_1 & w_2 & w_3 \end{vmatrix}$$

$$= \mathbf{u} \times \big[(v_2w_3 - w_2v_3)\mathbf{i} - (v_1w_3 - w_1v_3)\mathbf{j} + (v_1w_2 - v_2w_1)\mathbf{k} \big]$$

$$= \begin{vmatrix} \mathbf{i} & \mathbf{j} & \mathbf{k} \\ u_1 & u_2 & u_3 \\ (v_2w_3 - w_2v_3) & (w_1v_3 - v_1w_3) & (v_1w_2 - v_2w_1) \end{vmatrix}$$

$$= \big[(u_2(v_1w_2 - v_2w_1)) - u_3(w_1v_3 - v_1w_3) \big]\mathbf{i} - \big[u_1(v_1w_2 - v_2w_1) - u_3(v_2w_3 - w_2v_3) \big]\mathbf{j}$$

$$\quad + \big[u_1(w_1v_3 - v_1w_3) - u_2(v_2w_3 - w_2v_3) \big]\mathbf{k}$$

$$= (u_2w_2v_1 + u_3w_3v_1 - u_2v_2w_1 - u_3v_3w_1, u_1w_1v_2 + u_3w_3v_2 - u_1v_1w_2 - u_3v_3w_2,$$

$$\quad u_1w_1v_3 + u_2w_2v_3 - u_1v_1w_3 - u_2v_2w_3)$$

$$= (u_1w_1 + u_2w_2 + u_3w_3)(v_1, v_2, v_3) - (u_1v_1 + u_2v_2 + u_3v_3)(w_1, w_2, w_3)$$

$$= (\mathbf{u} \cdot \mathbf{w})\mathbf{v} - (\mathbf{u} \cdot \mathbf{v})\mathbf{w}$$

(b) Let

$$\mathbf{u} = (1, 0, 0), \mathbf{v} = (0, 1, 0) \quad \text{and} \quad \mathbf{w} = (1, 1, 1).$$

Then

$$\mathbf{v} \times \mathbf{w} = (1, 0, -1) \quad \text{and} \quad \mathbf{u} \times \mathbf{v} = (0, 0, 1).$$

So

$$\mathbf{u} \times (\mathbf{v} \times \mathbf{w}) = (1, 0, 0) \times (1, 0, -1) = (0, 1, 0),$$

while

$$(\mathbf{u} \times \mathbf{v}) \times \mathbf{w} = (0, 0, 1) \times (1, 1, 1) = (-1, 1, 0),$$

which are not equal.

51. (a) The standard basis for P_1 is $\{1, x\}$. Applying the Gram-Schmidt orthonormalization process produces the orthonormal basis

$$B = \{\mathbf{w}_1, \mathbf{w}_2\} = \left\{ \frac{1}{\sqrt{3}}, \frac{1}{3}(2x - 5) \right\}.$$

The least squares approximating function is given by $g(x) = \langle f, \mathbf{w}_1 \rangle \mathbf{w}_1 + \langle f, \mathbf{w}_2 \rangle \mathbf{w}_2$.

Find the inner products

$$\langle f_1, \mathbf{w}_1 \rangle = \int_1^4 \sqrt{x} \frac{1}{\sqrt{3}} dx = \frac{2}{3\sqrt{3}} x^{3/2} \Bigg]_1^4 = \frac{14}{3\sqrt{3}}$$

$$\langle f, \mathbf{w}_2 \rangle = \int_1^4 \sqrt{x} \left(\frac{1}{3} \right)(2x - 5) dx = \left[\frac{4}{15} x^{5/2} - \frac{10}{9} x^{3/2} \right]_1^4 = \frac{22}{45}$$

and conclude that

$$g(x) = \langle f, \mathbf{w}_1 \rangle \mathbf{w}_1 + \langle f, \mathbf{w}_2 \rangle \mathbf{w}_2 = \frac{14}{3\sqrt{3}} \frac{1}{\sqrt{3}} + \frac{22}{45} \left(\frac{1}{3} \right)(2x - 5) = \frac{44}{135} x + \frac{20}{27} = \frac{4}{135}(25 + 11x).$$

(b)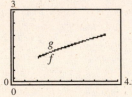

53. (a) The standard basis for P_1 is $\{1, x\}$. Applying the Gram-Schmidt orthonormalization process produces the orthonormal basis

$$B = \{\mathbf{w}_1, \mathbf{w}_2\} = \left\{ 1, \sqrt{3}(2x - 1) \right\}.$$

The least squares approximating function is then given by $g(x) = \langle f, \mathbf{w}_1 \rangle \mathbf{w}_1 + \langle f, \mathbf{w}_2 \rangle \mathbf{w}_2$.

Find the inner products

$$\langle f, \mathbf{w}_1 \rangle = \int_0^1 e^{-2x} dx = -\tfrac{1}{2} e^{-2x} \Bigg]_0^1 = -\tfrac{1}{2}(e^{-2} - 1)$$

$$\langle f, \mathbf{w}_2 \rangle = \int_0^1 e^{-2x} \sqrt{3}(2x - 1) dx = -\sqrt{3} \times e^{-2x} \Bigg]_0^1 = -\sqrt{3} e^{-2}$$

and conclude that

$$g(x) = \langle f, \mathbf{w}_1 \rangle \mathbf{w}_1 + \langle f, \mathbf{w}_2 \rangle \mathbf{w}_2 = -\tfrac{1}{2}(e^{-2} - 1) - \sqrt{3} e^{-2} \left(\sqrt{3}(2x - 1) \right) = -6e^{-2}x + \tfrac{1}{2}(5e^{-2} + 1) \approx -0.812x + 0.8383.$$

(b)

55. (a) The standard basis for P_1, is $\{1, x\}$. Applying the Gram-Schmidt orthonormalization process produces the orthonormal basis

$$B = \{\mathbf{w}_1, \mathbf{w}_2\} = \left\{\frac{\sqrt{2\pi}}{\pi}, \frac{\sqrt{6\pi}}{\pi^2}(4x - \pi)\right\}.$$

The least squares approximating function is then given by $g(x) = \langle f, \mathbf{w}_1\rangle\mathbf{w}_1 + \langle f, \mathbf{w}_2\rangle\mathbf{w}_2$.

Find the inner products

$$\langle f, \mathbf{w}_1\rangle = \int_0^{\pi/2} (\sin x)\left(\frac{\sqrt{2\pi}}{\pi}\right)dx = -\frac{\sqrt{2\pi}}{\pi}\cos x\bigg]_0^{\pi/2} = \frac{\sqrt{2\pi}}{\pi}$$

$$\langle f, \mathbf{w}_2\rangle = \int_0^{\pi/2} (\sin x)\left[\frac{\sqrt{6\pi}}{\pi^2}(4x - \pi)\right]dx = \frac{\sqrt{6\pi}}{\pi^2}[-4x\cos x + 4\sin x + \pi\cos x]_0^{\pi/2} = \frac{\sqrt{6\pi}}{\pi^2}(4 - \pi)$$

and conclude that

$$\begin{aligned}
g(x) &= \langle f, \mathbf{w}_1\rangle\mathbf{w}_1 + \langle f, \mathbf{w}_2\rangle\mathbf{w}_2 \\
&= \frac{\sqrt{2\pi}}{\pi}\left(\frac{\sqrt{2\pi}}{\pi}\right) + \frac{\sqrt{6\pi}}{\pi^2}(4 - \pi)\left[\frac{\sqrt{6\pi}}{\pi^2}(4x - \pi)\right] \\
&= \frac{2}{\pi} + \frac{6}{\pi^3}(4 - \pi)(4x - \pi) \\
&= \frac{24(4 - \pi)}{\pi^3}x - \frac{8(3 - \pi)}{\pi^2} \approx 0.6644x + 0.1148.
\end{aligned}$$

(b)

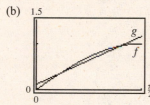

57. (a) The standard basis for P_2 is $\{1, x, x^2\}$. Applying the Gram-Schmidt orthonormalization process produces the orthonormal basis $B = \{\mathbf{w}_1, \mathbf{w}_2, \mathbf{w}_3\} = \{1, \sqrt{3}(2x - 1), \sqrt{5}(6x^2 - 6x + 1)\}$.

The least squares approximating function for f is given by $g(x) = \langle f, \mathbf{w}_1\rangle\mathbf{w}_1 + \langle f, \mathbf{w}_2\rangle\mathbf{w}_2 + \langle f, \mathbf{w}_3\rangle\mathbf{w}_3$.

Find the inner products

$$\langle f, \mathbf{w}_1\rangle = \int_0^1 x^3(1)dx = \frac{1}{4}x^4\bigg]_0^1 = \frac{1}{4}$$

$$\langle f, \mathbf{w}_2\rangle = \int_0^1 x^3[\sqrt{3}(2x - 1)]dx = \sqrt{3}\left[\frac{2}{5}x^5 - \frac{1}{4}x^4\right]_0^1 = \frac{3\sqrt{3}}{20}$$

$$\langle f, \mathbf{w}_3\rangle = \int_0^1 x^3[\sqrt{5}(6x^2 - 6x + 1)]dx = \sqrt{5}\left[x^6 - \frac{6}{5}x^5 + \frac{1}{4}x^4\right]_0^1 = \frac{\sqrt{5}}{20}$$

and conclude that

$$g(x) = \langle f, \mathbf{w}_1\rangle\mathbf{w}_1 + \langle f, \mathbf{w}_2\rangle\mathbf{w}_2 + \langle f, \mathbf{w}_3\rangle\mathbf{w}_3 = \frac{1}{4}(1) + \frac{3\sqrt{3}}{20}[\sqrt{3}(2x - 1)] + \frac{\sqrt{5}}{20}[\sqrt{5}(6x^2 - 6x + 1)] = \frac{3}{2}x^2 - \frac{3}{5}x + \frac{1}{20}.$$

(b)

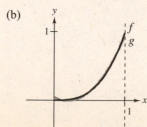

59. (a) The standard basis for P_2 is $\{1, x, x^2\}$. Applying the Gram-Schmidt orthonormalization process produces the orthonormal

basis $B = \{w_1, w_2, w_3\} = \left\{ \dfrac{1}{\sqrt{\pi}}, \dfrac{\sqrt{3}}{\pi\sqrt{\pi}}(2x - \pi), \dfrac{\sqrt{5}}{\pi^2\sqrt{\pi}}(6x^2 - 6\pi x + \pi^2) \right\}$.

The least squares approximating function for f is given by $g(x) = \langle f, w_1\rangle w_1 + \langle f, w_2\rangle w_2 + \langle f, w_3\rangle w_3$.

Find the inner products

$$\langle f, w_1\rangle = \int_0^{\pi} (\sin x)\left(\dfrac{1}{\sqrt{\pi}}\right)dx = -\dfrac{1}{\sqrt{\pi}}\cos x \Big]_0^{\pi} = \dfrac{2}{\sqrt{\pi}}$$

$$\langle f, w_2\rangle = \int_0^{\pi} (\sin x)\left[\dfrac{\sqrt{3}}{\pi\sqrt{\pi}}(2x - \pi)\right]dx = \dfrac{\sqrt{3}}{\pi\sqrt{\pi}}\left[-2x\cos x + 2\sin x + \pi\cos x\right]_0^{\pi} = 0$$

$$\langle f, w_3\rangle = \int_0^{\pi} (\sin x)\left[\dfrac{\sqrt{5}}{\pi^2\sqrt{\pi}}(6x^2 - 6\pi x + \pi^2)\right]dx$$

$$= \dfrac{\sqrt{5}}{\pi^2\sqrt{\pi}}\left[(12x - 6\pi)\sin x - (6x^2 - 6\pi x - 12 + \pi^2)\cos x\right]_0^{\pi} = \dfrac{2\sqrt{5}(\pi^2 - 12)}{\pi^2\sqrt{\pi}}$$

and conclude that

$$g(x) = \dfrac{2}{\sqrt{\pi}}\left(\dfrac{1}{\pi}\right) + \dfrac{2\sqrt{5}(\pi^2 - 12)}{\pi^2\sqrt{\pi}}\left[\dfrac{\sqrt{5}}{\pi^2\sqrt{\pi}}(6x^2 - 6\pi x + \pi^2)\right]$$

$$= \dfrac{12(\pi^2 - 10)}{\pi^3} - \dfrac{60(\pi^2 - 12)}{\pi^4}x + \dfrac{60(\pi^2 - 12)}{\pi^5}x^2 \approx -0.0505 + 1.3122x - 0.4177x^2.$$

(b)

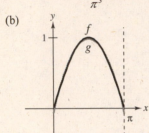

61. (a) The standard basis for P_2 is $\{1, x, x^2\}$. Applying the Gram-Schmidt orthonormalization process produces the orthonormal

basis $B = \{w_1, w_2, w_3\} = \left\{ \dfrac{1}{\sqrt{\pi}}, \dfrac{2\sqrt{3}}{\pi^{3/2}}x, \dfrac{6\sqrt{5}}{\pi^{5/2}}\left(x^2 - \dfrac{\pi^2}{12}\right) \right\}$.

The least squares approximating function is then given by $g(x) = \langle f, w\rangle w_1 + \langle f, w_2\rangle w_2 + \langle f, w_3\rangle w_3$.

Find the inner products

$$\langle f, w_1\rangle = \int_{-\pi/2}^{\pi/2} \dfrac{1}{\sqrt{\pi}}\cos x \, dx = \dfrac{\sin x}{\sqrt{\pi}}\Big]_{-\pi/2}^{\pi/2} = \dfrac{2}{\sqrt{\pi}}$$

$$\langle f, w_2\rangle = \int_{-\pi/2}^{\pi/2} \dfrac{2\sqrt{3}}{\pi^{3/2}}x\cos x \, dx = \left[\dfrac{2\sqrt{3}\cos x}{\pi^{3/2}} + \dfrac{2\sqrt{3}x\sin x}{\pi^{3/2}}\right]_{-\pi/2}^{\pi/2} = 0$$

$$\langle f, w_3\rangle = \int_{-\pi/2}^{\pi/2} \dfrac{6\sqrt{5}}{\pi^{3/2}}\left(x^2 - \dfrac{\pi^2}{12}\right)\cos x \, dx = \left[\dfrac{12\sqrt{5}x\cos x}{\pi^{5/2}} + \dfrac{\sqrt{5}(12x^2 - \pi^2 - 24)\sin x}{2\pi^{5/2}}\right]_{-\pi/2}^{\pi/2} = \dfrac{2\sqrt{5}(\pi^2 - 12)}{\pi^{5/2}}$$

and conclude that

$$g(x) = \langle f, w_1\rangle w_1 + \langle f, w_2\rangle w_2 + \langle f, w_3\rangle w_3$$

$$= \left(\dfrac{2}{\sqrt{\pi}}\right)\left(\dfrac{1}{\sqrt{\pi}}\right) + (0)\left(\dfrac{2\sqrt{3}}{\pi^{3/2}}x\right) + \left(\dfrac{2\sqrt{5}(\pi^2 - 12)}{\pi^{5/2}}\right)\left(\dfrac{6\sqrt{5}}{\pi^{5/2}}\left(x^2 - \dfrac{\pi^2}{12}\right)\right)$$

$$= \dfrac{2}{\pi} + \dfrac{60\pi^2 - 720}{\pi^5}\left(x^2 - \dfrac{\pi^2}{12}\right) = \left(\dfrac{60(\pi^2 - 12)}{\pi^5}\right)x^2 + \dfrac{60 - 3\pi^2}{\pi^3} \approx -0.4177x^2 + 0.9802.$$

(b)

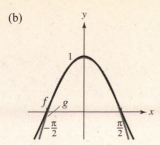

63. The third order Fourier approximation of $f(x) = \pi - x$ is of the form

$$g(x) = \frac{a_0}{2} + a_1 \cos x + b_1 \sin x + a_2 \cos 2x + b_2 \sin 2x + a_3 \cos 3x + b_3 \sin 3x.$$

Find the coefficients as follows.

$$a_0 = \frac{1}{\pi} \int_0^{2\pi} f(x)\, dx = \frac{1}{\pi} \int_0^{2\pi} (\pi - x)\, dx = -\frac{1}{2\pi}(\pi - x)^2 \Big]_0^{2\pi} = 0$$

$$a_j = \frac{1}{\pi} \int_0^{2\pi} f(x) \cos jx\, dx = \frac{1}{\pi} \int_0^{2\pi} (\pi - x) \cos jx\, dx = \left[-\frac{1}{j\pi} x \sin jx - \frac{1}{j^2 \pi} \cos jx + \frac{1}{j} \sin jx \right]_0^{2\pi} = 0, \ j = 1, 2, 3$$

$$b_j = \frac{1}{\pi} \int_0^{2\pi} f(x) \sin x\, jx\, dx = \frac{1}{\pi} \int_0^{2\pi} (\pi - x) \sin jx\, dx = \left[\frac{1}{j\pi} x \cos jx - \frac{1}{j^2 \pi} \sin jx - \frac{1}{j} \cos jx \right]_0^{2\pi} = \frac{2}{j}, \ j = 1, 2, 3$$

So, the approximation is $g(x) = 2 \sin x + \sin 2x + \dfrac{2}{3} \sin 3x.$

65. The third order Fourier approximation of $f(x) = (x - \pi)^2$ is of the form

$$g(x) = \frac{a_0}{2} + a_1 \cos x + b_1 \sin x + a_2 \cos 2x + b_2 \sin 2x + a_3 \cos 3x + b_3 \sin 3x.$$

Find the coefficients as follows.

$$a_0 = \frac{1}{\pi} \int_0^{2\pi} f(x)\, dx = \frac{1}{\pi} \int_0^{2\pi} (x - \pi)^2\, dx = \frac{1}{3\pi}(x - \pi)^3 \Big]_0^{2\pi} = \frac{2\pi^2}{3}$$

$$a_j = \frac{1}{\pi} \int_0^{2\pi} f(x) \cos jx\, dx$$

$$= \frac{1}{\pi} \int_0^{2\pi} (x - \pi)^2 \cos jx\, dx$$

$$= \left[\frac{\pi}{j} \sin jx - \frac{2}{j} x \sin jx + \frac{2}{j^2 \pi} x \cos jx + \frac{1}{j\pi} x^2 \sin jx - \frac{2}{j^3 \pi} \sin jx - \frac{2}{j^2} \cos jx \right]_0^{2\pi}$$

$$= \frac{4}{j^2}, \ j = 1, 2, 3$$

$$b_j = \frac{1}{\pi} \int_0^{2\pi} f(x) \sin jx\, dx$$

$$= \frac{1}{\pi} \int_0^{2\pi} (x - \pi)^2 \sin jx\, dx$$

$$= \left[-\frac{\pi}{j} \cos jx + \frac{2}{j} x \cos jx + \frac{2}{j^2 \pi} x \sin jx - \frac{1}{j\pi} x^2 \cos jx + \frac{2}{j^3 \pi} \cos jx - \frac{2}{j^2} \sin jx \right]_0^{2\pi}$$

$$= 0, \ j = 1, 2, 3$$

So, the approximation is $g(x) = \dfrac{\pi^2}{3} + 4 \cos x + \cos 2x + \dfrac{4}{9} \cos 3x.$

67. The first order Fourier approximation of $f(x) = e^{-x}$ is of the form $g(x) = \dfrac{a_0}{2} + a_1 \cos x + b_1 \sin x$.

Find the coefficients as follows.

$$a_0 = \frac{1}{\pi}\int_0^{2\pi} f(x)\,dx = \frac{1}{\pi}\int_0^{2\pi} e^{-x}\,dx = -\frac{1}{\pi}e^{-x}\Big]_0^{2\pi} = \frac{1 - e^{-2\pi}}{\pi}$$

$$a_1 = \frac{1}{\pi}\int_0^{2\pi} f(x)\cos x\,dx = \frac{1}{\pi}\int_0^{2\pi} e^{-x}\cos x\,dx = \left[-\frac{1}{2\pi}e^{-x}\cos x + \frac{1}{2\pi}e^{-x}\sin x\right]_0^{2\pi} = \frac{1 - e^{-2\pi}}{2\pi}$$

$$b_1 = \frac{1}{\pi}\int_0^{2\pi} f(x)\sin x\,dx = \frac{1}{\pi}\int_0^{2\pi} e^{-x}\sin x\,dx = \left[-\frac{1}{2\pi}e^{-x}\cos x - \frac{1}{2\pi}e^{-x}\sin x\right]_0^{2\pi} = \frac{1 - e^{-2\pi}}{2\pi}$$

So, the approximation is $g(x) = \dfrac{1}{2\pi}\left(1 - e^{-2\pi}\right)\left(1 + \cos x + \sin x\right)$.

69. The first order Fourier approximation of $f(x) = e^{-2x}$ is of the form $g(x) = \dfrac{a_0}{2} + a_1 \cos x + b_1 \sin x$.

Find the coefficients as follows.

$$a_0 = \frac{1}{\pi}\int_0^{2\pi} f(x)\,dx = \frac{1}{\pi}\int_0^{2\pi} e^{-2x}\,dx = \frac{1 - e^{-2x}}{2\pi}\Big]_0^{2\pi} = \frac{1 - e^{-4\pi}}{2\pi}$$

$$a_1 = \frac{1}{\pi}\int_0^{2\pi} f(x)\cos x\,dx = \frac{1}{\pi}\int_0^{2\pi} e^{-2x}\cos x\,dx = \left[-\frac{2}{5\pi}e^{-2x}\cos x + \frac{1}{5\pi}e^{-2x}\sin x\right]_0^{2\pi} = 2\left(\frac{1 - e^{-4\pi}}{5\pi}\right)$$

$$b_1 = \frac{1}{\pi}\int_0^{2\pi} f(x)\sin x\,dx = \frac{1}{\pi}\int_0^{2\pi} e^{-2x}\sin x\,dx = \left[-\frac{2}{5\pi}e^{-2x}\sin x - \frac{1}{5\pi}e^{-2x}\cos x\right]_0^{2\pi} = \frac{1 - e^{-4\pi}}{5\pi}$$

So, the approximation is

$$g(x) = \frac{1 - e^{-4\pi}}{4\pi} + 2\left(\frac{1 - e^{-4\pi}}{5\pi}\right)\cos x + \frac{1 - e^{-4\pi}}{5\pi}\sin x$$

$$= 5\left(\frac{1 - e^{-4\pi}}{20\pi}\right) + 8\left(\frac{1 - e^{-4\pi}}{20\pi}\right)\cos x + 4\left(\frac{1 - e^{-4\pi}}{20\pi}\right)\sin x$$

$$= \left(\frac{1 - e^{-4\pi}}{20\pi}\right)\left(5 + 8\cos x + 4\sin x\right).$$

71. The third order Fourier approximation of $f(x) = 1 + x$ is of the form

$$g(x) = \frac{a_0}{2} + a_1\cos x + b_1\sin x + a_2\cos 2x + b_2\sin 2x + a_3\cos 3x + b_3\sin 3x.$$

Find the coefficients as follows.

$$a_0 = \frac{1}{\pi}\int_0^{2\pi} f(x)\,dx = \frac{1}{\pi}\int_0^{2\pi} (1 + x)\,dx = \frac{1}{\pi}\left(x + \frac{x^2}{2}\right)\Big]_0^{2\pi} = 2 + 2\pi$$

$$a_j = \frac{1}{\pi}\int_0^{2\pi} f(x)\cos(jx)\,dx = \frac{1}{\pi}\int_0^{2\pi} (1 + x)\cos(jx)\,dx = \frac{1}{\pi}\left[\frac{1 + x}{j}\sin(jx) + \frac{1}{j^2}\cos(jx)\right]_0^{2\pi} = 0$$

$$b_j = \frac{1}{\pi}\int_0^{2\pi} f(x)\sin(jx)\,dx = \frac{1}{\pi}\int_0^{2\pi} (1 + x)\sin(jx)\,dx = \frac{1}{\pi}\left[\frac{-(1 + x)}{j}\cos(jx) + \frac{1}{j^2}\sin(jx)\right]_0^{2\pi} = \frac{-2}{j}$$

So, the approximation is $g(x) = (1 + \pi) - 2\sin x - \sin 2x - \dfrac{2}{3}\sin 3x$.

73. Because $f(x) = 2 \sin x \cos x = \sin 2x$, you see that the fourth order Fourier approximation is simply $g(x) = \sin 2x$.

75. Because $a_0 = 0$, $a_j = 0$ $(j = 1, 2, 3, \ldots, n)$, and $b_j = \dfrac{2}{j}$ $(j = 1, 2, 3, \ldots, n)$ the nth-order Fourier approximation is

$$g(x) = 2 \sin x + \sin 2x + \frac{2}{3} \sin 3x + \ldots + \frac{2}{n} \sin nx = \sum_{j=1}^{n} \frac{2}{j} \sin jx.$$

Review Exercises for Chapter 5

1. (a) $\|\mathbf{u}\| = \sqrt{1^2 + 2^2} = \sqrt{5}$

(b) $\|\mathbf{v}\| = \sqrt{4^2 + 1^2} = \sqrt{17}$

(c) $\mathbf{u} \cdot \mathbf{v} = 1(4) + 2(1) = 6$

(d) $d(\mathbf{u}, \mathbf{v}) = \|\mathbf{u} - \mathbf{v}\| = \|(-3, 1)\|$

$$= \sqrt{(-3)^2 + 1^2} = \sqrt{10}$$

3. (a) $\|\mathbf{u}\| = \sqrt{2^2 + 1^2 + 1^2} = \sqrt{6}$

(b) $\|\mathbf{v}\| = \sqrt{3^2 + 2^2 + (-1)^2} = \sqrt{14}$

(c) $\mathbf{u} \cdot \mathbf{v} = 2(3) + 1(2) + 1(-1) = 7$

(d) $d(\mathbf{u} \cdot \mathbf{v}) = \|\mathbf{u} - \mathbf{v}\| = \|(-1, -1, -2)\|$

$$= \sqrt{(-1)^2 + (-1)^2 + (-2)^2}$$

$$= \sqrt{6}$$

5. (a) $\|\mathbf{u}\| = \sqrt{1^2 + (-2)^2 + 0^2 + 1^2} = \sqrt{6}$

(b) $\|\mathbf{v}\| = \sqrt{1^2 + 1^2 + (-1)^2 + 0^2} = \sqrt{3}$

(c) $\mathbf{u} \cdot \mathbf{v} = 1(1) + (-2)(1) + 0(-1) + 1(0) = -1$

(d) $d(\mathbf{u} \cdot \mathbf{v}) = \|\mathbf{u} - \mathbf{v}\| = \|(0, -3, 1, 1)\|$

$$= \sqrt{0^2 + (-3)^2 + 1^2 + 1^2}$$

$$= \sqrt{11}$$

7. (a) $\|\mathbf{u}\| = \sqrt{0^2 + 1^2 + (-1)^2 + 1^2 + 2^2} = \sqrt{7}$

(b) $\|\mathbf{v}\| = \sqrt{0^2 + 1^2 + (-2)^2 + 1^2 + 1^2} = \sqrt{7}$

(c) $\mathbf{u} \cdot \mathbf{v} = 0(0) + 1(1) + (-1)(-2) + 1(1) + 2(1) = 6$

(d) $d(\mathbf{u} \cdot \mathbf{v}) = \|\mathbf{u} - \mathbf{v}\| = \|(0, 0, 1, 0, 1)\|$

$$= \sqrt{0^2 + 0^2 + 1^2 + 0^2 + 1^2}$$

$$= \sqrt{2}$$

9. The norm of $\mathbf{v}$ is

$$\|\mathbf{v}\| = \sqrt{5^2 + 3^2 + (-2)^2} = \sqrt{38}.$$

So, a unit vector in the direction of $\mathbf{v}$ is

$$\mathbf{u} = \frac{1}{\|\mathbf{v}\|} \mathbf{v} = \frac{1}{\sqrt{38}} (5, 3, -2) = \left(\frac{5}{\sqrt{38}}, \frac{3}{\sqrt{38}}, -\frac{2}{\sqrt{38}} \right).$$

11. The norm of $\mathbf{v}$ is

$$\|\mathbf{v}\| = \sqrt{1^2 + (-1)^2 + 2^2} = \sqrt{6}.$$

So, a unit vector in the direction of $\mathbf{v}$ is given by

$$\mathbf{u} = \frac{1}{\|\mathbf{v}\|} \mathbf{v} = \frac{1}{\sqrt{6}} (1, -1, 2) = \left(\frac{1}{\sqrt{6}}, \frac{-1}{\sqrt{6}}, \frac{2}{\sqrt{6}} \right).$$

13. The cosine of the angle θ between $\mathbf{u}$ and $\mathbf{v}$ is given by

$$\cos \theta = \frac{\mathbf{u} \cdot \mathbf{v}}{\|\mathbf{u}\|\|\mathbf{v}\|} = \frac{2(-3) + 2(3)}{\sqrt{2^2 + 2^2}\sqrt{(-3)^2 + 3^2}} = 0.$$

So, $\theta = \dfrac{\pi}{12}$ radians (90°).

15. The cosine of the angle θ between $\mathbf{u}$ and $\mathbf{v}$ is given by

$$\cos \theta = \frac{\mathbf{u} \cdot \mathbf{v}}{\|\mathbf{u}\|\|\mathbf{v}\|} = \frac{\cos \frac{3\pi}{4} \cos \frac{2\pi}{3} + \sin \frac{3\pi}{4} \sin \frac{2\pi}{3}}{\sqrt{\cos^2 \frac{3\pi}{4} + \sin^2 \frac{3\pi}{4}} \cdot \sqrt{\cos^2 \frac{2\pi}{3} + \sin^2 \frac{2\pi}{3}}} = \frac{\cos \left(\frac{3\pi}{4} - \frac{2\pi}{3} \right)}{1 \cdot 1} = \cos \frac{\pi}{12}$$

So, $\theta = \dfrac{\pi}{12}$ radians (15°).

17. The cosine of the angle θ between **u** and **v** is given by

$$\cos\theta = \frac{\mathbf{u}\cdot\mathbf{v}}{\|\mathbf{u}\|\|\mathbf{v}\|} = \frac{10(-2) + (-5)(1) + 15(-3)}{\sqrt{10^2 + (-5)^2 + 15^2}\sqrt{(-2)^2 + 1^2 + (-3)^2}} = -1.$$

So, $\theta = \pi$ radians (180°).

19. The projection of **u** onto **v** is given by

$$\text{proj}_\mathbf{v}\mathbf{u} = \frac{\mathbf{u}\cdot\mathbf{v}}{\mathbf{v}\cdot\mathbf{v}}\mathbf{v} = \frac{2(1) + 4(-5)}{1^2 + (-5)^2}(1, -5) = -\frac{9}{13}(1, -5) = \left(-\frac{9}{13}, \frac{45}{13}\right).$$

21. The projection of **u** onto **v** is given by

$$\text{proj}_\mathbf{v}\mathbf{u} = \frac{\mathbf{u}\cdot\mathbf{v}}{\mathbf{v}\cdot\mathbf{v}}\mathbf{v} = \frac{1(2) + 2(5)}{2^2 + 5^2}(2, 5) = \frac{12}{29}(2, 5) = \left(\frac{24}{29}, \frac{60}{29}\right).$$

23. The projection of **u** onto **v** is given by

$$\text{proj}_\mathbf{v}\mathbf{u} = \frac{\mathbf{u}\cdot\mathbf{v}}{\mathbf{v}\cdot\mathbf{v}}\mathbf{v} = \frac{0(3) + (-1)(2) + 2(4)}{3^2 + 2^2 + 4^2}(3, 2, 4) = \left(\frac{18}{29}, \frac{12}{29}, \frac{24}{29}\right).$$

25. (a) $\langle\mathbf{u}, \mathbf{v}\rangle = 2\left(\frac{3}{2}\right) + 2\left(-\frac{1}{2}\right)(2) + 3(1)(-1) = -2$

(b) $d(\mathbf{u}, \mathbf{v}) = \|\mathbf{u} - \mathbf{v}\| = \sqrt{\langle\mathbf{u} - \mathbf{v}, \mathbf{u} - \mathbf{v}\rangle} = \sqrt{\left(2 - \frac{3}{2}\right)^2 + 2\left(-\frac{1}{2} - 2\right)^2 + 3\left(1 - (-1)\right)^2} = \frac{3}{2}\sqrt{11}$

27. Verify the Triangle Inequality as follows.

$$\|\mathbf{u} + \mathbf{v}\| \le \|\mathbf{u}\| + \|\mathbf{v}\|$$

$$\left\|\left(\frac{7}{2}, \frac{3}{2}, 0\right)\right\| \le \sqrt{2^2 + 2\left(-\frac{1}{2}\right)^2 + 3(1)^2} + \sqrt{\left(\frac{3}{2}\right)^2 + 2(2)^2 + 3(-1)^2}$$

$$\sqrt{\left(\frac{7}{2}\right)^2 + 2\left(\frac{3}{2}\right)^2 + 0} \le \sqrt{\frac{15}{2}} + \frac{\sqrt{53}}{2}$$

$$\frac{\sqrt{67}}{2} \le \frac{\sqrt{30}}{2} + \frac{\sqrt{53}}{2}$$

$$4.093 \le 6.379$$

Verify the Cauchy-Schwarz Inequality as follows.

$$|\langle\mathbf{u}, \mathbf{v}\rangle| \le \|\mathbf{u}\|\|\mathbf{v}\|$$

$$\left|2\left(\frac{3}{2}\right) + 2\left(-\frac{1}{2}\right)(2) + 3(1)(-1)\right| \le \sqrt{\frac{15}{2}}\,\frac{\sqrt{53}}{2}$$

$$2 \le \frac{\sqrt{30}}{2}\,\frac{\sqrt{53}}{2}$$

$$2 \le 9.969$$

29. A vector $\mathbf{v} = (v_1, v_2, v_3)$ that is orthogonal to $\mathbf{u} = (0, -4, 3)$ must satisfy the equation

$$\mathbf{u}\cdot\mathbf{v} = (0, -4, 3)\cdot(v_1, v_2, v_3) = 0v_1 - 4v_2 + 3v_3 = 0.$$

This equation has solutions of the form $\mathbf{v} = (s, 3t, 4t)$, where s and t are any real numbers.

31. A vector $\mathbf{v} = (v_1, v_2, v_3, v_4)$ that is orthogonal to $\mathbf{u} = (1, -2, 2, 1)$ must satisfy the equation

$$\mathbf{u}\cdot\mathbf{v} = (1, -2, 2, 1)\cdot(v_1, v_2, v_3, v_4) = v_1 - 2v_2 + 2v_3 + v_4 = 0.$$

This equation has solutions of the form $(2r - 2s - t, r, s, t)$, where r, s, and t are any real numbers.

33. First orthogonalize the vectors in B.

$$\mathbf{w}_1 = (1, 1)$$

$$\mathbf{w}_2 = (0, 1) - \frac{1(0) + 1(1)}{1^2 + 1^2}(1, 1) = \left(-\frac{1}{2}, \frac{1}{2}\right)$$

Then normalize each vector.

$$\mathbf{u}_1 = \frac{1}{\|\mathbf{w}_1\|}\mathbf{w}_1 = \frac{1}{\sqrt{2}}(1, 1) = \left(\frac{1}{\sqrt{2}}, \frac{1}{\sqrt{2}}\right)$$

$$\mathbf{u}_2 = \frac{1}{\|\mathbf{w}_2\|}\mathbf{w}_2 = \sqrt{2}\left(-\frac{1}{2}, \frac{1}{2}\right) = \left(-\frac{1}{\sqrt{2}}, \frac{1}{\sqrt{2}}\right)$$

So, an orthonormal basis for R^2 is $\left\{\left(\frac{1}{\sqrt{2}}, \frac{1}{\sqrt{2}}\right), \left(-\frac{1}{\sqrt{2}}, \frac{1}{\sqrt{2}}\right)\right\}$.

35. $\mathbf{w}_1 = (0, 3, 4)$

$$\mathbf{w}_2 = (1, 0, 0) - \frac{1(0) + 0(3) + 0(4)}{0^2 + 3^2 + 4^2}(0, 3, 4) = (1, 0, 0)$$

$$\mathbf{w}_3 = (1, 1, 0) - \frac{1(1) + 1(0) + 0(0)}{1^2 + 0^2 + 0^2}(1, 0, 0) - \frac{1(0) + 1(3) + 0(4)}{0^2 + 3^2 + 4^2}(0, 3, 4) = \left(0, \frac{16}{25}, -\frac{12}{25}\right)$$

Then, normalize each vector.

$$\mathbf{u}_1 = \frac{1}{\|\mathbf{w}_1\|}\mathbf{w}_1 = \frac{1}{5}(0, 3, 4) = \left(0, \frac{3}{5}, \frac{4}{5}\right)$$

$$\mathbf{u}_2 = \frac{1}{\|\mathbf{w}_2\|}\mathbf{w}_2 = 1(1, 0, 0) = (1, 0, 0)$$

$$\mathbf{u}_3 = \frac{1}{\|\mathbf{w}_3\|}\mathbf{w}_3 = \frac{5}{4}\left(0, \frac{16}{25}, -\frac{12}{25}\right) = \left(0, \frac{4}{5}, -\frac{3}{5}\right)$$

So, an orthonormal basis for R^3 is $\left\{\left(0, \frac{3}{5}, \frac{4}{5}\right), (1, 0, 0), \left(0, \frac{4}{5}, -\frac{3}{5}\right)\right\}$.

37. (a) To find $\mathbf{x}$ as a linear combination of the vectors in B, solve the vector equation $c_1(0, 2, -2) + c_2(1, 0, -2) = (-1, 4, -2)$.

This produces the system of linear equations

$$\begin{aligned} c_2 &= -1 \\ 2c_1 &= 4 \\ -2c_1 - 2c_2 &= -2 \end{aligned}$$

which has the solution $c_1 = 2$ and $c_2 = -1$. So, $[\mathbf{x}]_B = (2, -1)$, and you can write $(-1, 4, -2) = 2(0, 2, -2) - (1, 0, -2)$.

(b) To apply the Gram-Schmidt orthonormalization process, first orthogonalize each vector in B.

$$\mathbf{w}_1 = (0, 2, -2)$$

$$\mathbf{w}_2 = (1, 0, -2) - \frac{1(0) + 0(2) + (-2)(-2)}{0^2 + 2^2 + (-2)^2}(0, 2, -2) = (1, -1, -1).$$

Then normalize $\mathbf{w}_1$ and $\mathbf{w}_2$ as follows.

$$\mathbf{u}_1 = \frac{1}{\|\mathbf{w}_1\|}\mathbf{w}_1 = \frac{1}{2\sqrt{2}}(0, 2, -2) = \left(0, \frac{1}{\sqrt{2}}, -\frac{1}{\sqrt{2}}\right)$$

$$\mathbf{u}_2 = \frac{1}{\|\mathbf{w}_2\|}\mathbf{w}_2 = \frac{1}{\sqrt{3}}(1, -1, -1) = \left(\frac{1}{\sqrt{3}}, -\frac{1}{\sqrt{3}}, -\frac{1}{\sqrt{3}}\right)$$

So, $B' = \left\{\left(0, \frac{1}{\sqrt{2}}, -\frac{1}{\sqrt{2}}\right), \left(\frac{1}{\sqrt{3}}, -\frac{1}{\sqrt{3}}, -\frac{1}{\sqrt{3}}\right)\right\}$.

(c) To find **x** as a linear combination of the vectors in B', solve the vector equation

$$c_1\left(0, \frac{1}{\sqrt{2}}, -\frac{1}{\sqrt{2}}\right) + c_2\left(\frac{1}{\sqrt{3}}, -\frac{1}{\sqrt{3}}, -\frac{1}{\sqrt{3}}\right) = (-1, 4, -2).$$

This produces the system of linear equations

$$\frac{1}{\sqrt{3}}c_2 = -1$$

$$\frac{1}{\sqrt{2}}c_1 - \frac{1}{\sqrt{3}}c_2 = 4$$

$$-\frac{1}{\sqrt{2}}c_1 - \frac{1}{\sqrt{3}}c_2 = -2$$

which has the solution $c_1 = 3\sqrt{2}$ and $c_2 = -\sqrt{3}$. So, $[\mathbf{x}]_{B'} = \left(3\sqrt{2}, -\sqrt{3}\right)$, and you can write

$$(-1, 4, -2) = 3\sqrt{2}\left(0, \frac{1}{\sqrt{2}}, -\frac{1}{\sqrt{2}}\right) - \sqrt{3}\left(\frac{1}{\sqrt{3}}, -\frac{1}{\sqrt{3}}, -\frac{1}{\sqrt{3}}\right).$$

39. (a) $\langle f, g \rangle = \int_0^1 f(x)g(x)\, dx = \int_0^1 x^3\, dx = \left.\frac{1}{4}x^4\right]_0^1 = \frac{1}{4}$

(b) Because

$$\langle g, g \rangle = \int_0^1 g(x)g(x)\, dx = \int_0^1 x^4\, dx = \left.\frac{1}{5}x^5\right]_0^1 = \frac{1}{5},$$

the norm of g is

$$\|g\| = \sqrt{\langle g, g \rangle} = \sqrt{\frac{1}{5}} = \frac{1}{\sqrt{5}}.$$

(c) Because

$$\langle f - g, f - g \rangle = \int_0^1 \left(x - x^2\right)^2 dx = \left[\frac{1}{3}x^3 - \frac{1}{2}x^4 + \frac{1}{5}x^5\right]_0^1 = \frac{1}{30},$$

the distance between f and g is

$$d(f, g) = \|f - g\| = \sqrt{\langle f - g, f - g \rangle} = \sqrt{\frac{1}{30}} = \frac{1}{\sqrt{30}}.$$

(d) First orthogonalize the vectors.

$$\mathbf{w}_1 = f = x$$

$$\mathbf{w}_2 = g - \frac{\langle g, \mathbf{w}_1 \rangle}{\langle \mathbf{w}_1, \mathbf{w}_1 \rangle}\mathbf{w}_1 = x^2 - \frac{\int_0^1 x^3\, dx}{\int_0^1 x^2\, dx}x = x^2 - \frac{3}{4}x$$

Then, normalize each vector. Because

$$\langle \mathbf{w}_1, \mathbf{w}_1 \rangle = \int_0^1 x^2\, dx = \left.\frac{1}{3}x^3\right]_0^1 = \frac{1}{3}$$

$$\langle \mathbf{w}_2, \mathbf{w}_2 \rangle = \int_0^1 \left(x^2 - \frac{3}{4}x\right)^2 dx = \left[\frac{1}{5}x^5 - \frac{3}{8}x^4 + \frac{3}{16}x^3\right]_0^1 = \frac{1}{80}$$

you have

$$\mathbf{u}_1 = \frac{1}{\|\mathbf{w}_1\|}\mathbf{w}_1 = \sqrt{3}x$$

$$\mathbf{u}_2 = \frac{1}{\|\mathbf{w}_2\|}\mathbf{w}_2 = 4\sqrt{5}x^2 - 3\sqrt{5}x = \sqrt{5}\left(4x^2 - 3x\right).$$

The orthonormal set is $B' = \left\{\sqrt{3}x, \sqrt{5}\left(4x^2 - 3x\right)\right\}$.

41. These functions are orthogonal because $\langle f, g \rangle = \int_{-1}^{1} \sqrt{1 - x^2} \, 2x\sqrt{1 - x^2} \, dx = \int_{-1}^{1} \left(2x - 2x^3\right) dx = \left[x^2 - \dfrac{x^4}{2}\right]_{-1}^{1} = 0.$

43. Vectors in W are of the form $(-s - t, s, t)$ where s and t are any real numbers. So, a basis for W is $\{(-1, 0, 1), (-1, 1, 0)\}$. Orthogonalize these vectors as follows.

$$\mathbf{w}_1 = (-1, 0, 1)$$

$$\mathbf{w}_2 = (-1, 1, 0) - \frac{-1(-1) + 1(0) + 0(1)}{(-1)^2 + 0^2 + 1^2}(-1, 0, 1) = \left(-\frac{1}{2}, 1, -\frac{1}{2}\right)$$

Finally, normalize $\mathbf{w}_1$ and $\mathbf{w}_2$ to obtain

$$\mathbf{u}_1 = \frac{1}{\|\mathbf{w}_1\|}\mathbf{w}_1 = \frac{1}{\sqrt{2}}(-1, 0, 1) = \left(-\frac{1}{\sqrt{2}}, 0, \frac{1}{\sqrt{2}}\right)$$

$$\mathbf{u}_2 = \frac{1}{\|\mathbf{w}_2\|}\mathbf{w}_2 = \frac{2}{\sqrt{6}}\left(-\frac{1}{2}, 1, -\frac{1}{2}\right) = \left(-\frac{1}{\sqrt{6}}, \frac{2}{\sqrt{6}}, -\frac{1}{\sqrt{6}}\right).$$

So, $W' = \left\{\left(-\dfrac{1}{\sqrt{2}}, 0, \dfrac{1}{\sqrt{2}}\right), \left(-\dfrac{1}{\sqrt{6}}, \dfrac{2}{\sqrt{6}}, -\dfrac{1}{\sqrt{6}}\right)\right\}.$

45. (a) $\langle f, g \rangle = \int_{-1}^{1} x\dfrac{1}{x^2 + 1}dx = \dfrac{1}{2}\ln(x^2 + 1)\Big]_{-1}^{1} = \dfrac{1}{2}\ln 2 - \dfrac{1}{2}\ln 2 = 0$

(b) The vectors are orthogonal.

(c) Because $\langle f, g \rangle = 0$, it follows that $|\langle f, g \rangle| \le \|f\| \, \|g\|$.

47. If $\|\mathbf{u}\| \le 1$ and $\|\mathbf{v}\| \le 1$, then the Cauchy-Schwarz Inequality implies that $|\langle \mathbf{u}, \mathbf{v} \rangle| \le \|\mathbf{u}\| \, \|\mathbf{v}\| \le 1$.

49. Let $\{\mathbf{v}_1, \dots, \mathbf{v}_m\}$ be a basis for V. You can extend this basis to one for R^n.

$$B = \{\mathbf{v}_1, \dots, \mathbf{v}_m, \mathbf{w}_{m+1}, \dots, \mathbf{w}_n\}$$

Now apply the Gram-Schmidt orthonormalization process to this basis, which results in the following basis for R^n.

$$B' = \{\mathbf{u}_1, \dots, \mathbf{u}_m, \mathbf{z}_{m+1}, \dots, \mathbf{z}_n\}$$

The first m vectors of B' still span V. Therefore, any vector $\mathbf{u} \in R^n$ is of the form

$$\mathbf{u} = c_1\mathbf{u}_1 + \cdots + c_m\mathbf{u}_m + c_{m+1}\mathbf{z}_{m+1} + \cdots + c_n\mathbf{z}_n = \mathbf{v} + \mathbf{w}$$

where $\mathbf{v} \in V$ and $\mathbf{w}$ is orthogonal to every vector in V.

51. First extend the set $\{\mathbf{u}_1, \dots, \mathbf{u}_m\}$ to an orthonormal basis for R^n.

$$B = \{\mathbf{u}_1, \dots, \mathbf{u}_m, \mathbf{u}_{m+1}, \dots, \mathbf{u}_n\}$$

If $\mathbf{v}$ is any vector in R^n, you have $\mathbf{v} = \displaystyle\sum_{i=1}^{n}(\mathbf{v} \cdot \mathbf{u}_i)\mathbf{u}_i$ which implies that $\|\mathbf{v}\|^2 = \langle \mathbf{v}, \mathbf{v} \rangle = \displaystyle\sum_{i=1}^{n}(\mathbf{v} \cdot \mathbf{u}_i)^2 \ge \sum_{i=1}^{m}(\mathbf{v} \cdot \mathbf{u}_i)^2.$

53. If $\mathbf{u}$ and $\mathbf{v}$ are orthogonal, then $\|\mathbf{u}\|^2 + \|\mathbf{v}\|^2 = \|\mathbf{u} + \mathbf{v}\|^2$ by the Pythagorean Theorem.

Furthermore, $\|\mathbf{u}\|^2 + \|-\mathbf{v}\|^2 = \|\mathbf{u}\|^2 + \|\mathbf{v}\|^2 = \|\mathbf{u} - \mathbf{v}\|^2$, which gives $\|\mathbf{u} + \mathbf{v}\|^2 = \|\mathbf{u} - \mathbf{v}\|^2 \Rightarrow \|\mathbf{u} + \mathbf{v}\| = \|\mathbf{u} - \mathbf{v}\|$.

On the other hand, if $\|\mathbf{u} + \mathbf{v}\| = \|\mathbf{u} - \mathbf{v}\|$, then $\langle \mathbf{u} + \mathbf{v}, \mathbf{u} + \mathbf{v} \rangle^2 = \langle \mathbf{u} - \mathbf{v}, \mathbf{u} - \mathbf{v} \rangle^2$, which implies that $\|\mathbf{u}\|^2 + \|\mathbf{v}\|^2 + 2\langle \mathbf{u}, \mathbf{v} \rangle = \|\mathbf{u}\|^2 + \|\mathbf{v}\|^2 - 2\langle \mathbf{u}, \mathbf{v} \rangle$, or $\langle \mathbf{u}, \mathbf{v} \rangle = 0$, and $\mathbf{u}$ and $\mathbf{v}$ are orthogonal.

55. $S^{\perp} = N(A^T)$, the orthogonal complement of S is the nullspace of A^T.

$$A = \begin{bmatrix} 1 & 2 & 0 \\ 2 & 1 & -1 \end{bmatrix} \Rightarrow \begin{bmatrix} 1 & 2 & 0 \\ 0 & -3 & -1 \end{bmatrix} \Rightarrow \begin{bmatrix} 1 & 0 & -\frac{2}{3} \\ 0 & 1 & \frac{1}{3} \end{bmatrix}$$

So, $S^{\perp}$ is spanned by $\mathbf{u} = \begin{bmatrix} 2 \\ -1 \\ 3 \end{bmatrix}$.

57. $A = \begin{bmatrix} 0 & 1 & 0 \\ 0 & -3 & 0 \\ 1 & 0 & 1 \end{bmatrix} \Rightarrow \begin{bmatrix} 1 & 0 & 1 \\ 0 & 1 & 0 \\ 0 & 0 & 0 \end{bmatrix}$

$A^T = \begin{bmatrix} 0 & 0 & 1 \\ 1 & -3 & 0 \\ 0 & 0 & 1 \end{bmatrix} \Rightarrow \begin{bmatrix} 1 & -3 & 0 \\ 0 & 0 & 1 \\ 0 & 0 & 0 \end{bmatrix}$

$R(A)$-basis: $\left\{ \begin{bmatrix} 0 \\ 0 \\ 1 \end{bmatrix}, \begin{bmatrix} 1 \\ -3 \\ 0 \end{bmatrix} \right\}$ $R(A^T)$-basis: $\left\{ \begin{bmatrix} 0 \\ 1 \\ 0 \end{bmatrix}, \begin{bmatrix} 1 \\ 0 \\ 1 \end{bmatrix} \right\}$ $N(A)$-basis: $\left\{ \begin{bmatrix} 1 \\ 0 \\ -1 \end{bmatrix} \right\}$ $N(A^T)$-basis: $\left\{ \begin{bmatrix} 3 \\ 1 \\ 0 \end{bmatrix} \right\}$

59. Use a graphing utility.

Least squares regression line: $y = 88.4t + 1592 \ (r^2 \approx 0.9619)$

Least squares cubic regression polynomial: $y = 1.778t^3 - 5.82t^2 + 77.9t + 1603 \ (r^2 \approx 0.9736)$

The cubic model is a better fit.

2010: For $t = 10 \Rightarrow y \approx \$3578$ million

61. Substitute the data points $(-1, 389.1)$, $(0, 399.5)$, $(1, 403.5)$, $(2, 409.7)$, $(3, 425.7)$, and $(4, 446.4)$ into the equation

$y = c_0 + c_1 t$ to obtain the following system.

$c_0 + c_1(-1) = 389.1$

$c_0 + c_1(0) = 399.5$

$c_0 + c_1(1) = 403.5$

$c_0 + c_1(2) = 409.7$

$c_0 + c_1(3) = 425.7$

$c_0 + c_1(4) = 446.4$

This produces the least squares problem

$$A\mathbf{x} = \mathbf{b}$$

$$\begin{bmatrix} 1 & -1 \\ 1 & 0 \\ 1 & 1 \\ 1 & 2 \\ 1 & 3 \\ 1 & 4 \end{bmatrix} \begin{bmatrix} c_0 \\ c_1 \end{bmatrix} = \begin{bmatrix} 389.1 \\ 399.5 \\ 403.5 \\ 409.7 \\ 425.7 \\ 446.4 \end{bmatrix}.$$

The normal equations are

$$A^T A\mathbf{x} = A^T \mathbf{b}$$

$$\begin{bmatrix} 6 & 9 \\ 9 & 31 \end{bmatrix} \begin{bmatrix} c_0 \\ c_1 \end{bmatrix} = \begin{bmatrix} 2473.9 \\ 3896.5 \end{bmatrix}$$

and the solution is $\mathbf{x} = \begin{bmatrix} c_0 \\ c_1 \end{bmatrix} \approx \begin{bmatrix} 396.4 \\ 10.61 \end{bmatrix}$. So, the least squares regression line is $y = 396.4 + 10.61t$.

63. Substitute the data points $(0, 431.4)$, $(1, 748.8)$, $(2, 1214.1)$, $(3, 2165.1)$, $(4, 3271.3)$, $(5, 4552.4)$, $(6, 5969.7)$, and $(7, 7150.0)$ into the quadratic polynomial $y = c_0 + c_1 t + c_2 t^2$ to obtain the following system.

$$c_0 + c_1(0) + c_2(0)^2 = 431.4$$
$$c_0 + c_1(1) + c_2(1)^2 = 748.8$$
$$c_0 + c_1(2) + c_2(2)^2 = 1214.1$$
$$c_0 + c_1(3) + c_2(3)^2 = 2165.1$$
$$c_0 + c_1(4) + c_2(4)^2 = 3271.3$$
$$c_0 + c_1(5) + c_2(5)^2 = 4552.4$$
$$c_0 + c_1(6) + c_2(6)^2 = 5969.7$$
$$c_0 + c_1(7) + c_2(7)^2 = 7150.0$$

This produces the least squares problem

$$A\mathbf{x} = \mathbf{b}$$

$$\begin{bmatrix} 1 & 0 & 0 \\ 1 & 1 & 1 \\ 1 & 2 & 4 \\ 1 & 3 & 9 \\ 1 & 4 & 16 \\ 1 & 5 & 25 \\ 1 & 6 & 36 \\ 1 & 7 & 49 \end{bmatrix} \begin{bmatrix} c_0 \\ c_1 \\ c_2 \end{bmatrix} = \begin{bmatrix} 431.4 \\ 748.8 \\ 1214.1 \\ 2165.1 \\ 3271.3 \\ 4552.4 \\ 5969.7 \\ 7150.0 \end{bmatrix}$$

The normal equations are

$$A^T A \mathbf{x} = A^T \mathbf{b}$$

$$\begin{bmatrix} 8 & 28 & 140 \\ 28 & 140 & 784 \\ 140 & 784 & 4676 \end{bmatrix} \begin{bmatrix} c_0 \\ c_1 \\ c_2 \end{bmatrix} = \begin{bmatrix} 25{,}502.8 \\ 131{,}387.7 \\ 756{,}501.1 \end{bmatrix}$$

and the solution is

$$\mathbf{x} = \begin{bmatrix} c_0 \\ c_1 \\ c_2 \end{bmatrix} \approx \begin{bmatrix} 315.0 \\ 365.26 \\ 91.112 \end{bmatrix}.$$

So, the least squares regression quadratic polynomial is $y = 315.0 + 365.26t + 91.112t^2$.

65. Substitute the data points $(0, 10{,}458.0)$, $(1, 9589.8)$, $(2, 9953.5)$, $(3, 9745.4)$, $(4, 10{,}472.0)$, $(5, 11{,}598.0)$, $(6, 12{,}670.0)$, and $(7, 13{,}680.0)$ into the quadratic polynomial $y = c_0 + c_1 t + c_2 t^2$ to obtain the following system.

$$c_0 + c_1(0) + c_2(0)^2 = 10{,}458.0$$
$$c_0 + c_1(1) + c_2(1)^2 = 9589.8$$
$$c_0 + c_1(2) + c_2(2)^2 = 9953.5$$
$$c_0 + c_1(3) + c_2(3)^2 = 9745.4$$
$$c_0 + c_1(4) + c_2(4)^2 = 10{,}472.0$$
$$c_0 + c_1(5) + c_2(5)^2 = 11{,}598.0$$
$$c_0 + c_1(6) + c_2(6)^2 = 12{,}670.0$$
$$c_0 + c_1(7) + c_2(7)^2 = 13{,}680.0$$

This produces the least squares problem

$$A\mathbf{x} = \mathbf{b}$$

$$\begin{bmatrix} 1 & 0 & 0 \\ 1 & 1 & 1 \\ 1 & 2 & 4 \\ 1 & 3 & 9 \\ 1 & 4 & 16 \\ 1 & 5 & 25 \\ 1 & 6 & 36 \\ 1 & 7 & 49 \end{bmatrix} \begin{bmatrix} c_0 \\ c_1 \\ c_2 \end{bmatrix} = \begin{bmatrix} 10{,}458.0 \\ 9589.8 \\ 9953.5 \\ 9745.4 \\ 10{,}472.0 \\ 11{,}598.0 \\ 12{,}670.0 \\ 13{,}680.0 \end{bmatrix}$$

The normal equations are

$$A^T A \mathbf{x} = A^T \mathbf{b}$$

$$\begin{bmatrix} 8 & 28 & 140 \\ 28 & 140 & 784 \\ 140 & 784 & 4676 \end{bmatrix} \begin{bmatrix} c_0 \\ c_1 \\ c_2 \end{bmatrix} = \begin{bmatrix} 88{,}166.7 \\ 330{,}391.0 \\ 1{,}721{,}054.4 \end{bmatrix}$$

and the solution is

$$\mathbf{x} = \begin{bmatrix} c_0 \\ c_1 \\ c_2 \end{bmatrix} \approx \begin{bmatrix} 10{,}265.4 \\ -542.62 \\ 151.692 \end{bmatrix}.$$

So, the least squares regression quadratic polynomial is $y = 10{,}265.4 - 542.62t + 151.692t^2$.

67. The cross product is

$$\mathbf{u} \times \mathbf{v} = \begin{vmatrix} \mathbf{i} & \mathbf{j} & \mathbf{k} \\ 1 & -1 & 1 \\ 0 & 1 & 1 \end{vmatrix} = -2\mathbf{i} - \mathbf{j} + \mathbf{k} = (-2, -1, 1).$$

Furthermore, $\mathbf{u} \times \mathbf{v}$ is orthogonal to both $\mathbf{u}$ and $\mathbf{v}$ because

$$\mathbf{u} \cdot (\mathbf{u} \times \mathbf{v}) = 1(-2) + (1)(-1) + 1(1) = 0$$

and

$$\mathbf{v} \cdot (\mathbf{u} \times \mathbf{v}) = 0(-2) + 1(-1) + 1(1) = 0.$$

69. The cross product is

$$\mathbf{u} \times \mathbf{v} = \begin{vmatrix} \mathbf{i} & \mathbf{j} & \mathbf{k} \\ 2 & 0 & -1 \\ 1 & 1 & -1 \end{vmatrix} = \mathbf{i} + \mathbf{j} + 2\mathbf{k} = (1, 1, 2).$$

Furthermore, $\mathbf{u} \times \mathbf{v}$ is orthogonal to both $\mathbf{u}$ and $\mathbf{v}$ because

$$\mathbf{u} \cdot (\mathbf{u} \times \mathbf{v}) = 2(1) + 0(1) + (-1)(2) = 0$$

and

$$\mathbf{v} \cdot (\mathbf{u} \times \mathbf{v}) = 1(1) + 1(1) + (-1)(2) = 0.$$

71. Because $\|\mathbf{u} \times \mathbf{v}\| = \|\mathbf{u}\| \|\mathbf{v}\| \sin \theta$, you see that $\mathbf{u}$ and $\mathbf{v}$ are orthogonal if and only if $\sin \theta = 1$, which means $\|\mathbf{u} \times \mathbf{v}\| = \|\mathbf{u}\| \|\mathbf{v}\|$.

73. Because

$$\mathbf{v} \times \mathbf{w} = \begin{vmatrix} \mathbf{i} & \mathbf{j} & \mathbf{k} \\ -1 & -1 & 0 \\ 3 & 4 & -1 \end{vmatrix} = \mathbf{i} - \mathbf{j} - \mathbf{k} = (1, -1, -1),$$

the volume is

$$|\mathbf{u} \cdot (\mathbf{v} \times \mathbf{w})| = |(1, 2, 1) \cdot (1, -1, -1)| = |-2| = 2.$$

75. The standard basis for P_1 is $\{1, x\}$. In the interval $[0, 2]$, the Gram-Schmidt orthonormalization process yields the orthonormal basis $\left\{ \dfrac{1}{\sqrt{2}}, \dfrac{\sqrt{3}}{\sqrt{2}}, (x - 1) \right\}$.

Because

$$\langle f, \mathbf{w}_1 \rangle = \int_0^2 x^3 \frac{1}{\sqrt{2}} dx = \frac{4}{\sqrt{2}}$$

$$\langle f, \mathbf{w}_2 \rangle = \int_0^2 x^3 \frac{\sqrt{3}}{\sqrt{2}} (x - 1) \, dx = \frac{\sqrt{3}}{\sqrt{2}} \int_0^2 \left(x^4 - x^3 \right) dx$$

$$= \frac{\sqrt{3}}{\sqrt{2}} \left(\frac{x^5}{5} - \frac{x^4}{5} \right) \Bigg]_0^2 = \frac{\sqrt{3}}{\sqrt{2}} \left(\frac{32}{5} - 4 \right) = \frac{\sqrt{3}}{\sqrt{2}} \left(\frac{12}{5} \right),$$

g is given by

$$g(x) = \langle f, \mathbf{w}_1 \rangle + \langle f, \mathbf{w}_2 \rangle \mathbf{w}_2$$

$$= \frac{4}{\sqrt{2}} \left(\frac{1}{\sqrt{2}} \right) + \frac{\sqrt{3}}{\sqrt{2}} \left(\frac{12}{5} \right) \frac{\sqrt{3}}{\sqrt{2}} (x - 1) = \frac{18}{5} x - \frac{8}{5}.$$

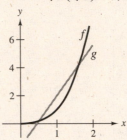

77. The standard basis for P_1 is $\{1, x\}$. In the interval $[0, \pi]$ the Gram-Schmidt orthonormalization process yields the orthonormal basis $\left\{ \dfrac{1}{\sqrt{\pi}}, \dfrac{\sqrt{3}}{\pi^{3/2}} (2x - \pi) \right\}$.

Because

$$\langle f, \mathbf{w}_1 \rangle = \int_0^\pi \sin x \cos x \left(\frac{1}{\sqrt{\pi}} \right) dx = 0$$

$$\langle f, \mathbf{w}_2 \rangle = \int_0^\pi \sin x \cos x \left(\frac{\sqrt{3}}{\pi^{3/2}} \right) (2x - \pi) \, dx = -\frac{\sqrt{3}}{2\pi^{1/2}},$$

g is given by

$$g(x) = \langle f, \mathbf{w}_1 \rangle \mathbf{w}_1 + \langle f, \mathbf{w}_2 \rangle \mathbf{w}_2$$

$$= 0 \left(\frac{1}{\sqrt{\pi}} \right) + \left(-\frac{\sqrt{3}}{2\pi^{1/2}} \right) \left(\frac{\sqrt{3}}{\pi^{3/2}} (2x - \pi) \right)$$

$$= -\frac{3x}{\pi^2} + \frac{3}{2\pi}.$$

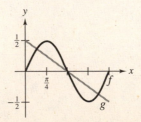

79. The standard basis for P_2 is $\{1, x, x^2\}$. In the interval $[1, 2]$, the Gram-Schmidt orthonormalization process yields the orthonormal basis

$$\left\{1, 2\sqrt{3}\left(x - \frac{3}{2}\right), \frac{30}{\sqrt{5}}\left(x^2 - 3x + \frac{13}{6}\right)\right\}.$$

Because

$$\langle f, \mathbf{w}_1 \rangle = \int_0^2 \frac{1}{x}\,dx = \ln 2$$

$$\langle f, \mathbf{w}_2 \rangle = \int_1^2 \frac{1}{x}2\sqrt{3}\left(x - \frac{3}{2}\right)dx = 2\sqrt{3}\int_1^2\left(1 - \frac{3}{2x}\right)dx = 2\sqrt{3}\left(1 - \frac{3}{2}\ln 2\right)$$

$$\langle f, \mathbf{w}_3 \rangle = \int_1^2 \frac{1}{x}\frac{30}{\sqrt{5}}\left(x^2 - 3x + \frac{13}{6}\right)dx = \frac{30}{\sqrt{5}}\int_1^2\left(x - 3 + \frac{13}{6x}\right)dx = \frac{30}{\sqrt{5}}\left(\frac{13}{6}\ln 2 - \frac{3}{2}\right),$$

g is given by $g(x) = \langle f, \mathbf{w}_1 \rangle \mathbf{w}_1 + \langle f, \mathbf{w}_2 \rangle \mathbf{w}_2 + \langle f, \mathbf{w}_3 \rangle \mathbf{w}_3$

$$= (\ln 2) + 2\sqrt{3}\left(1 - \frac{3}{2}\ln 2\right)2\sqrt{3}\left(x - \frac{3}{2}\right) + \frac{30}{\sqrt{5}}\left(\frac{13}{6}\ln 2 - \frac{3}{2}\right)\frac{30}{\sqrt{5}}\left(x^2 - 3x + \frac{13}{6}\right)$$

$$= \ln 2 + 12\left(1 - \frac{3}{2}\ln 2\right)\left(x - \frac{3}{2}\right) + 180\left(\frac{13}{6}\ln 2 - \frac{3}{2}\right)\left(x^2 - 3x + \frac{13}{6}\right) = .3274x^2 - 1.459x + 2.1175.$$

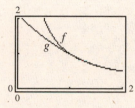

81. Find the coefficients as follows

$$a_0 = \frac{1}{\pi}\int_{-\pi}^{\pi} f(x)\,dx = \frac{1}{\pi}\int_{-\pi}^{\pi} x\,dx = 0$$

$$a_j = \frac{1}{\pi}\int_{-\pi}^{\pi} x\cos(jx)\,dx = \frac{1}{\pi}\left[\frac{1}{j^2}\cos(jx) + \frac{x}{j}\sin(jx)\right]_{-\pi}^{\pi} = 0,\ j = 1, 2\ldots$$

$$b_j = \frac{1}{\pi}\int_{-\pi}^{\pi} x\sin(jx) = \frac{1}{\pi}\left[\frac{1}{j^2}\sin(jx) - \frac{x}{j}\cos(jx)\right]_{-\pi}^{\pi} = -\frac{2}{j}\cos(\pi j)\ j = 1, 2, \ldots$$

So, the approximation is $g(x) = \dfrac{a_0}{2} + a_1\cos x + a_2\cos 2x + b_1\sin x + b_2\sin 2x = 2\sin x - \sin 2x.$

83. (a) True. See note following Theorem 5.17, page 338.

(b) True. See Theorem 5.18, part 3, page 339.

(c) True. See discussion starting on page 346.

Cumulative Test for Chapters 4 and 5

1. (a) $(1, -2) + (2, -5) = (3, -7)$

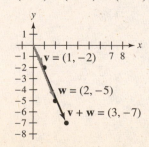

(b) $3(1, -2) = (3, -6)$

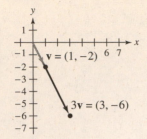

(c) $2(1, -2) - 4(2, -5) = (2, -4) - (8, -20) = (-6, 16)$

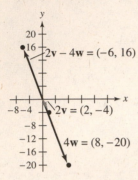

2. $\begin{bmatrix} 1 & -1 & 0 & 2 \\ 2 & 0 & 3 & 4 \\ 0 & 1 & 0 & 1 \end{bmatrix} \Rightarrow \begin{bmatrix} 1 & 0 & 0 & 3 \\ 0 & 1 & 0 & 1 \\ 0 & 0 & 1 & -\frac{2}{3} \end{bmatrix}$

$3(1, 2, 0) + (-1, 0, 1) - \frac{2}{3}(0, 3, 0) = (2, 4, 1)$

3. Not closed under addition: $\begin{bmatrix} 1 & 0 \\ 0 & 0 \end{bmatrix} + \begin{bmatrix} 0 & 0 \\ 0 & 1 \end{bmatrix} = \begin{bmatrix} 1 & 0 \\ 0 & 1 \end{bmatrix}$

4. Let $\mathbf{v} = (v_1, v_1 + v_2, v_2, v_2)$ and $\mathbf{u} = (u_1, u_1 + u_2, u_2, u_2)$ be two vectors in W.

$\mathbf{v} + \mathbf{u} = (v_1 + u_1, (v_1 + v_2) + (u_1 + u_2), v_2 + u_2, v_2 + u_2) = (v_1 + u_1, (v_1 + u_1) + (v_2 + u_2), v_2 + u_2, v_2 + u_2)$

$= (x_1, x_1 + x_2, x_2, x_2)$ where $x_1 = v_1 + u_1$ and $x_2 = v_2 + u_2$. So, $\mathbf{v} + \mathbf{u}$ is in W.

$c\mathbf{v} = c(v_1, v_1 + v_2, v_2, v_2) = (cv_1, c(v_1 + v_2), cv_2, cv_2) = (cv_1, cv_1 + cv_2, cv_2, cv_2) = (x_1, x_1 + x_2, x_2, x_2)$ where $x_1 = cv_1$ and

$v_2 = cv_2$. So, $c\mathbf{v}$ is in W. So, you can conclude that it is a subspace of R^4.

5. No: $(1, 1, 1) + (1, 1, 1) = (2, 2, 2)$

6. Yes, because $\begin{bmatrix} 1 & 2 & -1 & 0 \\ 1 & 3 & 0 & 2 \\ 0 & 0 & 1 & -1 \\ 1 & 0 & 0 & 1 \end{bmatrix}$ row reduces to I.

7. (a) See definition page 213.

(b) Linearly dependent

8. $B = \left\{ \begin{bmatrix} 1 & 0 & 0 \\ 0 & 0 & 0 \\ 0 & 0 & 0 \end{bmatrix}, \begin{bmatrix} 0 & 1 & 0 \\ 1 & 0 & 0 \\ 0 & 0 & 0 \end{bmatrix}, \begin{bmatrix} 0 & 0 & 1 \\ 0 & 0 & 0 \\ 1 & 0 & 0 \end{bmatrix}, \begin{bmatrix} 0 & 0 & 0 \\ 0 & 1 & 0 \\ 0 & 0 & 0 \end{bmatrix}, \begin{bmatrix} 0 & 0 & 0 \\ 0 & 0 & 1 \\ 0 & 1 & 0 \end{bmatrix}, \begin{bmatrix} 0 & 0 & 0 \\ 0 & 0 & 0 \\ 0 & 0 & 1 \end{bmatrix} \right\}$

Dimension 6

9. (a) A set of vectors $\{v_1, \ldots, v_n\}$ in a vector space V is a basis for **V** if the set is linearly independent and spans V.

(b) Yes. Because the set is linearly independent.

10.
$$\begin{bmatrix} 1 & 1 & 0 & 0 \\ -2 & -2 & 0 & 0 \\ 0 & 0 & 1 & 1 \\ 1 & 1 & 0 & 0 \end{bmatrix} \Rightarrow \begin{bmatrix} 1 & 1 & 0 & 0 \\ 0 & 0 & 1 & 1 \\ 0 & 0 & 0 & 0 \\ 0 & 0 & 0 & 0 \end{bmatrix} \begin{matrix} x_1 = -s \\ x_2 = s \\ x_3 = -t \\ x_4 = t \end{matrix} \quad \text{basis} \left\{ \begin{bmatrix} -1 \\ 1 \\ 0 \\ 0 \end{bmatrix}, \begin{bmatrix} 0 \\ 0 \\ -1 \\ 1 \end{bmatrix} \right\}$$

11.
$$\begin{bmatrix} 0 & 1 & 1 & 1 \\ 1 & 1 & 0 & 2 \\ 1 & 1 & 1 & -3 \end{bmatrix} \Rightarrow \begin{bmatrix} 1 & 0 & 0 & -4 \\ 0 & 1 & 0 & 6 \\ 0 & 0 & 1 & -5 \end{bmatrix}; [v]_B = \begin{bmatrix} -4 \\ 6 \\ -5 \end{bmatrix}$$

12. $[B' \vdots B] \Rightarrow [I \vdots P^{-1}];$
$$\begin{bmatrix} 1 & 1 & 0 & 2 & 1 & 0 \\ 1 & 1 & 1 & 1 & 0 & 1 \\ 2 & 1 & 2 & 0 & 0 & 1 \end{bmatrix} \Rightarrow \begin{bmatrix} 1 & 0 & 0 & 0 & 1 & -1 \\ 0 & 1 & 0 & 2 & 0 & 1 \\ 0 & 0 & 1 & -1 & -1 & 1 \end{bmatrix}$$

13. (a) $\|u\| = \sqrt{1^2 + 0^2 + 2^2} = \sqrt{5}$

(b) $\|u - v\| = \|(3, -1, -1)\|$
$$= \sqrt{3^2 + (-1)^2 + (-1)^2} = \sqrt{11}$$

(c) $u \cdot v = 1(-2) + 0(1) + 2(3) = 4$

(d) $\cos \theta = \dfrac{u \cdot v}{\|u\| \|v\|} = \dfrac{4}{\sqrt{5} \cdot \sqrt{14}} = \dfrac{4}{\sqrt{70}}$

$$\theta = \cos^{-1} \frac{4}{\sqrt{70}} \approx 1.0723 \text{ radians } (61.45°)$$

14. $\displaystyle\int_0^1 x^2(x + 2)\, dx = \left[\frac{x^4}{4} + \frac{2x^3}{3} \right]_0^1 = \frac{11}{12}$

15. $w_1 = (2, 0, 0)$

$w_2 = (1, 1, 1) - \dfrac{1}{2}(2, 0, 0) = (0, 1, 1)$

$w_3 = (0, 1, 2) - 0(2, 0, 0) - \dfrac{3}{2}(0, 1, 1) = \left(0, -\dfrac{1}{2}, \dfrac{1}{2}\right)$

Normalize each vector.

$u_1 = \dfrac{w_1}{\|w_1\|} = \dfrac{1}{2}(2, 0, 0) = (1, 0, 0)$

$u_2 = \dfrac{w_2}{\|w_2\|} = \dfrac{\sqrt{2}}{2}(0, 1, 1) = \left(0, \dfrac{\sqrt{2}}{2}, \dfrac{\sqrt{2}}{2}\right)$

$u_3 = \dfrac{w_3}{\|w_3\|} = \sqrt{2}\left(0, -\dfrac{1}{2}, \dfrac{1}{2}\right) = \left(0, -\dfrac{\sqrt{2}}{2}, \dfrac{\sqrt{2}}{2}\right)$

So, an orthonormal basis for R^3 is

$$\left\{ (1, 0, 0), \left(0, \frac{\sqrt{2}}{2}, \frac{\sqrt{2}}{2}\right), \left(0, -\frac{\sqrt{2}}{2}, \frac{\sqrt{2}}{2}\right) \right\}.$$

16. $\text{proj}_v u = \dfrac{u \cdot v}{v \cdot v} v = \dfrac{1}{13}(-3, 2)$

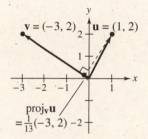

17. $A = \begin{bmatrix} 0 & 1 & 1 & 0 \\ -1 & 0 & 0 & 1 \\ 1 & 1 & 1 & 1 \end{bmatrix} \Rightarrow \begin{bmatrix} 1 & 0 & 0 & 0 \\ 0 & 1 & 1 & 0 \\ 0 & 0 & 0 & 1 \end{bmatrix}$

$A^T = \begin{bmatrix} 0 & -1 & 1 \\ 1 & 0 & 1 \\ 1 & 0 & 1 \\ 0 & 1 & 1 \end{bmatrix} \Rightarrow \begin{bmatrix} 1 & 0 & 0 \\ 0 & 1 & 0 \\ 0 & 0 & 1 \\ 0 & 0 & 0 \end{bmatrix}$

$R(A) = $ column space of $A = R^3$

$N(A)$-basis: $\left\{ \begin{bmatrix} 0 \\ 1 \\ -1 \\ 0 \end{bmatrix} \right\}$

$R(A^T)$-basis: $\left\{ \begin{bmatrix} 0 \\ 1 \\ 1 \\ 0 \end{bmatrix}, \begin{bmatrix} -1 \\ 0 \\ 0 \\ 1 \end{bmatrix}, \begin{bmatrix} 1 \\ 1 \\ 1 \\ 1 \end{bmatrix} \right\}$

$N(A^T) = \{0\}$

18. $S^\perp = N(A^T); \begin{bmatrix} 1 & 0 & 1 \\ -1 & 1 & 0 \end{bmatrix} \Rightarrow \begin{bmatrix} 1 & 0 & 1 \\ 0 & 1 & 1 \end{bmatrix} \Rightarrow S^\perp = \text{span} \left\{ \begin{bmatrix} -1 \\ -1 \\ 1 \end{bmatrix} \right\}$

19.
$$0\mathbf{v} = (0 + 0)\mathbf{v} = 0\mathbf{v} + 0\mathbf{v}$$
$$-0\mathbf{v} + 0\mathbf{v} = (-0\mathbf{v} + 0\mathbf{v}) + 0\mathbf{v}$$
$$\mathbf{0} = 0\mathbf{v}$$

20. Suppose $c_1\mathbf{x}_1 + \cdots + c_n\mathbf{x}_n + c\mathbf{y} = \mathbf{0}$

If $c = 0$, then $c_1\mathbf{x}_1 + \ldots + c_n\mathbf{x}_n = \mathbf{0}$ and
$\mathbf{x}_i$ independent $\Rightarrow c_i = 0$

If $c \neq 0$, then $\mathbf{y} = -(c_1/c)\mathbf{x}_1 + \cdots + -(c_n/c)\mathbf{x}_n$; a
contradiction.

21. Let $\mathbf{v}_1, \mathbf{v}_2 \in W^\perp$. $\langle\mathbf{v}_1, \mathbf{w}\rangle = 0, \langle\mathbf{v}_2, \mathbf{w}\rangle = 0$, for all
$\mathbf{w} \Rightarrow \mathbf{v}_1 + \mathbf{v}_2 \in W^\perp$ and $\langle c\mathbf{v}, \mathbf{w}\rangle = 0 \Rightarrow c\mathbf{v} \in W^\perp$.

Because $W^\perp$ is nonempty, and closed under addition and
scalar multiplication in $\mathbf{V}$ it is a subspace.

22. Substitute the points $(1, 1)$, $(2, 0)$, and $(5, -5)$ into the
equation $y = c_0 + c_1x$ to obtain the following system.

$$c_0 + c_1(1) = 1$$
$$c_0 + c_1(2) = 0$$
$$c_0 + c_1(5) = -5$$

This produces the least squares problem

$$A\mathbf{x} = \mathbf{b}$$

$$\begin{bmatrix} 1 & 1 \\ 1 & 2 \\ 1 & 5 \end{bmatrix} \begin{bmatrix} c_0 \\ c_1 \end{bmatrix} = \begin{bmatrix} 1 \\ 0 \\ -5 \end{bmatrix}.$$

The normal equations are

$$A^T A\mathbf{x} = A^T\mathbf{b}$$

$$\begin{bmatrix} 3 & 8 \\ 8 & 30 \end{bmatrix} \begin{bmatrix} c_0 \\ c_1 \end{bmatrix} = \begin{bmatrix} -4 \\ -24 \end{bmatrix}$$

and the solution is $\mathbf{x} = \begin{bmatrix} c_0 \\ c_1 \end{bmatrix} = \begin{bmatrix} \frac{36}{13} \\ -\frac{20}{13} \end{bmatrix}.$

So, the least squares regression line is $y = \frac{36}{13} - \frac{20}{13}x.$

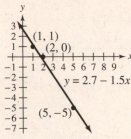

23. (a) rank $A = 3$

(b) first 3 rows of A

(c) columns 1, 3, 4 of A

(d)
$$\begin{aligned} x_1 &= 2r - 3s - 2t \\ x_2 &= r \\ x_3 &= 5s + 3t \\ x_4 &= -s - 7t \\ x_5 &= s \\ x_6 &= t \end{aligned} \left\{ \begin{bmatrix} 2 \\ 1 \\ 0 \\ 0 \\ 0 \\ 0 \end{bmatrix} \begin{bmatrix} -3 \\ 0 \\ 5 \\ -1 \\ 1 \\ 0 \end{bmatrix} \begin{bmatrix} -2 \\ 0 \\ 3 \\ -7 \\ 0 \\ 1 \end{bmatrix} \right\}$$

(e) no

(f) no

(g) yes

(h) no

24. $\|\mathbf{u} + \mathbf{v}\| = \|\mathbf{u} - \mathbf{v}\| \Leftrightarrow \|\mathbf{u} + \mathbf{v}\|^2 = \|\mathbf{u} - \mathbf{v}\|^2$

$$\Leftrightarrow (\mathbf{u} + \mathbf{v}) \cdot (\mathbf{u} + \mathbf{v}) = (\mathbf{u} - \mathbf{v}) \cdot (\mathbf{u} - \mathbf{v})$$
$$\Leftrightarrow \mathbf{u} \cdot \mathbf{u} + 2\mathbf{u} \cdot \mathbf{v} + \mathbf{v} \cdot \mathbf{v} = \mathbf{u} \cdot \mathbf{u} - 2\mathbf{u} \cdot \mathbf{v} + \mathbf{v} \cdot \mathbf{v}$$
$$\Leftrightarrow 2\mathbf{u} \cdot \mathbf{v} = -2\mathbf{u} \cdot \mathbf{v}$$
$$\Leftrightarrow \mathbf{u} \cdot \mathbf{v} = 0$$

CHAPTER 6
Linear Transformations

CHAPTER 6
Linear Transformations

Section 6.1   Introduction to Linear Transformations

1. (a) The image of **v** is

$$T(3, -4) = (3 + (-4), 3 - (-4)) = (-1, 7).$$

(b) If $T(v_1, v_2) = (v_1 + v_2, v_1 - v_2) = (3, 19)$, then

$$v_1 + v_2 = 3$$
$$v_1 - v_2 = 19,$$

which implies that $v_1 = 11$ and $v_2 = -8$. So, the preimage of **w** is $(11, -8)$.

3. (a) The image of **v** is

$$T(2, 3, 0) = (3 - 2, 2 + 3, 2(2)) = (1, 5, 4).$$

(b) If $T(v_1, v_2, v_3) = (v_2 - v_1, v_1 + v_2, 2v_1)$
$$= (-11, -1, 10),$$

then

$$v_2 - v_1 = -11$$
$$v_1 + v_2 = -1$$
$$2v_1 = 10$$

which implies that $v_1 = 5$ and $v_2 = -6$. So, the preimage of **w** is $\{(5, -6, t) : t \text{ is any real number}\}$.

5. (a) The image of **v** is

$$T(2, -3, -1) = (4(-3) - 2, 4(2) + 5(-3))$$
$$= (-14, -7).$$

(b) If $T(v_1, v_2, v_3) = (4v_2 - v_1, 4v_1 + 5v_2) = (3, 9)$,

then

$$-v_1 + 4v_2 = 3$$
$$4v_1 + 5v_2 = 9,$$

which implies that $v_1 = 1$, $v_2 = 1$, and $v_3 = t$,

where t is any real number. So, the preimage of **w** is
$\{(1, 1, t) : t \text{ is any real number}\}.$

7. (a) The image of **v** is

$$T(1, 1) = \left(\frac{\sqrt{2}}{2}(1) - \frac{\sqrt{2}}{2}(1), 1 + 1, 2(1) - 1 \right)$$
$$= (0, 2, 1).$$

(b) If $T(v_1, v_2) = \left(\frac{\sqrt{2}}{2}v_1 - \frac{\sqrt{2}}{2}v_2, v_1 + v_2, 2v_1 - v_2 \right)$
$$= (-5\sqrt{2}, -2, -16),$$

then

$$\frac{\sqrt{2}}{2}v_1 - \frac{\sqrt{2}}{2}v_2 = -5\sqrt{2}$$
$$v_1 + v_2 = -2$$
$$2v_1 - v_2 = -16$$

which implies that $v_1 = -6$ and $v_2 = 4$. So, the preimage of **w** is $(-6, 4)$.

9. T is *not* a linear transformation because it does not preserve addition nor scalar multiplication. For example,

$$T(1, 1) + T(1, 1) = (1, 1) + (1, 1)$$
$$= (2, 2) \neq (2, 1) = T(2, 2).$$

11. T preserves addition.

$$T(x_1, y_1, z_1) + T(x_2, y_2, z_2) = (x_1 + y_1, x_1 - y_1, z_1) + (x_2 + y_2, x_2 - y_2, z_2)$$
$$= (x_1 + y_1 + x_2 + y_2, x_1 - y_1 + x_2 - y_2, z_1 + z_2)$$
$$= ((x_1 + x_2) + (y_1 + y_2), (x_1 + x_2) - (y_1 + y_2), z_1 + z_2)$$
$$= T(x_1 + x_2, y_1 + y_2, z_1 + z_2)$$

T preserves scalar multiplication.

$$T(c(x, y, z)) = T(cx, cy, cz) = (cx + cy, cx - cy, cz) = c(x + y, x - y, z) = cT(x, y, z)$$

Therefore, T *is* a linear transformation.

13. T is *not* a linear transformation because it does not preserve addition nor scalar multiplication. For example,

$$T(0, 1) + T(1, 0) = (0, 0, 1) + (1, 0\ 0) = (1, 0, 1) \neq (1, 1, 1) = T(1, 1).$$

15. T is *not* a linear transformation because it does not preserve addition nor scalar multiplication. For example, $T(I_2) = 1$ but

$$T(2I_2) = 4 \neq 2T(I_2).$$

17. Let A and B be two elements of $M_{3,3}$ —two 3×3 matrices—and let c be a scalar. First

$$T(A + B) = \begin{bmatrix} 0 & 0 & 1 \\ 0 & 1 & 0 \\ 1 & 0 & 0 \end{bmatrix}(A + B) = \begin{bmatrix} 0 & 0 & 1 \\ 0 & 1 & 0 \\ 1 & 0 & 0 \end{bmatrix}A + \begin{bmatrix} 0 & 0 & 1 \\ 0 & 1 & 0 \\ 1 & 0 & 0 \end{bmatrix}B = T(A) + T(B)$$

by Theorem 2.3, part 2. And

$$T(cA) = \begin{bmatrix} 0 & 0 & 1 \\ 0 & 1 & 0 \\ 1 & 0 & 0 \end{bmatrix}(cA) = c\begin{bmatrix} 0 & 0 & 1 \\ 0 & 1 & 0 \\ 1 & 0 & 0 \end{bmatrix}A = cT(A)$$

by Theorem 2.3, part 4. So, T *is* a linear transformation.

19. T preserves addition.

$$T(A_1 + A_2) = (A_1 + A_2)^T = A_1^T + A_2^T = T(A_1) + T(A_2)$$

T preserves scalar multiplication.

$$T(cA) = (cA)^T = c(A^T) = cT(A).$$

Therefore, T *is* a linear transformation.

21. Let $\mathbf{u} = a_0 + a_1x + a_2x^2$, $\mathbf{v} = b_0 + b_1x + b_2x^2$.

Then $T(\mathbf{u} + \mathbf{v}) = (a_0 + b_0) + (a_1 + b_1) + (a_2 + b_2) + [(a_1 + b_1) + (a_2 + b_2)]x + (a_2 + b_2)x^2$

$$= (a_0 + a_1 + a_2) + (b_0 + b_1 + b_2) + [(a_1 + a_2) + (b_1 + b_2)]x + (a_2 + b_2)x^2$$

$$= T(\mathbf{u}) + T(\mathbf{v}), \text{ and}$$

$$T(c\mathbf{u}) = ca_0 + ca_1 + ca_2 + (ca_1 + ca_2)x + ca_2x^2 = cT(\mathbf{u}).$$

T *is* a linear transformation.

23. Because $(0, 3, -1)$ can be written as

$(0, 3, -1) = 0(1, 0, 0) + 3(0, 1, 0) - (0, 0, 1)$, you can use

Property 4 of Theorem 6.1 to write

$T(0, 3, -1) = 0 \cdot T(1, 0, 0) + 3T(0, 1, 0) - T(0, 0, 1)$

$\qquad = (0, 0, 0) + 3(1, 3, -2) - (0, -2, 2)$

$\qquad = (3, 11, -8).$

25. Because $(2, -4, 1)$ can be written as

$(2, -4, 1) = 2(1, 0, 0) - 4(0, 1, 0) + (0, 0, 1)$, you can use

Property 4 of Theorem 6.1 to write

$T(2, -4, 1) = 2T(1, 0, 0) - 4T(0, 1, 0) + T(0, 0, 1)$

$\qquad = 2(2, 4, -1) - 4(1, 3, -2) + (0, -2, 2)$

$\qquad = (0, -6, 8).$

27. Because $(2, 1, 0)$ can be written as

$(2, 1, 0) = 0(1, 1, 1) - (0, -1, 2) + 2(1, 0, 1)$, you can use

Property 4 of Theorem 6.1 to write

$T(2, 1, 0) = 0 \cdot T(1, 1, 1) - T(0, -1, 2) + 2T(1, 0, 1)$

$\qquad = (0, 0, 0) - (-3, 2, -1) + 2(1, 1, 0)$

$\qquad = (5, 0, 1).$

29. Because $(2, -1, 1)$ can be written as

$(2, -1, 1) = -\frac{3}{2}(1, 1, 1) - \frac{1}{2}(0, -1, 2) + \frac{7}{2}(1, 0, 1)$, you can

use Property 4 of Theorem 6.1 to write

$T(2, -1, 1) = -\frac{3}{2}T(1, 1, 1) - \frac{1}{2}T(0, -1, 2) + \frac{7}{2}T(1, 0, 1)$

$\qquad = -\frac{3}{2}(2, 0, -1) - \frac{1}{2}(-3, 2, -1) + \frac{7}{2}(1, 1, 0)$

$\qquad = \left(2, \frac{5}{2}, 2\right).$

31. Because the matrix has four columns, the dimension of R^n is 4. Because the matrix has three rows, the dimension of R^m is 3. So, $T: R^4 \to R^3$.

33. Because the matrix has five columns, the dimension of R^n is 5. Because the matrix has two rows, the dimension of R^m is 2. So, $T: R^5 \to R^2$.

35. Because the matrix has two columns, the dimension of R^n is 2. Because the matrix has two rows, the dimension of R^m is 2. So, $T: R^2 \to R^2$.

37. (a) $T(2, 4) = \begin{bmatrix} 1 & 2 \\ -2 & 4 \\ -2 & 2 \end{bmatrix} \begin{bmatrix} 2 \\ 4 \end{bmatrix} = \begin{bmatrix} 10 \\ 12 \\ 4 \end{bmatrix} = (10, 12, 4)$

  (b) The preimage of $(-1, 2, 2)$ is given by solving the equation

$$T(v_1, v_2) = \begin{bmatrix} 1 & 2 \\ -2 & 4 \\ -2 & 2 \end{bmatrix} \begin{bmatrix} v_1 \\ v_2 \end{bmatrix} = \begin{bmatrix} -1 \\ 2 \\ 2 \end{bmatrix}$$

  for $\mathbf{v} = (v_1, v_2)$. The equivalent system of linear equations

$$v_1 + 2v_2 = -1$$
$$-2v_1 + 4v_2 = 2$$
$$-2v_1 + 2v_2 = 2$$

  has the solution $v_1 = -1$ and $v_2 = 0$. So, $(-1, 0)$ is the preimage of $(-1, 2, 2)$ under T.

  (c) Because the system of linear equations represented by the equation

$$\begin{bmatrix} 1 & 2 \\ -2 & 4 \\ -2 & 2 \end{bmatrix} \begin{bmatrix} v_1 \\ v_2 \end{bmatrix} = \begin{bmatrix} 1 \\ 1 \\ 1 \end{bmatrix}$$

  has no solution, $(1, 1, 1)$ has no preimage under T.

39. (a)

$$T(1, 1, 1, 1) = \begin{bmatrix} -1 & 0 & 0 & 0 \\ 0 & 1 & 0 & 0 \\ 0 & 0 & 2 & 0 \\ 0 & 0 & 0 & 1 \end{bmatrix} \begin{bmatrix} 1 \\ 1 \\ 1 \\ 1 \end{bmatrix} = \begin{bmatrix} -1 \\ 1 \\ 2 \\ 1 \end{bmatrix} = (-1, 1, 2, 1).$$

  (b) The preimage of $(1, 1, 1, 1)$ is determined by solving the equation

$$T(v_1, v_2, v_3, v_4) = \begin{bmatrix} -1 & 0 & 0 & 0 \\ 0 & 1 & 0 & 0 \\ 0 & 0 & 2 & 0 \\ 0 & 0 & 0 & 1 \end{bmatrix} \begin{bmatrix} v_1 \\ v_2 \\ v_3 \\ v_4 \end{bmatrix} = \begin{bmatrix} 1 \\ 1 \\ 1 \\ 1 \end{bmatrix}$$

  for $\mathbf{v} = (v_1, v_2, v_3, v_4)$. The equivalent system of linear equations has solution $v_1 = -1, v_2 = 1, v_3 = \frac{1}{2}, v_4 = 1$. So, the preimage is $\left(-1, 1, \frac{1}{2}, 1\right)$.

41. (a) When $\theta = 45°$, $\cos \theta = \sin \theta = \frac{1}{\sqrt{2}}$, so $T(4, 4) = \left(4\left(\frac{1}{\sqrt{2}}\right) - 4\left(\frac{1}{\sqrt{2}}\right), 4\left(\frac{1}{\sqrt{2}}\right) + 4\left(\frac{1}{\sqrt{2}}\right)\right) = \left(0, 4\sqrt{2}\right)$.

  (b) When $\theta = 30°$, $\cos \theta = \frac{\sqrt{3}}{2}$ and $\sin \theta = \frac{1}{2}$, so $T(4, 4) = \left(4\left(\frac{\sqrt{3}}{2}\right) - 4\left(\frac{1}{2}\right), 4\left(\frac{1}{2}\right) + 4\left(\frac{\sqrt{3}}{2}\right)\right) = \left(2\sqrt{3} - 2, 2\sqrt{3} + 2\right)$.

  (c) When $\theta = 120°$, $\cos \theta = -\frac{1}{2}$ and $\sin \theta = \frac{\sqrt{3}}{2}$, so $T(5, 0) = \left(5\left(-\frac{1}{2}\right) - 0\left(\frac{\sqrt{3}}{2}\right), 5\left(\frac{\sqrt{3}}{2}\right) + 0\left(-\frac{1}{2}\right)\right) = \left(-\frac{5}{2}, \frac{5\sqrt{3}}{2}\right)$.

43. True. D_x is a linear transformation and therefore preserves addition and scalar multiplication.

45. False. $\sin 2x \neq 2 \sin x$ for all x.

47. If $D_x(g(x)) = 2x + 1$, then $g(x) = x^2 + x + C$.

49. If $D_x\big(g(x)\big) = \sin x$, then $g(x) = -\cos x + C$.

51. (a) $T\big(3x^2 - 2\big) = \int_0^1 \big(3x^2 - 2\big)\,dx = \big[x^3 - 2x\big]_0^1 = -1$

   (b) $T\big(x^3 - x^5\big) = \int_0^1 \big(x^3 - x^5\big)\,dx = \big[\tfrac{1}{4}x^4 - \tfrac{1}{6}x^6\big]_0^1 = \tfrac{1}{12}$

   (c) $T\big(4x - 6\big) = \int_0^1 \big(4x - 6\big)\,dx = \big[2x^2 - 6x\big]_0^1 = -4$

53. First express $(1, 0)$ in terms of $(1, 1)$ and $(1, -1)$: $(1, 0) = \tfrac{1}{2}(1, 1) + \tfrac{1}{2}(1, -1)$. Then,

$$T(1, 0) = T\big[\tfrac{1}{2}(1, 1) + \tfrac{1}{2}(1, -1)\big] = \tfrac{1}{2}T(1, 1) + \tfrac{1}{2}T(1, -1) = \tfrac{1}{2}(1, 0) + \tfrac{1}{2}(0, 1) = \big(\tfrac{1}{2}, \tfrac{1}{2}\big).$$

Similarly, express $(0, 2) = 1(1, 1) - 1(1, -1)$. Then,

$$T(0, 2) = T\big[(1, 1) - (1, -1)\big] = T(1, 1) - T(1, -1) = (1, 0) - (0, 1) = (1, -1).$$

55. $T\big(2 - 6x + x^2\big) = 2T(1) - 6T(x) + T\big(x^2\big) = 2x - 6(1 + x) + \big(1 + x + x^2\big) = -5 - 3x + x^2$

57. (a) True. See discussion before "Definition of a Linear Transformation," page 362.

   (b) False. $\cos(x_1 + x_2) \neq \cos x_1 + \cos x_2$

   (c) True. See Example 10, page 370.

59. (a) $T(x, y) = T\big[x(1, 0) + y(0, 1)\big] = xT(1, 0) + yT(0, 1) = x(1, 0) + y(0, 0) = (x, 0)$

   (b) T is the projection onto the x-axis.

61. (a) Because

$$\text{proj}_v\mathbf{u} = \frac{\mathbf{u} \cdot \mathbf{v}}{\mathbf{v} \cdot \mathbf{v}}\mathbf{v}$$

   and $T(\mathbf{u}) = \text{proj}_v\mathbf{u}$, you have $T(x, y) = \dfrac{x(1) + y(1)}{1^2 + 1^2}(1, 1) = \left(\dfrac{x + y}{2}, \dfrac{x + y}{2}\right)$.

   (b) From the result of part (a), where $(x, y) = (5, 0)$, $T(x, y) = \left(\dfrac{x + y}{2}, \dfrac{x + y}{2}\right) = \left(\dfrac{5}{2}, \dfrac{5}{2}\right)$.

   (c) From the result of part (a),

$$T(\mathbf{u} + \mathbf{w}) = T\big[(x_1, y_1) + (x_2, y_2)\big] = T(x_1 + x_2, y_1 + y_2)$$

$$= \left(\frac{x_1 + x_2 + y_1 + y_2}{2}, \frac{x_1 + x_2 + y_1 + y_2}{2}\right)$$

$$= \left(\frac{x_1 + y_1}{2}, \frac{x_1 + y_1}{2}\right) + \left(\frac{x_2 + y_2}{2}, \frac{x_2 + y_2}{2}\right)$$

$$= T(x_1, y_1) + T(x_2, y_2)$$

$$= T(\mathbf{u}) + T(\mathbf{w}).$$

   From the result of part (a),

$$T(c\mathbf{u}) = T\big[c(x, y)\big] = T(cx, cy) = \left(\frac{cx + cy}{2}, \frac{cx + cy}{2}\right) = c\left(\frac{x + y}{2}, \frac{x + y}{2}\right) = cT(x, y) = cT(\mathbf{u}).$$

63. Observe that $A\mathbf{u} = \begin{bmatrix} \tfrac{1}{2} & \tfrac{1}{2} \\ \tfrac{1}{2} & \tfrac{1}{2} \end{bmatrix}\begin{bmatrix} x \\ y \end{bmatrix} = \begin{bmatrix} \tfrac{1}{2}x + \tfrac{1}{2}y \\ \tfrac{1}{2}x + \tfrac{1}{2}y \end{bmatrix} = T(\mathbf{u}).$

65. (a) Because $T(0, 0) = (-h, -k) \neq (0, 0)$, a translation cannot be a linear transformation.

(b) $T(0, 0) = (0 - 2, 0 + 1) = (-2, 1)$

$T(2, -1) = (2 - 2, -1 + 1) = (0, 0)$

$T(5, 4) = (5 - 2, 4 + 1) = (3, 5)$

(c) Because $T(x, y) = (x - h, y - k) = (x, y)$ implies $x - h = x$ and $y - k = y$, a translation has no fixed points.

67. There are many possible examples. For instance, let $T: R^3 \to R^3$ be given by $T(x, y, z) = (0, 0, 0)$. Then if $\{\mathbf{v}_1, \mathbf{v}_2, \mathbf{v}_3\}$ is any set of linearly independent vectors, their images $T(\mathbf{v}_1), T(\mathbf{v}_2), T(\mathbf{v}_3)$ form a dependent set.

69. Let $T(\mathbf{v}) = \mathbf{v}$ be the identity transformation. Because

$T(\mathbf{u} + \mathbf{v}) = \mathbf{u} + \mathbf{v} = T(\mathbf{u}) + T(\mathbf{v})$ and

$T(c\mathbf{u}) = c\mathbf{u} = cT(\mathbf{u})$, T is a linear transformation.

71. T is a linear transformation because

$$T(A + B) = (a_{11} + b_{11}) + \cdots + (a_{nn} + b_{nn})$$
$$= (a_{11} + \cdots + a_{nn}) + (b_{11} + \cdots + b_{nn})$$
$$= T(A) + T(B)$$

and

$$T(cA) = ca_{11} + \cdots + ca_{nn}$$
$$= c(a_{11} + \cdots + a_{nn})$$
$$= cT(A).$$

73. Let $\mathbf{v} = c_1\mathbf{v}_1 + \cdots + c_n\mathbf{v}_n$ be an arbitrary vector in $\mathbf{v}$. Then,

$$T(\mathbf{v}) = T(c_1\mathbf{v}_1 + \cdots + c_n\mathbf{v}_n)$$
$$= c_1T(\mathbf{v}_1) + \cdots + c_nT(\mathbf{v}_n)$$
$$= \mathbf{0} + \cdots + \mathbf{0}$$
$$= \mathbf{0}.$$

Section 6.2　The Kernel and Range of a Linear Transformation

1. Because T sends every vector in R^3 to the zero vector, the kernel is R^3.

3. Solving the equation
$T(x, y, z, w) = (y, x, w, z) = (0, 0, 0, 0)$ yields the trivial solution $x = y = z = w = 0$.
So, $\ker(T) = \{(0, 0, 0, 0)\}$.

5. Solving the equation
$T(a_0 + a_1x + a_2x^2 + a_3x^3) = a_0 = 0$ yields solutions of the form $a_0 = 0$ and $a_1, a_2,$ and a_3 are any real numbers. So,
$\ker(T) = \{a_1x + a_2x^2 + a_3x^3: a_1, a_2, a_3 \in R\}$.

7. Solving the equation
$T(a_0 + a_1x + a_2x^2) = a_1 + 2a_2x = 0$ yields solutions of the form $a_1 = a_2 = 0$, and a_0 any real number. So,
$\ker(T) = \{a_0 : a_0 \in R\}$.

9. Solving the equation
$T(x, y) = (x + 2y, y - x) = (0, 0)$ yields the trivial solution $x = y = 0$. So,
$\ker(T) = \{(0, 0)\}$.

11. (a) Because

$$T(\mathbf{v}) = \begin{bmatrix} 1 & 2 \\ 3 & 4 \end{bmatrix}\begin{bmatrix} v_1 \\ v_2 \end{bmatrix} = \begin{bmatrix} 0 \\ 0 \end{bmatrix}$$

has only the trivial solution $v_1 = v_2 = 0$, the kernel is $\{(0, 0)\}$.

(b) Transpose A and find the equivalent reduced row-echelon form.

$$A^T = \begin{bmatrix} 1 & 3 \\ 2 & 4 \end{bmatrix} \Rightarrow \begin{bmatrix} 1 & 0 \\ 0 & 1 \end{bmatrix}$$

So, a basis for the range of A is $\{(1, 0), (0, 1)\}$.

13. (a) Because

$$T(\mathbf{v}) = \begin{bmatrix} 1 & -1 & 2 \\ 0 & 1 & 2 \end{bmatrix}\begin{bmatrix} v_1 \\ v_2 \\ v_3 \end{bmatrix} = \begin{bmatrix} 0 \\ 0 \end{bmatrix}$$

has solutions of the form $(-4t, -2t, t)$, where t is any real number, a basis for $\ker(T)$ is $\{(-4, -2, 1)\}$.

(b) Transpose A and find the equivalent reduced row-echelon form.

$$A^T = \begin{bmatrix} 1 & 0 \\ -1 & 1 \\ 2 & 2 \end{bmatrix} \Rightarrow \begin{bmatrix} 1 & 0 \\ 0 & 1 \\ 0 & 0 \end{bmatrix}$$

So, a basis for the range of A is $\{(1, 0), (0, 1)\}$.

15. (a) Because

$$T(\mathbf{v}) = \begin{bmatrix} 1 & 2 \\ -1 & -2 \\ 1 & 1 \end{bmatrix} \begin{bmatrix} v_1 \\ v_2 \end{bmatrix} = \begin{bmatrix} 0 \\ 0 \\ 0 \end{bmatrix}$$

has only the trivial solution $v_1 = v_2 = 0$, the kernel is $\{(0, 0)\}$.

(b) Transpose A and find the equivalent reduced row-echelon form.

$$A^T = \begin{bmatrix} 1 & -1 & 1 \\ 2 & -2 & 1 \end{bmatrix} \Rightarrow \begin{bmatrix} 1 & -1 & 0 \\ 0 & 0 & 1 \end{bmatrix}$$

So, a basis for the range of A is $\{(1, -1, 0), (0, 0, 1)\}$.

17. (a) Because

$$T(\mathbf{v}) = \begin{bmatrix} 1 & 2 & -1 & 4 \\ 3 & 1 & 2 & -1 \\ -4 & -3 & -1 & -3 \\ -1 & -2 & 1 & 1 \end{bmatrix} \begin{bmatrix} v_1 \\ v_2 \\ v_3 \\ v_4 \end{bmatrix} = \begin{bmatrix} 0 \\ 0 \\ 0 \\ 0 \end{bmatrix}$$

has solutions of the form $(-t, t, t, 0)$, where t is any real number, a basis for $\ker(T)$ is $\{(-1, 1, 1, 0)\}$.

(b) Transpose A and find the equivalent reduced row-echelon form.

$$A^T = \begin{bmatrix} 1 & 3 & -4 & -1 \\ 2 & 1 & -3 & -2 \\ -1 & 2 & -1 & 1 \\ 4 & -1 & -3 & 1 \end{bmatrix} \Rightarrow \begin{bmatrix} 1 & 0 & -1 & 0 \\ 0 & 1 & -1 & 0 \\ 0 & 0 & 0 & 1 \\ 0 & 0 & 0 & 0 \end{bmatrix}$$

So, a basis for the range of A is

$\{(1, 0, -1, 0), (0, 1, -1, 0), (0, 0, 0, 1)\}$.

Equivalently, you could use columns 1, 2 and 4 of the original matrix A.

19. (a) Because $T(\mathbf{x}) = \mathbf{0}$ has only the trivial solution $\mathbf{x} = (0, 0)$, the kernel of T is $\{(0, 0)\}$.

(b) $\text{nullity}(T) = \dim(\ker(T)) = 0$

(c) Transpose A and find the equivalent reduced row-echelon form.

$$A^T = \begin{bmatrix} -1 & 1 \\ 1 & 1 \end{bmatrix} \Rightarrow \begin{bmatrix} 1 & 0 \\ 0 & 1 \end{bmatrix}$$

So, $\text{range}(T) = R^2$.

(d) $\text{rank}(T) = \dim(\text{range}(T)) = 2$

21. (a) Because $T(\mathbf{x}) = \mathbf{0}$ has only the trivial solution $\mathbf{x} = (0, 0)$, the kernel of T is $\{(0, 0)\}$.

(b) $\text{nullity}(T) = \dim(\ker(T)) = 0$

(c) Transpose A and find the equivalent reduced row-echelon form.

$$A^T = \begin{bmatrix} 5 & 1 & 1 \\ -3 & 1 & -1 \end{bmatrix} \Rightarrow \begin{bmatrix} 1 & 0 & \frac{1}{4} \\ 0 & 1 & -\frac{1}{4} \end{bmatrix}$$

So, $\text{range}(T) = \{(4s, 4t, s - t) : s, t \in R\}$.

(d) $\text{rank}(T) = \dim(\text{range}(T)) = 2$

23. (a) The kernel of T is given by the solution to the equation $T(\mathbf{x}) = \mathbf{0}$. So,

$\ker(T) = \{(-11t, 6t, 4t) : t$ is any real number$\}$.

(b) $\text{nullity}(T) = \dim(\ker(T)) = 1$

(c) Transpose A and find the equivalent reduced row-echelon form.

$$A^T = \begin{bmatrix} 0 & 4 \\ -2 & 0 \\ 3 & 11 \end{bmatrix} \Rightarrow \begin{bmatrix} 1 & 0 \\ 0 & 1 \\ 0 & 0 \end{bmatrix}$$

So, $\text{range}(T) = R^2$.

(d) $\text{rank}(T) = \dim(\text{range}(T)) = 2$

25. (a) The kernel of T is given by the solution to the equation $T(\mathbf{x}) = \mathbf{0}$. So, $\ker(T) = \{(t, -3t) : t \in R\}$.

(b) $\text{nullity}(T) = \dim(\ker(T)) = 1$

(c) Transpose A and find its equivalent row-echelon form.

$$A^T = \begin{bmatrix} \frac{9}{10} & \frac{3}{10} \\ \frac{3}{10} & \frac{1}{10} \end{bmatrix} \Rightarrow \begin{bmatrix} 3 & 1 \\ 0 & 0 \end{bmatrix}$$

So, $\text{range}(T) = \{(3t, t) : t \in R\}$.

(d) $\text{rank}(T) = \dim(\text{range}(T)) = 1$

27. (a) The kernel of T is given by the solution to the equation $T(\mathbf{x}) = \mathbf{0}$.

So,

$\ker(T) = \{(s + t, s, -2t) : s \text{ and } t \text{ are real numbers}\}$.

(b) $\text{nullity}(T) = \dim(\ker(T)) = 2$

(c) Transpose A and find the equivalent reduced row-echelon form.

$$A^T = \begin{bmatrix} \frac{4}{9} & -\frac{4}{9} & \frac{2}{9} \\ -\frac{4}{9} & \frac{4}{9} & -\frac{2}{9} \\ \frac{2}{9} & -\frac{2}{9} & \frac{1}{9} \end{bmatrix} \Rightarrow \begin{bmatrix} 1 & -1 & \frac{1}{2} \\ 0 & 0 & 0 \\ 0 & 0 & 0 \end{bmatrix}$$

So,

$\text{range}(T) = \{(2t, -2t, t) : t \text{ is any real number}\}$.

(d) $\text{rank}(T) = \dim(\text{range}(T)) = 1$

29. (a) The kernel of T is given by the solution to the equation $T(\mathbf{x}) = \mathbf{0}$.

So, $\ker(T) = \{(2s - t, t, 4s, -5s, s) : s, t \in R\}$.

(b) $\text{nullity}(T) = \dim(\ker(T)) = 2$

(c) Transpose A and find its equivalent row-echelon form.

$$A^T = \begin{bmatrix} 2 & 1 & 3 & 6 \\ 2 & 1 & 3 & 6 \\ -3 & 1 & -5 & -2 \\ 1 & 1 & 0 & 4 \\ 13 & -1 & 14 & 16 \end{bmatrix} \Rightarrow \begin{bmatrix} 7 & 0 & 0 & 8 \\ 0 & 7 & 0 & 20 \\ 0 & 0 & 7 & 2 \\ 0 & 0 & 0 & 0 \\ 0 & 0 & 0 & 0 \end{bmatrix}$$

So,

$\text{range}(T) = \{7r, 7s, 7t, 8r + 20s + 2t) : r, s, t \in R\}$.

Equivalently, the range of T is spanned by columns 1, 3 and 4 of A.

(d) $\text{rank}(T) = \dim(\text{range}(T)) = 3$

39. $\text{rank}(T) + \text{nullity}(T) = \dim R^4 \Rightarrow \text{nullity}(T) = 4 - 2 = 2$

41. $\text{rank}(T) + \text{nullity}(T) = \dim R^4 \Rightarrow \text{nullity}(T) = 4 - 0 = 4$

43.

	Zero	**Standard Basis**
(a)	$(0, 0, 0, 0)$	$\{(1, 0, 0, 0), (0, 1, 0, 0), (0, 0, 1, 0), (0, 0, 0, 1)\}$

(b) $\begin{bmatrix} 0 \\ 0 \\ 0 \\ 0 \end{bmatrix}$ $\left\{ \begin{bmatrix} 1 \\ 0 \\ 0 \\ 0 \end{bmatrix}, \begin{bmatrix} 0 \\ 1 \\ 0 \\ 0 \end{bmatrix}, \begin{bmatrix} 0 \\ 0 \\ 1 \\ 0 \end{bmatrix}, \begin{bmatrix} 0 \\ 0 \\ 0 \\ 1 \end{bmatrix} \right\}$

31. Use Theorem 6.5 to find $\text{nullity}(T)$.

$\text{rank}(T) + \text{nullity}(T) = \dim(R^3)$

$\text{nullity}(T) = 3 - 2 = 1$

Because $\text{nullity}(T) = \dim(\ker(T)) = 1$, the kernel of T is a line in space. Furthermore, because $\text{rank}(T) = \dim(\text{range}(T)) = 2$, the range of T is a plane in space.

33. Because $\text{rank}(T) + \text{nullity}(T) = 3$ and you are given $\text{rank}(T) = 0$, it follows that $\text{nullity}(T) = 3$. So, the kernel of T is all of R^3, and the range is the single point $\{(0, 0, 0)\}$.

35. The preimage of $(0, 0, 0)$ is $\{(0, 0, 0)\}$.

So, $\text{nullity}(T) = 0$, and the rank of T is determined as follows.

$\text{rank}(T) + \text{nullity}(T) = \dim(R^3)$

$\text{rank}(T) = 3 - 0 = 3$

The kernel of T is the single point $(0, 0, 0)$. Because $\text{rank}(T) = \dim(\text{range}(T)) = 3$, the range of T is R^3.

37. The kernel of T is determined by solving

$T(x, y, z) = \dfrac{x + 2y + 2z}{9}(1, 2, 2) = (0, 0, 0)$, which

implies that $x + 2y + 2z = 0$. So, the nullity of T is 2, and the kernel is a plane. The range of T is found by observing that $\text{rank}(T) + \text{nullity}(T) = 3$. That is, the range of T is 1-dimensional, a line in

R^3 and $\text{range}(T) = \{(t, 2t, 2t) : t \in R\}$.

Zero	Standard Basis

(c) $\begin{bmatrix} 0 & 0 \\ 0 & 0 \end{bmatrix}$ $\left\{ \begin{bmatrix} 1 & 0 \\ 0 & 0 \end{bmatrix}, \begin{bmatrix} 0 & 1 \\ 0 & 0 \end{bmatrix}, \begin{bmatrix} 0 & 0 \\ 1 & 0 \end{bmatrix}, \begin{bmatrix} 0 & 0 \\ 0 & 1 \end{bmatrix} \right\}$

(d) $p(x) = 0$ $\{1, x, x^2, x^3\}$

(e) $(0, 0, 0, 0, 0)$ $\{(1, 0, 0, 0, 0), (0, 1, 0, 0, 0), (0, 0, 1, 0, 0), (0, 0, 0, 1, 0)\}$

45. Solve the equation $T(p) = \dfrac{d}{dx}\left(a_0 + a_1 x + a_2 x^2 + a_3 x^3 + a_4 x^4\right) = 0$ yielding $p = a_0$.

So, $\ker(T) = \{p(x) = a_0 : a_0 \text{ is a real number}\}$. (The constant polynomials)

47. First compute $T(\mathbf{u}) = \text{proj}_v \mathbf{u}$, for $\mathbf{u} = (x, y, z)$.

$$T(\mathbf{u}) = \text{proj}_v \mathbf{u} = \frac{(x, y, z) \cdot (2, -1, 1)}{(2, -1, 1) \cdot (2, -1, 1)}(2, -1, 1) = \frac{2x - y + z}{6}(2, -1, 1)$$

(a) Setting $T(\mathbf{u}) = \mathbf{0}$, you have $2x - y + z = 0$, so $\text{nullity}(T) = 2$, and $\text{rank}(T) = 3 - 2 = 1$.

(b) A basis for the kernel of T is obtained by solving $2x - y + z = 0$.

Letting $t = z$ and $s = y$, you have $x = \dfrac{1}{2}(y - z) = \dfrac{1}{2}s - \dfrac{1}{2}t$.

So, a basis for $\ker(T)$ is $\left\{ \left(\dfrac{1}{2}, 1, 0\right), \left(-\dfrac{1}{2}, 0, 1\right) \right\}$, or $\{(1, 2, 0), (1, 0, -2)\}$.

49. Because $|A| = -1 \neq 0$, the homogeneous equation $A\mathbf{x} = \mathbf{0}$ has only the trivial solution. So, $\ker(T) = \{(0, 0)\}$ and T is one-to-one (by Theorem 6.6). Furthermore, because $\text{rank}(T) = \dim(R^2) - \text{nullity}(T) = 2 - 0 = 2 = \dim(R^2)$, T is onto (by Theorem 6.7).

51. Because $|A| = -1 \neq 0$, the homogeneous equation $A\mathbf{x} = \mathbf{0}$ has only the trivial solution. So, $\ker(T) = \{(0, 0, 0)\}$ and T is one-to-one (by Theorem 6.6). Furthermore, because $\text{rank}(T) = \dim R^3 - \text{nullity}(T) = 3 - 0 = 3 = \dim(R^3)$, T is onto (by Theorem 6.7).

53. (a) False. See "Definition of Kernel of a Linear Transformation," page 374.

(b) False. See Theorem 6.4, page 378.

(c) True. See discussion before Theorem 6.6, page 382.

(d) True. See discussion before "Definition of Isomorphism," page 384.

55. (a) A is an $n \times n$ matrix and $\det(A) = \det(A^T) \neq 0$. So, the reduced row-echelon matrix equivalent to A^T has n nonzero rows and you can conclude that $\text{rank}(T) = n$.

(b) A is an $n \times n$ matrix and $\det(A) = \det(A^T) = 0$. So, the reduced row-echelon matrix equivalent to A^T has at least one row of zeros and you can conclude that $\text{rank}(T) < n$.

57. Theorem 6.9 tells you that if $M_{m,n}$ and $M_{j,k}$ are of the same dimension then they are isomorphic. So, you can conclude that $mn = jk$.

59. From Theorem 6.5,

$\text{rank}(T) + \text{nullity}(T) = n = \text{dimension of } V.$ T is one-to-one if and only if $\text{nullity}(T) = 0$ if and only if $\text{rank}(T) = \text{dimension of } V$.

61. Although they are not the same, they have the same dimension (4) and are isomorphic.

Section 6.3   Matrices for Linear Transformations

1. Because

$$T\left(\begin{bmatrix}1\\0\end{bmatrix}\right) = \begin{bmatrix}1\\1\end{bmatrix} \quad \text{and} \quad T\left(\begin{bmatrix}0\\1\end{bmatrix}\right) = \begin{bmatrix}2\\-2\end{bmatrix},$$

the standard matrix for T is $A = \begin{bmatrix}1 & 2\\1 & -2\end{bmatrix}$.

3. Because

$$T\left(\begin{bmatrix}1\\0\end{bmatrix}\right) = \begin{bmatrix}2\\1\\-4\end{bmatrix} \quad \text{and} \quad T\left(\begin{bmatrix}0\\1\end{bmatrix}\right) = \begin{bmatrix}-3\\1\\1\end{bmatrix},$$

the standard matrix for T is $A = \begin{bmatrix}2 & -3\\1 & -1\\-4 & 1\end{bmatrix}$.

5. Because

$$T\left(\begin{bmatrix}1\\0\\0\end{bmatrix}\right) = \begin{bmatrix}1\\1\\-1\end{bmatrix}, T\left(\begin{bmatrix}0\\1\\0\end{bmatrix}\right) = \begin{bmatrix}1\\-1\\0\end{bmatrix}, \text{ and } T\left(\begin{bmatrix}0\\0\\1\end{bmatrix}\right) = \begin{bmatrix}0\\0\\1\end{bmatrix},$$

the standard matrix for T is

$$A = \begin{bmatrix}1 & 1 & 0\\1 & -1 & 0\\-1 & 0 & 1\end{bmatrix}.$$

7. Because

$$T\left(\begin{bmatrix}1\\0\\0\end{bmatrix}\right) = \begin{bmatrix}0\\4\end{bmatrix}, T\left(\begin{bmatrix}0\\1\\0\end{bmatrix}\right) = \begin{bmatrix}-2\\0\end{bmatrix},$$

and $T\left(\begin{bmatrix}0\\0\\1\end{bmatrix}\right) = \begin{bmatrix}3\\11\end{bmatrix},$

the standard matrix for T is $A = \begin{bmatrix}0 & -2 & 3\\4 & 0 & 11\end{bmatrix}$.

9. Because

$$T\left(\begin{bmatrix}1\\0\\0\\0\end{bmatrix}\right) = \begin{bmatrix}0\\0\\0\\0\end{bmatrix}, \quad T\left(\begin{bmatrix}0\\1\\0\\0\end{bmatrix}\right) = \begin{bmatrix}0\\0\\0\\0\end{bmatrix}, \quad T\left(\begin{bmatrix}0\\0\\1\\0\end{bmatrix}\right) = \begin{bmatrix}0\\0\\0\\0\end{bmatrix},$$

and $T\left(\begin{bmatrix}0\\0\\0\\1\end{bmatrix}\right) = \begin{bmatrix}0\\0\\0\\0\end{bmatrix},$

the standard matrix for T is $A = \begin{bmatrix}0 & 0 & 0 & 0\\0 & 0 & 0 & 0\\0 & 0 & 0 & 0\\0 & 0 & 0 & 0\end{bmatrix}.$

11. Because

$$T\left(\begin{bmatrix}1\\0\\0\end{bmatrix}\right) = \begin{bmatrix}13\\6\end{bmatrix}, \quad T\left(\begin{bmatrix}0\\1\\0\end{bmatrix}\right) = \begin{bmatrix}-9\\5\end{bmatrix},$$

and $T\left(\begin{bmatrix}0\\0\\1\end{bmatrix}\right) = \begin{bmatrix}4\\-3\end{bmatrix},$

the standard matrix for T is $A = \begin{bmatrix}13 & -9 & 4\\6 & 5 & -3\end{bmatrix}$.

So, $T(\mathbf{v}) = \begin{bmatrix}13 & -9 & 4\\6 & 5 & -3\end{bmatrix}\begin{bmatrix}1\\-2\\1\end{bmatrix} = \begin{bmatrix}35\\-7\end{bmatrix}$

and $T(1, -2, 1) = (35, -7)$.

13. Because

$$T\left(\begin{bmatrix}1\\0\end{bmatrix}\right) = \begin{bmatrix}1\\1\\2\\0\end{bmatrix} \quad \text{and} \quad T\left(\begin{bmatrix}0\\1\end{bmatrix}\right) = \begin{bmatrix}1\\-1\\0\\2\end{bmatrix},$$

the standard matrix for T is

$$A = \begin{bmatrix}1 & 1\\1 & -1\\2 & 0\\0 & 2\end{bmatrix}.$$

So, $T(\mathbf{v}) = \begin{bmatrix}1 & 1\\1 & -1\\2 & 0\\0 & 2\end{bmatrix}\begin{bmatrix}3\\-3\end{bmatrix} = \begin{bmatrix}0\\6\\6\\-6\end{bmatrix}$

and $T(3, -3) = (0, 6, 6, -6).$

15. Because

$$T\left(\begin{bmatrix}1\\0\\0\\0\end{bmatrix}\right) = \begin{bmatrix}1\\0\end{bmatrix}, \quad T\left(\begin{bmatrix}0\\1\\0\\0\end{bmatrix}\right) = \begin{bmatrix}1\\0\end{bmatrix}, \quad T\left(\begin{bmatrix}0\\0\\1\\0\end{bmatrix}\right) = \begin{bmatrix}0\\1\end{bmatrix}, \text{ and}$$

$$T\left(\begin{bmatrix}0\\0\\0\\1\end{bmatrix}\right) = \begin{bmatrix}0\\1\end{bmatrix},$$

the standard matrix for T is

$$A = \begin{bmatrix}1 & 1 & 0 & 0\\0 & 0 & 1 & 1\end{bmatrix}.$$

So, $T(\mathbf{v}) = \begin{bmatrix}1 & 1 & 0 & 0\\0 & 0 & 1 & 1\end{bmatrix}\begin{bmatrix}1\\-1\\1\\-1\end{bmatrix} = \begin{bmatrix}0\\0\end{bmatrix}$

and $T(1, -1, 1, -1) = (0, 0)$.

17. (a) The matrix of the reflection through the origin,

$T(x, y) = (-x, -y)$, is given by

$$A = \begin{bmatrix}T(1, 0) \vdots T(0, 1)\end{bmatrix} = \begin{bmatrix}-1 & 0\\0 & -1\end{bmatrix}.$$

(b) The image of $\mathbf{v} = (3, 4)$ is given by

$$A\mathbf{v} = \begin{bmatrix}-1 & 0\\0 & -1\end{bmatrix}\begin{bmatrix}3\\4\end{bmatrix} = \begin{bmatrix}-3\\-4\end{bmatrix}.$$

So, $T(3, 4) = (-3, -4)$.

(c)

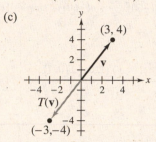

19. (a) The matrix of the reflection in the y-axis, $T(x, y) = (-x, y)$, is given by $A = \begin{bmatrix}T(1, 0) \vdots T(0, 1)\end{bmatrix} = \begin{bmatrix}-1 & 0\\0 & 1\end{bmatrix}.$

(b) The image of $\mathbf{v} = (2, -3)$ is given by $A\mathbf{v} = \begin{bmatrix}-1 & 0\\0 & 1\end{bmatrix}\begin{bmatrix}2\\-3\end{bmatrix} = \begin{bmatrix}-2\\-3\end{bmatrix}.$ So, $T(2, -3) = (-2, -3)$.

(c)

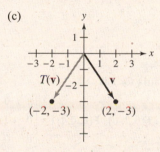

21. (a) The counterclockwise rotation of $135°$ in R^2 is given by

$$T(x, y) = \left(\cos(135)x - \sin(135)y, \sin(135)x + \cos(135)y\right) = \left(-\frac{\sqrt{2}}{2}x - \frac{\sqrt{2}}{2}y, \frac{\sqrt{2}}{2}x - \frac{\sqrt{2}}{2}y\right).$$

So, the matrix is $A = \begin{bmatrix}T(1, 0) \vdots T(0, 1)\end{bmatrix} = \begin{bmatrix}-\dfrac{\sqrt{2}}{2} & -\dfrac{\sqrt{2}}{2}\\[2mm] \dfrac{\sqrt{2}}{2} & -\dfrac{\sqrt{2}}{2}\end{bmatrix}.$

(b) The image of $\mathbf{v} = (4, 4)$ is given by $A\mathbf{v} = \begin{bmatrix}-\dfrac{\sqrt{2}}{2} & -\dfrac{\sqrt{2}}{2}\\[2mm] \dfrac{\sqrt{2}}{2} & -\dfrac{\sqrt{2}}{2}\end{bmatrix}\begin{bmatrix}4\\4\end{bmatrix} = \begin{bmatrix}-4\sqrt{2}\\0\end{bmatrix}.$ So, $T(4, 4) = \left(-4\sqrt{2}, 0\right)$.

(c)

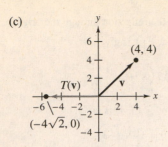

23. (a) The clockwise rotation of 60° is given by

$$T(x, \, y) = \big(\cos(-60)x - \sin(-60)y, \; \sin(-60)x + \cos(-60)y\big) = \left(\frac{1}{2}x + \frac{\sqrt{3}}{2}y, \; -\frac{\sqrt{3}}{2} + \frac{1}{2}y\right).$$

So, the matrix is $A = \big[\, T(1, 0) \vdots T(0, 1) \,\big] = \begin{bmatrix} \dfrac{1}{2} & \dfrac{\sqrt{3}}{2} \\[2mm] -\dfrac{\sqrt{3}}{2} & \dfrac{1}{2} \end{bmatrix}.$

(b) The image of $\mathbf{v} = (1, 2)$ is given by $A\mathbf{v} = \begin{bmatrix} \dfrac{1}{2} & \dfrac{\sqrt{3}}{2} \\[2mm] -\dfrac{\sqrt{3}}{2} & \dfrac{1}{2} \end{bmatrix} \begin{bmatrix} 1 \\ 2 \end{bmatrix} = \begin{bmatrix} \dfrac{1}{2} + \sqrt{3} \\[2mm] 1 - \dfrac{\sqrt{3}}{2} \end{bmatrix}.$ So, $T(1, 2) = \left(\dfrac{1}{2} + \sqrt{3}, 1 - \dfrac{\sqrt{3}}{2}\right).$

(c)

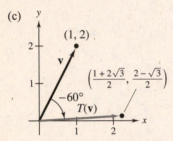

25. (a) The standard matrix for T is

$$A = \big[\, T(1, 0, 0) \vdots T(0, 1, 0) \vdots T(0, 0, 1) \,\big] = \begin{bmatrix} 1 & 0 & 0 \\ 0 & 1 & 0 \\ 0 & 0 & -1 \end{bmatrix}.$$

(b) The image of $\mathbf{v} = (3, 2, 2)$ is

$$A\mathbf{v} = \begin{bmatrix} 1 & 0 & 0 \\ 0 & 1 & 0 \\ 0 & 0 & -1 \end{bmatrix} \begin{bmatrix} 3 \\ 2 \\ 2 \end{bmatrix} = \begin{bmatrix} 3 \\ 2 \\ -2 \end{bmatrix}.$$

So, $T(3, 2, 2) = (3, 2, -2).$

(c)

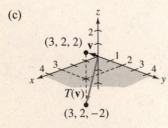

27. (a) The standard matrix for T is

$$A = \big[\, T(1, 0, 0) \vdots T(0, 1, 0) \vdots T(0, 0, 1) \,\big] = \begin{bmatrix} 1 & 0 & 0 \\ 0 & -1 & 0 \\ 0 & 0 & 1 \end{bmatrix}.$$

(b) The image of $\mathbf{v} = (1, 2, -1)$ is

$$A\mathbf{v} = \begin{bmatrix} 1 & 0 & 0 \\ 0 & -1 & 0 \\ 0 & 0 & 1 \end{bmatrix} \begin{bmatrix} 1 \\ 2 \\ -1 \end{bmatrix} = \begin{bmatrix} 1 \\ -2 \\ -1 \end{bmatrix}.$$

So, $T(1, 2, -1) = (1, -2, -1).$

(c)

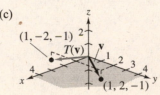

29. (a) The counterclockwise rotation of 30° is given by

$$T(x, y) = \left(\cos(30)x - \sin(30)y, \sin(30)x + \cos(30)y\right) = \left(\frac{\sqrt{3}}{2}x - \frac{1}{2}y, \frac{1}{2}x + \frac{\sqrt{3}}{2}y\right).$$

So, the matrix is $A = \left[T(1,0) \vdots T(0,1)\right] = \begin{bmatrix} \frac{\sqrt{3}}{2} & -\frac{1}{2} \\ \frac{1}{2} & \frac{\sqrt{3}}{2} \end{bmatrix}.$

(b) The image of $\mathbf{v} = (1, 2)$ is $A\mathbf{v} = \begin{bmatrix} \frac{\sqrt{3}}{2} & -\frac{1}{2} \\ \frac{1}{2} & \frac{\sqrt{3}}{2} \end{bmatrix}\begin{bmatrix} 1 \\ 2 \end{bmatrix} = \begin{bmatrix} \frac{\sqrt{3}}{2} - 1 \\ \frac{1}{2} + \sqrt{3} \end{bmatrix}.$ So, $T(1, 2) = \left(\frac{\sqrt{3}}{2} - 1, \frac{1}{2} + \sqrt{3}\right).$

(c)

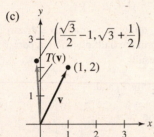

31. (a) The projection onto the vector $\mathbf{w} = (3, 1)$ is given by $T(\mathbf{v}) = \text{proj}_{\mathbf{w}}\mathbf{v} = \frac{3x + y}{10}(3, 1) = \left(\frac{3}{10}(3x + y), \frac{1}{10}(3x + y)\right).$

So, the matrix is $A = \left[T(1,0) \vdots T(0,1)\right] = \begin{bmatrix} \frac{9}{10} & \frac{3}{10} \\ \frac{3}{10} & \frac{1}{10} \end{bmatrix}.$

(b) The image of $\mathbf{v} = (1, 4)$ is given by $A\mathbf{v} = \begin{bmatrix} \frac{9}{10} & \frac{3}{10} \\ \frac{3}{10} & \frac{1}{10} \end{bmatrix}\begin{bmatrix} 1 \\ 4 \end{bmatrix} = \begin{bmatrix} \frac{21}{10} \\ \frac{7}{10} \end{bmatrix}.$ So, $T(1, 4) = \left(\frac{21}{10}, \frac{7}{10}\right).$

(c)

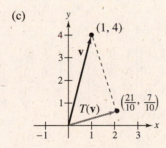

33. (a) The reflection of a vector **v** through **w** is given by

$$T(\mathbf{v}) = 2 \operatorname{proj}_{\mathbf{w}} \mathbf{v} - \mathbf{v}$$

$$T(x, y) = 2 \frac{3x + y}{10} (3, 1) - (x, y)$$

$$= \left(\frac{4}{5}x + \frac{3}{5}y, \frac{3}{5}x - \frac{4}{5}y \right).$$

The standard matrix for T is

$$A = \begin{bmatrix} T(1, 0) \vdots T(0, 1) \end{bmatrix} = \begin{bmatrix} \frac{4}{5} & \frac{3}{5} \\ \frac{3}{5} & -\frac{4}{5} \end{bmatrix}.$$

(b) The image of $\mathbf{v} = (1, 4)$ is

$$A\mathbf{v} = \begin{bmatrix} \frac{4}{5} & \frac{3}{5} \\ \frac{3}{5} & -\frac{4}{5} \end{bmatrix} \begin{bmatrix} 1 \\ 4 \end{bmatrix} = \begin{bmatrix} \frac{16}{5} \\ -\frac{13}{5} \end{bmatrix}.$$

So, $T(1, 4) = \left(\frac{16}{5}, -\frac{13}{5} \right)$.

(c)
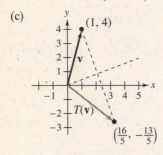

35. (a) The standard matrix for T is

$$A = \begin{bmatrix} 2 & 3 & -1 \\ 3 & 0 & -2 \\ 2 & -1 & 1 \end{bmatrix}$$

(b) The image of $\mathbf{v} = (1, 2, -1)$ is

$$A\mathbf{v} = \begin{bmatrix} 2 & 3 & -1 \\ 3 & 0 & -2 \\ 2 & -1 & 1 \end{bmatrix} \begin{bmatrix} 1 \\ 2 \\ -1 \end{bmatrix} = \begin{bmatrix} 9 \\ 5 \\ -1 \end{bmatrix}.$$

So, $T(1, 2, -1) = (9, 5, -1)$.

(c) Using a graphing utility or a computer software program to perform the multiplication in part (b) gives the same result.

37. (a) The standard matrix for T is

$$A = \begin{bmatrix} 1 & -1 & 0 & 0 \\ 0 & 0 & 1 & 0 \\ 1 & 2 & 0 & -1 \\ 0 & 0 & 0 & 1 \end{bmatrix}.$$

(b) The image of $\mathbf{v} = (1, 0, 1, -1)$ is

$$A\mathbf{v} = \begin{bmatrix} 1 & -1 & 0 & 0 \\ 0 & 0 & 1 & 0 \\ 1 & 2 & 0 & -1 \\ 0 & 0 & 0 & 1 \end{bmatrix} \begin{bmatrix} 1 \\ 0 \\ 1 \\ -1 \end{bmatrix} = \begin{bmatrix} 1 \\ 1 \\ 2 \\ -1 \end{bmatrix}.$$

So, $T(1, 0, 1, -1) = (1, 1, 2, -1)$.

(c) Using a graphing utility or a computer software program to perform the multiplication in part (b) gives the same result.

39. The standard matrices for T_1 and T_2 are

$$A_1 = \begin{bmatrix} 1 & -2 \\ 2 & 3 \end{bmatrix} \quad \text{and} \quad A_2 = \begin{bmatrix} 2 & 0 \\ 1 & -1 \end{bmatrix}.$$

The standard matrix for $T = T_2 \circ T_1$ is

$$A = A_2 A_1 = \begin{bmatrix} 2 & 0 \\ 1 & -1 \end{bmatrix} \begin{bmatrix} 1 & -2 \\ 2 & 3 \end{bmatrix} = \begin{bmatrix} 2 & -4 \\ -1 & -5 \end{bmatrix}$$

and the standard matrix for $T' = T_1 \circ T_2$ is

$$A' = A_1 A_2 = \begin{bmatrix} 1 & -2 \\ 2 & 3 \end{bmatrix} \begin{bmatrix} 2 & 0 \\ 1 & -1 \end{bmatrix} = \begin{bmatrix} 0 & 2 \\ 7 & -3 \end{bmatrix}.$$

41. The standard matrices for T_1 and T_2 are

$$A_1 = \begin{bmatrix} 1 & 0 & 0 \\ 0 & 1 & 0 \\ 0 & 0 & 1 \end{bmatrix} \quad \text{and} \quad A_2 = \begin{bmatrix} 0 & 0 & 0 \\ 1 & 0 & 0 \\ 0 & 0 & 0 \end{bmatrix}.$$

The standard matrix for $T = T_2 \circ T_1$ is

$$A = A_2 A_1 = \begin{bmatrix} 0 & 0 & 0 \\ 1 & 0 & 0 \\ 0 & 0 & 0 \end{bmatrix} \begin{bmatrix} 1 & 0 & 0 \\ 0 & 1 & 0 \\ 0 & 0 & 1 \end{bmatrix} = \begin{bmatrix} 0 & 0 & 0 \\ 1 & 0 & 0 \\ 0 & 0 & 0 \end{bmatrix} = A_2$$

and the standard matrix for $T' = T_1 \circ T_2$ is

$$A' = A_1 A_2 = \begin{bmatrix} 1 & 0 & 0 \\ 0 & 1 & 0 \\ 0 & 0 & 1 \end{bmatrix} \begin{bmatrix} 0 & 0 & 0 \\ 1 & 0 & 0 \\ 0 & 0 & 0 \end{bmatrix} = \begin{bmatrix} 0 & 0 & 0 \\ 1 & 0 & 0 \\ 0 & 0 & 0 \end{bmatrix} = A_2.$$

43. The standard matrices for T_1 and T_2 are

$$A_1 = \begin{bmatrix} -1 & 2 \\ 1 & 1 \\ 1 & -1 \end{bmatrix} \quad \text{and} \quad A_2 = \begin{bmatrix} 1 & -3 & 0 \\ 3 & 0 & 1 \end{bmatrix}.$$

The standard matrix for $T = T_2 \circ T_1$ is

$$A = A_2 A_1 = \begin{bmatrix} 1 & -3 & 0 \\ 3 & 0 & 1 \end{bmatrix}\begin{bmatrix} -1 & 2 \\ 1 & 1 \\ 1 & -1 \end{bmatrix} = \begin{bmatrix} -4 & -1 \\ -2 & 5 \end{bmatrix}$$

and the standard matrix for $T' = T_1 \circ T_2$ is

$$A' = A_1 A_2 = \begin{bmatrix} -1 & 2 \\ 1 & 1 \\ 1 & -1 \end{bmatrix}\begin{bmatrix} 1 & -3 & 0 \\ 3 & 0 & 1 \end{bmatrix} = \begin{bmatrix} 5 & 3 & 2 \\ 4 & -3 & 1 \\ -2 & -3 & -1 \end{bmatrix}.$$

45. The standard matrix for T is

$$A = \begin{bmatrix} 1 & 1 \\ 1 & -1 \end{bmatrix}.$$

Because $|A| = -2 \neq 0$, A is invertible.

$$A^{-1} = -\frac{1}{2}\begin{bmatrix} -1 & -1 \\ -1 & 1 \end{bmatrix} = \begin{bmatrix} \frac{1}{2} & \frac{1}{2} \\ \frac{1}{2} & -\frac{1}{2} \end{bmatrix}$$

So, $T^{-1}(x, y) = \left(\frac{1}{2}x + \frac{1}{2}y, \frac{1}{2}x - \frac{1}{2}y\right).$

47. The standard matrix for T is

$$A = \begin{bmatrix} 1 & 0 & 0 \\ 1 & 1 & 0 \\ 1 & 1 & 1 \end{bmatrix}.$$

Because $|A| = 1 \neq 0$, A is invertible. Calculate A^{-1} by Gauss-Jordan elimination

$$A^{-1} = \begin{bmatrix} 1 & 0 & 0 \\ -1 & 1 & 0 \\ 0 & -1 & 1 \end{bmatrix}$$

and conclude that

$$T^{-1}(x_1, x_2, x_3) = (x_1, -x_1 + x_2, -x_2 + x_3).$$

49. The standard matrix for T is

$$A = \begin{bmatrix} 2 & 0 \\ 0 & 0 \end{bmatrix}.$$

Because $|A| = 0$, A is not invertible, and so T is not invertible.

51. The standard matrix for T is

$$A = \begin{bmatrix} 1 & 1 \\ 3 & 3 \end{bmatrix}.$$

Because $|A| = 0$, A is not invertible, and so T is not invertible.

53. The standard matrix for T is $A = \begin{bmatrix} 5 & 0 \\ 0 & 5 \end{bmatrix}.$

Because $|A| = 25 \neq 0$, A is invertible.

$$A^{-1} = \begin{bmatrix} \frac{1}{5} & 0 \\ 0 & \frac{1}{5} \end{bmatrix} \quad \text{So, } T^{-1}(x, y) = \left(\frac{x}{5}, \frac{y}{5}\right).$$

55. The standard matrix for T is

$$A = \begin{bmatrix} 1 & -2 & 0 & 0 \\ 0 & 1 & 0 & 0 \\ 0 & 0 & 1 & 1 \\ 0 & 0 & 1 & 0 \end{bmatrix}.$$

Because $|A| = -1 \neq 0$, A is invertible. Calculate A^{-1} by Gauss-Jordan elimination

$$A^{-1} = \begin{bmatrix} 1 & 2 & 0 & 0 \\ 0 & 1 & 0 & 0 \\ 0 & 0 & 0 & 1 \\ 0 & 0 & 1 & -1 \end{bmatrix}$$

and conclude that

$$T^{-1}(x_1, x_2, x_3, x_4) = (x_1 + 2x_2, x_2, x_4, x_3 - x_4).$$

57. (a) The standard matrix for T is $A' = \begin{bmatrix} 1 & 1 \\ 1 & 0 \\ 0 & 1 \end{bmatrix}$

and the image of $\mathbf{v}$ under T is

$$A'\mathbf{v} = \begin{bmatrix} 1 & 1 \\ 1 & 0 \\ 0 & 1 \end{bmatrix}\begin{bmatrix} 5 \\ 4 \end{bmatrix} = \begin{bmatrix} 9 \\ 5 \\ 4 \end{bmatrix}. \quad \text{So, } T(\mathbf{v}) = (9, 5, 4).$$

(b) The image of each vector in B is as follows.

$$T(1, -1) = (0, 1, -1) = (1, 1, 0) + 0(0, 1, 1) - (1, 0, 1)$$
$$T(0, 1) = (1, 0, 1) = 0(1, 1, 0) + 0(0, 1, 1) + (1, 0, 1)$$

So, $[T(1, -1)]_{B'} = \begin{bmatrix} 1 \\ 0 \\ -1 \end{bmatrix}$ and $[T(0, 1)]_{B'} = \begin{bmatrix} 0 \\ 0 \\ 1 \end{bmatrix}$

which implies that $A = \begin{bmatrix} 1 & 0 \\ 0 & 0 \\ -1 & 1 \end{bmatrix}$. Then, because

$[\mathbf{v}]_B = \begin{bmatrix} 5 \\ 9 \end{bmatrix}$, you have

$$[T(\mathbf{v})]_{B'} = A[\mathbf{v}]_B = \begin{bmatrix} 1 & 0 \\ 0 & 0 \\ -1 & 1 \end{bmatrix}\begin{bmatrix} 5 \\ 9 \end{bmatrix} = \begin{bmatrix} 5 \\ 0 \\ 4 \end{bmatrix}.$$

So, $T(\mathbf{v}) = 5(1, 1, 0) + 4(1, 0, 1) = (9, 5, 4).$

59. (a) The standard matrix for T is $A' = \begin{bmatrix} 1 & -1 & 0 \\ 0 & 1 & -1 \end{bmatrix}$ and the image of $\mathbf{v}$ under T is $A'\mathbf{v} = \begin{bmatrix} 1 & -1 & 0 \\ 0 & 1 & -1 \end{bmatrix} \begin{bmatrix} 1 \\ 2 \\ -3 \end{bmatrix} = \begin{bmatrix} -1 \\ 5 \end{bmatrix}$.

So, $T(\mathbf{v}) = (-1, 5)$.

(b) The image of each vector in B is as follows.

$T(1, 1, 1) = (0, 0) = 0(1, 2) + 0(1, 1)$

$T(1, 1, 0) = (0, 1) = (1, 2) - (1, 1)$

$T(0, 1, 1) = (-1, 0) = (1, 2) - 2(1, 1)$

So, $\left[T(1, 1, 1) \right]_{B'} = \begin{bmatrix} 0 \\ 0 \end{bmatrix}$, $\left[T(1, 1, 0) \right]_{B'} = \begin{bmatrix} 1 \\ -1 \end{bmatrix}$, and $\left[T(0, 1, 1) \right]_{B'} = \begin{bmatrix} 1 \\ -2 \end{bmatrix}$, which implies that $A = \begin{bmatrix} 0 & 1 & 1 \\ 0 & -1 & -2 \end{bmatrix}$.

Then, because $\left[\mathbf{v} \right]_B = \begin{bmatrix} -4 \\ 5 \\ 1 \end{bmatrix}$, you have $\left[T(\mathbf{v}) \right]_{B'} = A\left[\mathbf{v} \right]_B = \begin{bmatrix} 0 & 1 & 1 \\ 0 & -1 & -2 \end{bmatrix} \begin{bmatrix} -4 \\ 5 \\ 1 \end{bmatrix} = \begin{bmatrix} 6 \\ -7 \end{bmatrix}$.

So, $T(\mathbf{v}) = 6(1, 2) - 7(1, 1) = (-1, 5)$.

61. (a) The standard matrix for T is $A' = \begin{bmatrix} 2 & 0 & 0 \\ 1 & 1 & 0 \\ 0 & 1 & 1 \\ 1 & 0 & 1 \end{bmatrix}$ and the image of $\mathbf{v} = (1, -5, 2)$ under T is $A'\mathbf{v} = \begin{bmatrix} 2 & 0 & 0 \\ 1 & 1 & 0 \\ 0 & 1 & 1 \\ 1 & 0 & 1 \end{bmatrix} \begin{bmatrix} 1 \\ -5 \\ 2 \end{bmatrix} = \begin{bmatrix} 2 \\ -4 \\ -3 \\ 3 \end{bmatrix}$.

So, $T(\mathbf{v}) = (2, -4, -3, 3)$.

(b) Because

$T(2, 0, 1) = (4, 2, 1, 3) = 2(1, 0, 0, 1) + (0, 1, 0, 1) + (1, 0, 1, 0) + (1, 1, 0, 0)$

$T(0, 2, 1) = (0, 2, 3, 1) = -2(1, 0, 0, 1) + 3(0, 1, 0, 1) + 3(1, 0, 1, 0) - (1, 1, 0, 0)$

$T(1, 2, 1) = (2, 3, 3, 2) = -(1, 0, 0, 1) + 3(0, 1, 0, 1) + 3(1, 0, 1, 0)$

the matrix for T relative to B and B' is

$A = \begin{bmatrix} 2 & -2 & -1 \\ 1 & 3 & 3 \\ 1 & 3 & 3 \\ 1 & -1 & 0 \end{bmatrix}$.

Because $\mathbf{v} = (1, -5, 2) = \frac{9}{2}(2, 0, 1) + \frac{11}{2}(0, 2, 1) - 8(1, 2, 1)$,

$\left[T(\mathbf{v}) \right]_{B'} = A\left[\mathbf{v} \right]_B = \begin{bmatrix} 2 & -2 & -1 \\ 1 & 3 & 3 \\ 1 & 3 & 3 \\ 1 & -1 & 0 \end{bmatrix} \begin{bmatrix} \frac{9}{2} \\ \frac{11}{2} \\ -8 \end{bmatrix} = \begin{bmatrix} 6 \\ -3 \\ -3 \\ -1 \end{bmatrix}$.

So, $T(1, -5, 2) = 6(1, 0, 0, 1) - 3(0, 1, 0, 1) - 3(1, 0, 1, 0) - (1, 1, 0, 0) = (2, -4, -3, 3)$.

63. (a) The standard matrix for T is $A' = \begin{bmatrix} 1 & 1 & 1 \\ -1 & 0 & 2 \\ 0 & 2 & -1 \end{bmatrix}$ and the image of $\mathbf{v} = (4, -5, 10)$ under T is

$A'\mathbf{v} = \begin{bmatrix} 1 & 1 & 1 \\ -1 & 0 & 2 \\ 0 & 2 & -1 \end{bmatrix} \begin{bmatrix} 4 \\ -5 \\ 10 \end{bmatrix} = \begin{bmatrix} 9 \\ 16 \\ -20 \end{bmatrix}$.

So, $T(\mathbf{v}) = (9, 16, -20)$.

(b) Because

$$T(2, 0, 1) = (3, 0, -1) = 2(1, 1, 1) + 1(1, 1, 0) - 3(0, 1, 1)$$

$$T(0, 2, 1) = (3, 2, 3) = 4(1, 1, 1) - (1, 1, 0) - (0, 1, 1)$$

$$T(1, 2, 1) = (4, 1, 3) = 6(1, 1, 1) - 2(1, 1, 0) - 3(0, 1, 1),$$

the matrix for T relative to B and B' is $A = \begin{bmatrix} 2 & 4 & 6 \\ 1 & -1 & -2 \\ -3 & -1 & -3 \end{bmatrix}$. Because $\mathbf{v} = (4, -5, 10) = \frac{25}{2}(2, 0, 1) + \frac{37}{2}(0, 2, 1) - 21(1, 2, 1),$

$$[T(\mathbf{v})]_{B'} = A[\mathbf{v}]_B = \begin{bmatrix} 2 & 4 & 6 \\ 1 & -1 & -2 \\ -3 & -1 & -3 \end{bmatrix} \begin{bmatrix} \frac{25}{2} \\ \frac{37}{2} \\ -21 \end{bmatrix} = \begin{bmatrix} -27 \\ 36 \\ 7 \end{bmatrix}.$$ So, $T(\mathbf{v}) = -27(1, 1, 1) + 36(1, 1, 0) + 7(0, 1, 1) = (9, 16, -20)$.

65. The image of each vector in B is as follows.

$$T(1) = x, \quad T(x) = x^2, \quad T(x^2) = x^3.$$

So, the matrix of T relative to B and B' is

$$A = \begin{bmatrix} 0 & 0 & 0 \\ 1 & 0 & 0 \\ 0 & 1 & 0 \\ 0 & 0 & 1 \end{bmatrix}.$$

67. The image of each vector in B is as follows.

$$D_x(1) = 0 = 0(1) + 0x + 0e^x + 0xe^x$$

$$D_x(x) = 1 = 1 + 0x + 0e^x + 0xe^x$$

$$D_x(e^x) = e^x = 0(1) + 0x + e^x + 0xe^x$$

$$D_x(xe^x) = e^x + xe^x = 0(1) + 0x + e^x + xe^x$$

So,

$$[D_x(1)]_B = \begin{bmatrix} 0 \\ 0 \\ 0 \\ 0 \end{bmatrix}, \quad [D_x(x)]_B = \begin{bmatrix} 1 \\ 0 \\ 0 \\ 0 \end{bmatrix},$$

$$[D_x(e^x)]_B = \begin{bmatrix} 0 \\ 0 \\ 1 \\ 0 \end{bmatrix}, \quad \text{and} \quad [D_x(xe^x)]_B = \begin{bmatrix} 0 \\ 0 \\ 1 \\ 1 \end{bmatrix},$$

which implies that

$$A = \begin{bmatrix} 0 & 1 & 0 & 0 \\ 0 & 0 & 0 & 0 \\ 0 & 0 & 1 & 1 \\ 0 & 0 & 0 & 1 \end{bmatrix}.$$

69. Because $3x - 2xe^x = 0(1) + 3(x) + 0(e^x) - 2(xe^x)$,

$$A[\mathbf{v}]_B = \begin{bmatrix} 0 & 1 & 0 & 0 \\ 0 & 0 & 0 & 0 \\ 0 & 0 & 1 & 1 \\ 0 & 0 & 0 & 1 \end{bmatrix} \begin{bmatrix} 0 \\ 3 \\ 0 \\ -2 \end{bmatrix} = \begin{bmatrix} 3 \\ 0 \\ -2 \\ -2 \end{bmatrix}.$$

So, $D_x(3x - 2xe^x) = 3 - 2e^x - 2xe^x$.

71. (a) The image of each vector in B is as follows.

$$T(1) = \int_0^x dt$$

$$= x = 0(1) + x + 0x^2 + 0x^3 + 0x^4$$

$$T(x) = \int_0^x t\, dt$$

$$= \frac{1}{2}x^2 = 0(1) + 0x + \frac{1}{2}x^2 + 0x^3 + 0x^4$$

$$T(x^2) = \int_0^x t^2\, dt$$

$$= \frac{1}{3}x^3 = 0(1) + 0x + 0x^2 + \frac{1}{3}x^3 + 0x^4$$

$$T(x^3) = \int_0^x t^3\, dt$$

$$= \frac{1}{4}x^4 = 0(1) + 0(x) + 0x^2 + 0x^3 + \frac{1}{4}x^4$$

So,

$$A = \begin{bmatrix} 0 & 0 & 0 & 0 \\ 1 & 0 & 0 & 0 \\ 0 & \frac{1}{2} & 0 & 0 \\ 0 & 0 & \frac{1}{3} & 0 \\ 0 & 0 & 0 & \frac{1}{4} \end{bmatrix}.$$

(b) The image of $p(x) = 6 - 2x + 3x^3$ under T relative to the basis of B' is given by

$$A[p]_B = \begin{bmatrix} 0 & 0 & 0 & 0 \\ 1 & 0 & 0 & 0 \\ 0 & \frac{1}{2} & 0 & 0 \\ 0 & 0 & \frac{1}{3} & 0 \\ 0 & 0 & 0 & \frac{1}{4} \end{bmatrix} \begin{bmatrix} 6 \\ -2 \\ 0 \\ 3 \end{bmatrix} = \begin{bmatrix} 0 \\ 6 \\ -1 \\ 0 \\ \frac{3}{4} \end{bmatrix}.$$

So, $T(p) = 0(1) + 6x - x^2 + 0x^3 + \frac{3}{4}x^4$

$$= 6x - x^2 + \frac{3}{4}x^4 = \int_0^x p(t)\, dt.$$

73. (a) True. See discussion, under "Composition of Linear Transformations," pages 390–391.

(b) False. See Example 3, page 392.

(c) False. See Theorem 6.12, page 393.

75. The standard basis for $M_{2,3}$ is

$$B = \left\{ \begin{bmatrix} 1 & 0 & 0 \\ 0 & 0 & 0 \end{bmatrix}, \begin{bmatrix} 0 & 1 & 0 \\ 0 & 0 & 0 \end{bmatrix}, \begin{bmatrix} 0 & 0 & 1 \\ 0 & 0 & 0 \end{bmatrix}, \begin{bmatrix} 0 & 0 & 0 \\ 1 & 0 & 0 \end{bmatrix}, \begin{bmatrix} 0 & 0 & 0 \\ 0 & 1 & 0 \end{bmatrix}, \begin{bmatrix} 0 & 0 & 0 \\ 0 & 0 & 1 \end{bmatrix} \right\}$$

and the standard basis for $M_{3,2}$ is

$$B' = \left\{ \begin{bmatrix} 1 & 0 \\ 0 & 0 \\ 0 & 0 \end{bmatrix}, \begin{bmatrix} 0 & 1 \\ 0 & 0 \\ 0 & 0 \end{bmatrix}, \begin{bmatrix} 0 & 0 \\ 1 & 0 \\ 0 & 0 \end{bmatrix}, \begin{bmatrix} 0 & 0 \\ 0 & 1 \\ 0 & 0 \end{bmatrix}, \begin{bmatrix} 0 & 0 \\ 0 & 0 \\ 1 & 0 \end{bmatrix}, \begin{bmatrix} 0 & 0 \\ 0 & 0 \\ 0 & 1 \end{bmatrix} \right\}.$$

By finding the image of each vector in B, you can find A.

$$T\left(\begin{bmatrix} 1 & 0 & 0 \\ 0 & 0 & 0 \end{bmatrix} \right) = \begin{bmatrix} 1 & 0 \\ 0 & 0 \\ 0 & 0 \end{bmatrix}, \quad T\left(\begin{bmatrix} 0 & 1 & 0 \\ 0 & 0 & 0 \end{bmatrix} \right) = \begin{bmatrix} 0 & 0 \\ 1 & 0 \\ 0 & 0 \end{bmatrix},$$

$$T\left(\begin{bmatrix} 0 & 0 & 1 \\ 0 & 0 & 0 \end{bmatrix} \right) = \begin{bmatrix} 0 & 0 \\ 0 & 0 \\ 1 & 0 \end{bmatrix}, \quad T\left(\begin{bmatrix} 0 & 0 & 0 \\ 1 & 0 & 0 \end{bmatrix} \right) = \begin{bmatrix} 0 & 1 \\ 0 & 0 \\ 0 & 0 \end{bmatrix},$$

$$T\left(\begin{bmatrix} 0 & 0 & 0 \\ 0 & 1 & 0 \end{bmatrix} \right) = \begin{bmatrix} 0 & 0 \\ 0 & 1 \\ 0 & 0 \end{bmatrix}, \quad T\left(\begin{bmatrix} 0 & 0 & 0 \\ 0 & 0 & 1 \end{bmatrix} \right) = \begin{bmatrix} 0 & 0 \\ 0 & 0 \\ 0 & 1 \end{bmatrix}.$$

So,

$$A = \begin{bmatrix} 1 & 0 & 0 & 0 & 0 & 0 \\ 0 & 0 & 0 & 1 & 0 & 0 \\ 0 & 1 & 0 & 0 & 0 & 0 \\ 0 & 0 & 0 & 0 & 1 & 0 \\ 0 & 0 & 1 & 0 & 0 & 0 \\ 0 & 0 & 0 & 0 & 0 & 1 \end{bmatrix}.$$

77. Let $(T_2 \circ T_1)(\mathbf{u}) = (T_2 \circ T_1)(\mathbf{v})$

$\qquad T_2(T_1(\mathbf{u})) = T_2(T_1(\mathbf{v}))$

$\qquad\quad T_1(\mathbf{u}) = T_1(\mathbf{v})$ $\qquad$ because T_2 one-to-one

$\qquad\qquad \mathbf{u} = \mathbf{v}$ $\qquad$ because T_1 one-to-one

Because $T_2 \circ T_1$ is one-to-one from V to V, it is also onto. The inverse is $T_1^{-1} \circ T_2^{-1}$ because

$$(T_2 \circ T_1) \circ (T_1^{-1} \circ T_2^{-1}) = T_2 \circ I \circ T_2^{-1} = I.$$

79. Sometimes it is preferable to use a nonstandard basis. If A_1 and A_2 are the standard matrices for T_1 and T_2 respectively, then the standard matrix for $T_2 \circ T_1$ is $A_2 A_1$ and the standard matrix for $T_1^{-1} \circ T_2^{-1}$ is $A_1^{-1} A_2^{-1}$. Because $(A_2 A_1)(A_1^{-1} A_2^{-1}) = A_2(I) = A_2^{-1} = I$, you have that the inverse of $T_2 \circ T_1$ is $T_1^{-1} \circ T_2^{-1}$.

Section 6.4 Transition Matrices and Similarity

1. (a) The standard matrix for T is

$$A = \begin{bmatrix} 2 & -1 \\ -1 & 1 \end{bmatrix}.$$

Furthermore, the transition matrix P from B' to the standard basis B, and its inverse, are

$$P = \begin{bmatrix} 1 & 0 \\ -2 & 3 \end{bmatrix} \quad \text{and} \quad P^{-1} = \begin{bmatrix} 1 & 0 \\ \frac{2}{3} & \frac{1}{3} \end{bmatrix}.$$

Therefore, the matrix for T relative to B' is

$$A' = P^{-1}AP = \begin{bmatrix} 1 & 0 \\ \frac{2}{3} & \frac{1}{3} \end{bmatrix}\begin{bmatrix} 2 & -1 \\ -1 & 1 \end{bmatrix}\begin{bmatrix} 1 & 0 \\ -2 & 3 \end{bmatrix} = \begin{bmatrix} 4 & -3 \\ \frac{5}{3} & -1 \end{bmatrix}.$$

(b) Because $A' = P^{-1}AP$, it follows that A and A' are similar.

3. (a) The standard matrix for T is

$$A = \begin{bmatrix} 1 & 1 \\ 0 & 4 \end{bmatrix}.$$

Furthermore, the transition matrix P from B' to the standard basis B, and its inverse, are

$$P = \begin{bmatrix} -4 & 1 \\ 1 & -1 \end{bmatrix} \quad \text{and} \quad P^{-1} = \begin{bmatrix} -\frac{1}{3} & -\frac{1}{3} \\ -\frac{1}{3} & -\frac{4}{3} \end{bmatrix}.$$

Therefore, the matrix for T relative to B' is

$$A' = P^{-1}AP = \begin{bmatrix} -\frac{1}{3} & -\frac{1}{3} \\ -\frac{1}{3} & -\frac{4}{3} \end{bmatrix}\begin{bmatrix} 1 & 1 \\ 0 & 4 \end{bmatrix}\begin{bmatrix} -4 & 1 \\ 1 & -1 \end{bmatrix} = \begin{bmatrix} -\frac{1}{3} & \frac{4}{3} \\ -\frac{13}{3} & \frac{16}{3} \end{bmatrix}.$$

(b) Because $A' = P^{-1}AP$, it follows that A and A' are similar.

5. (a) The standard matrix for T is

$$A = \begin{bmatrix} 1 & 0 & 0 \\ 0 & 1 & 0 \\ 0 & 0 & 1 \end{bmatrix}.$$

Furthermore, the transition matrix B' to the standard basis B, and its inverse, are

$$P = \begin{bmatrix} 1 & 1 & 0 \\ 1 & 0 & 1 \\ 0 & 1 & 1 \end{bmatrix} \quad \text{and} \quad P^{-1} = \begin{bmatrix} \frac{1}{2} & \frac{1}{2} & -\frac{1}{2} \\ \frac{1}{2} & -\frac{1}{2} & \frac{1}{2} \\ -\frac{1}{2} & \frac{1}{2} & \frac{1}{2} \end{bmatrix}.$$

Therefore, the matrix for T relative to B' is

$$A' = P^{-1}AP = \begin{bmatrix} \frac{1}{2} & \frac{1}{2} & -\frac{1}{2} \\ \frac{1}{2} & -\frac{1}{2} & \frac{1}{2} \\ -\frac{1}{2} & \frac{1}{2} & \frac{1}{2} \end{bmatrix}\begin{bmatrix} 1 & 0 & 0 \\ 0 & 1 & 0 \\ 0 & 0 & 1 \end{bmatrix}\begin{bmatrix} 1 & 1 & 0 \\ 1 & 0 & 1 \\ 0 & 1 & 1 \end{bmatrix} = \begin{bmatrix} 1 & 0 & 0 \\ 0 & 1 & 0 \\ 0 & 0 & 1 \end{bmatrix}.$$

(b) Because $A' = P^{-1}AP$, it follows that A and A' are similar.

7. (a) The standard matrix for T is

$$A = \begin{bmatrix} 1 & -1 & 2 \\ 2 & 1 & -1 \\ 1 & 2 & 1 \end{bmatrix}.$$

Furthermore, the transition matrix P from B' to the standard basis B, and its inverse, are

$$P = \begin{bmatrix} 1 & 0 & 1 \\ 0 & 2 & 2 \\ 1 & 2 & 0 \end{bmatrix} \quad \text{and} \quad P^{-1} = \begin{bmatrix} \frac{2}{3} & -\frac{1}{3} & \frac{1}{3} \\ -\frac{1}{3} & \frac{1}{6} & \frac{1}{3} \\ \frac{1}{3} & \frac{1}{3} & -\frac{1}{3} \end{bmatrix}.$$

Therefore, the matrix for T relative to B' is

$$A' = P^{-1}AP = \begin{bmatrix} \frac{2}{3} & -\frac{1}{3} & \frac{1}{3} \\ -\frac{1}{3} & \frac{1}{6} & \frac{1}{3} \\ \frac{1}{3} & \frac{1}{3} & -\frac{1}{3} \end{bmatrix} \begin{bmatrix} 1 & -1 & 2 \\ 2 & 1 & -1 \\ 1 & 2 & 1 \end{bmatrix} \begin{bmatrix} 1 & 0 & 1 \\ 0 & 2 & 2 \\ 1 & 2 & 0 \end{bmatrix} = \begin{bmatrix} \frac{7}{3} & \frac{10}{3} & -\frac{1}{3} \\ -\frac{1}{6} & \frac{4}{3} & \frac{8}{3} \\ \frac{2}{3} & -\frac{4}{3} & -\frac{2}{3} \end{bmatrix}.$$

(b) Because $A' = P^{-1}AP$, it follows that A and A' are similar.

9. (a) The transition matrix P from B' to B is found by row-reducing $[B \vdots B']$ to $[I \vdots P]$.

$$[B \vdots B'] = \begin{bmatrix} 1 & -2 & \vdots & -12 & -4 \\ 3 & -2 & \vdots & 0 & 4 \end{bmatrix} \Rightarrow [I \vdots P] = \begin{bmatrix} 1 & 0 & \vdots & 6 & 4 \\ 0 & 1 & \vdots & 9 & 4 \end{bmatrix}$$

So,

$$P = \begin{bmatrix} 6 & 4 \\ 9 & 4 \end{bmatrix}.$$

(b) The coordinate matrix for $\mathbf{v}$ relative to B is $[\mathbf{v}]_B = P[\mathbf{v}]_{B'} = \begin{bmatrix} 6 & 4 \\ 9 & 4 \end{bmatrix} \begin{bmatrix} -1 \\ 2 \end{bmatrix} = \begin{bmatrix} 2 \\ -1 \end{bmatrix}.$

Furthermore, the image of $\mathbf{v}$ under T relative to B is $[T(\mathbf{v})]_B = A[\mathbf{v}]_B = \begin{bmatrix} 3 & 2 \\ 0 & 4 \end{bmatrix} \begin{bmatrix} 2 \\ -1 \end{bmatrix} = \begin{bmatrix} 4 \\ -4 \end{bmatrix}.$

(c) The inverse of P is $P^{-1} = \begin{bmatrix} -\frac{1}{3} & \frac{1}{3} \\ \frac{3}{4} & -\frac{1}{2} \end{bmatrix}.$

The matrix of T relative to B' is then $A' = P^{-1}AP = \begin{bmatrix} -\frac{1}{3} & \frac{1}{3} \\ \frac{3}{4} & -\frac{1}{2} \end{bmatrix} \begin{bmatrix} 3 & 2 \\ 0 & 4 \end{bmatrix} \begin{bmatrix} 6 & 4 \\ 9 & 4 \end{bmatrix} = \begin{bmatrix} 0 & -\frac{4}{3} \\ 9 & 7 \end{bmatrix}.$

(d) The image of $\mathbf{v}$ under T relative to B' is $P^{-1}[T(\mathbf{v})]_B = \begin{bmatrix} -\frac{1}{3} & \frac{1}{3} \\ \frac{3}{4} & -\frac{1}{2} \end{bmatrix} \begin{bmatrix} 4 \\ -4 \end{bmatrix} = \begin{bmatrix} -\frac{8}{3} \\ 5 \end{bmatrix}.$

You can also find the image of $\mathbf{v}$ under T relative to B' by $A'[\mathbf{v}]_{B'} = \begin{bmatrix} 0 & -\frac{4}{3} \\ 9 & 7 \end{bmatrix} \begin{bmatrix} -1 \\ 2 \end{bmatrix} = \begin{bmatrix} -\frac{8}{3} \\ 5 \end{bmatrix}.$

11. (a) The transition matrix P from B' to B is found by row-reducing $[B \vdots B']$ to $[I \vdots P]$.

$$[B \vdots B'] = \begin{bmatrix} 1 & 1 & 0 & \vdots & 1 & 0 & 0 \\ 1 & 0 & 1 & \vdots & 0 & 1 & 0 \\ 0 & 1 & 1 & \vdots & 0 & 0 & 1 \end{bmatrix} \Rightarrow [I \vdots P] = \begin{bmatrix} 1 & 0 & 0 & \vdots & \frac{1}{2} & \frac{1}{2} & -\frac{1}{2} \\ 0 & 1 & 0 & \vdots & \frac{1}{2} & -\frac{1}{2} & \frac{1}{2} \\ 0 & 0 & 1 & \vdots & -\frac{1}{2} & \frac{1}{2} & \frac{1}{2} \end{bmatrix}$$

So,

$$P = \frac{1}{2}\begin{bmatrix} 1 & 1 & -1 \\ 1 & -1 & 1 \\ -1 & 1 & 1 \end{bmatrix}.$$

(b) The coordinate matrix for $\mathbf{v}$ relative to B is $[\mathbf{v}]_B = P[\mathbf{v}]_{B'} = \frac{1}{2}\begin{bmatrix} 1 & 1 & -1 \\ 1 & -1 & 1 \\ -1 & 1 & 1 \end{bmatrix}\begin{bmatrix} 1 \\ 0 \\ -1 \end{bmatrix} = \begin{bmatrix} 1 \\ 0 \\ -1 \end{bmatrix}.$

Furthermore, the image of $\mathbf{v}$ under T relative to B is $[T(\mathbf{v})]_B = A[\mathbf{v}]_B = \begin{bmatrix} \frac{3}{2} & -1 & -\frac{1}{2} \\ -\frac{1}{2} & 2 & \frac{1}{2} \\ \frac{1}{2} & 1 & \frac{5}{2} \end{bmatrix}\begin{bmatrix} 1 \\ 0 \\ -1 \end{bmatrix} = \begin{bmatrix} 2 \\ -1 \\ -2 \end{bmatrix}.$

(c) The matrix of T relative to B' is $A' = P^{-1}AP = \begin{bmatrix} 1 & 1 & 0 \\ 1 & 0 & 1 \\ 0 & 1 & 1 \end{bmatrix}\begin{bmatrix} \frac{3}{2} & -1 & -\frac{1}{2} \\ -\frac{1}{2} & 2 & \frac{1}{2} \\ \frac{1}{2} & 1 & \frac{5}{2} \end{bmatrix}\begin{bmatrix} \frac{1}{2} & \frac{1}{2} & -\frac{1}{2} \\ \frac{1}{2} & -\frac{1}{2} & \frac{1}{2} \\ -\frac{1}{2} & \frac{1}{2} & \frac{1}{2} \end{bmatrix} = \begin{bmatrix} 1 & 0 & 0 \\ 0 & 2 & 0 \\ 0 & 0 & 3 \end{bmatrix}.$

(d) The image of $\mathbf{v}$ under T relative to B' is $P^{-1}[T(\mathbf{v})]_B = \begin{bmatrix} 1 & 1 & 0 \\ 1 & 0 & 1 \\ 0 & 1 & 1 \end{bmatrix}\begin{bmatrix} 2 \\ -1 \\ -2 \end{bmatrix} = \begin{bmatrix} 1 \\ 0 \\ -3 \end{bmatrix}.$

You can also find the image of $\mathbf{v}$ under T relative to B' by $A'[\mathbf{v}]_{B'} = \begin{bmatrix} 1 & 0 & 0 \\ 0 & 2 & 0 \\ 0 & 0 & 3 \end{bmatrix}\begin{bmatrix} 1 \\ 0 \\ -1 \end{bmatrix} = \begin{bmatrix} 1 \\ 0 \\ -3 \end{bmatrix}.$

13. (a) The transition matrix P from B' to B is found by row-reducing $[B \vdots B']$ to $[I \vdots P]$.

$$[B \vdots B'] = \begin{bmatrix} 1 & -1 & \vdots & -4 & 0 \\ 2 & -1 & \vdots & 1 & 2 \end{bmatrix} \Rightarrow [I \vdots P] = \begin{bmatrix} 1 & 0 & \vdots & 5 & 2 \\ 0 & 1 & \vdots & 9 & 2 \end{bmatrix}$$

So,

$$P = \begin{bmatrix} 5 & 2 \\ 9 & 2 \end{bmatrix}.$$

(b) The coordinate matrix for $\mathbf{v}$ relative to B is $[\mathbf{v}]_B = P[\mathbf{v}]_{B'} = \begin{bmatrix} 5 & 2 \\ 9 & 2 \end{bmatrix}\begin{bmatrix} -1 \\ 4 \end{bmatrix} = \begin{bmatrix} 3 \\ -1 \end{bmatrix}.$

Furthermore, the image of $\mathbf{v}$ under T relative to B is $[T(\mathbf{v})]_B = A[\mathbf{v}]_B = \begin{bmatrix} 2 & 1 \\ 0 & -1 \end{bmatrix}\begin{bmatrix} 3 \\ -1 \end{bmatrix} = \begin{bmatrix} 5 \\ 1 \end{bmatrix}.$

(c) The matrix of T relative to B' is $A' = P^{-1}AP = \begin{bmatrix} -\frac{1}{4} & \frac{1}{4} \\ \frac{9}{8} & -\frac{5}{8} \end{bmatrix}\begin{bmatrix} 2 & 1 \\ 0 & -1 \end{bmatrix}\begin{bmatrix} 5 & 2 \\ 9 & 2 \end{bmatrix} = \begin{bmatrix} -7 & -2 \\ 27 & 8 \end{bmatrix}.$

(d) The image of $\mathbf{v}$ under T relative to B' is $P^{-1}[T(\mathbf{v})]_B = \begin{bmatrix} -\frac{1}{4} & \frac{1}{4} \\ \frac{9}{8} & -\frac{5}{8} \end{bmatrix}\begin{bmatrix} 5 \\ 1 \end{bmatrix} = \begin{bmatrix} -1 \\ 5 \end{bmatrix}.$

You can also find the image of $\mathbf{v}$ under T relative to B' by $A'[\mathbf{v}]_{B'} = \begin{bmatrix} -7 & -2 \\ 27 & 8 \end{bmatrix}\begin{bmatrix} -1 \\ 4 \end{bmatrix} = \begin{bmatrix} -1 \\ 5 \end{bmatrix}.$

15. If A and B are similar, then $B = P^{-1}AP$, for some nonsingular matrix P. So,

$|B| = |P^{-1}AP| = |P^{-1}||A||P| = |A|\dfrac{1}{|P|}|P| = |A|$. No, the converse is not true. For example, $\begin{vmatrix} 1 & 1 \\ 0 & 1 \end{vmatrix} = \begin{vmatrix} 1 & 0 \\ 0 & 1 \end{vmatrix}$, but these

matrices are not similar.

17. (a) $B = P^{-1}AP \Rightarrow B^T = (P^{-1}AP)^T = P^T A^T (P^{-1})^T = P^T A^T (P^T)^{-1}$, which shows that A^T and B^T are similar.

(b) If A is nonsingular, then so is $P^{-1}AP = B$, and

$$B = P^{-1}AP$$
$$B^{-1} = (P^{-1}AP)^{-1} = P^{-1}A^{-1}(P^{-1})^{-1} = P^{-1}A^{-1}P$$

which shows that A^{-1} and B^{-1} are similar.

(c) $B = P^{-1}AP \Rightarrow$

$$B^k = (P^{-1}AP)^k = (P^{-1}AP)(P^{-1}AP)\cdots(P^{-1}AP) \qquad (k \text{ times})$$
$$= P^{-1}A^k P.$$

19. Let A be an $n \times n$ matrix similar to I_n. Then there exists an invertible matrix P such that

$A = P^{-1}I_nP = P^{-1}P = I_n$. So, I_n is similar only to itself.

21. Let $A^2 = O$ and $B = P^{-1}AP$, then

$$B^2 = (P^{-1}AP)^2 = (P^{-1}AP)(P^{-1}AP) = P^{-1}A^2P = P^{-1}OP = O.$$

23. If A is similar to B, then $B = P^{-1}AP$. If B is similar to C, then $C = Q^{-1}BQ$.

So, $C = Q^{-1}BQ = Q^{-1}(P^{-1}AP)Q = (PQ)^{-1}A(PQ)$, which shows that A is similar to C.

25. Because $B = P^{-1}AP$, $B^2 = (P^{-1}AP)^2 = (P^{-1}AP)(P^{-1}AP) = P^{-1}A^2P$ which shows that A^2 is similar to B^2.

27. If $A = CD$ and C is nonsingular, then $C^{-1}A = D \Rightarrow C^{-1}AC = DC$, which shows that DC is similar to A.

29. The matrix for I relative to B and B' is the square matrix whose columns are the coordinates of $\mathbf{v}_1, \ldots \mathbf{v}_n$ relative to the standard basis. The matrix for I relative to B, or relative to B', is the identity matrix.

31. (a) True. See discussion, pages 399–400, and note that $A' = P^{-1}AP \Rightarrow PA'P^{-1} = PP^{-1}APP^{-1} = A$.

(b) False. Unless it is a diagonal matrix, see Example 5, pages 403–404.

Section 6.5 Applications of Linear Transformations

1. The standard matrix for T is

$$A = \begin{bmatrix} 1 & 0 \\ 0 & -1 \end{bmatrix}.$$

(a) $\begin{bmatrix} 1 & 0 \\ 0 & -1 \end{bmatrix}\begin{bmatrix} 3 \\ 5 \end{bmatrix} = \begin{bmatrix} 3 \\ -5 \end{bmatrix} \Rightarrow T(3, 5) = (3, -5)$

(b) $\begin{bmatrix} 1 & 0 \\ 0 & -1 \end{bmatrix}\begin{bmatrix} 2 \\ -1 \end{bmatrix} = \begin{bmatrix} 2 \\ 1 \end{bmatrix} \Rightarrow T(2, -1) = (2, 1)$

(c) $\begin{bmatrix} 1 & 0 \\ 0 & -1 \end{bmatrix}\begin{bmatrix} a \\ 0 \end{bmatrix} = \begin{bmatrix} a \\ 0 \end{bmatrix} \Rightarrow T(a, 0) = (a, 0)$

(d) $\begin{bmatrix} 1 & 0 \\ 0 & -1 \end{bmatrix}\begin{bmatrix} 0 \\ b \end{bmatrix} = \begin{bmatrix} 0 \\ -b \end{bmatrix} \Rightarrow T(0, b) = (0, -b)$

(e) $\begin{bmatrix} 1 & 0 \\ 0 & -1 \end{bmatrix}\begin{bmatrix} -c \\ d \end{bmatrix} = \begin{bmatrix} -c \\ -d \end{bmatrix} \Rightarrow T(-c, d) = (-c, -d)$

(f) $\begin{bmatrix} 1 & 0 \\ 0 & -1 \end{bmatrix}\begin{bmatrix} f \\ -g \end{bmatrix} = \begin{bmatrix} f \\ g \end{bmatrix} \Rightarrow T(f, -g) = (f, g)$

3. The standard matrix for T is

$$A = \begin{bmatrix} 0 & 1 \\ 1 & 0 \end{bmatrix}.$$

(a) $\begin{bmatrix} 0 & 1 \\ 1 & 0 \end{bmatrix}\begin{bmatrix} 0 \\ 1 \end{bmatrix} = \begin{bmatrix} 1 \\ 0 \end{bmatrix} \Rightarrow T(0, 1) = (1, 0)$

(b) $\begin{bmatrix} 0 & 1 \\ 1 & 0 \end{bmatrix}\begin{bmatrix} -1 \\ 3 \end{bmatrix} = \begin{bmatrix} 3 \\ -1 \end{bmatrix} \Rightarrow T(-1, 3) = (3, -1)$

(c) $\begin{bmatrix} 0 & 1 \\ 1 & 0 \end{bmatrix}\begin{bmatrix} a \\ 0 \end{bmatrix} = \begin{bmatrix} 0 \\ a \end{bmatrix} \Rightarrow T(a, 0) = (0, a)$

(d) $\begin{bmatrix} 0 & 1 \\ 1 & 0 \end{bmatrix}\begin{bmatrix} 0 \\ b \end{bmatrix} = \begin{bmatrix} b \\ 0 \end{bmatrix} \Rightarrow T(0, b) = (b, 0)$

(e) $\begin{bmatrix} 0 & 1 \\ 1 & 0 \end{bmatrix}\begin{bmatrix} -c \\ d \end{bmatrix} = \begin{bmatrix} d \\ -c \end{bmatrix} \Rightarrow T(-c, d) = (d, -c)$

(f) $\begin{bmatrix} 0 & 1 \\ 1 & 0 \end{bmatrix}\begin{bmatrix} f \\ -g \end{bmatrix} = \begin{bmatrix} -g \\ f \end{bmatrix} \Rightarrow T(f, -g) = (-g, f)$

5. (a) $T(x, y) = xT(1, 0) + yT(0, 1) = x(0, 1) + y(1, 0) = (y, x)$

(b) T is a reflection in the line $y = x$.

7. (a) Identify T as a vertical contraction from its standard matrix.

$$A = \begin{bmatrix} 1 & 0 \\ 0 & \frac{1}{2} \end{bmatrix}$$

(b)

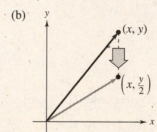

9. (a) Identify T as a horizontal expansion from its standard matrix

$$A = \begin{bmatrix} 4 & 0 \\ 0 & 1 \end{bmatrix}.$$

(b)

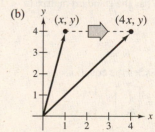

11. (a) Identify T as a horizontal shear from its matrix.

$$A = \begin{bmatrix} 1 & 3 \\ 0 & 1 \end{bmatrix}.$$

(b)

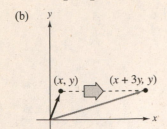

13. (a) Identify T as a vertical shear from its matrix

$$A = \begin{bmatrix} 1 & 0 \\ 2 & 1 \end{bmatrix}.$$

(b)

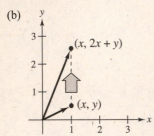

15. The reflection in the y-axis is given by
$T(x, y) = (-x, y)$. If (x, y) is a fixed point, then
$T(x, y) = (x, y) = (-x, y)$ which implies that
$x = 0$. So the set of fixed points is $\{(0, t) : t \in R\}$.

17. The reflection in the line $y = x$ is given by
$T(x, y) = (y, x)$. If (x, y) is a fixed point, then
$T(x, y) = (x, y) = (y, x)$ which implies that
$x = y$. So, the set of fixed points is $\{(t, t) : t \in R\}$.

19. A vertical contraction has the standard matrix $(k < 1)$

$$\begin{bmatrix} 1 & 0 \\ 0 & k \end{bmatrix}.$$

A fixed point of T satisfies the equation

$$T(\mathbf{v}) = \begin{bmatrix} 1 & 0 \\ 0 & k \end{bmatrix}\begin{bmatrix} v_1 \\ v_2 \end{bmatrix} = \begin{bmatrix} v_1 \\ kv_2 \end{bmatrix} = \begin{bmatrix} v_1 \\ v_2 \end{bmatrix} = \mathbf{v}.$$

So, the set of fixed points is $\{(t, 0) : t \text{ is a real number}\}$.

21. A horizontal shear has the form
$T(x, y) = (x + ky, y)$. If (x, y) is a fixed point,
then $T(x, y) = (x, y) = (x + ky, y)$ which implies that
$y = 0$. So, the set of fixed points is $\{(t, 0) : t \in R\}$.

23. Find the image of each vertex under $T(x, y) = (x, -y)$.

$$T(0, 0) = (0, 0), \qquad T(1, 0) = (1, 0),$$
$$T(1, 1) = (1, -1), \qquad T(0, 1) = (0, -1)$$

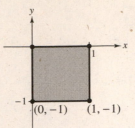

25. Find the image of each vertex under $T(x, y) = \left(\dfrac{x}{2}, y\right)$.

$$T(0, 0) = (0, 0), \qquad T(1, 0) = \left(\frac{1}{2}, 0\right),$$
$$T(1, 1) = \left(\frac{1}{2}, 1\right), \qquad T(0, 1) = (0, 1)$$

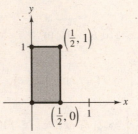

27. Find the image of each vertex under
$T(x, y) = (x + 2y, y)$.

$$T(0, 0) = (0, 0), \qquad T(1, 0) = (1, 0),$$
$$T(1, 1) = (3, 1), \qquad T(0, 1) = (2, 1)$$

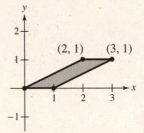

29. Find the image of each vertex under $T(x, y) = (-x, y)$.

$$T(0, 0) = (0, 0), \qquad T(0, 2) = (0, 2),$$
$$T(1, 2) = (-1, 2), \qquad T(1, 0) = (-1, 0)$$

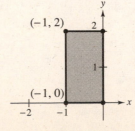

31. Find the image of each vertex under $T(x, y) = \left(x, \frac{1}{2}y\right)$.

$T(0, 0) = (0, 0), \qquad T(0, 2) = (0, 1),$

$T(1, 2) = (1, 1), \qquad T(1, 0) = (1, 0)$

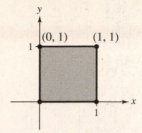

33. Find the image of each vertex under
$T(x, y) = (x + y, y)$.

$T(0, 0) = (0, 0), \qquad T(0, 2) = (2, 2),$

$T(1, 2) = (3, 2), \qquad T(1, 0) = (1, 0)$

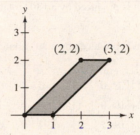

35. Find the image of each vertex under
$T(x, y) = (x + y, y)$.

(a) $T(0, 0) = (0, 0), \qquad T(1, 2) = (3, 2),$

$T(3, 6) = (9, 6), \qquad T(5, 2) = (7, 2),$

$T(6, 0) = (6, 0)$

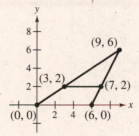

(b) $T(0, 0) = (0, 0) \qquad T(0, 6) = (6, 6)$

$T(6, 6) = (12, 6) \qquad T(6, 0) = (6, 0)$

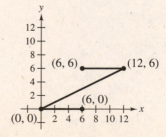

37. Find the image of each vertex under
$T(x, y) = \left(2x, \frac{1}{2}y\right).$

(a) $T(0, 0) = (0, 0) \qquad T(1, 2) = (2, 1)$

$T(3, 6) = (6, 3) \qquad T(5, 2) = (10, 1)$

$T(6, 0) = (12, 0)$

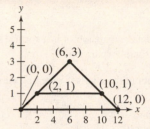

(b) $T(0, 0) = (0, 0) \qquad T(0, 6) = (0, 3)$

$T(6, 6) = (12, 3) \qquad T(6, 0) = (12, 0)$

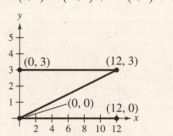

39. The images of the given vectors are as follows.

$T(1, 0) = \begin{bmatrix} 2 & 0 \\ 0 & 3 \end{bmatrix}\begin{bmatrix} 1 \\ 0 \end{bmatrix} = \begin{bmatrix} 2 \\ 0 \end{bmatrix} = (2, 0)$

$T(0, 1) = \begin{bmatrix} 2 & 0 \\ 0 & 3 \end{bmatrix}\begin{bmatrix} 0 \\ 1 \end{bmatrix} = \begin{bmatrix} 0 \\ 3 \end{bmatrix} = (0, 3)$

$T(2, 2) = \begin{bmatrix} 2 & 0 \\ 0 & 3 \end{bmatrix}\begin{bmatrix} 2 \\ 2 \end{bmatrix} = \begin{bmatrix} 4 \\ 6 \end{bmatrix} = (4, 6)$

The two triangles are shown in the following figure.

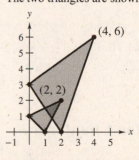

41. The linear transformation defined by A is a horizontal expansion.

43. The linear transformation defined by A is a reflection in the line $y = x$.

45. The linear transformation defined by A is a reflection in the x-axis followed by a vertical expansion.

47. Because

$$\begin{bmatrix} 1 & 0 \\ 2 & 1 \end{bmatrix}$$

represents a vertical shear and

$$\begin{bmatrix} 2 & 0 \\ 0 & 1 \end{bmatrix}$$

represents a horizontal expansion, A is a vertical shear *followed* by a horizontal expansion.

49. A rotation of $30°$ about the z-axis is given by the matrix

$$A = \begin{bmatrix} \cos 30° & -\sin 30° & 0 \\ \sin 30° & \cos 30° & 0 \\ 0 & 0 & 1 \end{bmatrix} = \begin{bmatrix} \frac{\sqrt{3}}{2} & -\frac{1}{2} & 0 \\ \frac{1}{2} & \frac{\sqrt{3}}{2} & 0 \\ 0 & 0 & 1 \end{bmatrix}.$$

51. A rotation of $60°$ about the y-axis is given by the matrix

$$A = \begin{bmatrix} \cos 60° & 0 & \sin 60° \\ 0 & 1 & 0 \\ -\sin 60° & 0 & \cos 60° \end{bmatrix} = \begin{bmatrix} \frac{1}{2} & 0 & \frac{\sqrt{3}}{2} \\ 0 & 1 & 0 \\ -\frac{\sqrt{3}}{2} & 0 & \frac{1}{2} \end{bmatrix}.$$

53. Using the matrix obtained in Exercise 49,

$$T(1,1,1) = \begin{bmatrix} \frac{\sqrt{3}}{2} & -\frac{1}{2} & 0 \\ \frac{1}{2} & \frac{\sqrt{3}}{2} & 0 \\ 0 & 0 & 1 \end{bmatrix}\begin{bmatrix} 1 \\ 1 \\ 1 \end{bmatrix} = \begin{bmatrix} \frac{(\sqrt{3}-1)}{2} \\ \frac{(1+\sqrt{3})}{2} \\ 1 \end{bmatrix}.$$

55. Using the matrix obtained in Exercise 51,

$$T(1,1,1) = \begin{bmatrix} \frac{1}{2} & 0 & \frac{\sqrt{3}}{2} \\ 0 & 1 & 0 \\ -\frac{\sqrt{3}}{2} & 0 & \frac{1}{2} \end{bmatrix}\begin{bmatrix} 1 \\ 1 \\ 1 \end{bmatrix} = \begin{bmatrix} \frac{1+\sqrt{3}}{2} \\ 1 \\ \frac{1-\sqrt{3}}{2} \end{bmatrix}.$$

57. The indicated tetrahedron is produced by a $90°$ rotation about the x-axis.

59. The indicated tetrahedron is produced by a $180°$ rotation about the y-axis.

61. The indicated tetrahedron is produced by a $90°$ rotation about the z-axis.

63. The matrix is $\begin{bmatrix} 0 & 0 & 1 \\ 0 & 1 & 0 \\ -1 & 0 & 0 \end{bmatrix}\begin{bmatrix} 1 & 0 & 0 \\ 0 & 0 & -1 \\ 0 & 1 & 0 \end{bmatrix} = \begin{bmatrix} 0 & 1 & 0 \\ 0 & 0 & -1 \\ -1 & 0 & 0 \end{bmatrix}.$

$T(0,0,0) = (0,0,0)$ and $T(1,1,1) = (1,-1,-1)$, so the line segment image runs from $(0,0,0)$ to $(1,-1,-1)$.

65. The matrix is $\begin{bmatrix} \cos 60° & 0 & \sin 60° \\ 0 & 1 & 0 \\ -\sin 60° & 0 & \cos 60° \end{bmatrix}\begin{bmatrix} \cos 30° & -\sin 30° & 0 \\ \sin 30° & \cos 30° & 0 \\ 0 & 0 & 1 \end{bmatrix} = \begin{bmatrix} \frac{1}{2} & 0 & \frac{\sqrt{3}}{2} \\ 0 & 1 & 0 \\ -\frac{\sqrt{3}}{2} & 0 & \frac{1}{2} \end{bmatrix}\begin{bmatrix} \frac{\sqrt{3}}{2} & -\frac{1}{2} & 0 \\ \frac{1}{2} & \frac{\sqrt{3}}{2} & 0 \\ 0 & 0 & 1 \end{bmatrix} = \begin{bmatrix} \frac{\sqrt{3}}{4} & -\frac{1}{4} & \frac{\sqrt{3}}{2} \\ \frac{1}{2} & \frac{\sqrt{3}}{2} & 0 \\ -\frac{3}{4} & \frac{\sqrt{3}}{4} & \frac{1}{2} \end{bmatrix}.$

$T(0,0,0) = (0,0,0)$ and $T(1,1,1) = \left(\frac{3\sqrt{3}-1}{4}, \frac{\sqrt{3}+1}{2}, \frac{\sqrt{3}-1}{4}\right)$, so the line segment runs from $(0,0,0)$ to

$\left(\frac{3\sqrt{3}-1}{4}, \frac{\sqrt{3}+1}{2}, \frac{\sqrt{3}-1}{4}\right)$.

Review Exercises for Chapter 6

1. (a) $T(\mathbf{v}) = T(2,-3) = (2,-4)$

(b) The preimage of $\mathbf{w}$ is given by solving the equation

$T(v_1, v_2) = (v_1, v_1 + 2v_2) = (4, 12)$.

The resulting system of linear equations

$$\begin{aligned} v_1 &= 4 \\ v_1 + 2v_2 &= 12 \end{aligned}$$

has the solution $v_1 = v_2 = 4$. So, the preimage of $\mathbf{w}$ is $(4, 4)$.

3. (a) $T(\mathbf{v}) = T(-3, 2, 5) = (0, -1, 7)$.

   (b) The preimage of $\mathbf{w}$ is given by solving the equation $T(v_1, v_2, v_3) = (0, v_1 + v_2, v_2 + v_3) = (0, 2, 5)$.

   The resulting system of linear equations has the solution of the form $v_1 = t - 3, v_2 = 5 - t, v_3 = t$, where t is any real

   number. So, the preimage of $\mathbf{w}$ is $\{(t - 3, 5 - t, t) : t \in R\}$.

5. T preserves addition.

$$T(x_1, x_2) + T(y_1, y_2) = (x_1 + 2x_2, -x_1 - x_2) + (y_1 + 2y_2, -y_1 - y_2)$$
$$= (x_1 + 2x_2 + y_1 + 2y_2, -x_1 - x_2 - y_1 - y_2)$$
$$= ((x_1 + y_1) + 2(x_2 + y_2), -(x_1 + y_1) - (x_2 + y_2))$$
$$= T(x_1 + y_1, x_2 + y_2)$$

T preserves scalar multiplication.

$$cT(x_1, x_2) = c(x_1 + 2x_2, -x_1 - x_2) = (cx_1 + 2(cx_2), -cx_1 - cx_2) = T(cx_1, cx_2)$$

So, T is a linear transformation with standard matrix $A = \begin{bmatrix} 1 & 2 \\ -1 & -1 \end{bmatrix}$.

7. T preserves addition.

$$T(x_1, y_1) + T(x_2, y_2) = (x_1 - 2y_1, 2y_1 - x_1) + (x_2 - 2y_2, 2y_2 - x_2)$$
$$= x_1 - 2y_1 + x_2 - 2y_2, 2y_1 - x_1 + 2y_2 - x_2$$
$$= (x_1 + x_2) - 2(y_1 + y_2), 2(y_1 + y_2) - (x_1 + x_2)$$
$$= T(x_1 + y_1, x_2 + y_2)$$

T preserves scalar multiplication.

$$cT(x, y) = c(x - 2y, 2y - x) = cx - 2cy, 2cy - cx = T(cx, cy)$$

So, T is a linear transformation with standard matrix $A = \begin{bmatrix} 1 & -2 \\ -1 & 2 \end{bmatrix}$.

9. T does not preserve addition or scalar multiplication, so, T is *not* a linear transformation. A counterexample is

$$T(1, 0) + T(0, 1) = (1 + h, k) + (h, 1 + k) = (1 + 2h, 1 + 2k) \neq T(1 + h, 1 + k) = T(1, 1).$$

11. T preserves addition.

$$T(x_1, x_2, x_3) + T(y_1, y_2, y_3)$$
$$= (x_1 - x_2, x_2 - x_3, x_3 - x_1) + (y_1 - y_2, y_2 - y_3, y_3 - y_1)$$
$$= (x_1 - x_2 + y_1 - y_2, x_2 - x_3 + y_2 - y_3, x_3 - x_1 + y_3 - y_1)$$
$$= ((x_1 + y_1) - (x_2 + y_2), (x_2 + y_2) - (x_3 + y_3), (x_3 + y_3) - (x_1 + y_1))$$
$$= T(x_1 + y_1, x_2 + y_2, x_3 + y_3)$$

T preserves scalar multiplication.

$$cT(x_1, x_2, x_3) = c(x_1 - x_2, x_2 - x_3, x_3 - x_1)$$
$$= (c(x_1 - x_2), c(x_2 - x_3), c(x_3 - x_1))$$
$$= (cx_1 - cx_2, cx_2 - cx_3, cx_3 - cx_1)$$
$$= T(cx_1, cx_2, cx_3)$$

So, T is a linear transformation with standard matrix $A = \begin{bmatrix} 1 & -1 & 0 \\ 0 & 1 & -1 \\ -1 & 0 & 1 \end{bmatrix}$.

13. Because $(1, 1) = \frac{1}{2}(2, 0) + \frac{1}{3}(0, 3)$,

$$T(1, 1) = \frac{1}{2}T(2, 0) + \frac{1}{3}T(0, 3)$$

$$= \frac{1}{2}(1, 1) + \frac{1}{3}(3, 3) = \left(\frac{3}{2}, \frac{3}{2}\right).$$

Because $(0, 1) = \frac{1}{3}(0, 3)$,

$$T(0, 1) = \frac{1}{3}T(0, 3)$$

$$= \frac{1}{3}(3, 3) = (1, 1).$$

15. Because $(0, -1) = -\frac{2}{3}(1, 1) - \frac{1}{3}(2, -1)$

$$T(0, -1) = -\frac{2}{3}T(1, 1) + \frac{1}{3}T(2, -1)$$

$$= -\frac{2}{3}(2, 3) + \frac{1}{3}(1, 0)$$

$$= \left(-\frac{4}{3}, -2\right) + \left(\frac{1}{3}, 0\right) = (-1, -2).$$

17. The standard matrix for T is

$$A = \begin{bmatrix} 1 & 0 & 0 \\ 0 & 1 & 0 \\ 0 & 0 & -1 \end{bmatrix}.$$

Therefore,

$$A^2 = \begin{bmatrix} 1^2 & 0 & 0 \\ 0 & 1^2 & 0 \\ 0 & 0 & (-1)^2 \end{bmatrix} = \begin{bmatrix} 1 & 0 & 0 \\ 0 & 1 & 0 \\ 0 & 0 & 1 \end{bmatrix} = I_3.$$

19. The standard matrix for T is

$$A = \begin{bmatrix} \cos\theta & -\sin\theta \\ \sin\theta & \cos\theta \end{bmatrix}.$$ Therefore,

$$A^3 = \begin{bmatrix} \cos 3\theta & -\sin 3\theta \\ \sin 3\theta & \cos 3\theta \end{bmatrix}.$$

21. (a) Because A is a 2×3 matrix, it maps R^3 into R^2, $(n = 3, m = 2)$.

(b) Because $T(\mathbf{v}) = A\mathbf{v}$ and

$$A\mathbf{v} = \begin{bmatrix} 0 & 1 & 2 \\ -2 & 0 & 0 \end{bmatrix}\begin{bmatrix} 6 \\ 1 \\ 1 \end{bmatrix} = \begin{bmatrix} 3 \\ -12 \end{bmatrix},$$

it follows that $T(6, 1, 1) = (3, -12)$.

(c) The preimage of $\mathbf{w}$ is given by the solution to the equation

$$T(v_1, v_2, v_3) = \mathbf{w} = (3, 5).$$

The equivalent system of linear equations

$$\begin{aligned} v_2 + 2v_3 &= 3 \\ -2v_1 \qquad\quad &= 5 \end{aligned}$$

has the solution

$$\left\{\left(-\frac{5}{2}, 3 - 2t, t\right) : t \text{ is a real number}\right\}.$$

23. (a) Because A is a 1×2 matrix, it maps R^2 into R^1, $(n = 2, m = 1)$.

(b) Because $T(\mathbf{v}) = A\mathbf{v}$ and $A\mathbf{v} = \begin{bmatrix} 1 & 1 \end{bmatrix}\begin{bmatrix} 2 \\ 3 \end{bmatrix} = 5$, it follows that $T(2, 3) = 5$.

(c) The preimage of $\mathbf{w} = (4)$ is given by the solution to this equation $T(v_1, v_2) = \mathbf{w} = (4)$.

The equivalent system of linear equations is $v_1 + v_2 = 4$, which has the solution

$$\{(4 - t, t) : t \in R\}.$$

25. (a) Because A is a 3×3 matrix, it maps R^3 into R^3, $(n = 3, m = 3)$.

(b) Because $T(\mathbf{v}) = A\mathbf{v}$ and

$$A\mathbf{v} = \begin{bmatrix} 1 & 1 & 1 \\ 0 & 1 & 1 \\ 0 & 0 & 1 \end{bmatrix}\begin{bmatrix} 2 \\ 1 \\ -5 \end{bmatrix} = \begin{bmatrix} -2 \\ -4 \\ -5 \end{bmatrix},$$

it follows that $T(2, 1, -5) = (-2, -4, -5)$.

(c) The preimage of $\mathbf{w} = (6, 4, 2)$ is given by the solution to the equation

$$T(v_1, v_2, v_3) = (6, 4, 2) = \mathbf{w}.$$

The equivalent system of linear equations has the solution $v_1 = v_2 = v_3 = 2$. So, the preimage is $(2, 2, 2)$.

27. (a) Because A is a 3×2 matrix, it maps R^2 into R^3, $(n = 2, m = 3)$.

(b) Because $T(\mathbf{v}) = A\mathbf{v}$ and

$$A\mathbf{v} = \begin{bmatrix} 4 & 0 \\ 0 & 5 \\ 1 & 1 \end{bmatrix}\begin{bmatrix} 2 \\ 2 \end{bmatrix} = \begin{bmatrix} 8 \\ 10 \\ 4 \end{bmatrix},$$

it follows that $T(2, 2) = (8, 10, 4)$.

(c) The preimage of $\mathbf{w} = (4, -5, 0)$ is given by the solution to the equation

$$T(v_1, v_2) = (4, -5, 0) = \mathbf{w}.$$

The equivalent system of linear equations has the solution $v_1 = 1$ and $v_2 = -1$. So, the preimage is $(1, -1)$.

29. (a) The standard matrix for T is

$$A = \begin{bmatrix} 2 & 4 & 6 & 5 \\ -1 & -2 & 2 & 0 \\ 0 & 0 & 8 & 4 \end{bmatrix}.$$

Solving $A\mathbf{v} = \mathbf{0}$ yields the solution

$$\{(-2s + 2t, s, t, -2t) : s \text{ and } t \text{ are real numbers}\}.$$

So, a basis for $\ker(T)$ is $\{(-2, 1, 0, 0), (2, 0, 1, -2)\}$.

(b) Use Gauss-Jordan elimination to reduce A^T as follows.

$$A^T = \begin{bmatrix} 2 & -1 & 0 \\ 4 & -2 & 0 \\ 6 & 2 & 8 \\ 5 & 0 & 4 \end{bmatrix} \Rightarrow \begin{bmatrix} 1 & 0 & \frac{4}{5} \\ 0 & 1 & \frac{8}{5} \\ 0 & 0 & 0 \\ 0 & 0 & 0 \end{bmatrix}$$

The nonzero row vectors form a basis for the range of T, $\left\{\left(1, 0, \frac{4}{5}\right), \left(0, 1, \frac{8}{5}\right)\right\}$.

31. (a) The standard matrix for T is

$$A = \begin{bmatrix} 1 & 1 & 0 \\ 0 & 1 & 1 \\ 1 & 0 & -1 \end{bmatrix}.$$

Solving $A\mathbf{v} = \mathbf{0}$ yields the solution $\{(t, -t, t) : t \in R\}$. So, a basis for $\ker(T)$ is $\{(1, -1, 1)\}$.

(b) Use Gauss-Jordan elimination to reduce A^T as follows.

$$A^T = \begin{bmatrix} 1 & 0 & 1 \\ 1 & 1 & 0 \\ 0 & 1 & -1 \end{bmatrix} \Rightarrow \begin{bmatrix} 1 & 0 & 1 \\ 0 & 1 & -1 \\ 0 & 0 & 0 \end{bmatrix}$$

The nonzero row vectors form a basis for the range of T, $\{(1, 0, 1), (0, 1, -1)\}$.

33. (a) To find the kernel of T, row-reduce A,

$$A = \begin{bmatrix} 1 & 2 \\ -1 & 0 \\ 1 & 1 \end{bmatrix} \Rightarrow \begin{bmatrix} 1 & 0 \\ 0 & 1 \\ 0 & 0 \end{bmatrix}$$

which shows that $\ker(T) = \{(0, 0)\}$.

(b) The range of T can be found by row-reducing the transpose of A.

$$A^T = \begin{bmatrix} 1 & -1 & 1 \\ 2 & 0 & 1 \end{bmatrix} \Rightarrow \begin{bmatrix} 1 & 0 & \frac{1}{2} \\ 0 & 1 & -\frac{1}{2} \end{bmatrix}$$

So, a basis for range (T) is $\left\{\left(1, 0, \frac{1}{2}\right), \left(0, 1, -\frac{1}{2}\right)\right\}$.

Or, use the 2 columns of A, $\{(1, -1, 1), (2, 0, 1)\}$.

(c) $\dim\big(\text{range}(T)\big) = \text{rank}(T) = 2$

(d) $\dim\big(\ker(T)\big) = \text{nullity}(T) = 0$

35. (a) To find the kernel of T, row-reduce A,

$$A = \begin{bmatrix} 2 & 1 & 3 \\ 1 & 1 & 0 \\ 0 & 1 & -3 \end{bmatrix} \Rightarrow \begin{bmatrix} 1 & 0 & 3 \\ 0 & 1 & -3 \\ 0 & 0 & 0 \end{bmatrix}$$

which shows that kernel $(T) = \{(-3t, 3t, t) : t \in R\}$. So, a basis for kernel$(T)$ is $\{(-3, 3, 1)\}$.

(b) The range of T can be found by row-reducing the transpose of A.

$$A^T = \begin{bmatrix} 2 & 1 & 0 \\ 1 & 1 & 1 \\ 3 & 0 & -3 \end{bmatrix} \Rightarrow \begin{bmatrix} 1 & 0 & -1 \\ 0 & 1 & 2 \\ 0 & 0 & 0 \end{bmatrix}$$

So, a basis for range(T) is $\{(1, 0, -1), (0, 1, 2)\}$.

(c) $\dim\big(\text{range}(T)\big) = \text{rank}(T) = 2$

(d) $\dim\big(\ker(T)\big) = \text{nullity}(T) = 1$

37. $\text{rank}(T) = \dim R^5 - \text{nullity}(T) = 5 - 2 = 3$

39. $\text{nullity}(T) = \dim(P_4) - \text{rank}(T) = 5 - 3 = 2$

41. The standard matrix for T is

$$A = \begin{bmatrix} 2 & 0 \\ 0 & 1 \end{bmatrix}.$$

A is invertible and its inverse is given by

$$A^{-1} = \begin{bmatrix} \frac{1}{2} & 0 \\ 0 & 1 \end{bmatrix}.$$

43. The standard matrix for T is

$$A = \begin{bmatrix} \cos \theta & -\sin \theta \\ \sin \theta & \cos \theta \end{bmatrix}.$$

A is invertible and its inverse is given by

$$A^{-1} = \begin{bmatrix} \cos \theta & \sin \theta \\ -\sin \theta & \cos \theta \end{bmatrix}.$$

45. The standard matrix of T is

$$A = \begin{bmatrix} 1 & 0 & 0 \\ 0 & 1 & 0 \\ 0 & 0 & 0 \end{bmatrix}.$$

Because A is *not* invertible, T has no inverse.

47. The standard matrix for T is

$$A = \begin{bmatrix} 1 & 1 & 0 \\ 0 & 1 & -1 \end{bmatrix}.$$

Because A is *not* invertible, T has no inverse.

49. The standard matrices for T_1 and T_2 are

$$A_1 = \begin{bmatrix} 1 & 0 \\ 1 & 1 \\ 0 & 1 \end{bmatrix} \quad \text{and} \quad A_2 = \begin{bmatrix} 0 & 0 & 0 \\ 0 & 1 & 0 \end{bmatrix}.$$

The standard matrix for $T = T_1 \circ T_2$ is

$$A = A_1 A_2 = \begin{bmatrix} 1 & 0 \\ 1 & 1 \\ 0 & 1 \end{bmatrix}\begin{bmatrix} 0 & 0 & 0 \\ 0 & 1 & 0 \end{bmatrix} = \begin{bmatrix} 0 & 0 & 0 \\ 0 & 1 & 0 \\ 0 & 1 & 0 \end{bmatrix}$$

and the standard matrix for $T' = T_2 \circ T_1$ is

$$A' = A_2 A_1 = \begin{bmatrix} 0 & 0 & 0 \\ 0 & 1 & 0 \end{bmatrix}\begin{bmatrix} 1 & 0 \\ 1 & 1 \\ 0 & 1 \end{bmatrix} = \begin{bmatrix} 0 & 0 \\ 1 & 1 \end{bmatrix}.$$

51. The standard matrix for the 90° counterclockwise notation is

$$A = \begin{bmatrix} \cos 90° & -\sin 90° \\ \sin 90° & \cos 90° \end{bmatrix} = \begin{bmatrix} 0 & -1 \\ 1 & 0 \end{bmatrix}.$$

Calculating the image of the three vertices,

$$\begin{bmatrix} 0 & -1 \\ 1 & 0 \end{bmatrix}\begin{bmatrix} 3 \\ 5 \end{bmatrix} = \begin{bmatrix} -5 \\ 3 \end{bmatrix},$$

$$\begin{bmatrix} 0 & -1 \\ 1 & 0 \end{bmatrix}\begin{bmatrix} 5 \\ 3 \end{bmatrix} = \begin{bmatrix} -3 \\ 5 \end{bmatrix},$$

$$\begin{bmatrix} 0 & -1 \\ 1 & 0 \end{bmatrix}\begin{bmatrix} 3 \\ 0 \end{bmatrix} = \begin{bmatrix} 0 \\ 3 \end{bmatrix},$$

you have the following graph.

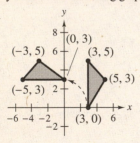

53. (a) Because $|A| = 6 \neq 0$, $\ker(T) = \{(0, 0)\}$ and the transformation is one-to-one.

   (b) Because the nullity of T is 0, the rank of T equals the dimension of the domain and the transformation is onto.

   (c) The transformation is one-to-one and onto (an isomorphism) and is, therefore, invertible.

55. (a) Because $|A| = 1 \neq 0$, $\ker(T) = \{(0, 0)\}$ and T is one-to-one.

   (b) Because $\text{rank}(A) = 2$, T is onto.

   (c) The transformation is one-to-one and onto, and is, therefore, invertible.

57. (a) The standard matrix for T is

$$A = \begin{bmatrix} -1 & 0 \\ 0 & 1 \\ 1 & 1 \end{bmatrix}$$

so it follows that

$$T(\mathbf{v}) = A(\mathbf{v}) = \begin{bmatrix} -1 & 0 \\ 0 & 1 \\ 1 & 1 \end{bmatrix}\begin{bmatrix} 0 \\ 1 \end{bmatrix} = \begin{bmatrix} 0 \\ 1 \\ 1 \end{bmatrix} = (0, 1, 1).$$

   (b) The image of each vector in B is as follows.

$$T(1, 1) = (-1, 1, 2) = (0, 1, 0) + 2(0, 0, 1) - (1, 0, 0)$$
$$T(1, -1) = (-1, -1, 0) = -(0, 1, 0) + 0(0, 0, 1) - (1, 0, 0)$$

Therefore,

$$\left[T(1, 1)\right]_{B'} = [1, 2, -1]^T$$

and

$$\left[T(1, -1)\right]_{B'} = [-1, 0, -1]^T$$

and

$$A' = \begin{bmatrix} 1 & -1 \\ 2 & 0 \\ -1 & -1 \end{bmatrix}.$$

Because

$$[\mathbf{v}]_B = \begin{bmatrix} \frac{1}{2} \\ -\frac{1}{2} \end{bmatrix},$$

the image of $\mathbf{v}$ under T relative to B' is

$$\left[T(\mathbf{v})\right]_{B'} = A'[\mathbf{v}]_B = \begin{bmatrix} 1 & -1 \\ 2 & 0 \\ -1 & -1 \end{bmatrix}\begin{bmatrix} \frac{1}{2} \\ -\frac{1}{2} \end{bmatrix} = \begin{bmatrix} 1 \\ 1 \\ 0 \end{bmatrix}.$$

So,

$$T(\mathbf{v}) = (0, 1, 0) + (0, 0, 1) + 0(1, 0, 0) = (0, 1, 1).$$

59. The standard matrix for T is

$$A = \begin{bmatrix} 1 & -3 \\ -1 & 1 \end{bmatrix}.$$

The transformation matrix from B' to the standard basis $B = \{(1, 0), (0, 1)\}$ is

$$P = \begin{bmatrix} 1 & 1 \\ -1 & 1 \end{bmatrix}.$$

The matrix A' for T relative to B' is

$$A' = P^{-1}AP = \begin{bmatrix} \frac{1}{2} & -\frac{1}{2} \\ \frac{1}{2} & \frac{1}{2} \end{bmatrix}\begin{bmatrix} 1 & -3 \\ -1 & 1 \end{bmatrix}\begin{bmatrix} 1 & 1 \\ -1 & 1 \end{bmatrix} = \begin{bmatrix} 3 & -1 \\ 1 & -1 \end{bmatrix}.$$

Because $A' = P^{-1}AP$, it follows that A and A' are similar.

61. (a) Because $T(\mathbf{v}) = T(x, y, z) = \text{proj}_{\mathbf{u}}\mathbf{v}$ where $\mathbf{u} = (0, 1, 2)$,

$$T(\mathbf{v}) = \frac{y + 2z}{5}(0, 1, 2).$$

So,

$$T(1, 0, 0) = (0, 0, 0), T(0, 1, 0) = \left(0, \tfrac{1}{5}, \tfrac{2}{5}\right), T(0, 0, 1) = \left(0, \tfrac{2}{5}, \tfrac{4}{5}\right)$$

and the standard matrix for T is

$$A = \begin{bmatrix} 0 & 0 & 0 \\ 0 & \tfrac{1}{5} & \tfrac{2}{5} \\ 0 & \tfrac{2}{5} & \tfrac{4}{5} \end{bmatrix}.$$

(b) $S = I - A$ satisfies $S(\mathbf{u}) = \mathbf{0}$. Letting $\mathbf{w}_1 = (1, 0, 0)$ and $\mathbf{w}_2 = (0, 2, -1)$ be two vectors orthogonal to $\mathbf{u}$,

$$\text{proj}_{\mathbf{w}_1}\mathbf{v} = \frac{x}{1}(1, 0, 0) \quad \Rightarrow \quad P_1 = \begin{bmatrix} 1 & 0 & 0 \\ 0 & 0 & 0 \\ 0 & 0 & 0 \end{bmatrix}$$

$$\text{proj}_{\mathbf{w}_2}\mathbf{v} = \frac{2y - z}{5}(0, 2, -1) \quad \Rightarrow \quad P_2 = \begin{bmatrix} 0 & 0 & 0 \\ 0 & \tfrac{4}{5} & -\tfrac{2}{5} \\ 0 & -\tfrac{2}{5} & \tfrac{1}{5} \end{bmatrix}.$$

So,

$$S = I - A = \begin{bmatrix} 1 & 0 & 0 \\ 0 & \tfrac{4}{5} & -\tfrac{2}{5} \\ 0 & -\tfrac{2}{5} & \tfrac{1}{5} \end{bmatrix} = P_1 + P_2$$

verifying that $S(\mathbf{v}) = \text{proj}_{\mathbf{w}_1}\mathbf{v} + \text{proj}_{\mathbf{w}_2}\mathbf{v}$.

(c) The kernel of T has basis $\{(1, 0, 0), (0, 2, -1)\}$, which is precisely the column space of S.

63. $S + T$ preserves addition.

$$(S + T)(\mathbf{v} + \mathbf{w}) = S(\mathbf{v} + \mathbf{w}) + T(\mathbf{v} + \mathbf{w})$$
$$= S(\mathbf{v}) + S(\mathbf{w}) + T(\mathbf{v}) + T(\mathbf{w})$$
$$= S(\mathbf{v}) + T(\mathbf{v}) + S(\mathbf{w}) + T(\mathbf{w})$$
$$= (S + T)\mathbf{v} + (S + T)(\mathbf{w})$$

$S + T$ preserves scalar multiplication.

$$(S + T)(c\mathbf{v}) = S(c\mathbf{v}) + T(c\mathbf{v})$$
$$= cS(\mathbf{v}) + cT(\mathbf{v})$$
$$= c(S(\mathbf{v}) + T(\mathbf{v}))$$
$$= c(S + T)(\mathbf{v})$$

kT preserves addition.

$$(kT)(\mathbf{v} + \mathbf{w}) = kT(\mathbf{v} + \mathbf{w}) = k(T(\mathbf{v}) + T(\mathbf{w}))$$
$$= kT(\mathbf{v}) + kT(\mathbf{w})$$
$$= (kT)(\mathbf{v}) + (kT)(\mathbf{w})$$

kT preserves scalar multiplication.

$$(kT)(c\mathbf{v}) = kT(c\mathbf{v}) = kcT(\mathbf{v}) = ckT(\mathbf{v}) = c(kT)(\mathbf{v}).$$

65. If S, T and $S + T$ are written as matrices, the number of linearly independent columns in $S + T$ cannot exceed the number of linearly independent columns in S plus the number of linearly independent columns in T, because the columns in $S + T$ are created by summing columns in S and T.

67. (a) T preserves addition.

$$T\left[\left(a_0 + a_1x + a_2x^2 + a_3x^3\right) + \left(b_0 + b_1x + b_2x^2 + b_3x^3\right)\right]$$

$$= T\left[\left(a_0 + b_0\right) + \left(a_1 + b_1\right)x + \left(a_2 + b_2\right)x^2 + \left(a_3 + b_3\right)x^3\right]$$

$$= \left(a_0 + b_0\right) + \left(a_1 + b_1\right) + \left(a_2 + b_2\right) + \left(a_3 + b_3\right)$$

$$= \left(a_0 + a_1 + a_2 + a_3\right) + \left(b_0 + b_1 + b_2 + b_3\right)$$

$$= T\left(a_0 + a_1x + a_2x^2 + a_3x^3\right) + T\left(b_0 + b_1x + b_2x^2 + b_3x^3\right)$$

T preserves scalar multiplication.

$$T\left(c\left(a_0 + a_1x + a_2x^2 + a_3x^3\right)\right) = T\left(ca_0 + ca_1x + ca_2x^2 + ca_3x^3\right)$$

$$= ca_0 + ca_1 + ca_2 + ca_3 = c\left(a_0 + a_1 + a_2 + a_3\right) = cT\left(a_0 + a_1x + a_2x^2 + a_3x^3\right)$$

(b) Because the range of T is R, rank$(T) = 1$. So, nullity$(T) = 4 - 1 = 3$.

(c) A basis for the kernel of T is obtained by solving $T\left(a_0 + a_1x + a_2x^2 + a_3x^3\right) = a_0 + a_1 + a_2 + a_3 = 0$.

Letting $a_3 = t$, $a_2 = s$, $a_1 = r$ be the free variables, $a_0 = -t - s - r$ and a basis is $\left\{-1 + x^3, -1 + x^2, -1 + x\right\}$.

69. Let B be a basis for V and let

$\left[\mathbf{v}_0\right]_B = \left[a_1, a_2, \ldots, a_n\right]^T$, where at least one $a_i \neq 0$ for

$i = 1, \ldots, n$. Then for $\left[\mathbf{v}\right]_B = \left[v_1, v_2, \ldots, v_n\right]^T$ you have

$\left[T(\mathbf{v})\right]_B = \left\langle \mathbf{v}, \mathbf{v}_0\right\rangle = a_1v_1 + a_2v_2 + \cdots + a_nv_n$.

The matrix for T relative to B is then

$A = \begin{bmatrix} a_1 & a_2 & \cdots & a_n \end{bmatrix}$.

Because A^T row-reduces to one nonzero row, the range of T is $\{t : t \in R\} = R$. So, the rank of T is 1 and nullity$(T) = n - 1$. Finally, $\ker(T) = \left\{\mathbf{v} : \left\langle \mathbf{v}, \mathbf{v}_0\right\rangle = 0\right\}$.

71. $M_{m,n}$ and $M_{p,q}$ will be isomorphic if they are of the same dimension. That is, $mn = pq$. Any function taking the standard basis of $M_{m,n}$ to the standard basis of $M_{p,q}$ will be an isomorphism.

73. (a) T is a vertical expansion.

(b)

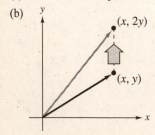

75. (a) T is a vertical shear.

(b)

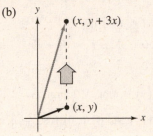

77. (a) T is a horizontal shear.

(b)

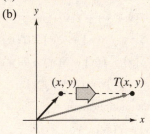

79. The image of each vertex is

$$T(0, 0) = (0, 0), T(1, 0) = (1, 0), T(0, 1) = (0, -1).$$

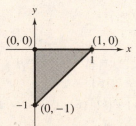

81. The image of each vertex is

$T(0, 0) = (0, 0)$, $T(1, 0) = (1, 0)$, and $T(0, 1) = (3, 1)$.

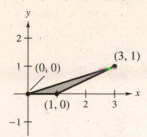

83. The transformation is a reflection in the line $y = x$

$$\begin{bmatrix} 0 & 1 \\ 1 & 0 \end{bmatrix}$$

followed by a horizontal expansion $\begin{bmatrix} 2 & 0 \\ 0 & 1 \end{bmatrix}$.

85. A rotation of 45° about the z-axis is given by

$$A = \begin{bmatrix} \cos 45° & -\sin 45° & 0 \\ \sin 45° & \cos 45° & 0 \\ 0 & 0 & 1 \end{bmatrix} = \begin{bmatrix} \frac{\sqrt{2}}{2} & -\frac{\sqrt{2}}{2} & 0 \\ \frac{\sqrt{2}}{2} & \frac{\sqrt{2}}{2} & 0 \\ 0 & 0 & 1 \end{bmatrix}.$$

Because

$$A\mathbf{v} = \begin{bmatrix} \frac{\sqrt{2}}{2} & -\frac{\sqrt{2}}{2} & 0 \\ \frac{\sqrt{2}}{2} & \frac{\sqrt{2}}{2} & 0 \\ 0 & 0 & 1 \end{bmatrix} \begin{bmatrix} 1 \\ -1 \\ 1 \end{bmatrix} = \begin{bmatrix} \sqrt{2} \\ 0 \\ 1 \end{bmatrix}$$

the image of $(1, -1, 1)$ is $(\sqrt{2}, 0, 1)$.

87. A rotation of 60° about the x-axis is given by

$$A = \begin{bmatrix} 1 & 0 & 0 \\ 0 & \cos 60° & -\sin 60° \\ 0 & \sin 60° & \cos 60° \end{bmatrix} = \begin{bmatrix} 1 & 0 & 0 \\ 0 & \frac{1}{2} & -\frac{\sqrt{3}}{2} \\ 0 & \frac{\sqrt{3}}{2} & \frac{1}{2} \end{bmatrix}.$$

Because

$$A\mathbf{v} = \begin{bmatrix} 1 & 0 & 0 \\ 0 & \frac{1}{2} & -\frac{\sqrt{3}}{2} \\ 0 & \frac{\sqrt{3}}{2} & \frac{1}{2} \end{bmatrix} \begin{bmatrix} 1 \\ -1 \\ 1 \end{bmatrix} = \begin{bmatrix} 1 \\ -\frac{1}{2} - \frac{\sqrt{3}}{2} \\ \frac{1}{2} - \frac{\sqrt{3}}{2} \end{bmatrix},$$

the image of $(1, -1, 1)$ is $\left(1, -\frac{1}{2} - \frac{\sqrt{3}}{2}, \frac{1}{2} - \frac{\sqrt{3}}{2}\right)$.

89. A rotation of 60° about the x-axis has the standard matrix

$$\begin{bmatrix} 1 & 0 & 0 \\ 0 & \cos 60° & -\sin 60° \\ 0 & \sin 60° & \cos 60° \end{bmatrix} = \begin{bmatrix} 1 & 0 & 0 \\ 0 & \frac{1}{2} & -\frac{\sqrt{3}}{2} \\ 0 & \frac{\sqrt{3}}{2} & \frac{1}{2} \end{bmatrix},$$

while a rotation of 30° about the z-axis has the standard matrix

$$\begin{bmatrix} \cos 30° & -\sin 30° & 0 \\ \sin 30° & \cos 30° & 0 \\ 0 & 0 & 1 \end{bmatrix} = \begin{bmatrix} \frac{\sqrt{3}}{2} & -\frac{1}{2} & 0 \\ \frac{1}{2} & \frac{\sqrt{3}}{2} & 0 \\ 0 & 0 & 1 \end{bmatrix}.$$

So, the pair of rotations is given by

$$\begin{bmatrix} \frac{\sqrt{3}}{2} & -\frac{1}{2} & 0 \\ \frac{1}{2} & \frac{\sqrt{3}}{2} & 0 \\ 0 & 0 & 1 \end{bmatrix} \begin{bmatrix} 1 & 0 & 0 \\ 0 & \frac{1}{2} & -\frac{\sqrt{3}}{2} \\ 0 & \frac{\sqrt{3}}{2} & \frac{1}{2} \end{bmatrix} = \begin{bmatrix} \frac{\sqrt{3}}{2} & -\frac{1}{4} & \frac{\sqrt{3}}{4} \\ \frac{1}{2} & \frac{\sqrt{3}}{4} & -\frac{3}{4} \\ 0 & \frac{\sqrt{3}}{2} & \frac{1}{2} \end{bmatrix}.$$

91. A rotation of 30° about the y-axis has the standard matrix

$$\begin{bmatrix} \cos 30° & 0 & \sin 30° \\ 0 & 1 & 0 \\ -\sin 30° & 0 & \cos 30° \end{bmatrix} = \begin{bmatrix} \frac{\sqrt{3}}{2} & 0 & \frac{1}{2} \\ 0 & 1 & 0 \\ -\frac{1}{2} & 0 & \frac{\sqrt{3}}{2} \end{bmatrix},$$

while a rotation of 45° about the z-axis has the standard matrix

$$\begin{bmatrix} \cos 45° & -\sin 45° & 0 \\ \sin 45° & \cos 45° & 0 \\ 0 & 0 & 1 \end{bmatrix} = \begin{bmatrix} \frac{\sqrt{2}}{2} & -\frac{\sqrt{2}}{2} & 0 \\ \frac{\sqrt{2}}{2} & \frac{\sqrt{2}}{2} & 0 \\ 0 & 0 & 1 \end{bmatrix}.$$

So, the pair of rotations is given by

$$\begin{bmatrix} \frac{\sqrt{2}}{2} & -\frac{\sqrt{2}}{2} & 0 \\ \frac{\sqrt{2}}{2} & \frac{\sqrt{2}}{2} & 0 \\ 0 & 0 & 1 \end{bmatrix} \begin{bmatrix} \frac{\sqrt{3}}{2} & 0 & \frac{1}{2} \\ 0 & 1 & 0 \\ -\frac{1}{2} & 0 & \frac{\sqrt{3}}{2} \end{bmatrix} = \begin{bmatrix} \frac{\sqrt{6}}{4} & -\frac{\sqrt{2}}{2} & \frac{\sqrt{2}}{4} \\ \frac{\sqrt{6}}{4} & \frac{\sqrt{2}}{2} & \frac{\sqrt{2}}{4} \\ -\frac{1}{2} & 0 & \frac{\sqrt{3}}{2} \end{bmatrix}.$$

93. The standard matrix for T is

$$\begin{bmatrix} \cos 45° & -\sin 45° & 0 \\ \sin 45° & \cos 45° & 0 \\ 0 & 0 & 1 \end{bmatrix} = \begin{bmatrix} \dfrac{\sqrt{2}}{2} & -\dfrac{\sqrt{2}}{2} & 0 \\ \dfrac{\sqrt{2}}{2} & \dfrac{\sqrt{2}}{2} & 0 \\ 0 & 0 & 1 \end{bmatrix}.$$

Therefore, T is given by

$$T(x, y, z) = \left(\frac{\sqrt{2}}{2}x - \frac{\sqrt{2}}{2}y, \frac{\sqrt{2}}{2}x + \frac{\sqrt{2}}{2}y, z \right).$$

The image of each vertex is as follows.

$T(0, 0, 0) = (0, 0, 0)$

$T(1, 0, 0) = \left(\dfrac{\sqrt{2}}{2}, \dfrac{\sqrt{2}}{2}, 0 \right)$

$T(1, 1, 0) = \left(0, \sqrt{2}, 0 \right)$

$T(0, 1, 0) = \left(-\dfrac{\sqrt{2}}{2}, \dfrac{\sqrt{2}}{2}, 0 \right)$

$T(0, 0, 1) = (0, 0, 1)$

$T(1, 0, 1) = \left(\dfrac{\sqrt{2}}{2}, \dfrac{\sqrt{2}}{2}, 1 \right)$

$T(1, 1, 1) = \left(0, \sqrt{2}, 1 \right)$

$T(0, 1, 1) = \left(-\dfrac{\sqrt{2}}{2}, \dfrac{\sqrt{2}}{2}, 1 \right)$

95. The standard matrix for T is

$$\begin{bmatrix} 1 & 0 & 0 \\ 0 & \cos 30° & -\sin 30° \\ 0 & \sin 30° & \cos 30° \end{bmatrix} = \begin{bmatrix} 1 & 0 & 0 \\ 0 & \dfrac{\sqrt{3}}{2} & -\dfrac{1}{2} \\ 0 & \dfrac{1}{2} & \dfrac{\sqrt{3}}{2} \end{bmatrix}.$$

Therefore, T is given by

$$T(x, y, z) = \left(x, \frac{\sqrt{3}}{2}y - \frac{1}{2}z, \frac{1}{2}y + \frac{\sqrt{3}}{2}z \right).$$

The image of each vertex is as follows.

$T(0, 0, 0) = (0, 0, 0)$

$T(1, 0, 0) = (1, 0, 0)$

$T(1, 1, 0) = \left(1, \dfrac{\sqrt{3}}{2}, \dfrac{1}{2} \right)$

$T(0, 1, 0) = \left(0, \dfrac{\sqrt{3}}{2}, \dfrac{1}{2} \right)$

$T(0, 0, 1) = \left(0, -\dfrac{1}{2}, \dfrac{\sqrt{3}}{2} \right)$

$T(1, 0, 1) = \left(1, -\dfrac{1}{2}, \dfrac{\sqrt{3}}{2} \right)$

$T(1, 1, 1) = \left(1, \dfrac{\sqrt{3}}{2} - \dfrac{1}{2}, \dfrac{1}{2} + \dfrac{\sqrt{3}}{2} \right)$

$T(0, 1, 1) = \left(0, \dfrac{\sqrt{3}}{2} - \dfrac{1}{2}, \dfrac{1}{2} + \dfrac{\sqrt{3}}{2} \right)$

97. (a) False. See "Elementary Matrices for Linear Transformations in the Plane," page 409.

    (b) True. See "Elementary Matrices for Linear Transformations in the Plane," page 407.

    (c) True. See discussion following Example 4, page 411.

99. (a) False. See "Remark," page 364.

    (b) False. See Theorem 6.7, page 383.

    (c) True. See discussion following Example 5, page 404.

C H A P T E R 7
Eigenvalues and Eigenvectors

C H A P T E R 7
Eigenvalues and Eigenvectors

Section 7.1   Eigenvalues and Eigenvectors

1. $A\mathbf{x}_1 = \begin{bmatrix} 1 & 0 \\ 0 & -1 \end{bmatrix}\begin{bmatrix} 1 \\ 0 \end{bmatrix} = \begin{bmatrix} 1 \\ 0 \end{bmatrix} = 1\begin{bmatrix} 1 \\ 0 \end{bmatrix} = \lambda_1\mathbf{x}_1$

$A\mathbf{x}_2 = \begin{bmatrix} 1 & 0 \\ 0 & -1 \end{bmatrix}\begin{bmatrix} 0 \\ 1 \end{bmatrix} = \begin{bmatrix} 0 \\ -1 \end{bmatrix} = -1\begin{bmatrix} 0 \\ 1 \end{bmatrix} = \lambda_2\mathbf{x}_2$

3. $A\mathbf{x}_1 = \begin{bmatrix} 1 & 1 \\ 1 & 1 \end{bmatrix}\begin{bmatrix} 1 \\ -1 \end{bmatrix} = \begin{bmatrix} 0 \\ 0 \end{bmatrix} = 0\begin{bmatrix} 1 \\ -1 \end{bmatrix} = \lambda_1\mathbf{x}_1$

$A\mathbf{x}_2 = \begin{bmatrix} 1 & 1 \\ 1 & 1 \end{bmatrix}\begin{bmatrix} 1 \\ 1 \end{bmatrix} = \begin{bmatrix} 2 \\ 2 \end{bmatrix} = 2\begin{bmatrix} 1 \\ 1 \end{bmatrix} = \lambda_2\mathbf{x}_2$

5. $A\mathbf{x}_1 = \begin{bmatrix} 2 & 3 & 1 \\ 0 & -1 & 2 \\ 0 & 0 & 3 \end{bmatrix}\begin{bmatrix} 1 \\ 0 \\ 0 \end{bmatrix} = \begin{bmatrix} 2 \\ 0 \\ 0 \end{bmatrix} = 2\begin{bmatrix} 1 \\ 0 \\ 0 \end{bmatrix} = \lambda_1\mathbf{x}_1$

$A\mathbf{x}_2 = \begin{bmatrix} 2 & 3 & 1 \\ 0 & -1 & 2 \\ 0 & 0 & 3 \end{bmatrix}\begin{bmatrix} 1 \\ -1 \\ 0 \end{bmatrix} = \begin{bmatrix} -1 \\ 1 \\ 0 \end{bmatrix} = -1\begin{bmatrix} 1 \\ -1 \\ 0 \end{bmatrix} = \lambda_2\mathbf{x}_2$

$A\mathbf{x}_3 = \begin{bmatrix} 2 & 3 & 1 \\ 0 & -1 & 2 \\ 0 & 0 & 3 \end{bmatrix}\begin{bmatrix} 5 \\ 1 \\ 2 \end{bmatrix} = \begin{bmatrix} 15 \\ 3 \\ 6 \end{bmatrix} = 3\begin{bmatrix} 5 \\ 1 \\ 2 \end{bmatrix} = \lambda_3\mathbf{x}_3$

7. $A\mathbf{x}_1 = \begin{bmatrix} 0 & 1 & 0 \\ 0 & 0 & 1 \\ 1 & 0 & 0 \end{bmatrix}\begin{bmatrix} 1 \\ 1 \\ 1 \end{bmatrix} = \begin{bmatrix} 1 \\ 1 \\ 1 \end{bmatrix} = 1\begin{bmatrix} 1 \\ 1 \\ 1 \end{bmatrix} = \lambda_1\mathbf{x}_1$

9. (a) $A(c\mathbf{x}_1) = \begin{bmatrix} 1 & 1 \\ 1 & 1 \end{bmatrix}\begin{bmatrix} c \\ -c \end{bmatrix} = \begin{bmatrix} 0 \\ 0 \end{bmatrix} = 0\begin{bmatrix} c \\ -c \end{bmatrix} = 0(c\mathbf{x}_1)$

(b) $A(c\mathbf{x}_2) = \begin{bmatrix} 1 & 1 \\ 1 & 1 \end{bmatrix}\begin{bmatrix} c \\ c \end{bmatrix} = \begin{bmatrix} 2c \\ 2c \end{bmatrix} = 2\begin{bmatrix} c \\ c \end{bmatrix} = 2(c\mathbf{x}_2)$

11. (a) Because

$A\mathbf{x} = \begin{bmatrix} 7 & 2 \\ 2 & 4 \end{bmatrix}\begin{bmatrix} 1 \\ 2 \end{bmatrix} = \begin{bmatrix} 11 \\ 10 \end{bmatrix} \neq \lambda\begin{bmatrix} 1 \\ 2 \end{bmatrix}$

$\mathbf{x}$ is *not* an eigenvector of A.

(b) Because

$A\mathbf{x} = \begin{bmatrix} 7 & 2 \\ 2 & 4 \end{bmatrix}\begin{bmatrix} 2 \\ 1 \end{bmatrix} = \begin{bmatrix} 16 \\ 8 \end{bmatrix} = 8\begin{bmatrix} 2 \\ 1 \end{bmatrix}$

$\mathbf{x}$ *is* an eigenvector of A (with a corresponding eigenvalue 8).

(c) Because

$A\mathbf{x} = \begin{bmatrix} 7 & 2 \\ 2 & 4 \end{bmatrix}\begin{bmatrix} 1 \\ -2 \end{bmatrix} = \begin{bmatrix} 3 \\ -6 \end{bmatrix} = 3\begin{bmatrix} 1 \\ -2 \end{bmatrix}$

$\mathbf{x}$ *is* an eigenvector of A (with a corresponding eigenvalue 3).

(d) Because

$A\mathbf{x} = \begin{bmatrix} 7 & 2 \\ 2 & 4 \end{bmatrix}\begin{bmatrix} -1 \\ 0 \end{bmatrix} = \begin{bmatrix} -7 \\ -2 \end{bmatrix} \neq \lambda\begin{bmatrix} -1 \\ 0 \end{bmatrix}$

$\mathbf{x}$ is *not* an eigenvector of A.

13. (a) Because

$A\mathbf{x} = \begin{bmatrix} -1 & -1 & 1 \\ -2 & 0 & -2 \\ 3 & -3 & 1 \end{bmatrix}\begin{bmatrix} 2 \\ -4 \\ 6 \end{bmatrix} = \begin{bmatrix} 8 \\ -16 \\ 24 \end{bmatrix} = 4\begin{bmatrix} 2 \\ -4 \\ 6 \end{bmatrix}$

$\mathbf{x}$ *is* an eigenvector of A (with a corresponding eigenvalue 4).

(b) Because

$A\mathbf{x} = \begin{bmatrix} -1 & -1 & 1 \\ -2 & 0 & -2 \\ 3 & -3 & 1 \end{bmatrix}\begin{bmatrix} 2 \\ 0 \\ 6 \end{bmatrix} = \begin{bmatrix} 4 \\ -16 \\ 12 \end{bmatrix} \neq \lambda\begin{bmatrix} 2 \\ 0 \\ 6 \end{bmatrix}$

$\mathbf{x}$ is *not* an eigenvector of A.

(c) Because

$A\mathbf{x} = \begin{bmatrix} -1 & -1 & 1 \\ -2 & 0 & -2 \\ 3 & -3 & 1 \end{bmatrix}\begin{bmatrix} 2 \\ 2 \\ 0 \end{bmatrix} = \begin{bmatrix} -4 \\ -4 \\ 0 \end{bmatrix} = -2\begin{bmatrix} 2 \\ 2 \\ 0 \end{bmatrix}$

$\mathbf{x}$ *is* an eigenvector of A (with a corresponding eigenvalue -2).

(d) Because

$A\mathbf{x} = \begin{bmatrix} -1 & -1 & 1 \\ -2 & 0 & -2 \\ 3 & -3 & 1 \end{bmatrix}\begin{bmatrix} -1 \\ 0 \\ 1 \end{bmatrix} = \begin{bmatrix} 2 \\ 0 \\ -2 \end{bmatrix} = -2\begin{bmatrix} -1 \\ 0 \\ 1 \end{bmatrix}$

$\mathbf{x}$ *is* an eigenvector of A (with a corresponding eigenvalue -2).

15. (a) The characteristic equation is $|\lambda I - A| = \begin{vmatrix} \lambda - 6 & 3 \\ 2 & \lambda - 1 \end{vmatrix} = \lambda^2 - 7\lambda = \lambda(\lambda - 7) = 0.$

(b) The eigenvalues are $\lambda_1 = 0$ and $\lambda_2 = 7$.

For $\lambda_1 = 0,$ $\begin{bmatrix} \lambda_1 - 6 & 3 \\ 2 & \lambda_1 - 1 \end{bmatrix}\begin{bmatrix} x_1 \\ x_2 \end{bmatrix} = \begin{bmatrix} 0 \\ 0 \end{bmatrix} \Rightarrow \begin{bmatrix} 2 & -1 \\ 0 & 0 \end{bmatrix}\begin{bmatrix} x_1 \\ x_2 \end{bmatrix} = \begin{bmatrix} 0 \\ 0 \end{bmatrix}.$

The solution is $\{(t, 2t) : t \in R\}$. So, an eigenvector corresponding to $\lambda_1 = 0$ is $(1, 2)$.

For $\lambda_2 = 7,$ $\begin{bmatrix} \lambda_2 - 6 & 3 \\ 2 & \lambda_2 - 1 \end{bmatrix}\begin{bmatrix} x_1 \\ x_2 \end{bmatrix} = \begin{bmatrix} 0 \\ 0 \end{bmatrix} \Rightarrow \begin{bmatrix} 1 & 3 \\ 0 & 0 \end{bmatrix}\begin{bmatrix} x_1 \\ x_2 \end{bmatrix} = \begin{bmatrix} 0 \\ 0 \end{bmatrix}.$

The solution is $\{(-3t, t) : t \in R\}$. So, an eigenvector corresponding to $\lambda_2 = 7$ is $(-3, 1)$.

17. (a) The characteristic equation is $|\lambda I - A| = \begin{vmatrix} \lambda - 1 & \frac{3}{2} \\ -\frac{1}{2} & \lambda + 1 \end{vmatrix} = \lambda^2 - \frac{1}{4} = 0.$

(b) The eigenvalues are $\lambda_1 = \frac{1}{2}$ and $\lambda_2 = -\frac{1}{2}$.

For $\lambda_1 = \frac{1}{2},$ $\begin{vmatrix} \lambda_1 - 1 & \frac{3}{2} \\ -\frac{1}{2} & \lambda_1 + 1 \end{vmatrix}\begin{bmatrix} x_1 \\ x_2 \end{bmatrix} = \begin{bmatrix} 0 \\ 0 \end{bmatrix} \Rightarrow \begin{bmatrix} 1 & -3 \\ 0 & 0 \end{bmatrix}\begin{bmatrix} x_1 \\ x_2 \end{bmatrix} = \begin{bmatrix} 0 \\ 0 \end{bmatrix}.$

The solution is $\{(-3t, t) : t \in R\}$. So, an eigenvector corresponding to $\lambda_1 = \frac{1}{2}$ is $(3, 1)$.

For $\lambda_2 = -\frac{1}{2},$ $\begin{vmatrix} \lambda_2 - 1 & \frac{3}{2} \\ -\frac{1}{2} & \lambda_2 + 1 \end{vmatrix}\begin{bmatrix} x_1 \\ x_2 \end{bmatrix} = \begin{bmatrix} 0 \\ 0 \end{bmatrix} \Rightarrow \begin{bmatrix} 1 & -1 \\ 0 & 0 \end{bmatrix}\begin{bmatrix} x_1 \\ x_2 \end{bmatrix} = \begin{bmatrix} 0 \\ 0 \end{bmatrix}.$

The solution is $\{(t, t) : t \in R\}$. So, an eigenvector corresponding to $\lambda_2 = -\frac{1}{2}$ is $(1, 1)$.

19. (a) The characteristic equation is $|\lambda I - A| = \begin{vmatrix} \lambda - 2 & 0 & -1 \\ 0 & \lambda - 3 & -4 \\ 0 & 0 & \lambda - 1 \end{vmatrix} = (\lambda - 2)(\lambda - 3)(\lambda - 1) = 0.$

(b) The eigenvalues are $\lambda_1 = 2, \lambda_2 = 3$ and $\lambda_3 = 1$.

For $\lambda_1 = 2,$ $\begin{bmatrix} \lambda_1 - 2 & 0 & -1 \\ 0 & \lambda_1 - 3 & -4 \\ 0 & 0 & \lambda_1 - 1 \end{bmatrix}\begin{bmatrix} x_1 \\ x_2 \\ x_3 \end{bmatrix} = \begin{bmatrix} 0 \\ 0 \\ 0 \end{bmatrix} \Rightarrow \begin{bmatrix} 0 & 1 & 0 \\ 0 & 0 & 1 \\ 0 & 0 & 0 \end{bmatrix}\begin{bmatrix} x_1 \\ x_2 \\ x_3 \end{bmatrix} = \begin{bmatrix} 0 \\ 0 \\ 0 \end{bmatrix}.$

The solution is $\{(t, 0, 0) : t \in R\}$. So, an eigenvector corresponding to $\lambda_1 = 2$ is $(1, 0, 0)$.

For $\lambda_2 = 3,$ $\begin{bmatrix} \lambda_2 - 2 & 0 & -1 \\ 0 & \lambda_2 - 3 & -4 \\ 0 & 0 & \lambda_2 - 1 \end{bmatrix}\begin{bmatrix} x_1 \\ x_2 \\ x_3 \end{bmatrix} = \begin{bmatrix} 0 \\ 0 \\ 0 \end{bmatrix} \Rightarrow \begin{bmatrix} 1 & 0 & 0 \\ 0 & 0 & 1 \\ 0 & 0 & 0 \end{bmatrix}\begin{bmatrix} x_1 \\ x_2 \\ x_3 \end{bmatrix} = \begin{bmatrix} 0 \\ 0 \\ 0 \end{bmatrix}.$

The solution is $\{(0, t, 0) : t \in R\}$. So, an eigenvector corresponding to $\lambda_2 = 3$ is $(0, 1, 0)$.

For $\lambda_3 = 1,$ $\begin{bmatrix} \lambda_3 - 2 & 0 & -1 \\ 0 & \lambda_3 - 3 & -4 \\ 0 & 0 & \lambda_3 - 1 \end{bmatrix}\begin{bmatrix} x_1 \\ x_2 \\ x_3 \end{bmatrix} = \begin{bmatrix} 0 \\ 0 \\ 0 \end{bmatrix} \Rightarrow \begin{bmatrix} 1 & 0 & 1 \\ 0 & 1 & 2 \\ 0 & 0 & 0 \end{bmatrix}\begin{bmatrix} x_1 \\ x_2 \\ x_3 \end{bmatrix} = \begin{bmatrix} 0 \\ 0 \\ 0 \end{bmatrix}.$

The solution is $\{(-t, -2t, t) : t \in R\}$. So, an eigenvector corresponding to $\lambda_3 = 1$ is $(-1, -2, 1)$.

21. (a) The characteristic equation is $|\lambda I - A| = \begin{vmatrix} \lambda - 2 & 2 & -3 \\ 0 & \lambda - 3 & 2 \\ 0 & 1 & \lambda - 2 \end{vmatrix} = (\lambda - 2)(\lambda - 4)(\lambda - 1) = 0.$

(b) The eigenvalues are $\lambda_1 = 1, \lambda_2 = 2,$ and $\lambda_3 = 4.$

For $\lambda_1 = 1, \begin{bmatrix} \lambda_1 - 2 & 2 & -3 \\ 0 & \lambda_1 - 3 & 2 \\ 0 & 1 & \lambda_1 - 2 \end{bmatrix} \begin{bmatrix} x_1 \\ x_2 \\ x_3 \end{bmatrix} = \begin{bmatrix} 0 \\ 0 \\ 0 \end{bmatrix} \Rightarrow \begin{bmatrix} -1 & 2 & -3 \\ 0 & -2 & 2 \\ 0 & 1 & -1 \end{bmatrix} \begin{bmatrix} x_1 \\ x_2 \\ x_3 \end{bmatrix} = \begin{bmatrix} 0 \\ 0 \\ 0 \end{bmatrix}.$

The solution is $\{(-t, t, t) : t \in R\}.$ So, an eigenvector corresponding to $\lambda_1 = 1$ is $(-1, 1, 1).$

For $\lambda_2 = 2, \begin{bmatrix} \lambda_2 - 2 & 2 & -3 \\ 0 & \lambda_2 - 3 & 2 \\ 0 & 1 & \lambda_2 - 2 \end{bmatrix} \begin{bmatrix} x_1 \\ x_2 \\ x_3 \end{bmatrix} = \begin{bmatrix} 0 \\ 0 \\ 0 \end{bmatrix} \Rightarrow \begin{bmatrix} 0 & 2 & -3 \\ 0 & -1 & 2 \\ 0 & 1 & 0 \end{bmatrix} \begin{bmatrix} x_1 \\ x_2 \\ x_3 \end{bmatrix} = \begin{bmatrix} 0 \\ 0 \\ 0 \end{bmatrix}.$

The solution is $\{(t, 0, 0) : t \in R\}.$ So, an eigenvector corresponding to $\lambda_2 = 2$ is $(1, 0, 0).$

For $\lambda_3 = 4, \begin{bmatrix} \lambda_3 - 2 & 2 & -3 \\ 0 & \lambda_3 - 3 & 2 \\ 0 & 1 & \lambda_3 - 2 \end{bmatrix} \begin{bmatrix} x_1 \\ x_2 \\ x_3 \end{bmatrix} = \begin{bmatrix} 0 \\ 0 \\ 0 \end{bmatrix} \Rightarrow \begin{bmatrix} 2 & 2 & -3 \\ 0 & 1 & 2 \\ 0 & 1 & 2 \end{bmatrix} \begin{bmatrix} x_1 \\ x_2 \\ x_3 \end{bmatrix} = \begin{bmatrix} 0 \\ 0 \\ 0 \end{bmatrix}.$

The solution is $\{(7t, -4t, 2t) : t \in R\}.$ So, an eigenvector corresponding to $\lambda_3 = 4$ is $(7, -4, 2).$

23. (a) The characteristic equation is $|\lambda I - A| = \begin{vmatrix} \lambda - 1 & -2 & 2 \\ 2 & \lambda - 5 & 2 \\ 6 & -6 & \lambda + 3 \end{vmatrix} = \lambda^3 - 3\lambda^2 - 9\lambda + 27 = (\lambda + 3)(\lambda - 3)^2 = 0.$

(b) The eigenvalues are $\lambda_1 = -3$ and $\lambda_2 = 3$ (repeated).

For $\lambda_1 = -3, \begin{bmatrix} \lambda_1 - 1 & -2 & 2 \\ 2 & \lambda_1 - 5 & 2 \\ 6 & -6 & \lambda_1 + 3 \end{bmatrix} \begin{bmatrix} x_1 \\ x_2 \\ x_3 \end{bmatrix} = \begin{bmatrix} 0 \\ 0 \\ 0 \end{bmatrix} \Rightarrow \begin{bmatrix} -4 & -2 & 2 \\ 2 & -8 & 2 \\ 6 & -6 & 0 \end{bmatrix} \begin{bmatrix} x_1 \\ x_2 \\ x_3 \end{bmatrix} = \begin{bmatrix} 0 \\ 0 \\ 0 \end{bmatrix}.$

The solution is $\{(t, t, 3t) : t \in R\}.$ So, an eigenvector corresponding to $\lambda_1 = -3$ is $(1, 1, 3).$

For $\lambda_2 = 3, \begin{bmatrix} \lambda_2 - 1 & -2 & 2 \\ 2 & \lambda_2 - 5 & 2 \\ 6 & -6 & \lambda_2 + 3 \end{bmatrix} \begin{bmatrix} x_1 \\ x_2 \\ x_3 \end{bmatrix} = \begin{bmatrix} 0 \\ 0 \\ 0 \end{bmatrix} \Rightarrow \begin{bmatrix} 2 & -2 & 2 \\ 2 & -2 & 2 \\ 6 & -6 & 6 \end{bmatrix} \begin{bmatrix} x_1 \\ x_2 \\ x_3 \end{bmatrix} = \begin{bmatrix} 0 \\ 0 \\ 0 \end{bmatrix}.$

The solution is $\{(s - t, s, t) : s, t \in R\}.$ So, two eigenvector corresponding to $\lambda_2 = 3$ are $(1, 1, 0)$ and $(1, 0, -1).$

25. (a) The characteristic equation is

$$|\lambda I - A| = \begin{vmatrix} \lambda & 3 & -5 \\ 4 & \lambda - 4 & 10 \\ 0 & 0 & \lambda - 4 \end{vmatrix} = (\lambda - 4)(\lambda^2 - 4\lambda - 12) = (\lambda - 4)(\lambda - 6)(\lambda + 2) = 0.$$

(b) The eigenvalues are $\lambda_1 = 4$, $\lambda_2 = 6$ and $\lambda_3 = -2$.

For $\lambda_1 = 4$, $\begin{bmatrix} \lambda_1 & 3 & -5 \\ 4 & \lambda_1 - 4 & 10 \\ 0 & 0 & \lambda_1 - 4 \end{bmatrix}\begin{bmatrix} x_1 \\ x_2 \\ x_3 \end{bmatrix} = \begin{bmatrix} 0 \\ 0 \\ 0 \end{bmatrix} \Rightarrow \begin{bmatrix} 2 & 0 & 5 \\ 0 & 1 & -5 \\ 0 & 0 & 0 \end{bmatrix}\begin{bmatrix} x_1 \\ x_2 \\ x_3 \end{bmatrix} = \begin{bmatrix} 0 \\ 0 \\ 0 \end{bmatrix}.$

The solution is $\{(-5t, 10t, 2t) : t \in R\}$. So, an eigenvector corresponding to $\lambda_1 = 4$ is $(-5, 10, 2)$.

For $\lambda_2 = 6$, $\begin{bmatrix} \lambda_2 & 3 & -5 \\ 4 & \lambda_2 - 4 & 10 \\ 0 & 0 & \lambda_2 - 4 \end{bmatrix}\begin{bmatrix} x_1 \\ x_2 \\ x_3 \end{bmatrix} = \begin{bmatrix} 0 \\ 0 \\ 0 \end{bmatrix} \Rightarrow \begin{bmatrix} 2 & 1 & 0 \\ 0 & 0 & 1 \\ 0 & 0 & 0 \end{bmatrix}\begin{bmatrix} x_1 \\ x_2 \\ x_3 \end{bmatrix} = \begin{bmatrix} 0 \\ 0 \\ 0 \end{bmatrix}.$

The solution is $\{(t, -2t, 0) : t \in R\}$. So, an eigenvector corresponding to $\lambda_2 = 6$ is $(1, -2, 0)$.

For $\lambda_3 = -2$, $\begin{bmatrix} \lambda_3 & 3 & -5 \\ 4 & \lambda_3 - 4 & 10 \\ 0 & 0 & \lambda_3 - 4 \end{bmatrix}\begin{bmatrix} x_1 \\ x_2 \\ x_3 \end{bmatrix} = \begin{bmatrix} 0 \\ 0 \\ 0 \end{bmatrix} \Rightarrow \begin{bmatrix} 2 & -3 & 0 \\ 0 & 0 & 1 \\ 0 & 0 & 0 \end{bmatrix}\begin{bmatrix} x_1 \\ x_2 \\ x_3 \end{bmatrix} = \begin{bmatrix} 0 \\ 0 \\ 0 \end{bmatrix}.$

The solution is $\{(3t, 2t, 0) : t \in R\}$. So, an eigenvector corresponding to $\lambda_3 = -2$ is $(3, 2, 0)$.

27. (a) The characteristic equation is

$$|\lambda I - A| = \begin{vmatrix} \lambda - 2 & 0 & 0 & 0 \\ 0 & \lambda - 2 & 0 & 0 \\ 0 & 0 & \lambda - 3 & 0 \\ 0 & 0 & -4 & \lambda \end{vmatrix} = \lambda(\lambda - 3)(\lambda - 2)^2 = 0.$$

(b) The eigenvalues are $\lambda_1 = 0$, $\lambda_2 = 3$ and $\lambda_3 = 2$ (repeated).

For $\lambda_1 = 0$, $\begin{bmatrix} -2 & 0 & 0 & 0 \\ 0 & -2 & 0 & 0 \\ 0 & 0 & -3 & 0 \\ 0 & 0 & -4 & 0 \end{bmatrix}\begin{bmatrix} x_1 \\ x_2 \\ x_3 \\ x_4 \end{bmatrix} = \begin{bmatrix} 0 \\ 0 \\ 0 \\ 0 \end{bmatrix} \Rightarrow \begin{bmatrix} 1 & 0 & 0 & 0 \\ 0 & 1 & 0 & 0 \\ 0 & 0 & 1 & 0 \\ 0 & 0 & 0 & 0 \end{bmatrix}\begin{bmatrix} x_1 \\ x_2 \\ x_3 \\ x_4 \end{bmatrix} = \begin{bmatrix} 0 \\ 0 \\ 0 \\ 0 \end{bmatrix}.$

The solution is $\{(0, 0, 0, t) : t \in R\}$. So, an eigenvector corresponding to $\lambda_1 = 0$ is $(0, 0, 0, 1)$.

For $\lambda_2 = 3$, $\begin{bmatrix} 1 & 0 & 0 & 0 \\ 0 & 1 & 0 & 0 \\ 0 & 0 & 0 & 0 \\ 0 & 0 & -4 & 3 \end{bmatrix}\begin{bmatrix} x_1 \\ x_2 \\ x_3 \\ x_4 \end{bmatrix} = \begin{bmatrix} 0 \\ 0 \\ 0 \\ 0 \end{bmatrix} \Rightarrow \begin{bmatrix} 1 & 0 & 0 & 0 \\ 0 & 1 & 0 & 0 \\ 0 & 0 & 1 & -\frac{3}{4} \\ 0 & 0 & 0 & 0 \end{bmatrix}\begin{bmatrix} x_1 \\ x_2 \\ x_3 \\ x_4 \end{bmatrix} = \begin{bmatrix} 0 \\ 0 \\ 0 \\ 0 \end{bmatrix}.$

The solution is $\{(0, 0, \frac{3}{4}t, t) : t \in R\}$. So, an eigenvector corresponding to $\lambda_2 = 3$ is $(0, 0, 3, 4)$.

For $\lambda_3 = 2$, $\begin{bmatrix} 0 & 0 & 0 & 0 \\ 0 & 0 & 0 & 0 \\ 0 & 0 & -1 & 0 \\ 0 & 0 & -4 & 2 \end{bmatrix}\begin{bmatrix} x_1 \\ x_2 \\ x_3 \\ x_4 \end{bmatrix} = \begin{bmatrix} 0 \\ 0 \\ 0 \\ 0 \end{bmatrix} \Rightarrow \begin{bmatrix} 0 & 0 & 1 & 0 \\ 0 & 0 & 0 & 1 \\ 0 & 0 & 0 & 0 \\ 0 & 0 & 0 & 0 \end{bmatrix}\begin{bmatrix} x_1 \\ x_2 \\ x_3 \\ x_4 \end{bmatrix} = \begin{bmatrix} 0 \\ 0 \\ 0 \\ 0 \end{bmatrix}.$

The solution is $\{(s, t, 0, 0) : s, t \in R\}$. So, two eigenvectors corresponding to $\lambda_3 = 2$ are $(1, 0, 0, 0)$ and $(0, 1, 0, 0)$.

29. Using a graphing utility: $\lambda = -2, 1$

31. Using a graphing utility: $\lambda = 5, 5$

33. Using a graphing utility: $\lambda = \frac{1}{3}, -\frac{1}{2}, 4$

35. Using a graphing utility: $\lambda = -1, 4, 4$

37. Using a graphing utility: $\lambda = 0, 3$

39. Using a graphing utility: $\lambda = 0, 0, 0, 21$

41. The characteristic equation is

$$|\lambda I - A| = \begin{vmatrix} \lambda - 4 & 0 \\ 3 & \lambda - 2 \end{vmatrix} = \lambda^2 - 6\lambda + 8 = 0.$$

Because

$$A^2 - 6A + 8I = \begin{bmatrix} 4 & 0 \\ -3 & 2 \end{bmatrix}^2 - 6\begin{bmatrix} 4 & 0 \\ -3 & 2 \end{bmatrix} + 8\begin{bmatrix} 1 & 0 \\ 0 & 1 \end{bmatrix}$$

$$= \begin{bmatrix} 16 & 0 \\ -18 & 4 \end{bmatrix} - \begin{bmatrix} 24 & 0 \\ -18 & 12 \end{bmatrix} + \begin{bmatrix} 8 & 0 \\ 0 & 8 \end{bmatrix}$$

$$= \begin{bmatrix} 0 & 0 \\ 0 & 0 \end{bmatrix}$$

the theorem holds for this matrix.

45. The characteristic equation is

$$|\lambda I - A| = \begin{vmatrix} \lambda & -2 & 1 \\ 1 & \lambda - 3 & -1 \\ 0 & 0 & \lambda + 1 \end{vmatrix} = \lambda^3 - 2\lambda^2 - \lambda + 2 = 0.$$

Because

$$A^3 - 2A^2 - A + 2 = \begin{bmatrix} 0 & 2 & -1 \\ -1 & 3 & 1 \\ 0 & 0 & -1 \end{bmatrix}^3 - 2\begin{bmatrix} 0 & 2 & -1 \\ -1 & 3 & 1 \\ 0 & 0 & -1 \end{bmatrix}^2 - \begin{bmatrix} 0 & 2 & -1 \\ -1 & 3 & 1 \\ 0 & 0 & -1 \end{bmatrix} + 2\begin{bmatrix} 1 & 0 & 0 \\ 0 & 1 & 0 \\ 0 & 0 & 1 \end{bmatrix}$$

$$= \begin{bmatrix} -6 & 14 & 5 \\ -7 & 15 & 7 \\ 0 & 0 & -1 \end{bmatrix} - 2\begin{bmatrix} -2 & 6 & 3 \\ -3 & 7 & 3 \\ 0 & 0 & 1 \end{bmatrix} - \begin{bmatrix} 0 & 2 & -1 \\ -1 & 3 & 1 \\ 0 & 0 & -1 \end{bmatrix} + \begin{bmatrix} 2 & 0 & 0 \\ 0 & 2 & 0 \\ 0 & 0 & 2 \end{bmatrix}$$

$$= \begin{bmatrix} 0 & 0 & 0 \\ 0 & 0 & 0 \\ 0 & 0 & 0 \end{bmatrix}$$

the theorem holds for this matrix.

43. The characteristic equation is

$$|\lambda I - A| = \begin{vmatrix} \lambda - 2 & 2 \\ -1 & \lambda - 5 \end{vmatrix} = \lambda^2 - 7\lambda + 12 = 0.$$

Because

$$A^2 - 7A + 12I = \begin{bmatrix} 2 & -2 \\ 1 & 5 \end{bmatrix}^2 - 7\begin{bmatrix} 2 & -2 \\ 1 & 5 \end{bmatrix} + 12\begin{bmatrix} 1 & 0 \\ 0 & 1 \end{bmatrix}$$

$$= \begin{bmatrix} 2 & -14 \\ 7 & 23 \end{bmatrix} - \begin{bmatrix} 14 & -14 \\ 7 & 35 \end{bmatrix} + \begin{bmatrix} 12 & 0 \\ 0 & 12 \end{bmatrix}$$

$$= \begin{bmatrix} 0 & 0 \\ 0 & 0 \end{bmatrix}$$

the theorem holds for this matrix.

47. The characteristic equation is

$$|\lambda I - A| = \begin{vmatrix} \lambda - 1 & 0 & 4 \\ 0 & \lambda - 3 & -1 \\ -2 & 0 & \lambda - 1 \end{vmatrix} = \lambda^3 - 5\lambda^2 + 15\lambda - 27 = 0.$$

Because

$$A^3 - 5A^2 + 15A - 27I = \begin{bmatrix} 1 & 0 & -4 \\ 0 & 3 & 1 \\ 2 & 0 & 1 \end{bmatrix}^3 - 5\begin{bmatrix} 1 & 0 & -4 \\ 0 & 3 & 1 \\ 2 & 0 & 1 \end{bmatrix}^2 + 15\begin{bmatrix} 1 & 0 & -4 \\ 0 & 3 & 1 \\ 2 & 0 & 1 \end{bmatrix} - 27\begin{bmatrix} 1 & 0 & 0 \\ 0 & 1 & 0 \\ 0 & 0 & 1 \end{bmatrix}$$

$$= \begin{bmatrix} -23 & 0 & 20 \\ 10 & 27 & 5 \\ -10 & 0 & -23 \end{bmatrix} - 5\begin{bmatrix} -7 & 0 & -8 \\ 2 & 9 & 4 \\ 4 & 0 & -7 \end{bmatrix} + 15\begin{bmatrix} 1 & 0 & -4 \\ 0 & 3 & 1 \\ 2 & 0 & 1 \end{bmatrix} - \begin{bmatrix} 27 & 0 & 0 \\ 0 & 27 & 0 \\ 0 & 0 & 27 \end{bmatrix}$$

$$= \begin{bmatrix} 0 & 0 & 0 \\ 0 & 0 & 0 \\ 0 & 0 & 0 \end{bmatrix}$$

the theorem holds for this matrix.

49. For the $n \times n$ matrix $A = \begin{bmatrix} a_{ij} \end{bmatrix}$, the sum of the diagonal entries, or the trace, of A is given by $\sum_{i=1}^{n} a_{ii}$.

Exercise 15: $\lambda_1 = 0, \lambda_2 = 7$

(a) $\sum_{i=1}^{2} \lambda_i = 7 = \sum_{i=1}^{2} a_{ii}$

(b) $|A| = \begin{vmatrix} 6 & -3 \\ -2 & 1 \end{vmatrix} = 0 = 0 \cdot 7 = \lambda_1 \cdot \lambda_2$

Exercise 17: $\lambda_1 = \frac{1}{2}, \lambda_2 = -\frac{1}{2}$

(a) $\sum_{i=1}^{2} \lambda_i = 0 = \sum_{i=1}^{2} a_{ii}$

(b) $|A| = \begin{vmatrix} 1 & -\frac{3}{2} \\ \frac{1}{2} & -1 \end{vmatrix} = -\frac{1}{4} = \frac{1}{2} \cdot \left(-\frac{1}{2}\right) = \lambda_1 \cdot \lambda_2$

Exercise 19: $\lambda_1 = 2, \lambda_2 = 3, \lambda_3 = 1$

(a) $\sum_{i=1}^{3} \lambda_i = 6 = \sum_{i=1}^{3} a_{ii}$

(b) $|A| = \begin{vmatrix} 2 & 0 & 1 \\ 0 & 3 & 4 \\ 0 & 0 & 1 \end{vmatrix} = 6 = 2 \cdot 3 \cdot 1 = \lambda_1 \cdot \lambda_2 \cdot \lambda_3$

Exercise 21: $\lambda_1 = 1, \lambda_2 = 2, \lambda_3 = 4$

(a) $\sum_{i=1}^{3} \lambda_i = 7 = \sum_{i=1}^{3} a_{ii}$

(b) $|A| = \begin{vmatrix} 2 & -2 & 3 \\ 0 & 3 & -2 \\ 0 & -1 & 2 \end{vmatrix} = 8 = 1 \cdot 2 \cdot 4 = \lambda_1 \cdot \lambda_2 \cdot \lambda_3$

Exercise 23: $\lambda_1 = -3, \lambda_2 = 3, \lambda_3 = 3$

(a) $\displaystyle\sum_{i=1}^{3}\lambda_i = 3 = \sum_{i=1}^{3} a_{ii}$

(b) $|A| = \begin{vmatrix} 1 & 2 & -2 \\ -2 & 5 & -2 \\ -6 & 6 & -3 \end{vmatrix} = -27 = -3 \cdot 3 \cdot 3 = \lambda_1 \cdot \lambda_2 \cdot \lambda_3$

Exercise 25: $\lambda_1 = 4, \lambda_2 = 6, \lambda_3 = -2$

(a) $\displaystyle\sum_{i=1}^{3}\lambda_i = 8 = \sum_{i=1}^{3} a_{ii}$

(b) $|A| = \begin{vmatrix} 0 & -3 & 5 \\ -4 & 4 & -10 \\ 0 & 0 & 4 \end{vmatrix} = -48 = 4 \cdot 6 \cdot (-2) = \lambda_1 \cdot \lambda_2 \cdot \lambda_3$

Exercise 27: $\lambda_1 = 0, \lambda_2 = 3, \lambda_3 = 2, \lambda_4 = 2$

(a) $\displaystyle\sum_{i=1}^{4}\lambda_i = 7 = \sum_{i=1}^{4} a_{ii}$

(b) $|A| = \begin{vmatrix} 2 & 0 & 0 & 0 \\ 0 & 2 & 0 & 0 \\ 0 & 0 & 3 & 0 \\ 0 & 0 & 4 & 0 \end{vmatrix} = 0 = 0 \cdot 3 \cdot 2 \cdot 2 = \lambda_1 \cdot \lambda_2 \cdot \lambda_3 \cdot \lambda_4$

51. Because the ith row of A is identical to the ith row of I, the ith row of A consists of zeros except for the main diagonal entry, which is 1. The ith row of $\lambda I - A$ then consists of zeros except for the main diagonal entry, which is $\lambda - 1$. Because

$$\det(A) = \sum_{j=1}^{n} a_{ij}C_{ij} = a_{i1}C_{i1} + a_{i2}C_{i2} + \cdots + a_{in}C_{in} \text{ and}$$

because each a_{ij} equals zero except for the main diagonal entry, $\det(\lambda I - A) = (\lambda - 1)c_{im} = 0$ where C_{im} is the cofactor determined for the main diagonal entry for this row. So, 1 is an eigenvalue of A.

53. If $A\mathbf{x} = \lambda\mathbf{x}$, then $A^{-1}A\mathbf{x} = A^{-1}\lambda\mathbf{x}$ and $\mathbf{x} = \lambda A^{-1}\mathbf{x}$. So,

$A^{-1}\mathbf{x} = \dfrac{1}{\lambda}\mathbf{x}$, which shows that $\mathbf{x}$ is an eigenvector of

A^{-1} with eigenvalue $\dfrac{1}{\lambda}$. The eigenvectors of A and

A^{-1} are the same.

55. The characteristic polynomial of A is $|\lambda I - A|$. The

constant term of this polynomial $(\text{in } \lambda)$ is obtained by

setting $\lambda = 0$. So, the constant term is
$|0I - A| = |-A| = \pm|A|$.

57. Assume that A is a real matrix with eigenvalues $\lambda_1, \lambda_2, \ldots, \lambda_n$ as its main diagonal entries. From Theorem 7.3, the eigenvalues of A are $\lambda_1, \lambda_2, \ldots, \lambda_n$. So, A has real eigenvalues. Because the determinant of A is $|A| = \lambda_1, \lambda_2, \ldots, \lambda_n$, it follows that A is nonsingular if and only if each λ is nonzero.

59. The characteristic equation of A is

$$\begin{vmatrix} \lambda - a & -b \\ 0 & \lambda - d \end{vmatrix} = (\lambda - a)(\lambda - d)$$
$$= \lambda^2 - (a + d)\lambda + ad = 0.$$

Because the given eigenvalues indicate a characteristic equation of $\lambda(\lambda - 1) = \lambda^2 - \lambda$,

$$\lambda^2 - (a + d)\lambda + ad = \lambda^2 - \lambda.$$

So, $a = 0$ and $d = 1$, or $a = 1$ and $d = 0$.

61. (a) False. See "Definition of Eigenvalue and Eigenvector," page 422.

(b) True. See discussion before Theorem 7.1, page 424.

(c) True. See Theorem 7.2, page 426.

63. Substituting the value $\lambda = 3$ yields the system

$$\begin{vmatrix} \lambda - 3 & 0 & 0 \\ 0 & \lambda - 3 & 0 \\ 0 & 0 & \lambda - 3 \end{vmatrix} \begin{bmatrix} x_1 \\ x_2 \\ x_3 \end{bmatrix} = \begin{bmatrix} 0 \\ 0 \\ 0 \end{bmatrix} \Rightarrow \begin{bmatrix} 0 & 0 & 0 \\ 0 & 0 & 0 \\ 0 & 0 & 0 \end{bmatrix} \begin{bmatrix} x_1 \\ x_2 \\ x_3 \end{bmatrix} = \begin{bmatrix} 0 \\ 0 \\ 0 \end{bmatrix}.$$

So, 3 has three linearly independent eigenvectors and the dimension of the eigenspace is 3.

65. Substituting the value $\lambda = 3$ yields the system

$$\begin{vmatrix} \lambda - 3 & 1 & 0 \\ 0 & \lambda - 3 & 1 \\ 0 & 0 & \lambda - 3 \end{vmatrix} \begin{bmatrix} x_1 \\ x_2 \\ x_3 \end{bmatrix} = \begin{bmatrix} 0 \\ 0 \\ 0 \end{bmatrix} \Rightarrow \begin{bmatrix} 0 & -1 & 0 \\ 0 & 0 & -1 \\ 0 & 0 & 0 \end{bmatrix} \begin{bmatrix} x_1 \\ x_2 \\ x_3 \end{bmatrix} = \begin{bmatrix} 0 \\ 0 \\ 0 \end{bmatrix}.$$

So, 3 has one linearly independent eigenvector, and the dimension of the eigenspace is 1.

67. $T(e^x) = \dfrac{d}{dx}[e^x] = e^x = 1(e^x)$

Therefore, $\lambda = 1$ is an eigenvalue.

69. The standard matrix for T is

$$A = \begin{bmatrix} 0 & -3 & 5 \\ -4 & 4 & -10 \\ 0 & 0 & 4 \end{bmatrix}.$$

The characteristic equation of A is

$$\begin{vmatrix} \lambda & 3 & -5 \\ 4 & \lambda - 4 & 10 \\ 0 & 0 & \lambda - 4 \end{vmatrix} = (\lambda + 2)(\lambda - 4)(\lambda - 6) = 0.$$

The eigenvalues of A are $\lambda_1 = -2$, $\lambda_2 = 4$, and $\lambda_3 = 6$. The corresponding eigenvectors are found by solving

$$\begin{bmatrix} \lambda_i & 3 & -5 \\ 4 & \lambda_i - 4 & 10 \\ 0 & 0 & \lambda_i - 4 \end{bmatrix} \begin{bmatrix} a_0 \\ a_1 \\ a_2 \end{bmatrix} = \begin{bmatrix} 0 \\ 0 \\ 0 \end{bmatrix}$$

for each λ_i. Thus,

$p_1(x) = 3 + 2x$, $p_2(x) = -5 + 10x + 2x^2$ and

$p_3(x) = -1 + 2x$ are eigenvectors corresponding to λ_1, λ_2, and λ_3.

71. The standard matrix for T is

$$A = \begin{bmatrix} 1 & 0 & -1 & 1 \\ 0 & 1 & 0 & 1 \\ -2 & 0 & 2 & -2 \\ 0 & 2 & 0 & 2 \end{bmatrix}.$$

Because the standard matrix is the same as that in Exercise 37, you know the eigenvalues are $\lambda_1 = 0$ and $\lambda_2 = 3$. So, two eigenvectors corresponding to $\lambda_1 = 0$ are

$$\mathbf{x}_1 = \begin{bmatrix} 1 \\ 0 \\ 1 \\ 0 \end{bmatrix} \quad \text{and} \quad \mathbf{x}_2 = \begin{bmatrix} 1 \\ 1 \\ 0 \\ -1 \end{bmatrix}$$

and an eigenvector corresponding to $\lambda_2 = 3$ is

$$\mathbf{x}_3 = \begin{bmatrix} 1 \\ 0 \\ -2 \\ 0 \end{bmatrix}.$$

73. 0 is the only eigenvalue of a nilpotent matrix. For if $A\mathbf{x} = \lambda\mathbf{x}$, then $A^2\mathbf{x} = A\lambda\mathbf{x} = \lambda^2\mathbf{x}$.

So,

$$A^k\mathbf{x} = \lambda^k\mathbf{x} = \mathbf{0} \Rightarrow \lambda^k = 0 \Rightarrow \lambda = 0$$

75. Let $\mathbf{x} = \begin{bmatrix} 1 \\ 1 \\ \vdots \\ 1 \end{bmatrix}$. Then $A\mathbf{x} = \begin{bmatrix} r \\ r \\ \vdots \\ r \end{bmatrix} = r\mathbf{x}$,

which shows that r is an eigenvalue of A with eigenvector $\mathbf{x}$.

For example, let $A = \begin{bmatrix} 1 & 2 \\ 3 & 0 \end{bmatrix}$.

Then $\begin{bmatrix} 1 & 2 \\ 3 & 0 \end{bmatrix}\begin{bmatrix} 1 \\ 1 \end{bmatrix} = \begin{bmatrix} 3 \\ 3 \end{bmatrix} = 3\begin{bmatrix} 1 \\ 1 \end{bmatrix}$.

Section 7.2   Diagonalization

1. $P^{-1} = \begin{bmatrix} 1 & -4 \\ -1 & 3 \end{bmatrix}$

$P^{-1}AP = \begin{bmatrix} 1 & -4 \\ -1 & 3 \end{bmatrix} \begin{bmatrix} -11 & 36 \\ -3 & 10 \end{bmatrix} \begin{bmatrix} -3 & -4 \\ -1 & -1 \end{bmatrix} = \begin{bmatrix} 1 & 0 \\ 0 & -2 \end{bmatrix}$

3. $P^{-1} = \begin{bmatrix} \frac{1}{5} & \frac{4}{5} \\ -\frac{1}{5} & \frac{1}{5} \end{bmatrix}$ and

$P^{-1}AP = \begin{bmatrix} \frac{1}{5} & \frac{4}{5} \\ -\frac{1}{5} & \frac{1}{5} \end{bmatrix} \begin{bmatrix} -2 & 4 \\ 1 & 1 \end{bmatrix} \begin{bmatrix} 1 & -4 \\ 1 & 1 \end{bmatrix} = \begin{bmatrix} 2 & 0 \\ 0 & -3 \end{bmatrix}$

5. $P^{-1} = \begin{bmatrix} \frac{2}{3} & -\frac{2}{3} & 1 \\ 0 & \frac{1}{4} & 0 \\ -\frac{1}{3} & \frac{1}{12} & 0 \end{bmatrix}$ and

$P^{-1}AP = \begin{bmatrix} \frac{2}{3} & -\frac{2}{3} & 1 \\ 0 & \frac{1}{4} & 0 \\ -\frac{1}{3} & \frac{1}{12} & 0 \end{bmatrix} \begin{bmatrix} -1 & 1 & 0 \\ 0 & 3 & 0 \\ 4 & -2 & 5 \end{bmatrix} \begin{bmatrix} 0 & 1 & -3 \\ 0 & 4 & 0 \\ 1 & 2 & 2 \end{bmatrix} = \begin{bmatrix} 5 & 0 & 0 \\ 0 & 3 & 0 \\ 0 & 0 & -1 \end{bmatrix}$

7. $P^{-1} = \begin{bmatrix} 1 & -\frac{1}{2} & \frac{5}{2} \\ 0 & \frac{1}{2} & -\frac{1}{2} \\ 0 & 0 & 1 \end{bmatrix}$ and

$P^{-1}AP = \begin{bmatrix} 1 & -\frac{1}{2} & \frac{5}{2} \\ 0 & \frac{1}{2} & -\frac{1}{2} \\ 0 & 0 & 1 \end{bmatrix} \begin{bmatrix} 4 & -1 & 3 \\ 0 & 2 & 1 \\ 0 & 0 & 3 \end{bmatrix} \begin{bmatrix} 1 & 1 & -2 \\ 0 & 2 & 1 \\ 0 & 0 & 1 \end{bmatrix} = \begin{bmatrix} 4 & 0 & 0 \\ 0 & 2 & 0 \\ 0 & 0 & 3 \end{bmatrix}$

9. A has only one eigenvalue, $\lambda = 0$, and a basis for the eigenspace is $\{(0, 1)\}$. So, A does not satisfy Theorem 7.5 (it does not have two linearly independent eigenvectors) and is not diagonalizable.

11. The matrix has eigenvalues $\lambda = 1$ (repeated), and a basis for the eigenspace is $\{(1, 0)\}$. So, A does not satisfy Theorem 7.5 (it does not have two linearly independent eigenvectors) and it is not diagonalizable.

13. The matrix has eigenvalues $\lambda = 1$ (repeated) and $\lambda_2 = 2$. A basis for the eigenspace associated with $\lambda = 1$ is $\{(1, 0, 0)\}$. So, the matrix has only two linearly independent eigenvectos and, by Theorem 7.5 it is not diagonalizable.

15. From Exercise 37, Section 7.1, A has only three linearly independent eigenvectors. So, A does not satisfy Theorem 7.5 and is not diagonalizable.

17. The eigenvalues of A are $\lambda_1 = 0$ and $\lambda_2 = 2$. Because A has two distinct eigenvalues, it is diagonalizable (by Theorem 7.6).

19. The eigenvalues of A are $\lambda = 0$ and $\lambda = 2$ (repeated). Because A does not have three <u>distinct</u> eigenvalues, Theorem 7.6 does not guarantee that A is diagonalizable.

21. The eigenvalues of A are $\lambda_1 = 0$, $\lambda_2 = 7$ (see Exercise 15, Section 7.1). The corresponding eigenvectors $(1, 2)$ and $(-3, 1)$ are used to form the columns of P. So,

$P = \begin{bmatrix} 1 & -3 \\ 2 & 1 \end{bmatrix} \Rightarrow P^{-1} = \begin{bmatrix} \frac{1}{7} & \frac{3}{7} \\ -\frac{2}{7} & \frac{1}{7} \end{bmatrix}$

and

$P^{-1}AP = \begin{bmatrix} \frac{1}{7} & \frac{3}{7} \\ -\frac{2}{7} & \frac{1}{7} \end{bmatrix} \begin{bmatrix} 6 & -3 \\ -2 & 1 \end{bmatrix} \begin{bmatrix} 1 & -3 \\ 2 & 1 \end{bmatrix} = \begin{bmatrix} 0 & 0 \\ 0 & 7 \end{bmatrix}.$

23. The eigenvalues of A are $\lambda_1 = \frac{1}{2}$ and $\lambda_2 = -\frac{1}{2}$ (see Exercise 17, Section 7.1). The corresponding eigenvectors $(3, 1)$ and $(1, 1)$ are used to form the columns of P. So,

$$P = \begin{bmatrix} 3 & 1 \\ 1 & 1 \end{bmatrix} \Rightarrow P^{-1} = \begin{bmatrix} \frac{1}{2} & -\frac{1}{2} \\ -\frac{1}{2} & \frac{3}{2} \end{bmatrix}$$

and

$$P^{-1}AP = \begin{bmatrix} \frac{1}{2} & -\frac{1}{2} \\ -\frac{1}{2} & \frac{3}{2} \end{bmatrix} \begin{bmatrix} 1 & -\frac{3}{2} \\ \frac{1}{2} & -1 \end{bmatrix} \begin{bmatrix} 3 & 1 \\ 1 & 1 \end{bmatrix} = \begin{bmatrix} \frac{1}{2} & 0 \\ 0 & -\frac{1}{2} \end{bmatrix}.$$

25. The eigenvalues of A are $\lambda_1 = 1$, $\lambda_2 = 2$, $\lambda_3 = 4$ (see Exercise 21, Section 7.1). The corresponding eigenvectors $(-1, 1, 1)$, $(1, 0, 0)$, $(7, -4, 2)$ are used to form the columns of P. So,

$$P = \begin{bmatrix} -1 & 1 & 7 \\ 1 & 0 & -4 \\ 1 & 0 & 2 \end{bmatrix} \Rightarrow P^{-1} = \begin{bmatrix} 0 & \frac{1}{3} & \frac{2}{3} \\ 1 & \frac{3}{2} & -\frac{1}{2} \\ 0 & -\frac{1}{6} & \frac{1}{6} \end{bmatrix}$$

and

$$P^{-1}AP = \begin{bmatrix} 0 & \frac{1}{3} & \frac{2}{3} \\ 1 & \frac{3}{2} & -\frac{1}{2} \\ 0 & -\frac{1}{6} & \frac{1}{6} \end{bmatrix} \begin{bmatrix} 2 & -2 & 3 \\ 0 & 3 & -2 \\ 0 & -1 & 2 \end{bmatrix} \begin{bmatrix} -1 & 1 & 7 \\ 1 & 0 & -4 \\ 1 & 0 & 2 \end{bmatrix} = \begin{bmatrix} 1 & 0 & 0 \\ 0 & 2 & 0 \\ 0 & 0 & 4 \end{bmatrix}.$$

27. The eigenvalues of A are $\lambda_1 = -3$ and $\lambda_2 = 3$ (repeated) (see Exercise 23, Section 7.1).

The corresponding eigenvectors $(1, 1, 3)$, $(1, 1, 0)$, and $(1, 0, -1)$ are used to form the columns of P. So,

$$P = \begin{bmatrix} 1 & 1 & 1 \\ 1 & 1 & 0 \\ 3 & 0 & -1 \end{bmatrix} \Rightarrow P^{-1} = \begin{bmatrix} \frac{1}{3} & -\frac{1}{3} & \frac{1}{3} \\ -\frac{1}{3} & \frac{4}{3} & -\frac{1}{3} \\ 1 & -1 & 0 \end{bmatrix}$$

and

$$P^{-1}AP = \begin{bmatrix} \frac{1}{3} & -\frac{1}{3} & \frac{1}{3} \\ -\frac{1}{3} & \frac{4}{3} & -\frac{1}{3} \\ 1 & -1 & 0 \end{bmatrix} \begin{bmatrix} 1 & 2 & -2 \\ -2 & 5 & -2 \\ -6 & 6 & -3 \end{bmatrix} \begin{bmatrix} 1 & 1 & 1 \\ 1 & 1 & 0 \\ 3 & 0 & -1 \end{bmatrix} = \begin{bmatrix} -3 & 0 & 0 \\ 0 & 3 & 0 \\ 0 & 0 & 3 \end{bmatrix}.$$

29. The eigenvalues of A are $\lambda_1 = -2$, $\lambda_2 = 6$ and $\lambda_3 = 4$ (see Exercise 25, Section 7.1).

The corresponding eigenvectors $(3, 2, 0)$, $(-1, 2, 0)$ and $(-5, 10, 2)$ are used to form the columns of P. So,

$$P = \begin{bmatrix} 3 & -1 & -5 \\ 2 & 2 & 10 \\ 0 & 0 & 2 \end{bmatrix} \Rightarrow P^{-1} = \begin{bmatrix} \frac{1}{4} & \frac{1}{8} & 0 \\ -\frac{1}{4} & \frac{3}{8} & -\frac{5}{2} \\ 0 & 0 & \frac{1}{2} \end{bmatrix}$$

and

$$P^{-1}AP = \begin{bmatrix} \frac{1}{4} & \frac{1}{8} & 0 \\ -\frac{1}{4} & \frac{3}{8} & -\frac{5}{2} \\ 0 & 0 & \frac{1}{2} \end{bmatrix} \begin{bmatrix} 0 & -3 & 5 \\ -4 & 4 & -10 \\ 0 & 0 & 4 \end{bmatrix} \begin{bmatrix} 3 & -1 & -5 \\ 2 & 2 & 10 \\ 0 & 0 & 2 \end{bmatrix} = \begin{bmatrix} -2 & 0 & 0 \\ 0 & 6 & 0 \\ 0 & 0 & 4 \end{bmatrix}.$$

31. The eigenvalues of A are $\lambda_1 = 1$ and $\lambda_2 = 2$. Furthermore, there are just two linearly independent eigenvectors of A, $\mathbf{x}_1 = (-1, 0, 1)$ and $\mathbf{x}_2 = (0, 1, 0)$. So, A is not diagonalizable.

33. The eigenvalues of A are $\lambda_1 = 2$, $\lambda_2 = 1$, $\lambda_3 = -1$ and $\lambda_{4'} = -2$. The corresponding eigenvectors $(4, 4, 4, 1)$, $(0, 0, 3, 1)$, $(0, -2, 1, 1)$ and $(0, 0, 0, 1)$ are used to form the columns of P. So,

$$P = \begin{bmatrix} 4 & 0 & 0 & 0 \\ 4 & 0 & -2 & 0 \\ 4 & 3 & 1 & 0 \\ 1 & 1 & 1 & 1 \end{bmatrix} \quad \Rightarrow \quad P^{-1} = \begin{bmatrix} \frac{1}{4} & 0 & 0 & 0 \\ -\frac{1}{2} & \frac{1}{6} & \frac{1}{3} & 0 \\ \frac{1}{2} & -\frac{1}{2} & 0 & 0 \\ -\frac{1}{4} & \frac{1}{3} & -\frac{1}{3} & 1 \end{bmatrix}$$

and

$$P^{-1}AP = \begin{bmatrix} \frac{1}{4} & 0 & 0 & 0 \\ -\frac{1}{2} & \frac{1}{6} & \frac{1}{3} & 0 \\ \frac{1}{2} & -\frac{1}{2} & 0 & 0 \\ -\frac{1}{4} & \frac{1}{3} & -\frac{1}{3} & 1 \end{bmatrix} \begin{bmatrix} 2 & 0 & 0 & 0 \\ 3 & -1 & 0 & 0 \\ 0 & 1 & 1 & 0 \\ 0 & 0 & 1 & -2 \end{bmatrix} \begin{bmatrix} 4 & 0 & 0 & 0 \\ 4 & 0 & -2 & 0 \\ 4 & 3 & 1 & 0 \\ 1 & 1 & 1 & 1 \end{bmatrix} = \begin{bmatrix} 2 & 0 & 0 & 0 \\ 0 & 1 & 0 & 0 \\ 0 & 0 & -1 & 0 \\ 0 & 0 & 0 & -2 \end{bmatrix}$$

35. The standard matrix for T is

$$A = \begin{bmatrix} 1 & 1 \\ 1 & 1 \end{bmatrix}$$

which has eigenvalues $\lambda_1 = 0$ and $\lambda_2 = 2$ and corresponding eigenvectors $(1, -1)$ and $(1, 1)$. Let $B = \{(1, -1), (1, 1)\}$ and find the image of each vector in B.

$$\left[T(1, -1)\right]_B = \left[(0, 0)\right]_B = (0, 0)$$
$$\left[T(1, 1)\right]_B = \left[(2, 2)\right]_B = (0, 2)$$

The matrix of T relative to B is then

$$A' = \begin{bmatrix} 0 & 0 \\ 0 & 2 \end{bmatrix}.$$

37. The standard matrix for T is

$$A = \begin{bmatrix} 1 & 0 \\ 1 & 2 \end{bmatrix}.$$

which has eigenvalues $\lambda_1 = 1$ and $\lambda_2 = 2$ and corresponding eigenvectors $-1 + x$ and x. Let $B = \{-1 + x, x\}$ and find the image of each vector in B.

$$\left[T(-1 + x)\right]_B = \left[-1 + x\right]_B = (1, 0)$$
$$\left[T(x)\right]_B = \left[2x\right]_B = (0, 2)$$

The matrix of T relative to B is then

$$A' = \begin{bmatrix} 1 & 0 \\ 0 & 2 \end{bmatrix}.$$

39. (a) $B^k = \left(P^{-1}AP\right)^k = \left(P^{-1}AP\right)\left(P^{-1}AP\right) \dots \left(P^{-1}AP\right)$ (k times)

$\qquad\qquad = P^{-1}A^k P$

(b) $B = P^{-1}AP \quad \Rightarrow \quad A = PBP^{-1} \quad \Rightarrow \quad A^k = PB^k P^{-1}$ from part (a).

41. The eigenvalues and corresponding eigenvectors of A are $\lambda_1 = -2$, $\lambda_2 = 1$, $\mathbf{x}_1 = \left(-\frac{3}{2}, 1\right)$, and $\mathbf{x}_2 = (-2, 1)$. Construct a nonsingular matrix P from the eigenvectors of A,

$$P = \begin{bmatrix} -\frac{3}{2} & -2 \\ 1 & 1 \end{bmatrix}$$

and find a diagonal matrix B similar to A.

$$B = P^{-1}AP = \begin{bmatrix} 2 & 4 \\ -2 & -3 \end{bmatrix}\begin{bmatrix} 10 & 18 \\ -6 & -11 \end{bmatrix}\begin{bmatrix} -\frac{3}{2} & -2 \\ 1 & 1 \end{bmatrix} = \begin{bmatrix} -2 & 0 \\ 0 & 1 \end{bmatrix}$$

Then

$$A^6 = PB^6P^{-1} = \begin{bmatrix} -\frac{3}{2} & -2 \\ 1 & 1 \end{bmatrix}\begin{bmatrix} 64 & 0 \\ 0 & 1 \end{bmatrix}\begin{bmatrix} 2 & 4 \\ -2 & -3 \end{bmatrix} = \begin{bmatrix} -188 & -378 \\ 126 & 253 \end{bmatrix}.$$

43. The eigenvalues and corresponding eigenvectors of A are $\lambda_1 = 0$, $\lambda_2 = 2$ (repeated), $\mathbf{x}_1 = (-1, 3, 1)$, $\mathbf{x}_2 = (3, 0, 1)$ and $\mathbf{x}_3 = (-2, 1, 0)$. Construct a nonsingular matrix P from the eigenvectors of A.

$$P = \begin{bmatrix} -1 & 3 & -2 \\ 3 & 0 & 1 \\ 1 & 1 & 0 \end{bmatrix}$$

and find a diagonal matrix B similar to A.

$$B = P^{-1}AP = \begin{bmatrix} \frac{1}{2} & 1 & -\frac{3}{2} \\ -\frac{1}{2} & -1 & \frac{5}{2} \\ -\frac{3}{2} & -2 & \frac{9}{2} \end{bmatrix}\begin{bmatrix} 3 & 2 & -3 \\ -3 & -4 & 9 \\ -1 & -2 & 5 \end{bmatrix}\begin{bmatrix} -1 & 3 & -2 \\ 3 & 0 & 1 \\ 1 & 1 & 0 \end{bmatrix} = \begin{bmatrix} 0 & 0 & 0 \\ 0 & 2 & 0 \\ 0 & 0 & 2 \end{bmatrix}$$

Then,

$$A^8 = PB^8P^{-1} = P\begin{bmatrix} 0 & 0 & 0 \\ 0 & 256 & 0 \\ 0 & 0 & 256 \end{bmatrix}P^{-1} = \begin{bmatrix} 384 & 256 & -384 \\ -384 & -512 & 1152 \\ -128 & -256 & 640 \end{bmatrix}.$$

45. (a) True. See the proof of Theorem 7.4, pages 436–437.

(b) False. See Theorem 7.6, page 442.

47. Yes, the order of the elements on the main diagonal may change. For instance,

$$\begin{bmatrix} 1 & 0 \\ 0 & 2 \end{bmatrix} \text{ and } \begin{bmatrix} 2 & 0 \\ 0 & 1 \end{bmatrix} \text{ are similar.}$$

49. Assume that A is diagonalizable, $P^{-1}AP = D$, where D is diagonal. Then

$$D^T = \left(P^{-1}AP\right)^T = P^T A^T \left(P^{-1}\right)^T = P^T A^T \left(P^T\right)^{-1}$$

is diagonal, which shows that A^T is diagonalizable.

51. Assume that A is diagonalizable with n real eigenvalues $\lambda_1, \ldots, \lambda_n$. Then if $PAP^{-1} = D$, D is diagonal,

$$|A| = |P^{-1}AP| = \begin{vmatrix} \lambda_1 & 0 & \cdots & 0 \\ 0 & \lambda_2 & & \vdots \\ \vdots & & & 0 \\ 0 & \cdots & 0 & \lambda_n \end{vmatrix} = \lambda_1\lambda_2 \ldots \lambda_n.$$

53. Let the eigenvalues of the diagonalizable matrix A be all ± 1. Then there exists an invertible matrix P such that

$$P^{-1}AP = D$$

where D is diagonal with ± 1 along the main diagonal. So $A = PDP^{-1}$ and because $D^{-1} = D$,

$$A^{-1} = \left(PDP^{-1}\right)^{-1} = \left(P^{-1}\right)^{-1}D^{-1}P^{-1} = PDP^{-1} = A.$$

55. Given that $P^{-1}AP = D$, where D is diagonal,

$$A = PDP^{-1} \text{ and } A^{-1} = \left(PDP^{-1}\right)^{-1} = \left(P^{-1}\right)^{-1}D^{-1}P^{-1} = PD^{-1}P^{-1} \implies P^{-1}A^{-1}P = D^{-1},$$

which shows that A^{-1} is diagonalizable.

57. A is triangular, so the eigenvalues are simply the entries on the main diagonal. So, the only eigenvalue is $\lambda = k$, and a basis for the eigenspace is $\{(0, 0)\}$.

Because matrix A does not have two linearly independent eigenvectors, it does not satisfy Theorem 7.5 and it is not diagonalizable.

Section 7.3   Symmetric Matrices and Orthogonal Diagonalization

1. Because $\begin{bmatrix} 1 & 3 \\ 3 & -1 \end{bmatrix}^T = \begin{bmatrix} 1 & 3 \\ 3 & -1 \end{bmatrix}$

the matrix *is* symmetric.

3. Because

$$\begin{bmatrix} 4 & -2 & 1 \\ 3 & 1 & 2 \\ 1 & 2 & 1 \end{bmatrix}^T \neq \begin{bmatrix} 4 & 3 & 1 \\ -2 & 1 & 2 \\ 1 & 2 & 1 \end{bmatrix}$$

the matrix is *not* symmetric.

5. Because

$$\begin{bmatrix} 0 & 1 & 2 & -1 \\ 1 & 0 & -3 & 2 \\ 2 & -3 & 0 & 1 \\ -1 & 2 & 1 & -2 \end{bmatrix}^T = \begin{bmatrix} 0 & 1 & 2 & -1 \\ 1 & 0 & -3 & 2 \\ 2 & -3 & 0 & 1 \\ -1 & 2 & 1 & -2 \end{bmatrix}$$

the matrix *is* symmetric.

7. The characteristic equation of A is

$$|\lambda I - A| = \begin{vmatrix} \lambda - 3 & -1 \\ -1 & \lambda - 3 \end{vmatrix} = (\lambda - 4)(\lambda - 2) = 0$$

Therefore, the eigenvalues of A are $\lambda_1 = 4$ and $\lambda_2 = 2$. The dimension of the corresponding eigenspace of each eigenvalue is 1 (by Theorem 7.7).

9. The characteristic equation of A is

$$|\lambda I - A| = \begin{vmatrix} \lambda - 3 & 0 & 0 \\ 0 & \lambda - 2 & 0 \\ 0 & 0 & \lambda - 2 \end{vmatrix}$$

$$= (\lambda - 3)(\lambda - 2)^2 = 0.$$

Therefore, the eigenvalues of A are $\lambda_1 = 3$ and $\lambda_2 = 2$. The dimension of the eigenspace corresponding $\lambda_1 = 3$ is 1. The multiplicity of $\lambda_2 = 2$ is 2, so the dimension of the corresponding eigenspace is 2 (by Theorem 7.7).

11. The characteristic equation of A is

$$|\lambda I - A| = \begin{vmatrix} \lambda & -2 & -2 \\ -2 & \lambda & -2 \\ -2 & -2 & \lambda \end{vmatrix} = (\lambda + 2)^2(\lambda - 4) = 0.$$

The eigenvalues of A are $\lambda_1 = -2$ and $\lambda_2 = 4$. The multiplicity of $\lambda_1 = -2$ is 2, so the dimension of the corresponding eigenspace is 2 (by Theorem 7.7). The dimension for the eigenspace corresponding to $\lambda_2 = 4$ is 1.

13. The characteristic equation of A is

$$|\lambda I - A| = \begin{vmatrix} \lambda & -1 & -1 \\ -1 & \lambda & -1 \\ -1 & -1 & \lambda - 1 \end{vmatrix}$$

$$= (\lambda + 1)(\lambda^2 - 2\lambda - 1) = 0.$$

Therefore, the eigenvalues are $\lambda_1 = -1$, $\lambda_2 = 1 - \sqrt{2}$, and $\lambda_3 = 1 + \sqrt{2}$. The dimension of the eigenspace corresponding to each eigenvalue is 1.

15. Because the column vectors of the matrix form an orthonormal set, the matrix *is* orthogonal.

17. Because the column vectors of the matrix do not form an orthonormal set $[(-4, 0, 3)$ and $(3, 0, 4)$ are not unit vectors], the matrix is *not* orthogonal.

19. Because the column vectors of the matrix form an orthonormal set, the matrix *is* orthogonal.

21. Because the column vectors of the matrix form an orthonormal set, the matrix *is* orthogonal.

23. The eigenvalues of A are $\lambda_1 = 0$ and $\lambda_2 = 2$, with corresponding eigenvectors $(1, -1)$ and $(1, 1)$, respectively.

Normalize each eigenvector to form the columns of P. Then

$$P = \begin{bmatrix} \dfrac{\sqrt{2}}{2} & \dfrac{\sqrt{2}}{2} \\ -\dfrac{\sqrt{2}}{2} & \dfrac{\sqrt{2}}{2} \end{bmatrix}$$

and

$$P^T AP = \begin{bmatrix} \dfrac{\sqrt{2}}{2} & -\dfrac{\sqrt{2}}{2} \\ \dfrac{\sqrt{2}}{2} & \dfrac{\sqrt{2}}{2} \end{bmatrix} \begin{bmatrix} 1 & 1 \\ 1 & 1 \end{bmatrix} \begin{bmatrix} \dfrac{\sqrt{2}}{2} & \dfrac{\sqrt{2}}{2} \\ -\dfrac{\sqrt{2}}{2} & \dfrac{\sqrt{2}}{2} \end{bmatrix} = \begin{bmatrix} 0 & 0 \\ 0 & 2 \end{bmatrix}.$$

25. The eigenvalues of A are $\lambda_1 = 0$ and $\lambda_2 = 3$, with corresponding eigenvectors $\left(\dfrac{\sqrt{2}}{2}, -1\right)$ and $\left(\sqrt{2}, 1\right)$, respectively.

Normalize each eigenvector to form the columns of P. Then

$$P = \begin{bmatrix} \dfrac{\sqrt{3}}{3} & \dfrac{\sqrt{6}}{3} \\ -\dfrac{\sqrt{6}}{3} & \dfrac{\sqrt{3}}{3} \end{bmatrix}$$

and

$$P^T AP = \begin{bmatrix} \dfrac{\sqrt{3}}{3} & -\dfrac{\sqrt{6}}{3} \\ \dfrac{\sqrt{6}}{3} & \dfrac{\sqrt{3}}{3} \end{bmatrix} \begin{bmatrix} 2 & \sqrt{2} \\ \sqrt{2} & 1 \end{bmatrix} \begin{bmatrix} \dfrac{\sqrt{3}}{3} & \dfrac{\sqrt{6}}{3} \\ -\dfrac{\sqrt{6}}{3} & \dfrac{\sqrt{3}}{3} \end{bmatrix} = \begin{bmatrix} 0 & 0 \\ 0 & 3 \end{bmatrix}.$$

27. The eigenvalues of A are $\lambda_1 = -15$ and $\lambda_2 = 0$, and $\lambda_3 = 15$, with corresponding eigenvectors $(-2, 1, 2)$, $(-1, 2, -2)$ and $(2, 2, 1)$, respectively. Normalize each eigenvector to form the columns of P. Then

$$P = \begin{bmatrix} -\dfrac{2}{3} & -\dfrac{1}{3} & \dfrac{2}{3} \\ \dfrac{1}{3} & \dfrac{2}{3} & \dfrac{2}{3} \\ \dfrac{2}{3} & -\dfrac{2}{3} & \dfrac{1}{3} \end{bmatrix}$$

and

$$P^T AP = \begin{bmatrix} -\dfrac{2}{3} & \dfrac{1}{3} & \dfrac{2}{3} \\ -\dfrac{1}{3} & \dfrac{2}{3} & -\dfrac{2}{3} \\ \dfrac{2}{3} & \dfrac{2}{3} & \dfrac{1}{3} \end{bmatrix} \begin{bmatrix} 0 & 10 & 10 \\ 10 & 5 & 0 \\ 10 & 0 & -5 \end{bmatrix} \begin{bmatrix} -\dfrac{2}{3} & -\dfrac{1}{3} & \dfrac{2}{3} \\ \dfrac{1}{3} & \dfrac{2}{3} & \dfrac{2}{3} \\ \dfrac{2}{3} & -\dfrac{2}{3} & \dfrac{1}{3} \end{bmatrix} = \begin{bmatrix} -15 & 0 & 0 \\ 0 & 0 & 0 \\ 0 & 0 & 15 \end{bmatrix}.$$

29. The eigenvalues of A are $\lambda_1 = -2$, $\lambda_2 = 2$, and $\lambda_3 = 4$, with corresponding eigenvectors $(-1, -1, 1)$, $(-1, 1, 0)$, and $(1, 1, 2)$, respectively. Normalize each eigenvector to form the columns of P. Then

$$P = \begin{bmatrix} -\dfrac{\sqrt{3}}{3} & -\dfrac{\sqrt{2}}{2} & \dfrac{\sqrt{6}}{6} \\[2mm] -\dfrac{\sqrt{3}}{2} & \dfrac{\sqrt{2}}{2} & \dfrac{\sqrt{6}}{6} \\[2mm] \dfrac{\sqrt{3}}{3} & 0 & \dfrac{\sqrt{6}}{3} \end{bmatrix}$$

and

$$P^T A P = \begin{bmatrix} -\dfrac{\sqrt{3}}{3} & -\dfrac{\sqrt{3}}{3} & \dfrac{\sqrt{3}}{3} \\[2mm] -\dfrac{\sqrt{2}}{2} & \dfrac{\sqrt{2}}{2} & 0 \\[2mm] \dfrac{\sqrt{6}}{6} & \dfrac{\sqrt{6}}{6} & \dfrac{\sqrt{6}}{3} \end{bmatrix} \begin{bmatrix} 1 & -1 & 2 \\ -1 & 1 & 2 \\ 2 & 2 & 2 \end{bmatrix} \begin{bmatrix} -\dfrac{\sqrt{3}}{3} & -\dfrac{\sqrt{2}}{2} & \dfrac{\sqrt{6}}{6} \\[2mm] -\dfrac{\sqrt{3}}{3} & \dfrac{\sqrt{2}}{2} & \dfrac{\sqrt{6}}{6} \\[2mm] \dfrac{\sqrt{3}}{2} & 0 & \dfrac{\sqrt{6}}{3} \end{bmatrix} = \begin{bmatrix} -2 & 0 & 0 \\ 0 & 2 & 0 \\ 0 & 0 & 4 \end{bmatrix}.$$

31. The eigenvalues of A are $\lambda_1 = 2$ and $\lambda_2 = 6$, with corresponding eigenvectors $(1, -1, 0, 0)$ and $(0, 0, 1, -1)$ for λ_1 and $(1, 1, 0, 0)$ and $(0, 0, 1, 1)$ for λ_2. Normalize each eigenvector to form the columns of P. Then

$$P = \begin{bmatrix} \dfrac{\sqrt{2}}{2} & 0 & \dfrac{\sqrt{2}}{2} & 0 \\[2mm] -\dfrac{\sqrt{2}}{2} & 0 & \dfrac{\sqrt{2}}{2} & 0 \\[2mm] 0 & \dfrac{\sqrt{2}}{2} & 0 & \dfrac{\sqrt{2}}{2} \\[2mm] 0 & -\dfrac{\sqrt{2}}{2} & 0 & \dfrac{\sqrt{2}}{2} \end{bmatrix}$$

and

$$P^T A P = \begin{bmatrix} \dfrac{\sqrt{2}}{2} & -\dfrac{\sqrt{2}}{2} & 0 & 0 \\[2mm] 0 & 0 & \dfrac{\sqrt{2}}{2} & -\dfrac{\sqrt{2}}{2} \\[2mm] \dfrac{\sqrt{2}}{2} & \dfrac{\sqrt{2}}{2} & 0 & 0 \\[2mm] 0 & 0 & \dfrac{\sqrt{2}}{2} & \dfrac{\sqrt{2}}{2} \end{bmatrix} \begin{bmatrix} 4 & 2 & 0 & 0 \\ 2 & 4 & 0 & 0 \\ 0 & 0 & 4 & 2 \\ 0 & 0 & 2 & 4 \end{bmatrix} \begin{bmatrix} \dfrac{\sqrt{2}}{2} & 0 & \dfrac{\sqrt{2}}{2} & 0 \\[2mm] -\dfrac{\sqrt{2}}{2} & 0 & \dfrac{\sqrt{2}}{2} & 0 \\[2mm] 0 & \dfrac{\sqrt{2}}{2} & 0 & \dfrac{\sqrt{2}}{2} \\[2mm] 0 & -\dfrac{\sqrt{2}}{2} & 0 & \dfrac{\sqrt{2}}{2} \end{bmatrix}$$

$$= \begin{bmatrix} 2 & 0 & 0 & 0 \\ 0 & 2 & 0 & 0 \\ 0 & 0 & 6 & 0 \\ 0 & 0 & 0 & 6 \end{bmatrix}$$

33. (a) True. See Theorem 7.10, page 453.

(b) True. See Theorem 7.9, page 452.

35. $\left(A^T A\right)^T = A^T \left(A^T\right)^T = A^T A \Rightarrow A^T A$ is symmetric.

$\left(A A^T\right)^T = \left(A^T\right)^T A^T = A A^T \Rightarrow A A^T$ is symmetric.

37. If A is orthogonal, then $A A^T = I$. So,

$$1 = \left| A A^T \right| = \left| A \right| \left| A^T \right| = \left| A \right|^2 \Rightarrow \left| A \right| = \pm 1.$$

39. Observe that A is orthogonal because

$$A^{-1} = \begin{bmatrix} \cos \theta & \sin \theta \\ -\sin \theta & \cos \theta \end{bmatrix} \frac{1}{\cos^2 \theta + \sin^2 \theta} = A^T.$$

41. Let A be orthogonal, $A^{-1} = A^T$.

Then $\left(A^T\right)^{-1} = \left(A^{-1}\right)^{-1} = A = \left(A^T\right)^T \Rightarrow A^T$ is orthogonal.

Furthermore,

$\left(A^{-1}\right)^{-1} = \left(A^T\right)^{-1} = \left(A^{-1}\right)^T \Rightarrow A^{-1}$ is orthogonal.

Section 7.4   Applications of Eigenvalues and Eigenvectors

1. $\mathbf{x}_2 = A\mathbf{x}_1 = \begin{bmatrix} 0 & 2 \\ \frac{1}{2} & 0 \end{bmatrix}\begin{bmatrix} 10 \\ 10 \end{bmatrix} = \begin{bmatrix} 20 \\ 5 \end{bmatrix}$

$\mathbf{x}_3 = A\mathbf{x}_2 = \begin{bmatrix} 0 & 2 \\ \frac{1}{2} & 0 \end{bmatrix}\begin{bmatrix} 20 \\ 5 \end{bmatrix} = \begin{bmatrix} 10 \\ 10 \end{bmatrix}$

3. $\mathbf{x}_2 = A\mathbf{x}_1 = \begin{bmatrix} 0 & 3 & 4 \\ 1 & 0 & 0 \\ 0 & \frac{1}{2} & 0 \end{bmatrix}\begin{bmatrix} 12 \\ 12 \\ 12 \end{bmatrix} = \begin{bmatrix} 84 \\ 12 \\ 6 \end{bmatrix}$

$\mathbf{x}_3 = A\mathbf{x}_2 = \begin{bmatrix} 0 & 3 & 4 \\ 1 & 0 & 0 \\ 0 & \frac{1}{2} & 0 \end{bmatrix}\begin{bmatrix} 84 \\ 12 \\ 6 \end{bmatrix} = \begin{bmatrix} 60 \\ 84 \\ 6 \end{bmatrix}$

5. The eigenvalues are 1 and -1. Choosing the positive eigenvalue, $\lambda = 1$, the corresponding eigenvector is found by row-reducing $\lambda I - A = I - A$.

$\begin{bmatrix} 1 & -2 \\ -\frac{1}{2} & 1 \end{bmatrix} \Rightarrow \begin{bmatrix} 1 & -2 \\ 0 & 0 \end{bmatrix}$

So, an eigenvector is $(2, 1)$, and the stable age

distribution vector is $\mathbf{x} = t\begin{bmatrix} 2 \\ 1 \end{bmatrix}$.

7. The eigenvalues of A are -1 and 2. Choosing the positive eigenvalue, let $\lambda = 2$.

An eigenvector corresponding to $\lambda = 2$ is found by row-reducing $2I - A$.

$\begin{bmatrix} 2 & -3 & -4 \\ -1 & 2 & 0 \\ 0 & -\frac{1}{2} & 2 \end{bmatrix} \Rightarrow \begin{bmatrix} 1 & 0 & -8 \\ 0 & 1 & -4 \\ 0 & 0 & 0 \end{bmatrix}$

So, an eigenvector is $(8, 4, 1)$ and stable age distribution

vector is $\mathbf{x} = t\begin{bmatrix} 8 \\ 4 \\ 1 \end{bmatrix}$.

9. Construct the age transition matrix.

$A = \begin{bmatrix} 2 & 4 & 2 \\ 0.75 & 0 & 0 \\ 0 & 0.25 & 0 \end{bmatrix}$

The current age distribution vector is

$\mathbf{x}_1 = \begin{bmatrix} 120 \\ 120 \\ 120 \end{bmatrix}$.

In one year, the age distribution vector will be

$\mathbf{x}_2 = A\mathbf{x}_1 = \begin{bmatrix} 2 & 4 & 2 \\ 0.75 & 0 & 0 \\ 0 & 0.25 & 0 \end{bmatrix}\begin{bmatrix} 120 \\ 120 \\ 120 \end{bmatrix} = \begin{bmatrix} 960 \\ 90 \\ 30 \end{bmatrix}$.

In two years, the age distribution vector will be

$\mathbf{x}_3 = A\mathbf{x}_2 = \begin{bmatrix} 2 & 4 & 2 \\ 0.75 & 0 & 0 \\ 0 & 0.25 & 0 \end{bmatrix}\begin{bmatrix} 960 \\ 90 \\ 30 \end{bmatrix} = \begin{bmatrix} 2340 \\ 720 \\ 22.5 \end{bmatrix}$.

11. Construct the age transition matrix.

$A = \begin{bmatrix} 2 & 5 & 2 \\ 0.6 & 0 & 0 \\ 0 & 0.5 & 0 \end{bmatrix}$

The current age distribution vector is

$\mathbf{x}_1 = \begin{bmatrix} 100 \\ 100 \\ 100 \end{bmatrix}$.

In one year, the age distribution vector will be

$\mathbf{x}_2 = A\mathbf{x}_1 = \begin{bmatrix} 2 & 5 & 2 \\ 0.6 & 0 & 0 \\ 0 & 0.5 & 0 \end{bmatrix}\begin{bmatrix} 100 \\ 100 \\ 100 \end{bmatrix} = \begin{bmatrix} 900 \\ 60 \\ 50 \end{bmatrix}$.

In two years, the age distribution vector will be

$\mathbf{x}_3 = A\mathbf{x}_2 = \begin{bmatrix} 2 & 5 & 2 \\ 0.6 & 0 & 0 \\ 0 & 0.5 & 0 \end{bmatrix}\begin{bmatrix} 900 \\ 60 \\ 50 \end{bmatrix} = \begin{bmatrix} 2200 \\ 540 \\ 30 \end{bmatrix}$.

13. The solution to the differential equation $y' = ky$ is

$y = Ce^{kt}$. So, $y_1 = C_1e^{2t}$ and $y_2 = C_2e^{t}$.

15. The solution to the differential equation $y' = ky$ is

$y = Ce^{kt}$. So, $y_1 = C_1e^{-t}$, $y_2 = C_2e^{6t}$ and $y_3 = C_3e^t$.

17. The solution to the differential equation $y' = ky$ is

$y = Ce^{kt}$. So, $y_1 = C_1e^{2t}$, $y_2 = C_2e^{-t}$, and $y_3 = C_3e^t$.

19. This system has the matrix form

$$\mathbf{y}' = \begin{bmatrix} y_1' \\ y_2' \end{bmatrix} = \begin{bmatrix} 1 & -4 \\ 0 & 2 \end{bmatrix}\begin{bmatrix} y_1 \\ y_2 \end{bmatrix} = A\mathbf{y}.$$

The eigenvalues of A are $\lambda_1 = 1$ and $\lambda_2 = 2$, with corresponding eigenvectors $(1, 0)$ and

$(-4, 1)$, respectively. So, diagonalize A using a matrix P whose columns are the eigenvectors of A.

$$P = \begin{bmatrix} 1 & -4 \\ 0 & 1 \end{bmatrix} \quad \text{and} \quad P^{-1}AP = \begin{bmatrix} 1 & 0 \\ 0 & 2 \end{bmatrix}$$

The solution of the system $\mathbf{w}' = P^{-1}AP\mathbf{w}$ is

$w_1 = C_1e^t$ and $w_2 = C_2e^{2t}$. Return to the original system by applying the substitution $\mathbf{y} = P\mathbf{w}$.

$$\mathbf{y} = \begin{bmatrix} y_1 \\ y_2 \end{bmatrix} = \begin{bmatrix} 1 & -4 \\ 0 & 1 \end{bmatrix}\begin{bmatrix} w_1 \\ w_2 \end{bmatrix} = \begin{bmatrix} w_1 - 4w_2 \\ w_2 \end{bmatrix}.$$

So, the solution is

$y_1 = C_1e^t - 4C_2e^{2t}$

$y_2 = C_2e^{2t}$.

21. This system has the matrix form

$$\mathbf{y}' = \begin{bmatrix} y_1' \\ y_2' \end{bmatrix} = \begin{bmatrix} 1 & 2 \\ 2 & 1 \end{bmatrix}\begin{bmatrix} y_1 \\ y_2 \end{bmatrix} = A\mathbf{y}.$$

The eigenvalues of A are $\lambda_1 = -1$ and $\lambda_2 = 3$ with corresponding eigenvectors $\mathbf{x}_1 = (1, -1)$ and

$\mathbf{x}_2 = (1, 1)$, respectively. So, diagonalize A using a matrix P whose columns vectors are the eigenvectors of A.

$$P = \begin{bmatrix} 1 & 1 \\ -1 & 1 \end{bmatrix} \quad \text{and} \quad P^{-1}AP = \begin{bmatrix} -1 & 0 \\ 0 & 3 \end{bmatrix}.$$

The solution of the system $\mathbf{w}' = P^{-1}AP\mathbf{w}$ is

$w_1 = C_1e^{-t}$ and $w_2 = C_2e^{3t}$. Return to the original system by applying the substitution $\mathbf{y} = P\mathbf{w}$.

$$\mathbf{y} = \begin{bmatrix} y_1 \\ y_2 \end{bmatrix} = \begin{bmatrix} 1 & 1 \\ -1 & 1 \end{bmatrix}\begin{bmatrix} w_1 \\ w_2 \end{bmatrix} = \begin{bmatrix} w_1 + w_2 \\ -w_1 + w_2 \end{bmatrix}$$

So, the solution is

$y_1 = C_1e^{-t} + C_2e^{3t}$

$y_2 = -C_1e^{-t} + C_2e^{3t}$.

23. This system has the matrix form

$$\mathbf{y}' = \begin{bmatrix} y_1' \\ y_2' \\ y_3' \end{bmatrix} = \begin{bmatrix} 0 & -3 & 5 \\ -4 & 4 & -10 \\ 0 & 0 & 4 \end{bmatrix}\begin{bmatrix} y_1 \\ y_2 \\ y_3 \end{bmatrix} = A\mathbf{y}.$$

The eigenvalues of A are $\lambda_1 = -2$, $\lambda_2 = 6$ and $\lambda_3 = 4$, with corresponding eigenvectors $(3, 2, 0)$, $(-1, 2, 0)$ and

$(-5, 10, 2)$, respectively. So, diagonalize A using a matrix P whose column vectors are the eigenvectors of A.

$$P = \begin{bmatrix} 3 & -1 & -5 \\ 2 & 2 & 10 \\ 0 & 0 & 2 \end{bmatrix} \quad \text{and} \quad P^{-1}AP = \begin{bmatrix} -2 & 0 & 0 \\ 0 & 6 & 0 \\ 0 & 0 & 4 \end{bmatrix}$$

The solution of the system $\mathbf{w}' = P^{-1}AP\mathbf{w}$ is $w_1 = C_1e^{-2t}$, $w_2 = C_2e^{6t}$ and $w_3 = C_3e^{4t}$.

Return to the original system by applying the substitution $\mathbf{y} = P\mathbf{w}$.

$$\mathbf{y} = \begin{bmatrix} y_1 \\ y_2 \\ y_3 \end{bmatrix} = \begin{bmatrix} 3 & -1 & -5 \\ 2 & 2 & 10 \\ 0 & 0 & 2 \end{bmatrix}\begin{bmatrix} w_1 \\ w_2 \\ w_3 \end{bmatrix} = \begin{bmatrix} 3w_1 - w_2 - 5w_3 \\ 2w_1 + 2w_2 + 10w_3 \\ 2w_3 \end{bmatrix}$$

So, the solution is

$y_1 = 3C_1e^{-2t} - C_2e^{6t} - 5C_3e^{4t}$

$y_2 = 2C_1e^{-2t} + 2C_2e^{6t} + 10C_3e^{4t}$

$y_3 = 2C_3e^{4t}$.

25. This system has the matrix form

$$\mathbf{y}' = \begin{bmatrix} y_1' \\ y_2' \\ y_3' \end{bmatrix} = \begin{bmatrix} 1 & -2 & 1 \\ 0 & 2 & 4 \\ 0 & 0 & 3 \end{bmatrix} \begin{bmatrix} y_1 \\ y_2 \\ y_3 \end{bmatrix} = A\mathbf{y}.$$

The eigenvalues of A are $\lambda_1 = 1$, $\lambda_2 = 2$, and $\lambda_3 = 3$ with corresponding eigenvectors $\mathbf{x}_1 = (1, 0, 0)$, $\mathbf{x}_2 = (-2, 1, 0)$, and $\mathbf{x}_3 = (-7, 8, 2)$. So, diagonalize A using a matrix P whose column vectors are the eigenvectors of A.

$$P = \begin{bmatrix} 1 & -2 & -7 \\ 0 & 1 & 8 \\ 0 & 0 & 2 \end{bmatrix} \quad \text{and} \quad P^{-1}AP = \begin{bmatrix} 1 & 0 & 0 \\ 0 & 2 & 0 \\ 0 & 0 & 3 \end{bmatrix}$$

The solution of the system $\mathbf{w}' = P^{-1}AP\mathbf{w}$ is $w_1 = C_1e^t$, $w_2 = C_2e^{2t}$, and $w_3 = C_3e^{3t}$.

Return to the original system by applying the substitution $\mathbf{y} = P\mathbf{w}$.

$$\mathbf{y} = \begin{bmatrix} y_1 \\ y_2 \\ y_3 \end{bmatrix} = \begin{bmatrix} 1 & -2 & -7 \\ 0 & 1 & 8 \\ 0 & 0 & 2 \end{bmatrix} \begin{bmatrix} w_1 \\ w_2 \\ w_3 \end{bmatrix} = \begin{bmatrix} w_1 - 2w_2 - 7w_3 \\ w_2 + 8w_3 \\ 2w_3 \end{bmatrix}$$

So, the solution is

$$y_1 = C_1e^t - 2C_2e^{2t} - 7C_3e^{3t}$$
$$y_2 = \qquad\quad C_2e^{2t} + 8C_3e^{3t}$$
$$y_3 = \qquad\qquad\qquad 2C_3e^{3t}.$$

27. Because

$$\mathbf{y}' = \begin{bmatrix} y_1' \\ y_2' \end{bmatrix} = \begin{bmatrix} 1 & 1 \\ 0 & 1 \end{bmatrix} \begin{bmatrix} y_1 \\ y_2 \end{bmatrix} = A\mathbf{y},$$

the system represented by $\mathbf{y}' = A\mathbf{y}$ is

$$y_1' = y_1 + y_2$$
$$y_2' = \qquad y_2.$$

Note that

$$y_1' = C_1e^t + C_2te^t + C_2e^t = y_1 + y_2$$
$$y_2' = C_2e^t = y_2.$$

29. Because

$$\mathbf{y}' = \begin{bmatrix} y_1' \\ y_2' \\ y_3' \end{bmatrix} = \begin{bmatrix} 0 & 1 & 0 \\ 0 & 0 & 1 \\ 0 & -4 & 0 \end{bmatrix} \begin{bmatrix} y_1 \\ y_2 \\ y_3 \end{bmatrix} = A\mathbf{y},$$

the system represented by $\mathbf{y}' = A\mathbf{y}$ is

$$y_1' = y_2$$
$$y_2' = \quad y_3$$
$$y_3' = -4y_2.$$

Note that

$$y_1' = -2C_2 \sin 2t + 2C_3 \cos 2t = y_2$$
$$y_2' = -4C_3 \sin 2t - 4C_2 \cos 2t = y_3$$
$$y_3' = 8C_2 \sin 2t - 8C_3 \cos 2t = -4y_2.$$

31. The matrix of the quadratic form is

$$A = \begin{bmatrix} a & \dfrac{b}{2} \\ \dfrac{b}{2} & c \end{bmatrix} = \begin{bmatrix} 1 & 0 \\ 0 & 1 \end{bmatrix}.$$

33. The matrix of the quadratic form is

$$A = \begin{bmatrix} a & \dfrac{b}{2} \\ \dfrac{b}{2} & c \end{bmatrix} = \begin{bmatrix} 9 & 5 \\ 5 & -4 \end{bmatrix}.$$

35. The matrix of the quadratic form is

$$A = \begin{bmatrix} a & \dfrac{b}{2} \\ \dfrac{b}{2} & c \end{bmatrix} = \begin{bmatrix} 0 & 5 \\ 5 & -10 \end{bmatrix}.$$

37. The matrix of the quadratic form is

$$A = \begin{bmatrix} a & \dfrac{b}{2} \\ \dfrac{b}{2} & c \end{bmatrix} = \begin{bmatrix} 2 & -\dfrac{3}{2} \\ -\dfrac{3}{2} & -2 \end{bmatrix}.$$

The eigenvalues of A are $\lambda_1 = -\dfrac{5}{2}$ and $\lambda_2 = \dfrac{5}{2}$ with corresponding eigenvectors $\mathbf{x}_1 = (1, 3)$ and $\mathbf{x}_2 = (-3, 1)$ respectively.

Using unit vectors in the direction of $\mathbf{x}_1$ and $\mathbf{x}_2$ to form the columns of P yields

$$P = \begin{bmatrix} \dfrac{1}{\sqrt{10}} & -\dfrac{3}{\sqrt{10}} \\ \dfrac{3}{\sqrt{10}} & \dfrac{1}{\sqrt{10}} \end{bmatrix}.$$

Note that

$$P^T A P = \begin{bmatrix} \dfrac{1}{\sqrt{10}} & \dfrac{3}{\sqrt{10}} \\ -\dfrac{3}{\sqrt{10}} & \dfrac{1}{\sqrt{10}} \end{bmatrix} \begin{bmatrix} 2 & -\dfrac{3}{2} \\ -\dfrac{3}{2} & -2 \end{bmatrix} \begin{bmatrix} \dfrac{1}{\sqrt{10}} & -\dfrac{3}{\sqrt{10}} \\ \dfrac{3}{\sqrt{10}} & \dfrac{1}{\sqrt{10}} \end{bmatrix} = \begin{bmatrix} -\dfrac{5}{2} & 0 \\ 0 & \dfrac{5}{2} \end{bmatrix}.$$

39. The matrix of the quadratic form is

$$A = \begin{bmatrix} a & \dfrac{b}{2} \\ \dfrac{b}{2} & c \end{bmatrix} = \begin{bmatrix} 13 & 3\sqrt{3} \\ 3\sqrt{3} & 7 \end{bmatrix}.$$

The eigenvalues of A are $\lambda_1 = 4$ and $\lambda_2 = 16$, with corresponding eigenvectors $\mathbf{x}_1 = \left(1, -\sqrt{3}\right)$ and $\mathbf{x}_2 = \left(\sqrt{3}, 1\right)$, respectively. Using unit vectors in the direction of $\mathbf{x}_1$ and $\mathbf{x}_2$ to form the columns of P,

$$P = \begin{bmatrix} \dfrac{1}{2} & \dfrac{\sqrt{3}}{2} \\ -\dfrac{\sqrt{3}}{2} & \dfrac{1}{2} \end{bmatrix} \quad \text{and} \quad P^T A P = \begin{bmatrix} 4 & 0 \\ 0 & 16 \end{bmatrix}.$$

41. The matrix of the quadratic form is

$$A = \begin{bmatrix} a & \dfrac{b}{2} \\ \dfrac{b}{2} & c \end{bmatrix} = \begin{bmatrix} 16 & -12 \\ -12 & 9 \end{bmatrix}.$$

The eigenvalues of A are $\lambda_1 = 0$ and $\lambda_2 = 25$, with corresponding eigenvectors $\mathbf{x}_1 = (3, 4)$ and $\mathbf{x}_2 = (-4, 3)$ respectively. Using unit vectors in the direction of $\mathbf{x}_1$ and $\mathbf{x}_2$ to form the columns of P,

$$P = \begin{bmatrix} \dfrac{3}{5} & -\dfrac{4}{5} \\ \dfrac{4}{5} & \dfrac{3}{5} \end{bmatrix} \quad \text{and} \quad P^T A P = \begin{bmatrix} 0 & 0 \\ 0 & 25 \end{bmatrix}.$$

43. The matrix of the quadratic form is

$$A = \begin{bmatrix} a & \dfrac{b}{2} \\ \dfrac{b}{2} & c \end{bmatrix} = \begin{bmatrix} 13 & -4 \\ -4 & 7 \end{bmatrix}.$$

This matrix has eigenvalues of 5 and 15 with corresponding unit eigenvectors $\left(\dfrac{1}{\sqrt{5}}, \dfrac{2}{\sqrt{5}}\right)$ and $\left(-\dfrac{2}{\sqrt{5}}, \dfrac{1}{\sqrt{5}}\right)$ respectively. Let

$$P = \begin{bmatrix} \dfrac{1}{\sqrt{5}} & -\dfrac{2}{\sqrt{5}} \\ \dfrac{2}{\sqrt{5}} & \dfrac{1}{\sqrt{5}} \end{bmatrix} \quad \text{and} \quad P^T A P = \begin{bmatrix} 5 & 0 \\ 0 & 15 \end{bmatrix}.$$

This implies that the rotated conic is an ellipse with equation

$$5(x')^2 + 15(y')^2 = 45.$$

45. The matrix of the quadratic form is $A = \begin{bmatrix} a & \dfrac{b}{2} \\ \dfrac{b}{2} & c \end{bmatrix} = \begin{bmatrix} 7 & 16 \\ 16 & -17 \end{bmatrix}$.

This matrix has eigenvalues of -25 and 15, with corresponding unit eigenvectors $\left(\dfrac{1}{\sqrt{5}}, -\dfrac{2}{\sqrt{5}} \right)$ and $\left(\dfrac{2}{\sqrt{5}}, \dfrac{1}{\sqrt{5}} \right)$ respectively.

Let $P = \begin{bmatrix} \dfrac{1}{\sqrt{5}} & \dfrac{2}{\sqrt{5}} \\ -\dfrac{2}{\sqrt{5}} & \dfrac{1}{\sqrt{5}} \end{bmatrix}$ and $P^T A P = \begin{bmatrix} -25 & 0 \\ 0 & 15 \end{bmatrix}$.

This implies that the rotated conic is a hyperbola with equation $-25(x')^2 + 15(y')^2 - 50 = 0$.

47. The matrix of the quadratic form is $A = \begin{bmatrix} a & \dfrac{b}{2} \\ \dfrac{b}{2} & c \end{bmatrix} = \begin{bmatrix} 2 & 2 \\ 2 & 2 \end{bmatrix}$.

This matrix has eigenvalues of 0 and 4, with corresponding unit eigenvectors $\left(\dfrac{1}{\sqrt{2}}, -\dfrac{1}{\sqrt{2}} \right)$ and

$\left(\dfrac{1}{\sqrt{2}}, \dfrac{1}{\sqrt{2}} \right)$ respectively. Let $P = \begin{bmatrix} \dfrac{1}{\sqrt{2}} & \dfrac{1}{\sqrt{2}} \\ -\dfrac{1}{\sqrt{2}} & \dfrac{1}{\sqrt{2}} \end{bmatrix}$ and $P^T A P = \begin{bmatrix} 0 & 0 \\ 0 & 4 \end{bmatrix}$.

This implies that the rotated conic is a parabola. Furthermore, $[d \quad e]P = [6\sqrt{2} \quad 2\sqrt{2}] \begin{bmatrix} \dfrac{1}{\sqrt{2}} & \dfrac{1}{\sqrt{2}} \\ -\dfrac{1}{\sqrt{2}} & \dfrac{1}{\sqrt{2}} \end{bmatrix} = [4 \quad 8] = [d' \quad e']$.

So, the equation in the $x'y'$-coordinate system is, $4(y')^2 + 4x' + 8y' + 4 = 0$.

49. The matrix of the quadratic form is $A = \begin{bmatrix} 0 & \dfrac{1}{2} \\ \dfrac{1}{2} & 0 \end{bmatrix}$.

This matrix has eigenvalues of $-\dfrac{1}{2}$ and $\dfrac{1}{2}$ with corresponding unit eigenvectors $\left(-\dfrac{1}{\sqrt{2}}, \dfrac{1}{\sqrt{2}} \right)$ and

$\left(\dfrac{1}{\sqrt{2}}, \dfrac{1}{\sqrt{2}} \right)$ respectively. Let $P = \begin{bmatrix} -\dfrac{1}{\sqrt{2}} & \dfrac{1}{\sqrt{2}} \\ \dfrac{1}{\sqrt{2}} & \dfrac{1}{\sqrt{2}} \end{bmatrix}$ and $P^T A P = \begin{bmatrix} -\dfrac{1}{2} & 0 \\ 0 & \dfrac{1}{2} \end{bmatrix}$.

This implies that the rotated conic is a hyperbola. Furthermore,

$[d \quad e]P = [1 \quad -2] \begin{bmatrix} -\dfrac{1}{\sqrt{2}} & \dfrac{1}{\sqrt{2}} \\ \dfrac{1}{\sqrt{2}} & \dfrac{1}{\sqrt{2}} \end{bmatrix} = \left[-\dfrac{3}{\sqrt{2}} \quad -\dfrac{1}{\sqrt{2}} \right] = [d' \quad e']$,

so, the equation in the $x'y'$-coordinate system is $-\dfrac{1}{2}(x')^2 + \dfrac{1}{2}(y')^2 - \dfrac{3}{\sqrt{2}}x' - \dfrac{1}{\sqrt{2}}y' + 3 = 0$.

51. The matrix of the quadratic form is $A = \begin{bmatrix} 3 & -1 & 0 \\ -1 & 3 & 0 \\ 0 & 0 & 8 \end{bmatrix}$.

The eigenvalues of A are 2, 4, and 8 with corresponding unit eigenvectors $\left(\frac{1}{\sqrt{2}}, \frac{1}{\sqrt{2}}, 0 \right), \left(-\frac{1}{\sqrt{2}}, \frac{1}{\sqrt{2}}, 0 \right)$ and

$(0, 0, 1)$ respectively. Then let $P = \begin{bmatrix} \frac{1}{\sqrt{2}} & -\frac{1}{\sqrt{2}} & 0 \\ \frac{1}{\sqrt{2}} & \frac{1}{\sqrt{2}} & 0 \\ 0 & 0 & 1 \end{bmatrix}$ and $P^T AP = \begin{bmatrix} 2 & 0 & 0 \\ 0 & 4 & 0 \\ 0 & 0 & 8 \end{bmatrix}$.

Furthermore,

$$\begin{bmatrix} g & h & i \end{bmatrix} P = \begin{bmatrix} 0 & 0 & 0 \end{bmatrix} \begin{bmatrix} \frac{1}{\sqrt{2}} & -\frac{1}{\sqrt{2}} & 0 \\ \frac{1}{\sqrt{2}} & \frac{1}{\sqrt{2}} & 0 \\ 0 & 0 & 1 \end{bmatrix} = \begin{bmatrix} 0 & 0 & 0 \end{bmatrix} = \begin{bmatrix} g' & h' & i' \end{bmatrix}.$$

So, the equation of the rotated quadratic surface is $2(x')^2 + 4(y')^2 + 8(z')^2 - 16 = 0$.

53. The matrix of the quadratic form is $A = \begin{bmatrix} 1 & 0 & 0 \\ 0 & 2 & 1 \\ 0 & 1 & 2 \end{bmatrix}$.

The eigenvalues of A are 1, 1, and 3 with corresponding unit eigenvectors $(1, 0, 0), (0, -1, 1)$, and $(0, 1, 1)$ respectively.

Then let $P = \begin{bmatrix} 1 & 0 & 0 \\ 0 & -\frac{\sqrt{2}}{2} & \frac{\sqrt{2}}{2} \\ 0 & \frac{\sqrt{2}}{2} & \frac{\sqrt{2}}{2} \end{bmatrix}$ and $P^T AP = \begin{bmatrix} 1 & 0 & 0 \\ 0 & 1 & 0 \\ 0 & 0 & 3 \end{bmatrix}$.

Furthermore,

$$\begin{bmatrix} g & h & i \end{bmatrix} P = \begin{bmatrix} 0 & 0 & 0 \end{bmatrix} \begin{bmatrix} 1 & 0 & 0 \\ 0 & -\frac{\sqrt{2}}{2} & \frac{\sqrt{2}}{2} \\ 0 & \frac{\sqrt{2}}{2} & \frac{\sqrt{2}}{2} \end{bmatrix} = \begin{bmatrix} 0 & 0 & 0 \end{bmatrix} = \begin{bmatrix} g' & h' & i' \end{bmatrix}.$$

So, the equation of the rotated quadratic surface is $(x')^2 + (y')^2 + 3(z')^2 - 1 = 0$.

55. Let $P = \begin{bmatrix} a & c \\ c & d \end{bmatrix}$ be a 2×2 orthogonal matrix such that $|P| = 1$. Define $\theta \in [0, 2\pi]$ as follows.

(i) If $a = 1$, then $c = 0, b = 0$ and $d = 1$, so let $\theta = 0$.

(ii) If $a = -1$, then $c = 0, b = 0$ and $d = -1$, so let $\theta = \pi$.

(iii) If $a \geq 0$ and $c > 0$, let $\theta = \arccos(a), 0 < \theta \leq \pi/2$.

(iv) If $a \geq 0$ and $c < 0$, let $\theta = 2\pi - \arccos(a), 3\pi/2 \leq \theta < 2\pi$.

(v) If $a \leq 0$ and $c > 0$, let $\theta = \arccos(a), \pi/2 \leq \theta < \pi$.

(vi) If $a \leq 0$ and $c < 0$, let $\theta = 2\pi - \arccos(a), \pi < \theta \leq 3\pi/2$.

In each of these cases, you can confirm that $P = \begin{bmatrix} a & b \\ c & d \end{bmatrix} = \begin{bmatrix} \cos \theta & -\sin \theta \\ \sin \theta & \cos \theta \end{bmatrix}$.

Review Exercises for Chapter 7

1. (a) The characteristic equation of A is given by

$$|\lambda I - A| = \begin{vmatrix} \lambda - 2 & -1 \\ -5 & \lambda + 2 \end{vmatrix} = \lambda^2 - 9 = 0.$$

(b) The eigenvalues of A are $\lambda_1 = -3$ and $\lambda_2 = 3$.

(c) To find the eigenvectors corresponding to $\lambda_1 = -3$, solve the matrix equation $(\lambda_1 I - A)\mathbf{x} = \mathbf{0}$. Row-reduce the augmented matrix to yield

$$\begin{bmatrix} -5 & -1 & : & 0 \\ -5 & -1 & : & 0 \end{bmatrix} \Rightarrow \begin{bmatrix} 1 & \frac{1}{5} & : & 0 \\ 0 & 0 & : & 0 \end{bmatrix}.$$

So, $\mathbf{x}_1 = (1, -5)$ is an eigenvector and $\{(1, -5)\}$ is a basis for the eigenspace corresponding to $\lambda_1 = -3$. Similarly, solve $(\lambda_2 I - A)\mathbf{x} = \mathbf{0}$ for $\lambda_2 = 3$. So, $\mathbf{x}_2 = (1, 1)$ is an eigenvector and $\{(1, 1)\}$ is a basis for the eigenspace corresponding to $\lambda_2 = 3$.

3. (a) The characteristic equation of A is given by

$$|\lambda I - A| = \begin{vmatrix} \lambda - 9 & -4 & 3 \\ 2 & \lambda & -6 \\ 1 & 4 & \lambda - 11 \end{vmatrix} = \lambda^3 - 20\lambda^2 + 128\lambda - 256 = (\lambda - 4)(\lambda - 8)^2.$$

(b) The eigenvalues of A are $\lambda_1 = 4$ and $\lambda_2 = 8$ (repeated).

(c) To find the eigenvectors corresponding to $\lambda_1 = 4$, solve the matrix equation $(\lambda_1 I - A)\mathbf{x} = \mathbf{0}$. Row-reducing the augmented matrix,

$$\begin{bmatrix} -5 & -4 & 3 & : & 0 \\ 2 & 4 & -6 & : & 0 \\ 1 & 4 & -7 & : & 0 \end{bmatrix} \Rightarrow \begin{bmatrix} 1 & 0 & 1 & : & 0 \\ 0 & 1 & -2 & : & 0 \\ 0 & 0 & 0 & : & 0 \end{bmatrix}$$

you can see that a basis for the eigenspace of $\lambda_1 = 4$ is $\{(-1, 2, 1)\}$. Similarly, solve $(\lambda_2 I - A)\mathbf{x} = \mathbf{0}$ for $\lambda_2 = 8$. So, a basis for the eigenspace of $\lambda_2 = 8$ is $\{(3, 0, 1), (-4, 1, 0)\}$.

5. (a) The characteristic equation of A is given by

$$|\lambda I - A| = \begin{vmatrix} \lambda - 2 & 0 & -1 \\ 0 & \lambda - 3 & -4 \\ 0 & 0 & \lambda - 1 \end{vmatrix} = (\lambda - 2)(\lambda - 3)(\lambda - 1) = 0.$$

(b) The eigenvalues of A are $\lambda_1 = 1$, $\lambda_2 = 2$ and $\lambda_3 = 3$.

(c) To find the eigenvectors corresponding to $\lambda_1 = 1$, solve the matrix equation $(\lambda_1 I - A)\mathbf{x} = \mathbf{0}$. Row-reducing the augmented matrix,

$$\begin{bmatrix} -1 & 0 & -1 & : & 0 \\ 0 & -2 & -4 & : & 0 \\ 0 & 0 & 0 & : & 0 \end{bmatrix} \Rightarrow \begin{bmatrix} 1 & 0 & 1 & : & 0 \\ 0 & 1 & 2 & : & 0 \\ 0 & 0 & 0 & : & 0 \end{bmatrix}$$

you can see that a basis for the eigenspace of $\lambda_1 = 1$ is $\{(-1, -2, 1)\}$. Similarly, solve $(\lambda_2 I - A)\mathbf{x} = \mathbf{0}$ for $\lambda_2 = 2$, and you see that $\{(1, 0, 0)\}$ is a basis for the eigenspace of $\lambda_2 = 2$. Finally, solve $(\lambda_3 I - A)\mathbf{x} = \mathbf{0}$ for $\lambda_3 = 3$, and you discover that $\{(0, 1, 0)\}$ is a basis for its eigenspace.

7. (a) The characteristic equation of A is given by

$$|\lambda I - A| = \begin{vmatrix} \lambda - 2 & -1 & 0 & 0 \\ -1 & \lambda - 2 & 0 & 0 \\ 0 & 0 & \lambda - 2 & -1 \\ 0 & 0 & -1 & \lambda - 2 \end{vmatrix} = (\lambda - 1)^2(\lambda - 3)^2 = 0.$$

(b) The eigenvalues of A are $\lambda_1 = 1$ (repeated) and $\lambda_2 = 3$ (repeated).

(c) To find the eigenvectors corresponding to $\lambda_1 = 1$, solve the matrix equation $(\lambda_1 I - A)\mathbf{x} = \mathbf{0}$ for $\lambda_1 = 1$. Row reducing the augmented matrix,

$$\begin{bmatrix} -1 & -1 & 0 & 0 & \vdots & 0 \\ -1 & -1 & 0 & 0 & \vdots & 0 \\ 0 & 0 & -1 & -1 & \vdots & 0 \\ 0 & 0 & -1 & -1 & \vdots & 0 \end{bmatrix} \Rightarrow \begin{bmatrix} 1 & 1 & 0 & 0 & \vdots & 0 \\ 0 & 0 & 1 & 1 & \vdots & 0 \\ 0 & 0 & 0 & 0 & \vdots & 0 \\ 0 & 0 & 0 & 0 & \vdots & 0 \end{bmatrix}$$

you see that a basis for the eigenspace of $\lambda_1 = 1$ is $\{(1, -1, 0, 0), (0, 0, 1, -1)\}$. Similarly, solve $(\lambda_2 I - A)\mathbf{x} = \mathbf{0}$ for $\lambda_2 = 3$, and discover that a basis for the eigenspace of $\lambda_2 = 3$ is $\{(1, 1, 0, 0), (0, 0, 1, 1)\}$.

9. The eigenvalues of A are the solutions of

$$|\lambda I - A| = \begin{vmatrix} \lambda + 2 & 1 & -3 \\ 0 & \lambda - 1 & -2 \\ 0 & 0 & \lambda - 1 \end{vmatrix} = (\lambda + 2)(\lambda - 1)^2 = 0.$$

The eigenspace corresponding to the repeated eigenvalue $\lambda = 1$ has dimension 1, and so A is *not* diagonalizable.

11. The eigenvalues of A are the solutions to

$$|\lambda I - A| = \begin{vmatrix} \lambda - 1 & 0 & -2 \\ 0 & \lambda - 1 & 0 \\ -2 & 0 & \lambda - 1 \end{vmatrix} = (\lambda - 3)(\lambda - 1)(\lambda + 1) = 0.$$

Therefore, the eigenvalues are 3, 1, and -1. The corresponding eigenvectors are the solutions of $(\lambda I - A)\mathbf{x} = \mathbf{0}$. So, an eigenvector corresponding to 3 is $(1, 0, 1)$, an eigenvector corresponding to 1 is $(0, 1, 0)$, an eigenvector corresponding to -1 is $(1, 0, -1)$. Now form P using the eigenvectors of A as column vectors.

$$P = \begin{vmatrix} 1 & 0 & 1 \\ 0 & 1 & 0 \\ 1 & 0 & -1 \end{vmatrix}$$

Note that

$$P^{-1}AP = \begin{bmatrix} \frac{1}{2} & 0 & \frac{1}{2} \\ 0 & 1 & 0 \\ \frac{1}{2} & 0 & -\frac{1}{2} \end{bmatrix}\begin{bmatrix} 1 & 0 & 2 \\ 0 & 1 & 0 \\ 2 & 0 & 1 \end{bmatrix}\begin{bmatrix} 1 & 0 & 1 \\ 0 & 1 & 0 \\ 1 & 0 & -1 \end{bmatrix} = \begin{bmatrix} 3 & 0 & 0 \\ 0 & 1 & 0 \\ 0 & 0 & -1 \end{bmatrix}.$$

13. Consider the characteristic equation $|\lambda I - A| = \begin{vmatrix} \lambda - \cos\theta & \sin\theta \\ -\sin\theta & \lambda - \cos\theta \end{vmatrix} = \lambda^2 - 2\cos\theta \cdot \lambda + 1 = 0.$

The discriminant of this quadratic equation in λ is $b^2 - 4ac = 4\cos^2\theta - 4 = -4\sin^2\theta.$

Because $0 < \theta < \pi$, this discriminant is always negative, and the characteristic equation has no real roots.

15. The eigenvalue is $\lambda = 0$ (repeated). To find its corresponding eigenspace, solve

$$\begin{bmatrix} \lambda & -2 & \vdots & 0 \\ 0 & \lambda & \vdots & 0 \end{bmatrix} = \begin{bmatrix} 0 & -2 & \vdots & 0 \\ 0 & 0 & \vdots & 0 \end{bmatrix} \Rightarrow \begin{bmatrix} 0 & 1 & \vdots & 0 \\ 0 & 0 & \vdots & 0 \end{bmatrix}.$$

Because the eigenspace is only one-dimensional, the matrix A is not diagonalizable.

17. The eigenvalue is $\lambda = 3$ (repeated). To find its corresponding eigenspace, solve $(\lambda I - A)\mathbf{x} = \mathbf{0}$ with $\lambda = 3$.

$$\begin{bmatrix} \lambda - 3 & 0 & 0 & \vdots & 0 \\ -1 & \lambda - 3 & 0 & \vdots & 0 \\ 0 & 0 & \lambda - 3 & \vdots & 0 \end{bmatrix} = \begin{bmatrix} 0 & 0 & 0 & \vdots & 0 \\ -1 & 0 & 0 & \vdots & 0 \\ 0 & 0 & 0 & \vdots & 0 \end{bmatrix} \Rightarrow \begin{bmatrix} 1 & 0 & 0 & \vdots & 0 \\ 0 & 0 & 0 & \vdots & 0 \\ 0 & 0 & 0 & \vdots & 0 \end{bmatrix}$$

Because the eigenspace is only two-dimensional, the matrix A is not diagonalizable.

19. The eigenvalues of B are 1 and 2 with corresponding eigenvectors $(0, 1)$ and $(1, 0)$, respectively. Form the columns of P from the eigenvectors of B. So,

$$P = \begin{bmatrix} 0 & 1 \\ 1 & 0 \end{bmatrix}$$

$$P^{-1}BP = \begin{bmatrix} 0 & 1 \\ 1 & 0 \end{bmatrix}\begin{bmatrix} 2 & 0 \\ 0 & 1 \end{bmatrix}\begin{bmatrix} 0 & 1 \\ 1 & 0 \end{bmatrix} = \begin{bmatrix} 1 & 0 \\ 0 & 2 \end{bmatrix} = A.$$

Therefore, A and B are similar.

21. Because the eigenspace corresponding to $\lambda = 1$ of matrix A has dimension 1, while that of matrix B has dimension 2, the matrices are not similar.

23. Because

$$A^T = \begin{bmatrix} -\dfrac{\sqrt{2}}{2} & \dfrac{\sqrt{2}}{2} \\ \dfrac{\sqrt{2}}{2} & \dfrac{\sqrt{2}}{2} \end{bmatrix} = A,$$

A is symmetric. Furthermore, the column vectors of A form an orthonormal set. So, A is both symmetric and orthogonal.

25. Because

$$A^T = \begin{bmatrix} 0 & 0 & 1 \\ 0 & 1 & 0 \\ 1 & 0 & 1 \end{bmatrix} = A,$$

A is symmetric. However, column 3 is not a unit vector, so A is *not* orthogonal.

27. Because

$$A^T = \begin{bmatrix} -\dfrac{2}{3} & \dfrac{2}{3} & \dfrac{1}{3} \\ \dfrac{1}{3} & \dfrac{2}{3} & -\dfrac{2}{3} \\ -\dfrac{2}{3} & -\dfrac{1}{3} & \dfrac{2}{3} \end{bmatrix} \neq A,$$

A is *not* symmetric. Because the column vectors of A do not form an orthonormal set (columns 2 and 3 are not orthogonal), A is *not* orthogonal.

29. The eigenvalues of A are 5 and -5 with corresponding unit eigenvectors $\left(\dfrac{2}{\sqrt{5}}, \dfrac{1}{\sqrt{5}}\right)$ and $\left(-\dfrac{1}{\sqrt{5}}, \dfrac{2}{\sqrt{5}}\right)$, respectively. Form the columns of P with the eigenvectors of A.

$$P = \begin{bmatrix} \dfrac{2}{\sqrt{5}} & -\dfrac{1}{\sqrt{5}} \\ \dfrac{1}{\sqrt{5}} & \dfrac{2}{\sqrt{5}} \end{bmatrix}.$$

31. The eigenvalues of A are 3 and 1 (repeated), with corresponding unit eigenvectors $\left(\dfrac{1}{\sqrt{2}}, 0, -\dfrac{1}{\sqrt{2}}\right), \left(\dfrac{1}{\sqrt{2}}, 0, \dfrac{1}{\sqrt{2}}\right)$ and $(0, 1, 0)$. Form the columns of P from the eigenvectors of A.

$$P = \begin{bmatrix} \dfrac{1}{\sqrt{2}} & \dfrac{1}{\sqrt{2}} & 0 \\ 0 & 0 & 1 \\ -\dfrac{1}{\sqrt{2}} & \dfrac{1}{\sqrt{2}} & 0 \end{bmatrix}$$

33. The eigenvalues of A are $\dfrac{1}{6}$ and 1. The eigenvectors corresponding to $\lambda = 1$ are $\mathbf{x} = t(3, 2)$. By choosing $t = \dfrac{1}{5}$, you can find the steady state probability vector for A to be $\mathbf{v} = \left(\dfrac{3}{5}, \dfrac{2}{5}\right)$. Note that

$$A\mathbf{v} = \begin{bmatrix} \dfrac{2}{3} & \dfrac{1}{2} \\ \dfrac{1}{3} & \dfrac{1}{2} \end{bmatrix}\begin{bmatrix} \dfrac{3}{5} \\ \dfrac{2}{5} \end{bmatrix} = \begin{bmatrix} \dfrac{3}{5} \\ \dfrac{2}{5} \end{bmatrix} = \mathbf{v}.$$

35. The eigenvalues of A are $\dfrac{1}{2}$ and 1. The eigenvectors corresponding to $\lambda = 1$ are $\mathbf{x} = t(3, 2)$. By choosing $t = \dfrac{1}{5}$, you can find the steady state probability vector for A to be $\mathbf{v} = \left(\dfrac{3}{5}, \dfrac{2}{5}\right)$. Note that

$$A\mathbf{v} = \begin{bmatrix} 0.8 & 0.3 \\ 0.2 & 0.7 \end{bmatrix}\begin{bmatrix} \dfrac{3}{5} \\ \dfrac{2}{5} \end{bmatrix} = \begin{bmatrix} \dfrac{3}{5} \\ \dfrac{2}{5} \end{bmatrix} = \mathbf{v}.$$

37. The eigenvalues of A are $0, \frac{1}{2}$ and 1. The eigenvectors corresponding to $\lambda = 1$ are $\mathbf{x} = t(1, 2, 1)$. By choosing $t = \frac{1}{4}$, you can find the steady state probability vector for A to be $\mathbf{v} = \left(\frac{1}{4}, \frac{1}{2}, \frac{1}{4}\right)$. Note that

$$A\mathbf{v} = \begin{bmatrix} \frac{1}{2} & \frac{1}{4} & 0 \\ \frac{1}{2} & \frac{1}{2} & \frac{1}{2} \\ 0 & \frac{1}{4} & \frac{1}{2} \end{bmatrix} \begin{bmatrix} \frac{1}{4} \\ \frac{1}{2} \\ \frac{1}{4} \end{bmatrix} = \begin{bmatrix} \frac{1}{4} \\ \frac{1}{2} \\ \frac{1}{4} \end{bmatrix} = \mathbf{v}.$$

39. The eigenvalues of A are 0.6 and 1. The eigenvectors corresponding to $\lambda = 1$ are $\mathbf{x} = t(4, 5, 7)$. By choosing $t = \frac{1}{16}$, you can find the steady state probability vector for A to be $\mathbf{v} = \left(\frac{1}{4}, \frac{5}{16}, \frac{7}{16}\right)$.

Note that

$$A\mathbf{v} = \begin{bmatrix} 0.7 & 0.1 & 0.1 \\ 0.2 & 0.7 & 0.1 \\ 0.1 & 0.2 & 0.8 \end{bmatrix} \begin{bmatrix} 0.25 \\ 0.3125 \\ 0.4375 \end{bmatrix} = \begin{bmatrix} 0.25 \\ 0.3125 \\ 0.4375 \end{bmatrix} = \mathbf{v}.$$

45. $A^2 = 10A - 24I_2 = \begin{bmatrix} 80 & -40 \\ 20 & 20 \end{bmatrix} - \begin{bmatrix} 24 & 0 \\ 0 & 24 \end{bmatrix} = \begin{bmatrix} 56 & -40 \\ 20 & -4 \end{bmatrix}$

$A^3 = 10A^2 - 24A = \begin{bmatrix} 560 & -400 \\ 200 & -40 \end{bmatrix} - \begin{bmatrix} 192 & -96 \\ 48 & 48 \end{bmatrix} = \begin{bmatrix} 368 & -304 \\ 152 & -88 \end{bmatrix}$

47. (a) True. If $A\mathbf{x} = \lambda\mathbf{x}$, then $A^2\mathbf{x} = A(A\mathbf{x}) = A(\lambda\mathbf{x}) = \lambda(A\mathbf{x}) = \lambda^2\mathbf{x}$ showing that $\mathbf{x}$ is an eigenvector of A^2.

(b) False. For example, $(1, 0)$ is an eigenvector of $A^2 = \begin{bmatrix} 1 & 0 \\ 0 & 1 \end{bmatrix}$, but not of $A = \begin{bmatrix} 0 & 1 \\ 1 & 0 \end{bmatrix}$.

49. Because $A^{-1}(AB)A = BA$, you can see that AB and BA are similar.

51. The eigenvalues of A are $a + b$ and $a - b$, with corresponding unit eigenvectors $\left(\frac{1}{\sqrt{2}}, \frac{1}{\sqrt{2}}\right)$ and $\left(-\frac{1}{\sqrt{2}}, \frac{1}{\sqrt{2}}\right)$, respectively. So, $P = \begin{bmatrix} \frac{1}{\sqrt{2}} & -\frac{1}{\sqrt{2}} \\ \frac{1}{\sqrt{2}} & \frac{1}{\sqrt{2}} \end{bmatrix}$. Note that

$$P^{-1}AP = \begin{bmatrix} \frac{1}{\sqrt{2}} & \frac{1}{\sqrt{2}} \\ -\frac{1}{\sqrt{2}} & \frac{1}{\sqrt{2}} \end{bmatrix} \begin{bmatrix} a & b \\ b & a \end{bmatrix} \begin{bmatrix} \frac{1}{\sqrt{2}} & -\frac{1}{\sqrt{2}} \\ \frac{1}{\sqrt{2}} & \frac{1}{\sqrt{2}} \end{bmatrix} = \begin{bmatrix} a + b & 0 \\ 0 & a - b \end{bmatrix}.$$

53. (a) A is diagonalizable if and only if $a = b = c = 0$.

(b) If exactly two of a, b, and c are zero, then the eigenspace of 2 has dimension 3. If exactly one of a, b, c is zero, then the dimension of the eigenspace is 2. If none of a, b, c is zero, the eigenspace is dimension 1.

41. $(P^T A P)^T = P^T A^T (P^T)^T = P^T A P$ (because A is symmetric), which shows that $P^T A P$ is symmetric.

43. From the form $p(\lambda) = a_0 + a_1\lambda + a_2\lambda^2$, you have $a_0 = 0$, $a_1 = -9$, and $a_2 = 4$. This implies that the companion matrix of p is

$$A = \begin{bmatrix} 0 & 1 \\ -\dfrac{a_0}{a_2} & -\dfrac{a_1}{a_2} \end{bmatrix} \begin{bmatrix} 0 & 1 \\ 0 & \dfrac{9}{4} \end{bmatrix}.$$

The eigenvalues of A are 0 and $\dfrac{9}{4}$, the zeros of p.

55. (a) True. See "Definitions of Eigenvalue and Eigenvector," page 422.

(b) False. See Theorem 7.4, page 436.

(c) True. See "Definition of a Diagonalizable Matrix," page 435.

57. The population after one transition is

$$\mathbf{x}_2 = A\mathbf{x}_1 = \begin{bmatrix} 0 & 1 \\ \frac{1}{4} & 0 \end{bmatrix} \begin{bmatrix} 100 \\ 100 \end{bmatrix} = \begin{bmatrix} 100 \\ 25 \end{bmatrix}$$

and after two transitions is

$$\mathbf{x}_3 = A\mathbf{x}_2 = \begin{bmatrix} 0 & 1 \\ \frac{1}{4} & 0 \end{bmatrix} \begin{bmatrix} 100 \\ 25 \end{bmatrix} = \begin{bmatrix} 25 \\ 25 \end{bmatrix}.$$

The eigenvalues of A are $-\frac{1}{2}$ and $\frac{1}{2}$. Choose the positive eigenvalue and find the corresponding eigenvectors to be multiples of $(2, 1)$. So, the stable age distribution vector is $\mathbf{x} = t\begin{bmatrix} 2 \\ 1 \end{bmatrix}$.

59. The population after one transition is

$$\mathbf{x}_2 = \begin{bmatrix} 0 & 3 & 12 \\ 1 & 0 & 0 \\ 0 & \frac{1}{6} & 0 \end{bmatrix} \begin{bmatrix} 300 \\ 300 \\ 300 \end{bmatrix} = \begin{bmatrix} 4500 \\ 300 \\ 50 \end{bmatrix}$$

and after two transitions is

$$\mathbf{x}_3 = \begin{bmatrix} 0 & 3 & 12 \\ 1 & 0 & 0 \\ 0 & \frac{1}{6} & 0 \end{bmatrix} \begin{bmatrix} 4500 \\ 300 \\ 50 \end{bmatrix} = \begin{bmatrix} 1500 \\ 4500 \\ 50 \end{bmatrix}.$$

The positive eigenvalue 2 has corresponding eigenvector $(24, 12, 1)$, which is a stable distribution.

So, the stable age distribution vector is $\mathbf{x} = t\begin{bmatrix} 24 \\ 12 \\ 1 \end{bmatrix}$.

61. Construct the age transition matrix.

$$A = \begin{bmatrix} 4 & 6 & 2 \\ 0.9 & 0 & 0 \\ 0 & 0.75 & 0 \end{bmatrix}$$

The current age distribution vector is

$$\mathbf{x}_1 = \begin{bmatrix} 120 \\ 120 \\ 120 \end{bmatrix}.$$

In one year, the age distribution vector will be

$$\mathbf{x}_2 = A\mathbf{x}_1 = \begin{bmatrix} 4 & 6 & 2 \\ 0.9 & 0 & 0 \\ 0 & 0.75 & 0 \end{bmatrix} \begin{bmatrix} 120 \\ 120 \\ 120 \end{bmatrix} = \begin{bmatrix} 1440 \\ 108 \\ 90 \end{bmatrix}.$$

In two years, the age distribution vector will be

$$\mathbf{x}_3 = A\mathbf{x}_2 = \begin{bmatrix} 4 & 6 & 2 \\ 0.9 & 0 & 0 \\ 0 & 0.75 & 0 \end{bmatrix} \begin{bmatrix} 1440 \\ 108 \\ 90 \end{bmatrix} = \begin{bmatrix} 6588 \\ 1296 \\ 81 \end{bmatrix}.$$

63. The matrix corresponding to the system $\mathbf{y}' = A\mathbf{y}$ is

$$A = \begin{bmatrix} 1 & 2 \\ 0 & 0 \end{bmatrix}.$$

This matrix has eigenvalues of 0 and 1, with corresponding eigenvectors $(-2, 1)$ and $(1, 0)$, respectively. So, a matrix P that diagonalizes A is

$$P = \begin{bmatrix} -2 & 1 \\ 1 & 0 \end{bmatrix} \quad \text{and} \quad P^{-1}AP = \begin{bmatrix} 0 & 0 \\ 0 & 1 \end{bmatrix}.$$

The system represented by $\mathbf{w}' = P^{-1}AP\mathbf{w}$ yields the solution $w_1' = 0$ and $w_2' = w_2$.

So $w_1 = C_1$ and $w_2 = C_2 e^t$. Substitute $\mathbf{y} = P\mathbf{w}$ and write

$$\begin{bmatrix} y_1 \\ y_2 \end{bmatrix} = \begin{bmatrix} -2 & 1 \\ 1 & 0 \end{bmatrix} \begin{bmatrix} w_1 \\ w_2 \end{bmatrix} = \begin{bmatrix} -2w_1 + w_2 \\ w_1 \end{bmatrix}.$$

This implies that the solution is

$$y_1 = -2C_1 + C_2 e^t$$
$$y_2 = C_1.$$

65. The matrix corresponding to the system $\mathbf{y}' = A\mathbf{y}$ is

$$A = \begin{bmatrix} 0 & 1 & 0 \\ 1 & 0 & 0 \\ 0 & 0 & 0 \end{bmatrix}.$$

This matrix has eigenvalues 1, -1, and 0 with corresponding eigenvectors $(1, 1, 0)$, $(1, -1, 0)$, and $(0, 0, 1)$.

So, a matrix P that diagonalizes A is

$$P = \begin{bmatrix} 1 & 1 & 0 \\ 1 & -1 & 0 \\ 0 & 0 & 1 \end{bmatrix} \quad \text{and} \quad P^{-1}AP = \begin{bmatrix} 1 & 0 & 0 \\ 0 & -1 & 0 \\ 0 & 0 & 0 \end{bmatrix}.$$

The system represented by $\mathbf{w}' = P^{-1}AP\mathbf{w}$ has solutions $w_1 = C_1 e^t$, $w_2 = C_2 e^{-t}$, and $w_3 = C_3 e^0 = C_3$

Substitute $\mathbf{y} = P\mathbf{w}$ and obtain

$$\begin{bmatrix} y_1 \\ y_2 \\ y_3 \end{bmatrix} = \begin{bmatrix} 1 & 1 & 0 \\ 1 & -1 & 0 \\ 0 & 0 & 1 \end{bmatrix} \begin{bmatrix} w_1 \\ w_2 \\ w_3 \end{bmatrix} = \begin{bmatrix} w_1 + w_2 \\ w_1 - w_2 \\ w_3 \end{bmatrix},$$

which yields the solution

$$y_1 = C_1 e^t + C_2 e^{-t}$$
$$y_2 = C_1 e^t - C_2 e^{-t}$$
$$y_3 = C_3.$$

67. The matrix of the quadratic form is

$$A = \begin{bmatrix} a & \dfrac{b}{2} \\ \dfrac{b}{2} & c \end{bmatrix} = \begin{bmatrix} 1 & \dfrac{3}{2} \\ \dfrac{3}{2} & 1 \end{bmatrix}.$$

The eigenvalues are $\dfrac{5}{2}$ and $-\dfrac{1}{2}$ with corresponding unit

eigenvectors $\left(\dfrac{1}{\sqrt{2}}, \dfrac{1}{\sqrt{2}}\right)$ and $\left(-\dfrac{1}{\sqrt{2}}, \dfrac{1}{\sqrt{2}}\right)$,

respectively.

Then form the columns of P from the eigenvectors of A.

$$P = \begin{bmatrix} \dfrac{1}{\sqrt{2}} & -\dfrac{1}{\sqrt{2}} \\ \dfrac{1}{\sqrt{2}} & \dfrac{1}{\sqrt{2}} \end{bmatrix} \quad \text{and} \quad P^T A P = \begin{bmatrix} \dfrac{5}{2} & 0 \\ 0 & -\dfrac{1}{2} \end{bmatrix}.$$

This implies that the equation of the rotated conic is

$$\tfrac{5}{2}(x')^2 - \tfrac{1}{2}(y')^2 - 3 = 0.$$

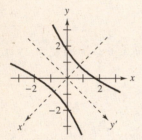

69. The matrix of the quadratic form is

$$A = \begin{bmatrix} a & \dfrac{b}{2} \\ \dfrac{b}{2} & c \end{bmatrix} = \begin{bmatrix} 0 & \dfrac{1}{2} \\ \dfrac{1}{2} & 0 \end{bmatrix}.$$

The eigenvalues are $\dfrac{1}{2}$ and $-\dfrac{1}{2}$, with corresponding unit

eigenvectors $\left(\dfrac{1}{\sqrt{2}}, \dfrac{1}{\sqrt{2}}\right)$ and $\left(-\dfrac{1}{\sqrt{2}}, \dfrac{1}{\sqrt{2}}\right)$. Use these

eigenvectors to form the columns of P.

$$P = \begin{bmatrix} \dfrac{1}{\sqrt{2}} & -\dfrac{1}{\sqrt{2}} \\ \dfrac{1}{\sqrt{2}} & \dfrac{1}{\sqrt{2}} \end{bmatrix} \quad \text{and} \quad P^T A P = \begin{bmatrix} \dfrac{1}{2} & 0 \\ 0 & -\dfrac{1}{2} \end{bmatrix}$$

This implies that the equation of the rotated conic is
$\tfrac{1}{2}(x')^2 - \tfrac{1}{2}(y')^2 - 2 = 0$, a hyperbola.

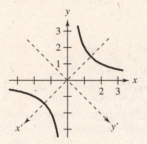

Cumulative Test for Chapters 6 and 7

1. T preserves addition.

$$\begin{aligned} T(x_1, y_1, z_1) + T(x_2, y_2, z_2) &= (2x_1, x_1 + y_1) + (2x_2, x_2 + y_2) \\ &= (2x_1 + 2x_2, x_1 + y_1 + x_2 + y_2) \\ &= \big(2(x_1 + x_2), (x_1 + x_2) + (y_1 + y_2)\big) \\ &= T(x_1 + x_2, y_1 + y_2, z_1 + z_2) \end{aligned}$$

T preserves scalar multiplication.

$$\begin{aligned} T(c(x, y, z)) &= T(cx, cy, cz) \\ &= (2cx, cx + cy) \\ &= c(2x, x + y) \\ &= cT(x, y, z) \end{aligned}$$

Therefore, T is a linear transformation.

2. No, T is not a linear transformation. For example,

$$T\left[2\begin{bmatrix} 1 & 0 \\ 0 & 1 \end{bmatrix}\right] = T\begin{bmatrix} 2 & 0 \\ 0 & 2 \end{bmatrix} = \left\|\begin{bmatrix} 2 & 0 \\ 0 & 2 \end{bmatrix} + \begin{bmatrix} 2 & 0 \\ 0 & 2 \end{bmatrix}\right\| = \begin{vmatrix} 4 & 0 \\ 0 & 4 \end{vmatrix} = 16$$

$$2T\begin{bmatrix} 1 & 0 \\ 0 & 1 \end{bmatrix} = 2\left\|\begin{bmatrix} 1 & 0 \\ 0 & 1 \end{bmatrix} + \begin{bmatrix} 1 & 0 \\ 0 & 1 \end{bmatrix}\right\| = 2\begin{vmatrix} 2 & 0 \\ 0 & 2 \end{vmatrix} = 2 \cdot 4 = 8$$

3. (a) $T(1, -2) = \begin{bmatrix} 1 & 0 \\ -1 & 0 \\ 0 & 0 \end{bmatrix}\begin{bmatrix} 1 \\ -2 \end{bmatrix} = \begin{bmatrix} 1 \\ -1 \\ 0 \end{bmatrix}$

(b) $\begin{bmatrix} 1 & 0 \\ -1 & 0 \\ 0 & 0 \end{bmatrix}\begin{bmatrix} x \\ y \end{bmatrix} = \begin{bmatrix} x \\ -x \\ 0 \end{bmatrix} = \begin{bmatrix} 5 \\ -5 \\ 0 \end{bmatrix} \quad \Rightarrow \quad x = 5, y = t$

The preimage of $(5, -5, 0)$ is $(5, t)$, where t is any real number.

4. The kernel is the solution space of the homogeneous system

$$\begin{aligned} x_1 - x_2 \qquad\qquad &= 0 \\ -x_1 + x_2 \qquad\qquad &= 0 \\ x_3 + x_4 &= 0. \end{aligned}$$

$$\begin{bmatrix} 1 & -1 & 0 & 0 \\ -1 & 1 & 0 & 0 \\ 0 & 0 & 1 & 1 \end{bmatrix} \Rightarrow \begin{bmatrix} 1 & -1 & 0 & 0 \\ 0 & 0 & 0 & 0 \\ 0 & 0 & 1 & 1 \end{bmatrix} \Rightarrow \begin{aligned} x_1 &= x_2 \\ x_3 &= -x_4 \end{aligned}$$

So, $\ker(T) = \{(s, s, -t, t) : s, t \in R\}$.

5. $A = \begin{bmatrix} 1 & 0 & 1 & 0 \\ 0 & -1 & 0 & -1 \end{bmatrix} \Rightarrow \begin{bmatrix} 1 & 0 & 1 & 0 \\ 0 & 1 & 0 & 1 \end{bmatrix}$

(a) basis for kernel: $\left\{ \begin{bmatrix} 0 \\ -1 \\ 0 \\ 1 \end{bmatrix}, \begin{bmatrix} 1 \\ 0 \\ -1 \\ 0 \end{bmatrix} \right\}$

(b) basis for range (column space of A): $\left\{ \begin{pmatrix} 1 \\ 0 \end{pmatrix}, \begin{pmatrix} 0 \\ 1 \end{pmatrix} \right\}$

(c) rank $= 2$, nullity $= 2$

6. Because

$$T\left(\begin{bmatrix} 1 \\ 0 \\ 0 \end{bmatrix}\right) = \begin{bmatrix} 1 \\ 0 \\ 1 \end{bmatrix}, \quad T\left(\begin{bmatrix} 0 \\ 1 \\ 0 \end{bmatrix}\right) = \begin{bmatrix} 1 \\ 1 \\ 0 \end{bmatrix} \quad \text{and} \quad T\left(\begin{bmatrix} 0 \\ 0 \\ 1 \end{bmatrix}\right) = \begin{bmatrix} 0 \\ 1 \\ -1 \end{bmatrix},$$

the standard matrix for T is

$$A = \begin{bmatrix} 1 & 1 & 0 \\ 0 & 1 & 1 \\ 1 & 0 & -1 \end{bmatrix}.$$

7. $T\left(\begin{bmatrix} 1 \\ 0 \end{bmatrix}\right) = \begin{bmatrix} \frac{1}{2} \\ -\frac{1}{2} \end{bmatrix}, T\left(\begin{bmatrix} 0 \\ 1 \end{bmatrix}\right) = \begin{bmatrix} -\frac{1}{2} \\ \frac{1}{2} \end{bmatrix}, A = \begin{bmatrix} \frac{1}{2} & -\frac{1}{2} \\ -\frac{1}{2} & \frac{1}{2} \end{bmatrix}$

$T\left(\begin{bmatrix} 1 \\ 1 \end{bmatrix}\right) = \begin{bmatrix} \frac{1}{2} & -\frac{1}{2} \\ -\frac{1}{2} & \frac{1}{2} \end{bmatrix}\begin{bmatrix} 1 \\ 1 \end{bmatrix} = \begin{bmatrix} 0 \\ 0 \end{bmatrix}$

$T\left(\begin{bmatrix} -2 \\ 2 \end{bmatrix}\right) = \begin{bmatrix} \frac{1}{2} & -\frac{1}{2} \\ -\frac{1}{2} & \frac{1}{2} \end{bmatrix}\begin{bmatrix} -2 \\ 2 \end{bmatrix} = \begin{bmatrix} -2 \\ 2 \end{bmatrix}$

8. Matrix of T is $A = \begin{bmatrix} 1 & -1 \\ 2 & 1 \end{bmatrix}$.

$A^{-1} = \frac{1}{3}\begin{bmatrix} 1 & 1 \\ -2 & 1 \end{bmatrix} \Rightarrow T^{-1}(x, y) = \left(\frac{1}{3}x + \frac{1}{3}y, -\frac{2}{3}x + \frac{1}{3}y\right)$

$\left(T^{-1} \circ T\right)(3, -2) = T^{-1}(T(3, -2)) = T^{-1}(5, 4) = (3, -2)$

9. $T(1, 1) = (1, 2, 2) = -1(1, 0, 0) + 0(1, 1, 0) + 2(1, 1, 1)$

$T(1, 0) = (0, 2, 1) = -2(1, 0, 0) + 1(1, 1, 0) + 1(1, 1, 1)$

$A = \begin{bmatrix} -1 & -2 \\ 0 & 1 \\ 2 & 1 \end{bmatrix}$

$T(0, 1) = A[\mathbf{v}]_B = \begin{bmatrix} -1 & -2 \\ 0 & 1 \\ 2 & 1 \end{bmatrix}\begin{bmatrix} 1 \\ -1 \end{bmatrix} = \begin{bmatrix} 1 \\ -1 \\ 1 \end{bmatrix} = (1, 0, 0) - (1, 1, 0) + (1, 1, 1) = (1, 0, 1)$

10. (a) $A = \begin{bmatrix} 1 & -2 \\ 1 & 4 \end{bmatrix}$

(b) $[B : B'] \Rightarrow [I : P] \Rightarrow P = \begin{bmatrix} 1 & 1 \\ 1 & 2 \end{bmatrix}$

(c) $P^{-1} = \begin{bmatrix} 2 & -1 \\ -1 & 1 \end{bmatrix} \Rightarrow A' = P^{-1}AP = \begin{bmatrix} -7 & -15 \\ 6 & 12 \end{bmatrix}$

(d) $\left[T(\mathbf{v})\right]_{B'} = A'[\mathbf{v}]_{B'} = \begin{bmatrix} -7 & -15 \\ 6 & 12 \end{bmatrix}\begin{bmatrix} 3 \\ -2 \end{bmatrix} = \begin{bmatrix} 9 \\ -6 \end{bmatrix}$

(e) $[\mathbf{v}]_B = P[\mathbf{v}]_{B'} = \begin{bmatrix} 1 & 1 \\ 1 & 2 \end{bmatrix}\begin{bmatrix} 3 \\ -2 \end{bmatrix} = \begin{bmatrix} 1 \\ -1 \end{bmatrix}$

$T(\mathbf{v}) = P\left[T(\mathbf{v})\right]_{B'} = \begin{bmatrix} 1 & 1 \\ 1 & 2 \end{bmatrix}\begin{bmatrix} 9 \\ -6 \end{bmatrix} = \begin{bmatrix} 3 \\ -3 \end{bmatrix}$

$\left[T(\mathbf{v})\right]_B = A[\mathbf{v}]_B = \begin{bmatrix} 1 & -2 \\ 1 & 4 \end{bmatrix}\begin{bmatrix} 1 \\ -1 \end{bmatrix} = \begin{bmatrix} 3 \\ -3 \end{bmatrix}$

11. The characteristic equation is

$$|\lambda I - A| = \begin{vmatrix} \lambda - 1 & -2 & -1 \\ 0 & \lambda - 3 & -1 \\ 0 & 3 & \lambda + 1 \end{vmatrix}$$

$$= (\lambda - 1)(\lambda^2 - 2\lambda)$$

$$= \lambda(\lambda - 1)(\lambda - 2) = 0.$$

For $\lambda_1 = 1$:

$$\begin{bmatrix} 0 & -2 & -1 \\ 0 & -2 & -1 \\ 0 & 3 & 2 \end{bmatrix} \begin{bmatrix} x_1 \\ x_2 \\ x_3 \end{bmatrix} = \begin{bmatrix} 0 \\ 0 \\ 0 \end{bmatrix}$$

The solution is $\{(t, 0, 0) : t \in R\}$ and an eigenvector is $(1, 0, 0)$.

For $\lambda_2 = 0$:

$$\begin{bmatrix} -1 & -2 & -1 \\ 0 & -3 & -1 \\ 0 & 3 & 1 \end{bmatrix} \begin{bmatrix} x_1 \\ x_2 \\ x_3 \end{bmatrix} = \begin{bmatrix} 0 \\ 0 \\ 0 \end{bmatrix}$$

The solution is $\{(-t, -t, 3t) : t \in R\}$ and an eigenvector is $(-1, -1, 3)$.

For $\lambda_3 = 2$:

$$\begin{bmatrix} 1 & -2 & -1 \\ 0 & -1 & -1 \\ 0 & 3 & 3 \end{bmatrix} \begin{bmatrix} x_1 \\ x_2 \\ x_3 \end{bmatrix} = \begin{bmatrix} 0 \\ 0 \\ 0 \end{bmatrix}$$

The solution is $\{(t, t, -t) : t \in R\}$ and an eigenvector is $(1, 1, -1)$.

12. Because A is a triangular matrix, $\lambda = 1$ (repeated).

For $\lambda = 1$:

$$\begin{bmatrix} 0 & 1 & -1 \\ 0 & 0 & -2 \\ 0 & 0 & 0 \end{bmatrix} \begin{bmatrix} x_1 \\ x_2 \\ x_3 \end{bmatrix} = \begin{bmatrix} 0 \\ 0 \\ 0 \end{bmatrix}$$

The solution is $\{(t, 0, 0) : t \in R\}$ and an eigenvector is $(1, 0, 0)$.

13. The eigenvalues of A are $\lambda_1 = 2$ and $\lambda_2 = 4$. The corresponding eigenvectors $(3, 1)$ and $(1, 1)$ are used to form the columns of P. So,

$$P = \begin{bmatrix} 3 & 1 \\ 1 & 1 \end{bmatrix} \Rightarrow P^{-1} = \begin{bmatrix} \frac{1}{2} & -\frac{1}{2} \\ -\frac{1}{2} & \frac{3}{2} \end{bmatrix}$$

and

$$P^{-1}AP = \begin{bmatrix} \frac{1}{2} & -\frac{1}{2} \\ -\frac{1}{2} & \frac{3}{2} \end{bmatrix} \begin{bmatrix} 1 & 3 \\ -1 & 5 \end{bmatrix} \begin{bmatrix} 3 & 1 \\ 1 & 1 \end{bmatrix} = \begin{bmatrix} 2 & 0 \\ 0 & 4 \end{bmatrix}.$$

14. The standard matrix for T is

$$A = \begin{bmatrix} 2 & 0 & -2 \\ 0 & 2 & -2 \\ 3 & 0 & -3 \end{bmatrix}$$

which has eigenvalues $\lambda_1 = 2$, $\lambda_2 = 0$, and $\lambda_3 = -1$ and corresponding eigenvectors $(0, 1, 0)$, $(1, 1, 1)$, and $(2, 2, 3)$.

So, $B = \{(0, 1, 0), (1, 1, 1), (2, 2, 3)\}$ and

$$P = \begin{bmatrix} 0 & 1 & 2 \\ 1 & 1 & 2 \\ 0 & 1 & 3 \end{bmatrix} \Rightarrow P^{-1} \begin{bmatrix} -2 & 2 & 0 \\ 0 & 0 & 0 \\ 1 & 0 & -1 \end{bmatrix}$$

and

$$P^{-1}AP = \begin{bmatrix} 2 & 0 & 0 \\ 0 & 0 & 0 \\ 0 & 0 & -1 \end{bmatrix}.$$

15. The eigenvalues of A are $\lambda_1 = -2$ and $\lambda_2 = 4$, with corresponding eigenvectors $(1, -1)$ and $(1, 1)$, respectively. Normalize each eigenvector to form the columns of P. Then

$$P = \begin{bmatrix} \frac{1}{\sqrt{2}} & \frac{1}{\sqrt{2}} \\ -\frac{1}{\sqrt{2}} & \frac{1}{\sqrt{2}} \end{bmatrix}$$

and

$$P^T AP = \begin{bmatrix} \frac{1}{\sqrt{2}} & -\frac{1}{\sqrt{2}} \\ \frac{1}{\sqrt{2}} & \frac{1}{\sqrt{2}} \end{bmatrix} \begin{bmatrix} 1 & 3 \\ 3 & 1 \end{bmatrix} \begin{bmatrix} \frac{1}{\sqrt{2}} & \frac{1}{\sqrt{2}} \\ -\frac{1}{\sqrt{2}} & \frac{1}{\sqrt{2}} \end{bmatrix} = \begin{bmatrix} -2 & 0 \\ 0 & 4 \end{bmatrix}.$$

16. Eigenvalues and eigenvectors of A are

$$\lambda = 2, \begin{bmatrix} 1 \\ 1 \\ 1 \end{bmatrix}; \lambda = -1, \begin{bmatrix} 1 \\ 0 \\ -1 \end{bmatrix}, \begin{bmatrix} 1 \\ -1 \\ 0 \end{bmatrix}.$$

Using the Gram-Schmidt orthonormalization process, you obtain

$$P = \begin{bmatrix} \frac{1}{\sqrt{3}} & \frac{1}{\sqrt{2}} & \frac{1}{\sqrt{6}} \\ \frac{1}{\sqrt{3}} & 0 & -\frac{2}{\sqrt{6}} \\ \frac{1}{\sqrt{3}} & -\frac{1}{\sqrt{2}} & \frac{1}{\sqrt{6}} \end{bmatrix}$$

and $P^T AP = \begin{bmatrix} 2 & 0 & 0 \\ 0 & -1 & 0 \\ 0 & 0 & -1 \end{bmatrix}.$

17. The solution to the differential equation $y' = Ky$ is
$y = Ce^{Kt}$. So, $y_1 = C_1e^t$ and $y_2 = C_2e^{3t}$.

18. Because $a = 4$, $b = -8$, and $c = 4$, the matrix is

$$A = \begin{bmatrix} a & \dfrac{b}{2} \\ \dfrac{b}{2} & c \end{bmatrix} = \begin{bmatrix} 4 & -4 \\ -4 & 4 \end{bmatrix}.$$

19. Construct the age transition matrix.

$$A = \begin{bmatrix} 3 & 6 & 3 \\ 0.8 & 0 & 0 \\ 0 & 0.4 & 0 \end{bmatrix}$$

The current age distribution vector is

$$\mathbf{x}_1 = \begin{bmatrix} 150 \\ 150 \\ 150 \end{bmatrix}.$$

In one year, the age distribution vector will be

$$\mathbf{x}_2 = A\mathbf{x}_1 = \begin{bmatrix} 3 & 6 & 3 \\ 0.8 & 0 & 0 \\ 0 & 0.4 & 0 \end{bmatrix} \begin{bmatrix} 150 \\ 150 \\ 150 \end{bmatrix} = \begin{bmatrix} 1800 \\ 120 \\ 60 \end{bmatrix}.$$

In two years, the age distribution vector will be

$$\mathbf{x}_3 = A\mathbf{x}_2 = \begin{bmatrix} 3 & 6 & 3 \\ 0.8 & 0 & 0 \\ 0 & 0.4 & 0 \end{bmatrix} \begin{bmatrix} 1800 \\ 120 \\ 60 \end{bmatrix} = \begin{bmatrix} 6300 \\ 1440 \\ 48 \end{bmatrix}.$$

20. λ is an <u>eigenvalue</u> of A if there exists a nonzero vector $\mathbf{x}$ such that $A\mathbf{x} = \lambda\mathbf{x}$. $\mathbf{x}$ is called an <u>eigenvector</u> of A. If A is $n \times n$, A can have n eigenvalues, possibly complex and possibly repeated.

21. P is <u>orthogonal</u> if $P^{-1} = P^T$.

$1 = \det\left(P \cdot P^{-1}\right)$

$\quad = \det\left(PP^T\right) = \left(\det P\right)^2 \Rightarrow \det P = \pm 1$

22. There exists P such that $P^{-1}AP = D$. A and B are similar implies that there exists Q such that

$A = Q^{-1}BQ.$ Then

$D = P^{-1}AP = P^{-1}\left(Q^{-1}BQ\right)P = \left(QP\right)^{-1}B\left(QP\right).$

23. If 0 is an eigenvalue of A, then $\left|A - \lambda I\right| = \left|A\right| = 0$, and A is singular.

24. See proof of Theorem 7.9, page 452.

25. The range of T is nonempty because it contains $\mathbf{0}$. Let $T(\mathbf{u})$ and $T(\mathbf{v})$ be two vectors in the range of T. Then $T(\mathbf{u}) + T(\mathbf{v}) = T(\mathbf{u} + \mathbf{v})$. But because $\mathbf{u}$ and $\mathbf{v}$ are in V, it follows that $\mathbf{u} + \mathbf{v}$ is also in V, which in turn implies that $T(\mathbf{u} + \mathbf{v})$ is in the range.

Similarly, let $T(\mathbf{u})$ be in the range of T, and let c be a scalar. Then $cT(\mathbf{u}) = T(c\mathbf{u})$. But because $c\mathbf{u}$ is in V, this implies that $T(c\mathbf{u})$ is in the range.

26. If T is one-to-one and $\mathbf{v} \in$ kernel, then
$T(\mathbf{v}) = T(\mathbf{0}) = \mathbf{0} \Rightarrow \mathbf{v} = \mathbf{0}$.

If kernel $= \{\mathbf{0}\}$ and $T(\mathbf{v}_1) = T(\mathbf{v}_2)$, then

$T(\mathbf{v}_1 - \mathbf{v}_2) = \mathbf{0} \Rightarrow \mathbf{v}_1 - \mathbf{v}_2 \in$ kernel $\Rightarrow \mathbf{v}_1 = \mathbf{v}_2$.

27. If λ is an eigenvalue of A and $A^2 = O$, then

$A\mathbf{x} = \lambda\mathbf{x}, \mathbf{x} \neq \mathbf{0}$

$A^2\mathbf{x} = A(\lambda\mathbf{x}) = \lambda^2\mathbf{x}$

$\mathbf{0} = \lambda^2\mathbf{x} \Rightarrow \lambda^2 = 0 \Rightarrow \lambda = 0.$